The Enzymes

VOLUME XXIV

PROTEIN METHYLTRANSFERASES

THE ENZYMES

Edited by

Steven G. Clarke

Department of Chemistry and Biochemistry
Molecular Biology Institute
University of California, Los Angeles
Los Angeles, CA 90095, USA

Fuyuhiko Tamanoi

Department of Microbiology,
Immunology, and Molecular Genetics
Molecular Biology Institute
University of California, Los Angeles
Los Angeles, CA 90095, USA

Volume XXIV

PROTEIN METHYLTRANSFERASES

AMSTERDAM • BOSTON • HEIDELBERG • LONDON
NEW YORK • OXFORD • PARIS • SAN DIEGO
SAN FRANCISCO • SYDNEY • TOKYO

ELSEVIER

Academic Press is an imprint of Elsevier

Academic Press is an imprint of Elsevier
30 Corporate Drive, Suite 400, Burlington, MA 01803, USA
525 B Street, Suite 1900, San Diego, California 92101-4495, USA
84 Theobald's Road, London WC1X 8RR, UK

Library of Congress Cataloging-in-Publication Data
An application has been submitted.

British Library Cataloguing-in-Publication Data
A catalogue record for this book is available from the British Library.

ISBN 13: 978-0-12-122725-8
ISBN 10: 0-12-122725-1

For information on all Academic Press publications
visit our Web site at www.books.elsevier.com

Printed and bound by CPI Group (UK) Ltd, Croydon, CR0 4YY
Transferred to Digital Printing, 2013

Contents

Part I: Overview of Protein Methyltransferases

1. Protein Methyltransferases: Their Distribution Among the Five Structural Classes of AdoMet-Dependent Methyltransferases

HEIDI L. SCHUBERT, ROBERT M. BLUMENTHAL, AND XIAODONG CHENG

Part II: Modification of Lysine and Arginine Residues in Signal Transduction, Transcription, Translation, and Other Functions

2. The Family of Protein Arginine Methyltransferases

MARK T. BEDFORD

3. Diverse Roles of Protein Arginine Methyltransferases

ANNE E. MCBRIDE

4. Structure of Protein Arginine Methyltransferases

XING ZHANG AND XIAODONG CHENG

5. Methylation and Demethylation of Histone Arg and Lys Residues in Chromatin Structure and Function

YANMING WANG

6. Structure of SET Domain Protein Lysine Methyltransferases

BING XIAO, STEVEN J. GAMBLIN, AND JONATHAN R. WILSON

7. Non-Histone Protein Lysine Methyltransferases: Structure and Catalytic Roles

LYNNETTE M. A. DIRK, RAYMOND C. TRIEVEL, AND ROBERT L. HOUTZ

8. Demethylation Pathways for Histone Methyllysine Residues

FEDERICO FORNERIS, CLAUDIA BINDA, MARIA ANTONIETTA VANONI, ANDREA MATTEVI, AND ELENA BATTAGLIOLI

Part III: Biological Regulation by Protein Methyl Ester Formation

9. Structure and Function of Isoprenylcysteine Carboxylmethyltransferase (Icmt): A Key Enzyme in CaaX Processing

JESSICA L. ANDERSON AND CHRISTINE A. HRYCYNA

10. Genetic Approaches to Understanding the Physiologic Importance of the Carboxyl Methylation of Isoprenylated Proteins

STEPHEN G. YOUNG, STEVEN G. CLARKE, MARTIN O. BERGO, MARK PHILIPS, AND LOREN G. FONG

11. Reversible Methylation of Protein Phosphatase 2A

SARI LONGIN AND JOZEF GORIS

12. Reversible Methylation of Glutamate Residues in the Receptor Proteins of Bacterial Sensory Systems

FRANCES M. ANTOMMATTEI AND ROBERT M. WEIS

Part IV: Recognition of Damaged Proteins in Aging by Protein Methyltransferases

13. Protein L-Isoaspartyl, D-Aspartyl O-Methyltransferases: Catalysts for Protein Repair

CLARE M. O'CONNOR

Part V: Modification of Proteins by Methylation of Glutamine and Asparagine Residues

14. Modification of Glutamine Residues in Proteins Involved in Translation

HEIDI L. SCHUBERT

15. Modification of Phycobiliproteins at Asparagine Residues

ALAN V. KLOTZ

Part VI: Inhibition of Methyltransferases by Metabolites

16. Inhibition of Mammalian Protein Methyltransferases by 5′-Methylthioadenosine (MTA): A Mechanism of Action of Dietary SAMe?

STEVEN G. CLARKE

Preface

The third edition of *The Enzymes* strives to distill our wealth of knowledge about the catalysts that make life possible. This collection brings together investigators from various fields, highlighting how common threads connect seemingly disparate studies—cancer biologists with protein chemists, neuroscientists with enzymologists, and gerontologists with biochemists. One area of particular interest, especially for understanding the regulation of physiological function, concerns enzymes involved in the posttranslational modification of proteins.

With the determination of the human genome sequence, we have been learning how important posttranscriptional events are in providing diversity to gene products. Posttranslational modifications can both directly regulate protein function and play an important role in determining their intracellular localization and membrane associations. Furthermore, the stability and turnover of proteins are influenced by their modification. Finally, signal transduction reactions are largely dependent on protein covalent modification.

Protein methylation has recently emerged as one of the most exciting areas of study on posttranslational modification. For example, a search of the PubMed database in December of 2005 revealed that 250 of 491 papers retrieved for "protein arginine methylation" were published since 2000–more than for all of the previous 36 years combined. A search for "histone methylation" yielded 482 papers in 2005, with the year then yet to close! Other topics in protein methylation have gained similar prominence as their importance has been recognized. We are confident that the chapters in this volume will offer neophytes to the field a rapid and authoritative entry and will provide scientists already reaping the scientific fruits in these areas a broad and comprehensive view of the larger picture.

A large family of protein methyltransferases has been identified and their structural properties have been characterized. These studies have provided novel insights into how methylation regulates a variety of biological functions. This volume is intended to capture these recent developments concerning protein methyltansferases. We open this volume with the overview of diverse structures of the methyltransferase family (Chapter 1). We then focus on protein arginine and protein lysine methyltansferases that modify proteins involved in a number of cellular processes such as transcription, RNA metabolism, and DNA repair (Chapters 2 through 8).

Of particular interest is the regulation of chromosome function by methylating histones. Recently, the identification of the first lysine-specific histone demethylase LSD1 was reported. Discovery of this enzyme provided an answer to a long-standing dispute concerning the presence of histone demethylating enzymes. It is now clear that histone methylation is a dynamic reversible event. Methylation also plays critical roles in signal transduction involving cell proliferation. This is exemplified by the characterization of isomethylcysteine methyltransferases (Icmt). The significance of this membrane-bound enzyme in the function of Ras and other prenylated proteins is discussed in Chapters 9 and 10.

Regulation of signal transduction events by modifying protein phosphatase 2A or by affecting bacterial chemotaxis proteins is discussed in Chapters 11 and 12. Methylation plays important roles in aging, explored in Chapter 13. Methylation of the polypepetide release factor leading to the regulation of protein synthesis as well as regulation of spectroscopic properties of phycobiliprotein complex is discussed in Chapters 14 and 15.

The last chapter discusses regulation of protein methylation by metabolites, including a novel suggestion for the effectiveness of the commonly used nutriceutical SAMe, or S-adenosylmethionine, the methyl group donor of all of these reactions. Small molecule inhibitors of protein methyltransferases are beginning to be identified, and the discussion on this topic can be found in a number of chapters.

In this volume, we focus on enzymes catalyzing the methylation of proteins. However, there is a large number of enzymes that catalyze methylation on nucleic acids such as DNA and RNA methyltransferases. In addition, lipids and small molecules are also targets of methylation. Because of page restrictions, we have not been able to include these enzymes in this volume. These topics will be the subject of future volumes.

Fuyuhiko Tamanoi
Steven G. Clarke
Molecular Biology Institute, UCLA
December 8, 2005

Part I

Overview of Protein Methyltransferases

1

Protein Methyltransferases: Their Distribution Among the Five Structural Classes of AdoMet-Dependent Methyltransferases[1]

HEIDI L. SCHUBERT[a] • ROBERT M. BLUMENTHAL[b] • XIAODONG CHENG[c]

[a]*Department of Biochemistry*
University of Utah
15 North Medical Drive East
Salt Lake City, UT 84112, USA

[b]*Department of Medical Microbiology and Immunology and Program in Bioinformatics and Proteomics/Genomics*
Medical University of Ohio
3000 Arlington Avenue
Toledo, OH 43614, USA

[c]*Department of Biochemistry*
Emory University School of Medicine
1510 Clifton Road Northeast
Atlanta, GA 30322, USA

---

[1]The review is modified and updated from the article "Many Paths to Methyltransferase: A Chronicle of Convergence" (originally published in *TRENDS in Biochemical Sciences*, June 2003, vol. 28, no. 6, pp. 329–335) with permission of Elsevier Science Ltd.

THE ENZYMES, Vol. XXIV
Copyright © 2006 by Elsevier Inc.

I.   Abstract

S-adenosyl-L-methionine (AdoMet) dependent methyltransferases (MTases) are involved in biosynthesis, signal transduction, protein repair, chromatin regulation, and gene silencing. Five different structural folds (designated I through V) have been described that bind AdoMet and catalyze methyltransfer to diverse substrates, although the great majority of known MTases have the Class I fold. Even within a particular MTase class the amino-acid sequence similarity can be as low as 10%. Thus, the structural and catalytic requirements for methyltransfer from AdoMet appear to be remarkably flexible. MTases that act on protein substrates have been found to date among three of the five structural classes (I, the classical fold; III, the corrin MTase fold; and V, the SET fold).

"There are many paths to the top of the mountain, but the view is always the same."—Chinese proverb *The Columbia World of Quotations,* New York, Columbia University Press, 1996

II.   Introduction

Following ATP, *S*-adenosyl-L-methionine (AdoMet) is the second most widely used enzyme substrate [1]. The majority of AdoMet-dependent reactions involve methyltransfer, leaving the product *S*-adenosyl-L-homocysteine (AdoHcy). The huge preference for AdoMet over other methyl donors, such as folate, reflects favorable energetics resulting from the charged methylsulfonium center – the ΔG° for (AdoMet + Hcy $\rightarrow$ AdoHcy + Met) is -17 kcal mol^{-1}, over double that for (ATP $\rightarrow$ ADP + P$_i$) [1]. Methylation substrates range in size from arsenite to DNA and proteins, and the atomic targets can be carbon, oxygen, nitrogen, sulfur, or even halides [2, 3].

The first structure of an AdoMet-dependent methyltransferases (MTase), determined in 1993, was for the DNA C5-cytosine MTase M.HhaI [4]. For several years thereafter, a variety of additional MTases (with a wide range of different substrates) were found to share the same basic structure. More recently, however, AdoMet-dependent methylation has been found to be the target of functional convergence that is catalyzed by enzymes with remarkably distinct structures. The Protein Data Bank (PDB) currently (as of June 2005) includes >245 structures (with MTase in the title) for >126 distinct AdoMet-dependent MTases (structures with MTase in the title and sequences 90% similar removed) from >30 different classes of enzymes as defined by the Enzyme Classification (EC) system (Table 1.1; [5]).

The purpose of this review is to compare and contrast the five known structurally distinct families of AdoMet-dependent MTases (Classes I through V), focusing in particular on MTases with protein substrates. The phenomenon of

TABLE 1.1

FIVE CLASSES OF METHYLTRANSFERASE STRUCTURES (2003)

EC Classification	Name	PDB Code
Class I		
2.1.1.2	Guanidinoacetate N-MTase	1KHH, 1P1B, 1P1C, 1XCJ
2.1.1.5	Betaine homocysteine MTase	1LT7, 1LT8, 1YMY
2.1.1.6	Catechol O-Mtase	1VID, 1JR4, 1HID
2.1.1.8	Histamine N-MTase	1JQD, 1JQE, 1ICZ
2.1.1.20	Glycine N-Mtase	1BHJ, 1D2C, 1D2G, 1D2H, 1KIA, 1D2G, 1D2H, 1NBH, 1NBI, 1R74, 1XVA,
2.1.1.28	Phenylethanolamine N-MTase	1HNN, 1N7I, 1N7J
2.1.1.37	DNA (cytosine) C5-MTase	1G55
2.1.1.43	Dot1p-like histone-lysine N-MTase	1NW3, 1U2Z
2.1.1.46	Isoflavone O4'-MTase	1FP2, 1FPX
2.1.1.48	rRNA (adenine) N6-MTase	1QAM, 1QAN, 1QAO, 1QAQ, 1YUB, 2ERC
2.1.1.63	O6-Alkylguanine-DNA MTase	1EH6, 1EH7, 1EH8, 1MGT, 1QNT, 1SFE, 1T38, 1T39
2.1.1.68	Caffeate O-Mtase	1KYW, 1KYZ, 1KNS, 1L0U
2.1.1.72	Adenine-specific DNA N6-MTase	1G38, 1ADM, 1AQI, 1AQJ, 2ADM, 1EG2, 1G60, 2DPM, 1NW5, 1NW6, 1NW7, 1NW8, 1Q0S, 1Q0T, 1YF3, 1YFL
2.1.1.73	Cytosine-specific DNA C5-MTase	1OMH, 1DCT, 1FJX, 1HMY, 1MHT, 2HMY, 3MHT, 4MHT, 5MHT, 6MHT, 7MHT, 8MHT, 9MHT, 1M0E, 1SKM, 1SVU
2.1.1.77	Protein-L-isoaspartate O-MTase	1DL5, 1L1N, 1JG1, 1JG2, 1JG3, 1JG4, 1KR5, 1R18, 1VE
2.1.1.79	Cyclopropane-fatty-acyl-phospholipid synthase	1KP9, 1KPG, 1KPH, 1KPI, 1L1E, 1TPY
2.1.1.80	Protein-glutamate O-MTase	1AF7, 1BC5
2.1.1.104	Caffeoyl Coenzyme A 3-O-MTase	1SUI, 1SUS
2.1.1.113	Cytosine-N4-DNA-MTase	1BOO
2.1.1.125	Protein Arginine N-MTase	1F3L, 1G6Q, 1OR8, 1ORH, 1ORI
2.1.1.132	Precorrin-6y C-MTase	1KXZ, 1L3B, 1L3C, 1L3I
2.7.7.19	Polynucleotide adenylyltransferase	1VPT, 3MAG, 1B42, 1BKY, 1EAM, 1EQA, 1JSZ, 1JTE, 1JTF, 1P39, 1VP3, 1VP9, 1VPT, 2VP3, 3MAG, 3MCT, 4DCG, 1AV6, 1L2Q, 1V39
2.1.1.?	Isoliquiritigenin (chalcone) O2'-MTase	1FP1, 1FPQ
2.1.1.?	Salicylic Acid Carboxyl MTase	1M6E
2.1.1.?	Protein-glutamine N5-MTase	1NV8, 1NV9, 1T43, 1SG9, 1VQ1
2.1.1.?	FtsJ RNA MTase	1EIZ, 1EJ0
2.1.1.?	Ribosomal RNA uracil-C5-MTase	1UWV, 2BH2
2.1.1.?	Ribosomal RNA cytosine-C5-MTase	1SQF, 1SQG
2.1.1.?	Hypothetical MTase	1IM8, 1I9G, 1IXK, 1M6Y, 1DUS, 1WXW, 1WXX, 1WY7, 1WZN

Continued

TABLE 1.1

FIVE CLASSES OF METHYLTRANSFERASE STRUCTURES (2003)—cont'd

EC Classification	Name	PDB Code
Class II		
2.1.1.13	Reactivation domain of MetH	1MSK, 1J6R
Class III		
2.1.1.133	Precorrin-4 C11-MTase	1CBF, 2CBF, 2cbf
2.1.1.98	Diphthine synthase	1VCE, 1VHV, 1WDE
Class IV		
2.1.1.?	Ribosomal RNA 2′O-MTase	1GZO, 1IPA, 1X7O, 1X7P
2.1.1.31	tRNA guanine-N1-MTase	1OY5, 1P9P, 1UAJ
2.1.1.34	tRNA guanosine-2′-O-MTase	1J85, 1MXI, 1V2X
2.1.1.?	Hypothetical MTase	1NS5, 1O6D, 1VHK, 1K3R, 1NXZ
Class V		
2.1.1.43	Histone-lysine N-MTase	1ML9, 1MT6, 1H3I, 1MUF, 1N6A, 1N6C, 1O9S, 1MVX, 1MVH, 1PEG, 1XQH, 1ZKK, 2BQZ
2.1.1.127	[Ribulose-bisphosphate-carboxylase]-lysine N-MTase	1MLV, 1OZV, 1P0Y, 1ML9

enzymes from distinct structural families catalyzing the same reaction, termed enzyme analogy, has been noted for several decades [6]. Many enzymes exhibit "catalytic promiscuity," resulting in a pluripotency that can be shaped by mutation and selection [7, 8]. This can lead to a given protein structure playing several distinct catalytic roles [9], but also results in distinct protein structures playing a common catalytic role. Perhaps such flexibility is particularly easy where highly exergonic reactions are involved. As ATP is the only enzyme substrate more widely used than AdoMet, it seems logical that the current champion for greatest number of analogous families is the ATP-dependent protein phosphoryltransferases (protein kinases), with seven known structurally distinct families [10]. Nonetheless, this degree of analogy appears to be quite rare, and the five known families of AdoMet-dependent MTases also provide an impressive example of functional convergence.

III. Class I MTases: In the Beginning, an MTase Was an MTase Was an MTase

Starting with the M.HhaI DNA-MTase structure in 1993 [4], a continuing string of structures for AdoMet-dependent MTases have been reported. These structures are remarkably similar, comprising a seven-stranded β sheet with a central topological switch point and characteristic reversed β hairpin at the

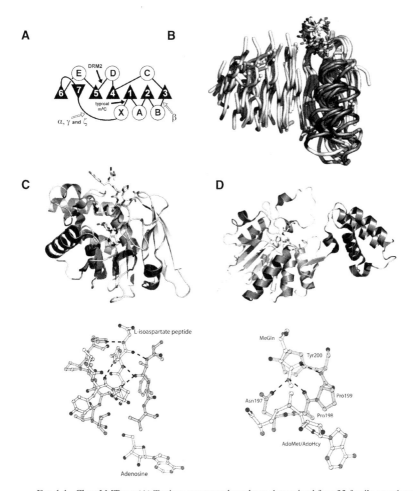

FIG. 1.1. Class I MTases (A) Tertiary structures have been determined for >33 family members, most containing a seven-stranded β sheet flanked by α helices. In this representation of circular permutation, the arrows indicate topological breakpoints at which the structure is opened in various DNA amino-MTases [60], generating the amino and carboxyl termini. *(Reprinted with permission [20].)* (B) The strong structural similarity within the Class I methyltransferases (MTases) can be seen in the alignment of 14 structures via their β strands. For clarity, only the peptide backbone of the catalytic core containing the main β sheet and the first α helix is shown. The cofactor (AdoMet or AdoHcy) is shown in stick-model form. Structures include protein isoaspartate O-MTase (PDB code 1I1N), protein arginine N-MTase (1F3L), cyclopropane-fatty-acyl-phospholipid synthase (1KPH), catechol O-MTase (1VID), M.TaqI (2ADM), M.HhaI (6MHT), CheR (1AF7), FtsJ (1EIZ), isoliquiritigenin O-MTase (1FP1), ErmC′ (1QAN), VP39 (1VPT), Hemk/PrmC (1NV8), mj0882 (1DUS), and hypothetical rv2118c (1I9G). (C) Protein isoaspartate O-MTase facilitates protein repair by binding the residues surrounding the mutant isoaspartate as if it were a β strand within the MTase structure (1JG3). (D) Protein glutamine N5-MTase PrmC/HemK bound to an AdoMet-AdoHcy mixture and partially methylated glutamine (1NV8). Atoms are colored as follows: oxygen atoms (dark), nitrogen atoms (medium gray), protein carbon atoms (shaded), and carbon atoms of AdoMet/AdoHcy and substrate (white).

carboxyl end of the sheet ($6\uparrow$ $7\downarrow$ $5\uparrow$ $4\uparrow$ $1\uparrow$ $2\uparrow$ $3\uparrow$; Figure 1.1a). This sheet is flanked by α helices to form a doubly wound open $\alpha\beta\alpha$ sandwich, and is henceforth referred to as the *Class I MTase* structure (Figure 1.1a). The first β strand typically ends in a GxGxG (or at least GxG) motif (the hallmark of a nucleotide-binding site) bending sharply underneath the AdoMet to initiate the first α helix. The only other strongly conserved position is an acidic residue at the end of $\beta2$ that forms hydrogen bonds to both hydroxyls of the AdoMet ribose [11]. The spatial conservation of this class of MTase is so pronounced that superimposing the cores (150 amino acids on average) of 20 MTases gives root-mean-square deviations for the Cα atoms of just 3.6 ± 0.5 Å when 190 pairwise comparisons are averaged. Figure 1.1b displays the alignment of 14 unique structures displaying their β sheet and first α helix. Several groups have noted that with the exception of strand 7 this structure is quite similar to that of the NAD(P)-binding Rossmann-fold domains [11]. In both groups of proteins, the central topological switch point results in a deep cleft in which the AdoMet or NAD(P) binds.

Although highly conserved, the Class I family contains several members with variant structural features [5]. Some Class I MTases are homodimeric [12], tetrameric [13], or even oligomeric [14, 15], but most are monomers. One of the smallest MTases (catechol O-MTase [16]) consists of just the consensus structural core, whereas most MTases contain auxiliary domains that are inserted throughout the MTase fold and appear to play roles in substrate recognition. Strands 6 and 7 are reversed in the primary sequence of protein isoaspartyl MTase [17], and are absent from protein arginine (R) MTase (PRMT) [12, 14, 15]. Finally, circular permutation of the overall topology has been proven for some DNA MTases [18, 19] and is predicted for some RNA MTases [20] (Figure 1.1a). Together, these differences suggest structural flexibility, but it (incorrectly) appeared for a time that AdoMet-dependent MTases were variations on a single basic theme.

Many of the known Class I MTases act on DNA to regulate gene expression, to repair mutations, or to protect against bacterial restriction enzymes. Initially, it was a mystery as to how MTases acted on nucleotides that are held inside the DNA duplex by base pairing and stacking (seemingly inaccessible to an enzyme's active site). The answer came from the M.HhaI DNA C5-MTase complexed with a synthetic DNA duplex [21]. In a process termed *base flipping*, the enzyme simply rotates the target DNA base ~180 degrees on its flanking phosphodiester bonds such that the base projects into the catalytic pocket [22]. This strategy helps explain the tremendous diversity of substrates accommodated by MTases sharing the Class I structure.

For the great majority of protein MTases characterized to date, information on substrate recognition is limited to an enzyme-peptide substrate complex [14, 23] (Figure 1.1c) or an enzyme with a free residue in the active site [24] (Figure 1.1d), as opposed to an enzyme-protein substrate complex. Whereas methylated amino acids often project from the surface of their protein substrates and may lie on a helix, strand, or loop, structural loops containing modified amino acids probably adopt alternate conformations depending on their status. For example, the GGQ loop of the translation release factor (the HemK substrate) is methylated on glutamine, but the Gln residue is not particularly accessible in the independent crystal structures of the release factor [25], and it is not visible in structures of complexes with the ribosome [26].

IV. Class II MTases: A Lesson in Cobalamin

As early as 1996, there was a hint that not all AdoMet-dependent MTases would follow the same structural theme. The *Escherichia coli* cobalamin (vitamin B_{12})-dependent methionine synthase, MetH, generates methionine from homocysteine, transferring a methyl group from a folate derivative to the bound cobalamin and thence to homocysteine. Periodically, the B_{12} cobalt is oxidized to a dead-end form, and reactivation requires reductive methylation using AdoMet, flavodoxin, and an additional structural domain (Figure 1.2a) [27]. The MetH reactivation domain (*Class II MTase*)—dominated by a long, central, antiparallel β sheet flanked by groups of helices at either end (Fig. 2b)—looks nothing like the Class I MTases in either overall architecture or interactions with AdoMet [28]. AdoMet is bound in an extended conformation to a shallow groove along the edges of the β strands, forming hydrogen bonds to a conserved RxxxGY motif [28] (Figure 1.2c). Large conformational changes are required to position the reactivation domain near the cobalamin substrate within the main catalytic domain [28, 29]. However, interesting from a biochemical standpoint the flavodoxin requirement and subordinate role of this reactivation domain led many to underestimate its relevance to understanding the structural diversity possible among independent MTases. It turned out that this "black sheep" was just the beginning.

V. Class III MTases: Another Lesson in Cobalamin

By 1998 the MTase field was again surprised; this time by the homodimeric structure of CbiF, an MTase that acts on ring carbons of the large planar precorrin substrates during cobalamin biosynthesis [30]. The active site in this structural family (*Class III MTases*) is tucked into a cleft between two αβα domains, each containing five strands and four helices (Figure 1.3a). A GxGxG motif occurs at the C-terminal end of the first β strand, similar to the Class I enzymes, but surprisingly does not contact AdoMet (at least in the absence of the precorrin substrate). Instead, the AdoMet is tightly folded, and binds between the two domains (Figure 1.3b).

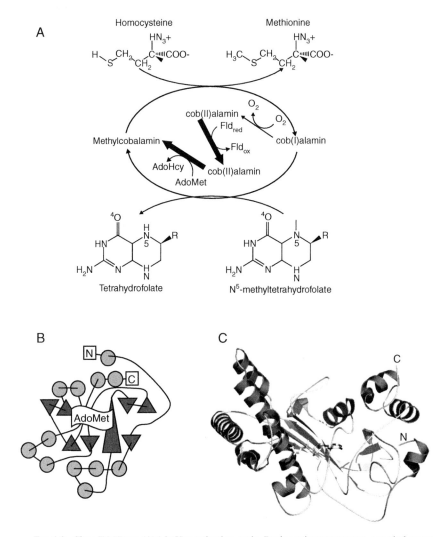

FIG. 1.2. Class II MTases (A) MetH reactivation cycle. During primary turnover, a methyl group is transferred from CH_3-H_4 folate to the enzyme-bound cob(I)alamin cofactor, forming H_4-folate and methylcobalamin. Methylcobalamin is then demethylated by Hcy, forming methionine (Met) and regenerating cob(I)alamin [80]. In reactivation, cob(II)alamin is reduced by one-electron transfer from flavodoxin and the resulting cob(I)alamin is methylated by AdoMet [80]. *(Reprinted with permission [81]. Copyright 2004 National Academy of Sciences, USA.)* (B) The reactivation domain of methionine synthase contains a series of long β strands. (C) AdoMet binds to in a shallow groove on the surface of the MetH domain (1MSK).

Class III MTases also include enzymes that act on amino acids within proteins. During the synthesis of diphthamide (the amino acid targeted by diphtheria toxin), AdoMet is utilized in two different ways. Initially the 3-amino-3-carboxypropyl group from AdoMet is transferred to a histidine residue on elongation factor 2 to generate a modified amino acid, which is later tri-methylated on the amino group by diphthine synthase [31]. Diphthine synthase adopts the Class III MTase structure (Figure 1.3c) (PDB 1VHV [32], 1WDE [33], 1VCE [34]). It is interesting that all described Class III MTases act on substrates that contain imidazole rings. However, diphthine synthases methylate the amino carboxyl modification

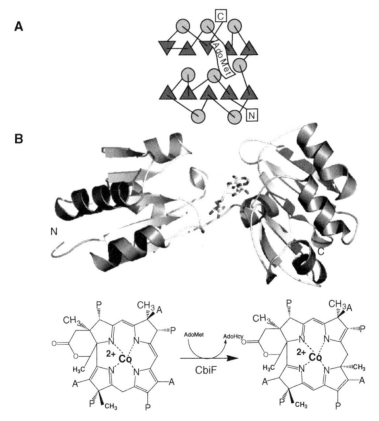

FIG. 1.3. Class III MTases (A) The bilobal structure of CbiF contains two αβα domains. (B) (Top) AdoMet binds between the two domains of CbiF and a groove in the N-terminal domain is proposed to be the active-site cleft (1CBF). (Bottom) CbiF catalyzes one of eight methylations in the formation of cobalamin (vitamin B_{12}). The majority of MTases are predicted to adopt the Class III fold and methylate a ring carbon in the presence or absence of the central cobalt ion. CbiF specifically methylates carbon C11, though a later enzyme transfers the methyl group to C12 where it is found in cobalamin.

Continued

C

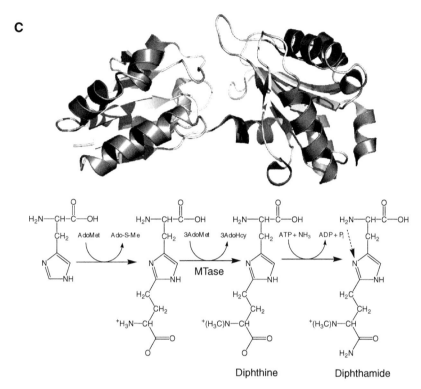

Diphthine   Diphthamide

FIG. 1.3. Class III MTases—cont'd (C) (Top) Structure of diphthine synthase (1VHV). (Bottom) Diphthamide is modified histidine residue in EF-2, and the target of the Diphtheria toxin which goes on to ADP-ribosylate the residue on an imidazole nitrogen (dashed arrow). A multi-step synthesis begins with a 3-amino-3-carboxypropyl transfer from AdoMet to the imidazole side chain, followed by tri-methylation of the free amine group by a Class III MTase, Dph5. A final ATP-dependent amidation of the carboxyl group yields diphthamide.

of histidine, while corrin MTases act on the imidazole ring itself. Thus the diphtine synthases and the corrin MTases are unlikely to have very similar substrate recognition processes.

VI.   Class IV MTases: Knotty New Structure SPOUTs

Two additional disparate MTase structural classes were recognized in the same year (2002). The SPOUT family of RNA MTases provides the only known cases of *Class IV* structure [35–37]. These enzymes are unique in three ways. First, they include a six-stranded parallel β sheet flanked by seven α helices, of which the first three strands form half a Rossman fold (Figure 1.4a). Second, the

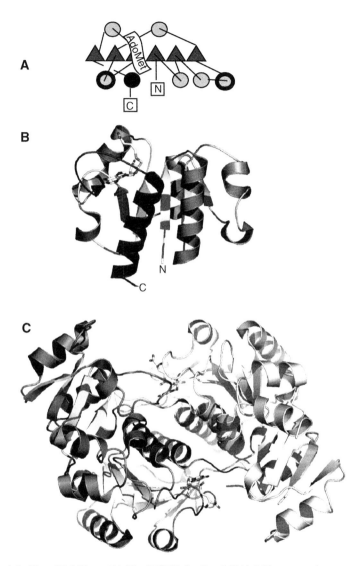

FIG. 1.4. Class IV MTases (A) The SPOUT family of RNA MTases contains a novel knot structure at the C-terminus. Helices that are not conserved between family members are shown with a bold circle in the topology diagram. (B) AdoHcy binds above strands 4 to 6 of YibK in a bent conformation (1MXI). (C) The active site(s) of AviRb from *Streptomyces viridochromogenes* is located near the subunit interface of a homodimer (1X7P).

topology of the structure is such that a significant portion of the C terminus is tucked back into the structure in a "knot" (Figure 1.4b). This rare substructure is formed by the last ~30 residues, including the last α helix, and contains several catalytic residues that confirm its structural importance [36, 37]. Third, their active site is located near the subunit interface of a homodimer, and may encompass residues extending from both subunits (Figure 1.4c). The structure of a SpoU homolog, *Haemophilus influenzae* (HI0766) YibK, has been determined in the presence of AdoHcy [38] (Figure 1.4b), whereas the structure of the antibiotic resistance-mediating MTase AviRb from *Streptomyces viridochromogenes* has been determined in the presence of AdoMet [39] (Figure 1.4c). Both structures revealed the AdoMet/AdoHcy to be bound above strands 4 to 6 in a bent conformation (see material following).

VII.   Class V MTases: Pseudo Knotty Structures Come in SETs

The most recently described structural family of AdoMet-dependent MTases is the SET-domain proteins [40–45], which constitute *Class V*. These Class V MTases contain a series of eight curved β strands forming three small sheets (Figure 1.5a), with the C-terminus tucked underneath a surface loop forming a knot-like structure similar to the Class IV MTases but constructed on a totally different topology (Figure 1.5b)[40–45]. Flanking the SET domain are diverse sequences termed the pre- and post-SET regions, which are often essential for MTase activity and might participate in substrate recognition and specificity [46]. AdoHcy bound to the SET domain is kinked in a manner similar to that of the Class III CbiF-bound AdoHcy and binds on a concave surface of the enzyme (Figure 1.5c) near an invariant tyrosine residue that has been implicated in the catalytic reaction.

Amino acid sequence comparison [47] suggests that there are hundreds of these proteins, and several have been shown to methylate lysines in the flexible

FIG. 1.5. Class V MTases (A) The SET domain containing a protein (histone)-lysine N-MTase family is formed by the combination of three small β sheets. (B) Representative examples of SET domain containing structures: (*left-hand panel*) *Neurospora* DIM-5 (PDB 1PEG) and (*right-hand panel*) human SET7/9 (PDB 1O9S). (C) Active site of SET domain: (*left-hand panel*) H3 peptide-binding site in DIM-5 with the target Lys-9 inserted into a channel (PDB 1PEG) and (*right-hand panel*) the AdoHcy-binding site in SET7/9, located at the opposite end of the target lysine-binding channel (PDB 1O9S). (D) The active sites in DIM-5 (PDB 1PEG) (*left-hand panel*) and SET7/9 (PDB 1MT6) (*right-hand panel*). The arrow indicates the movement of the methyl group transferred from the AdoMet methylsulfonium group to the target amino group. (E) Structural comparison of active sites in DIM-5 and SET7/9. Either two tyrosines and one phenylalanine (DIM-5) or three tyrosines (SET7/9) surround the target lysine. (*Panels B through E reprinted with permission from the Annual Review of Biophysics and Biomolecular Structure, Volume 34(c), 2005, by Annual Reviews, www.annualreviews.org.*)

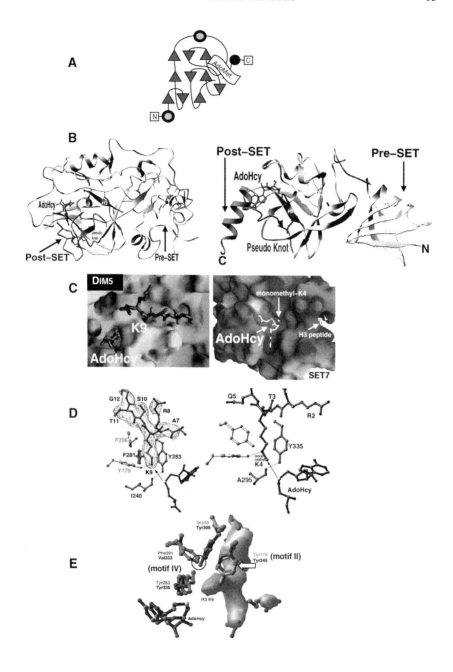

tails of histones or in Rubisco (ribulose-1,5-bisphosphate carboxylase) [48]. Five ternary structures for Class V MTases have been solved: human SET8 (also called Pr-SET7) in complex with a peptide containing histone H4 Lys-20 [49, 50], human SET7/9 in complex with either a peptide containing histone H3 Lys-4 [51] (Figure 1.5c, right-hand panel) or a peptide derived from tumour suppressor p53 [52], *Neurospora crassa* DIM-5 in complex with histone H3 Lys-9 peptide [53] (Figure 1.5c, left-hand panel), and a nonhistone Rubisco MTase in complex with a free lysine [54]. These structures reveal that the target lysine is inserted into a narrow channel such that the target nitrogen lies in close proximity to the methyl-donor AdoMet at the opposite end of the channel (Figure 1.5c). Aromatic side chains form the channel wall, and make van der Waals contacts to the methylene part of the target lysine side chain. At the bottom of the channel, the terminal ε-amino group of the substrate lysine hydrogen bonds the hydroxyl of a catalytic Tyr (Y178 in DIM-5, Y245 in SET7/9) (Figure 1.5d). However, in Rubisco MTase the ε-amino is hydrogen bound to the main chain carbonyl of Arg222.

SET-domain histone lysine MTases differ both in their substrate specificity for the various acceptor lysines and in their product specificity for the number of methyl groups (one, two, or three) they transfer. DIM-5 and SET7/9 generate distinct products: DIM-5 forms trimethyl-lysine [53, 55] and SET7/9 forms only monomethyl-lysine [51, 53]. A likely structural explanation for their different product specificities is that residues in the lysine-binding channel of SET7/9 sterically exclude the target lysine side chain with methyl group(s). Comparison of the two active sites pinpointed the difference to a single amino acid that occupies a structurally similar position in both enzymes (F281 of DIM-5 and Y305 of SET7/9). Although the two residues are not aligned at the primary sequence level, the edge of the F281 phenyl ring in DIM-5 points to the same position as the Y305 hydroxyl in SET7/9, both in close proximity to the terminal ε-amino group of target lysine (Figure 1.5e). It was hypothesized that the Y305 hydroxyl in SET7/9 may be the source of steric hindrance limiting methylation [53]. Therefore, DIM-5 and SET7/9 mutants that swapped these residues were created. Remarkably, this swap almost completely inverts methylation product specificity [53]. Importantly, neither the lysine target specificity (Lys-9 versus Lys-4) nor the overall reaction rate for each enzyme was changed. Thus, DIM-5 was converted from a Lys-9 tri-MTase to a Lys-9 mono/di-MTase by F281Y mutation, whereas SET7/9 Y305F generated dimethylated instead of monomethylated Lys-4.

VIII.   Conformations of AdoMet and AdoHcy

The bound AdoMet or AdoHcy ligand exhibits significantly different conformations in the five structural classes, which emphasizes its flexibility. Figure 1.6 compares the prototypical AdoMet and AdoHcy conformations of each structural class by aligning the molecules via their ribose moieties. The ribose ring of

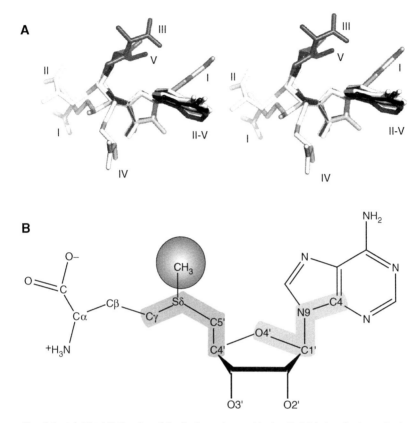

FIG. 1.6. AdoMet (A) Overlay of the *S*-adenosyl-L-methionine (AdoMet) or *S*-adenosyl-L-homocysteine (AdoHcy) conformations in different methyltransferase (MTase) classes. The molecules are aligned via the ribose rings, and show varying degrees of bend at the intersection between the adenosine and methionine moieties. Described in order of structural class, indicating enzyme/PDB code: Class I, M.HhaI/ 6MHT; Class II, MetH/1MSK; Class III, CbiF/1CBF; Class IV, YbiK/1K2X; and Class V, Rubisco/ 1MVL. (B) AdoMet atomic nomenclature. The O4′-C1′-N9-C4 and C4′-C5′-Sδ-Cγ dihedrals are highlighted light and dark gray, respectively.

AdoMet in Class I adopts an envelope 2′-endo conformation, with its base in the anti position at ~135 degrees (defined by the O4′-C1′-N9-C4 dihedral; Figure 1.6b). The ribose rings of the other AdoMet classes adopt an envelope 2′-exo conformation with the base in the anti position at ~180 degrees. These differences, although significant, are small compared with the differences in the C4′-C5′-Sδ-Cγ dihedral angle, which begins to define whether the nucleotide is extended or folded. An extended *trans* conformation (~180 degrees) is adopted by the Class I MTases; an angle of approximately –90 degrees is adopted by Classes II, III and V; and the AdoHcy twists in the opposite direction with an angle of 80 degrees in Class IV. The next dihedral angle (C4′-C5′-Sδ-Cγ; Figure 1.6b) further separates the classes such that overall the Class I and Class II ligands are relatively extended but Class III through V MTases bind the methyl donor in tightly folded conformations (Figure 1.6a). Because AdoMet is an exchanged substrate, its solvent accessibility might be directly related to the rate of catalytic turnover, or could at least provide an indication of structural flexibility of the catalytic core. The classes of MTase surround their methyl donor to differing degrees such that in catechol-O-MTase (Class I) <1% of the AdoMet surface area is exposed to solvent, whereas this exposure is 8 to 21% in the other classes.

IX.   Diverse Set of Mechanisms for a Conserved Class of MTase

Substrate-bound complexes have been determined mainly for Class I structures, although as noted previously several Class V (SET) MTases have been characterized in complex with lysine-containing peptides [51–53]. All MTases are thought to proceed with direct transfer of the methyl group to substrate with inversion of symmetry in an S_N2-like mechanism [56, 57]. This reaction also requires that a proton be removed before, concurrent with, or after methyl transfer (Figure 1.7a). Even within the structurally conserved family of Class I MTases, a wide variety of mechanisms have evolved to activate the catalytic nucleophile, dependent on the polarizability of the target atom.

A.   N-METHYLATION

Nitrogen appears to be the dominant target among protein MTases. A common mode of substrate binding is the use of an [D/N/S]PP[Y/F] motif (DPPY, for brevity), employed by the protein N5-glutamine MTase, PrmC/HemK [24], and by DNA N6-adenine [58, 59] (Figure 1.7b) and DNA N4-cytosine MTases [18]. This set of diverse substrates indicates that the DPPY motif is not nucleotide specific but is selective for nitrogens conjugated to a planar system such as an amide moiety or a nucleotide base. DPPY motifs extend from the C-terminus of β strand 4 in these Class I structures (see Figure 1.1a), in which the di-proline bends the polypeptide toward the surface of the active site. Two hydrogen bonds are formed between the nucleophilic nitrogen and both the oxygen atom of the [D/N/S] side chain and the carbonyl oxygen of the first

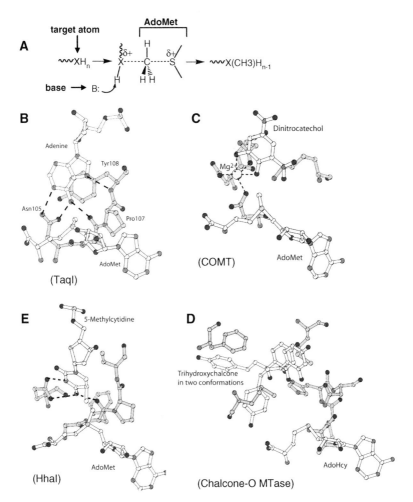

FIG. 1.7. Mechanism and substrate interactions of AdoMet-dependent MTases (A) General mechanism of methyl transfer from AdoMet to a target atom (X), and proton abstraction via a base (B:). (B) DNA adenine N6-MTase M.TaqI bound to adenosine (a combination of two PDB files: 2ADM and 1G38). (C) Catechol O-MTase bound to magnesium and 3,5-dinitrocatechol (1VID). (D) Chalcone O-MTase bound to two alternate conformations of 2′,4,4′-trihydroxychalcone (1FP1). (E) DNA C5-cytosine MTase M.HhaI bound to a mixture of AdoMet and AdoHcy and partially methylated cytosine (6MHT).

proline (Figure 1.7b and Figure 1.1d, bottom panel). These hydrogen bonds position the substrate such that the lone-pair electrons on the nitrogen nucleophile point toward the incoming methyl group [60]. A charged methylamine intermediate results after methyl transfer, but resolves into the neutral sp^2-hybridized amide upon proton loss to solvent during product release [24, 57, 61]. Other Class I N-MTases (including protein arginine MTase [12, 14, 15], small molecule glycine-N MTase [13], and phenylethanolamine-N MTase [62]) do not contain the DPPY motif but instead use acidic residue(s) to neutralize the positive charge on the substrate amino group.

Using entirely different structural scaffolding, the Class V SET-domain MTases bind to a kinked AdoMet molecule on the opposite side of a small channel from the N5 nitrogen of a peptide lysine substrate [51–54] (Figure 1.5d). The C-terminal tail of the domain (post-SET, Figure 1.5b) forms a pseudo knot and provides an integral part of the hydrophobic active-site pocket, either by packing an α-helix (SET7/9; Figure 1.5, right-hand panel) or a metal center (DIM-5; Figure 1.5b, left-hand panel) onto the active site.

B. O-METHYLATION

Several O-MTase structures have been determined, including catechol O-MTase (COMT) [16], the dimeric structures of two plant O-MTases [63], and the protein glutamate O-MTase CheR [64]. A Mg^{2+} ion is required to bind and orient the two catechol hydroxyls in COMT (Figure 1.7c). Molecular simulation and pK_a studies suggest that the Mg^{2+} ion acts primarily to organize the substrate-binding site and not as a general base [65, 66]. Instead, a nearby Lys appears to deprotonate the substrate hydroxyl before attack on the AdoMet methyl group. The chalcone (Figure 1.7d) and isoflavone O-MTases do not require a metal ion but do require a histidine residue to deprotonate the hydroxyls of these plant metabolites [63].

Glutamate O-methylation by CheR differs in that the methyl group is transferred to a carboxylic acid rather than to a hydroxyl moiety. The active site of CheR contains the side chains of arginine and tyrosine residues [64], which might position the glutamic acid substrate and facilitate the methyl-transfer reaction. The negatively charged carboxyl group might attack the methyl group unassisted. The isoapartate O-MTase recognizes damaged asparagine residues as part of an essential repair process. This enzyme binds a VYP(L-isoAsp)HA peptide by forming a series of hydrogen bonds to the surrounding peptide main chain as if the substrate were a β strand within the enzyme itself [23] (Figure 1.1c). The other protein MTases may form similar interactions with their substrates: PRMT, CheR, HemK/PrmC, and the Class V SET-domain lysine N-MTases [51–53] (see Figure 1.5b).

C.   C-METHYLATION

Very little is known about protein MTases that directly target a carbon atom. However, methyl groups attached to carbon atoms have been observed on a glutamine and an arginine within the methylcoenzyme M reductase structure [67]. The novel modifications are at the Cα carbon of glutamine, resulting in 2-(S)-methylglutamine and Cδ carbon of arginine, in turn resulting in 5-(S)-methylarginine. The methyl groups do derive from AdoMet, but this happens at a stage in protein synthesis prior to quaternary structure formation as the residues are buried within the core of the protein near the active site.

Many substrate-bound structures are known for DNA C5-cytosine MTases, including M.HhaI [21, 68]. A ProCys dipeptide is universally conserved within the active site of C5-cytosine MTases, and structurally resembles the PY portion of the DPPY motif (see section on N-methylation), whereas an aspartic acid residue from a neighboring portion of the enzyme functionally replaces the [D/N/S] residue of the DPPY motif. In M.HhaI, the N4 nitrogen of the target cytosine is positioned farther away from the AdoMet by hydrogen bonds such that the C5 atom is presented as a methylation target (Figure 1.7e). Because methylation on carbon atoms is more difficult than on polarizable nitrogens, the nucleotide must first be activated by covalent-bond formation between the conserved Cys thiol and carbon C6 [69]. This generates a negative charge on C5 that facilitates methyl transfer.

The functional flexibility of the Class I structure is further illustrated by a large family of RNA C5-cytosine MTases [70]. These enzymes adopt a Class I fold. However, the cysteine nucleophile [71] in the active site is contributed by the end of β strand 5 rather than the end of strand 4, as in M.HhaI and many other C5-cytosine MTases (Figure 1.1a). The functional significance of this surprising variation in linear (as opposed to 3D) position of the active site nucleophile, confirmed by the structures of *E. coli* Fmu [72] (an rRNA C5-cytosine MTase) and RumA [73, 74] (an rRNA 5-methyluridine MTase), is that Cys389-SG in RumA is covalently attached to the C6 of target uridine through a thioether linkage [74].

X.   Conclusions

Evolution has independently achieved AdoMet-dependent MTase activity at least five times, producing five unique structural MTase classes. These are summarized and compared in Figure 1.8 (color figures at end of book). Most of the other examples of analogous enzyme families also use substrates, such as ATP or NAD, which include a nucleotide "handle" for binding. The Class I and Class IV MTases are plausibly derived from Rossmann-fold proteins, and even

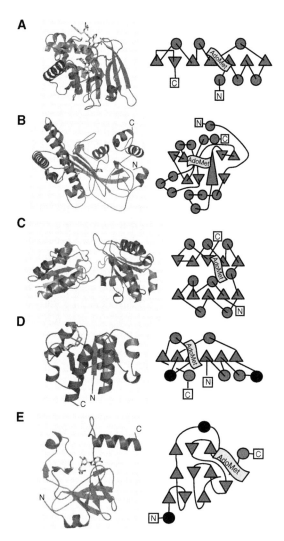

FIG. 1.8. The tertiary structures of the five classes of *S*-adenosyl-ʟ-methionine (AdoMet)-dependent methyltransferases (MTases). In each case a representative tertiary structure (left) and topology diagram (right) is shown. (A) Protein isoaspartate O-MTase (1JG3). (B) The reactivation domain MetH of methionine synthase (1MSK). (C) Diphthine synthase (1VHV). (D) SpoU homolog, *Haemophilus influenzae* (HI0766) YibK (1MXI). (E) SET7/9 (1O9S). (See color plate.)

the class III CbiF structure contains a GxGxG nucleotide-binding motif, but uses it unconventionally. The Class II and Class V MTases do not appear to have structural analogs, and thus their evolutionary history is not yet clear. If the Class I MTases are actually derived from the ubiquitous Rossmann-fold proteins, multiple independent evolutionary sublineages might explain the predominance of the Class I enzyme family. The limited sequence similarity between Class I proteins could even be consistent with independent evolution from a generic GxGxG-containing nucleotide-binding domain.

Catalysis of AdoMet-dependent methyltransfer does not appear to be rigidly restricted by tertiary structure or local spatial requirements. Even within the structural constraints of the Class I family, many different methods of binding substrate and activating a nucleophile have been described. One important question that remains is whether chance or functional constraints define which reactions are carried out by which classes of MTase. It could be argued that precorrin MTases have adopted a novel conformation because of their rigid, planar substrates. However, it appears that not all tetrapyrrole biosynthetic MTases are in Class III. Precorrin-C6 MTase CbiT has [75], and protoporphyrin IX O-MTase is predicted to have, a Class I structure [76]. We now know, further-more, that not all Class III MTases act on corrins, as illustrated by the diphthine synthases (see previous discussion). Similarly, not all histone-lysine N-MTases contain the Class V SET domain. For example, the Dot1 histone H3-Lys79 N-MTase belongs to Class I [77, 78]. At this point, it is not yet clear whether the (incompletely) distinct roles played by the various MTase classes is a result of functional differences resulting from the various structural frameworks or is just an accident of evolutionary history.

New methyltransferase activities are still being described, and although the genome projects are providing large lists of enzymes orthologous to members of the five known structural classes it would not be a great shock to find members of MTase Class VI and up among uncharacterized open reading frames [79]. In fact, given what we now know about enzyme evolution these new MTase classes might already have been (mis)annotated based on their similarity to characterized enzymes [9]. The challenge will be for biochemists to test annotation claims, or to discover the real identity of these new genes, whereas structural biologists can begin to address substrate recognition and catalytic mechanisms, particularly in the newer structural classes.

Protein MTases play strikingly diverse roles (as outlined in the other chapters of this volume), and even protein MTases playing essentially the same role can belong to distinct functional classes (as illustrated by the Dot1p and SET histone-lysine N-MTases mentioned previously). Protein MTases structurally character-ized to date fall into classes I, III, and V, but there is no reason to doubt that our appreciation of protein MTase structural diversity will continue to parallel our growing understanding of their substantial functional diversity.

ACKNOWLEDGMENTS

H.L.S. was supported by a grant from NIH (GM56775), R.M.B. was supported by a grant from the U.S. NSF (MCB-0516692), and X.C. was supported by a grant from NIH (GM068680) and the Georgia Research Alliance.

REFERENCES

1. Cantoni, G.L. (1975). Biological methylation: selected aspects. *Annu Rev Biochem* 44:435–451.
2. Ohsawa, N., Tsujita, M., Morikawa, S., and Itoh, N. (2001). Purification and characterization of a monohalomethane-producing enzyme S-adenosyl-L-methionine: halide ion methyltransferase from a marine microalga, Pavlova pinguis. *Biosci Biotechnol Biochem* 65:2397–2404.
3. Attieh, J.M., Hanson, A.D., and Saini, H.S. (1995). Purification and characterization of a novel methyltransferase responsible for biosynthesis of halomethanes and methanethiol in Brassica oleracea. *J Biol Chem* 270:9250–9257.
4. Cheng, X., Kumar, S., Posfai, J., Pflugrath, J.W., and Roberts, R.J. (1993). Crystal structure of the HhaI DNA methyltransferase complexed with S-adenosyl-L-methionine. *Cell* 74:299–307.
5. Martin, J.L., and McMillan, F.M. (2002). SAM (dependent) I AM: the S-adenosylmethionine-dependent methyltransferase fold. *Curr Opin Struct Biol* 12:783–793.
6. Galperin, M.Y., Walker, D.R., and Koonin, E.V. (1998). Analogous enzymes: independent inventions in enzyme evolution. *Genome Res* 8:779–790.
7. O'Brien, P.J., and Herschlag, D. (1999). Catalytic promiscuity and the evolution of new enzymatic activities. *Chem Biol* 6:R91–R105.
8. Anantharaman, V., Aravind, L., and Koonin, E.V. (2003). Emergence of diverse biochemical activities in evolutionarily conserved structural scaffolds of proteins. *Curr Opin Chem Biol* 7:12–20.
9. Gerlt, J.A., and Babbitt, P.C. (2001). Divergent evolution of enzymatic function: mechanistically diverse superfamilies and functionally distinct suprafamilies. *Annu Rev Biochem* 70:209–246.
10. Cheek, S., Zhang, H., and Grishin, N.V. (2002). Sequence and structure classification of kinases. *J Mol Biol* 320:855–881.
11. Fauman, E.B., Blumenthal, R.M., and Cheng, X. (1999). Structure and evolution of AdoMet-dependent methyltransferases. In S-adenosylmethionine-dependent Methyltransferases: Structures and Functions. Cheng, X., and Blumenthal, R.M. (eds.), pp 1–38, World Scientific Publishing, River Edge, New York.
12. Zhang, X., Zhou, L., and Cheng, X. (2000). Crystal structure of the conserved core of protein arginine methyltransferase PRMT3. *Embo J* 19:3509–3519.
13. Huang, Y., Komoto, J., Konishi, K., Takata, Y., Ogawa, H., Gomi, T., Fujioka, M., and Takusagawa, F. (2000). Mechanisms for auto-inhibition and forced product release in glycine N-methyltransferase: crystal structures of wild-type, mutant R175K and S-adenosylhomocysteine-bound R175K enzymes. *J Mol Biol* 298:149–162.
14. Zhang, X., and Cheng, X. (2003). Structure of the predominant protein arginine methyltransferase PRMT1 and analysis of its binding to substrate peptides. *Structure* 11:509–520.
15. Weiss, V.H., McBride, A.E., Soriano, M.A., Filman, D.J., Silver, P.A., and Hogle, J.M. (2000). The structure and oligomerization of the yeast arginine methyltransferase, Hmt1. *Nat Struct Biol* 7:1165–1171.
16. Vidgren, J., Svensson, L.A., and Liljas, A. (1994). Crystal structure of catechol O-methyltransferase. *Nature* 368:354–358.

17. Skinner, M.M., Puvathingal, J.M., Walter, R.L., and Friedman, A.M. (2000). Crystal structure of protein isoaspartyl methyltransferase: a catalyst for protein repair. *Structure* 8:1189–1201.
18. Gong, W., O'Gara, M., Blumenthal, R.M., and Cheng, X. (1997). Structure of pvu II DNA-(cytosine N4) methyltransferase, an example of domain permutation and protein fold assignment. *Nucleic Acids Res* 25:2702–2715.
19. Scavetta, R.D., Thomas, C.B., Walsh, M.A., Szegedi, S., Joachimiak, A., Gumport, R.I., and Churchill, M.E. (2000). Structure of RsrI methyltransferase, a member of the N6-adenine beta class of DNA methyltransferases. *Nucleic Acids Res* 28:3950–3961.
20. Bujnicki, J.M., Feder, M., Radlinska, M., and Blumenthal, R.M. (2002). Structure prediction and phylogenetic analysis of a functionally diverse family of proteins homologous to the MT-A70 subunit of the human mRNA:m(6)A methyltransferase. *J Mol Evol* 55:431–444.
21. Klimasauskas, S., Kumar, S., Roberts, R.J., and Cheng, X. (1994). HhaI methyltransferase flips its target base out of the DNA helix. *Cell* 76:357–369.
22. Cheng, X., and Roberts, R.J. (2001). AdoMet-dependent methylation, DNA methyltransferases and base flipping. *Nucleic Acids Res* 29:3784–3795.
23. Griffith, S.C., Sawaya, M.R., Boutz, D.R., Thapar, N., Katz, J.E., Clarke, S., and Yeates, T.O. (2001). Crystal structure of a protein repair methyltransferase from Pyrococcus furiosus with its L-isoaspartyl peptide substrate. *J Mol Biol* 313:1103–1116.
24. Schubert, H.L., Phillips, J.D., and Hill, C.P. (2003). Structures along the catalytic pathway of PrmC/HemK, an N5-glutamine AdoMet-dependent methyltransferase. *Biochemistry* 42:5592–5599.
25. Vestergaard, B., Van, L.B., Andersen, G.R., Nyborg, J., Buckingham, R.H., and Kjeldgaard, M. (2001). Bacterial polypeptide release factor RF2 is structurally distinct from eukaryotic eRF1. *Mol Cell* 8:1375–1382.
26. Rawat, U.B., Zavialov, A.V., Sengupta, J., Valle, M., Grassucci, R.A., Linde, J., Vestergaard, B., Ehrenberg, M., and Frank, J. (2003). A cryo-electron microscopic study of ribosome-bound termination factor RF2. *Nature* 421:87–90.
27. Drummond, J.T., Huang, S., Blumenthal, R.M., and Matthews, R.G. (1993). Assignment of enzymatic function to specific protein regions of cobalamin-dependent methionine synthase from Escherichia coli. *Biochemistry* 32:9290–9295.
28. Dixon, M.M., Huang, S., Matthews, R.G., and Ludwig, M. (1996). The structure of the C-terminal domain of methionine synthase: presenting S-adenosylmethionine for reductive methylation of B12. *Structure* 4:1263–1275.
29. Jarrett, J.T., Huang, S., and Matthews, R.G. (1998). Methionine synthase exists in two distinct conformations that differ in reactivity toward methyltetrahydrofolate, adenosylmethionine, and flavodoxin. *Biochemistry* 37:5372–5382.
30. Schubert, H.L., Wilson, K.S., Raux, E., Woodcock, S.C., and Warren, M.J. (1998). The X-ray structure of a cobalamin biosynthetic enzyme, cobalt-precorrin-4 methyltransferase. *Nat Struct Biol* 5:585–592.
31. Chen, J.Y., and Bodley, J.W. (1988). Biosynthesis of diphthamide in Saccharomyces cerevisiae. Partial purification and characterization of a specific S-adenosylmethionine: elongation factor 2 methyltransferase. *J Biol Chem* 263:11692–11696.
32. Genomix, S. (2003). Crystal Structure Of Diphthine Synthase.
33. Kishishita, S., Murayama, K., Shirouzu, M., and Yokoyama, S. (2004). Crystal structure of the conserved hypothetical protein ape0931 from aeropyrum pernix K1. Protein Data Bank (*www.rscb.org/pdb*).
34. Kunishima, N., and Shimizu, K. (2005). Crystal structure of diphthine synthase from pyrococcus horikoshii Ot3. Protein Data Bank (*www.rcsb.org/pdb*).
35. Anantharaman, V., Koonin, E.V., and Aravind, L. (2002). SPOUT: a class of methyltransferases that includes spoU and trmD RNA methylase superfamilies, and novel superfamilies of predicted prokaryotic RNA methylases. *J Mol Microbiol Biotechnol* 4:71–75.

36. Michel, G., Sauve, V., Larocque, R., Li, Y., Matte, A., and Cygler, M. (2002). The structure of the RlmB 23S rRNA methyltransferase reveals a new methyltransferase fold with a unique knot. *Structure* 10:1303–1315.
37. Nureki, O., Shirouzu, M., Hashimoto, K., Ishitani, R., Terada, T., Tamakoshi, M., Oshima, T., Chijimatsu, M., Takio, K., Vassylyev, D.G., Shibata, T., Inoue, Y., Kuramitsu, S., and Yokoyama, S. (2002). An enzyme with a deep trefoil knot for the active-site architecture. *Acta Crystallogr D Biol Crystallogr* 58:1129–1137.
38. Lim, K., Zhang, H., Tempczyk, A., Krajewski, W., Bonander, N., Toedt, J., Howard, A., Eisenstein, E., and Herzberg, O. (2003). Structure of the YibK methyltransferase from Haemophilus influenzae (HI0766): a cofactor bound at a site formed by a knot. *Proteins* 51:56–67.
39. Mosbacher, T.G., Bechthold, A., and Schulz, G.E. (2005). Structure and function of the antibiotic resistance-mediating methyltransferase AviRb from Streptomyces viridochromogenes. *J Mol Biol* 345:535–545.
40. Zhang, X., Tamaru, H., Khan, S.I., Horton, J.R., Keefe, L.J., Selker, E.U., and Cheng, X. (2002). Structure of the Neurospora SET domain protein DIM-5, a histone H3 lysine methyltransferase. *Cell* 111:117–127.
41. Trievel, R.C., Beach, B.M., Dirk, L.M., Houtz, R.L., and Hurley, J.H. (2002). Structure and catalytic mechanism of a SET domain protein methyltransferase. *Cell* 111:91–103.
42. Wilson, J.R., Jing, C., Walker, P.A., Martin, S.R., Howell, S.A., Blackburn, G.M., Gamblin, S.J., and Xiao, B. (2002). Crystal structure and functional analysis of the histone methyltransferase SET7/9. *Cell* 111:105–115.
43. Min, J., Zhang, X., Cheng, X., Grewal, S.I., and Xu, R.M. (2002). Structure of the SET domain histone lysine methyltransferase Clr4. *Nat Struct Biol* 9:828–832.
44. Jacobs, S.A., Harp, J.M., Devarakonda, S., Kim, Y., Rastinejad, F., and Khorasanizadeh, S. (2002). The active site of the SET domain is constructed on a knot. *Nat Struct Biol* 9:833–838.
45. Kwon, T., Chang, J.H., Kwak, E., Lee, C.W., Joachimiak, A., Kim, Y.C., Lee, J., and Cho, Y. (2003). Mechanism of histone lysine methyl transfer revealed by the structure of SET7/9-AdoMet. *Embo J* 22:292–303.
46. Cheng, X., Collins, R.E., and Zhang, X. (2005). Structural and sequence motifs of protein (histone) methylation enzymes. Annu Rev Biophys Biomol Struct *34*:267–294.
47. Doerks, T., Copley, R.R., Schultz, J., Ponting, C.P., and Bork, P. (2002). Systematic identification of novel protein domain families associated with nuclear functions. *Genome Res* 12:47–56.
48. Yeates, T.O. (2002). Structures of SET domain proteins: protein lysine methyltransferases make their mark. *Cell* 111:5–7.
49. Couture, J.F., Collazo, E., Brunzelle, J.S., and Trievel, R.C. (2005). Structural and functional analysis of SET8, a histone H4 Lys-20 methyltransferase. *Genes Dev* 19:1455–1465.
50. Xiao, B., Jing, C., Kelly, G., Walker, P.A., Muskett, F.W., Frenkiel, T.A., Martin, S.R., Sarma, K., Reinberg, D., Gamblin, S.J., and Wilson, J.R. (2005). Specificity and mechanism of the histone methyltransferase Pr-Set7. *Genes Dev* 19:1444–1454.
51. Xiao, B., Jing, C., Wilson, J.R., Walker, P.A., Vasisht, N., Kelly, G., Howell, S., Taylor, I.A., Blackburn, G.M., and Gamblin, S.J. (2003). Structure and catalytic mechanism of the human histone methyltransferase SET7/9. *Nature* 421:652–656.
52. Chuikov, S., Kurash, J.K., Wilson, J.R., Xiao, B., Justin, N., Ivanov, G.S., McKinney, K., Tempst, P., Prives, C., Gamblin, S.J., Barlev, N.A., and Reinberg, D. (2004). Regulation of p53 activity through lysine methylation. *Nature* 432:353–360.
53. Zhang, X., Yang, Z., Khan, S.I., Horton, J.R., Tamaru, H., Selker, E.U., and Cheng, X. (2003). Structural basis for the product specificity of histone lysine methyltransferases. *Mol Cell* 12:177–185.

54. Trievel, R.C., Flynn, E.M., Houtz, R.L., and Hurley, J.H. (2003). Mechanism of multiple lysine methylation by the SET domain enzyme Rubisco LSMT. *Nat Struct Biol* 10:545–552.
55. Tamaru, H., Zhang, X., McMillen, D., Singh, P.B., Nakayama, J., Grewal, S.I., Allis, C.D., Cheng, X., and Selker, E.U. (2003). Trimethylated lysine 9 of histone H3 is a mark for DNA methylation in Neurospora crassa. *Nat Genet* 34:75–79.
56. Woodard, R.W., Tsai, M.D., Floss, H.G., Crooks, P.A., and Coward, J.K. (1980). Stereochemical course of the transmethylation catalyzed by catechol O-methyltransferase. *J Biol Chem* 255:9124–9127.
57. Coward, J.K. (1977). Chemical mechanisms of methyl transfer reactions: comparison of methylases with nonenzymic "model reactions." In *The biochemistry of adenosylmethionine*, (Salvatore, F., ed.), pp. 127–144, New York: Columbia University Press.
58. Goedecke, K., Pignot, M., Goody, R.S., Scheidig, A.J., and Weinhold, E. (2001). Structure of the N6-adenine DNA methyltransferase M.TaqI in complex with DNA and a cofactor analog. *Nat Struct Biol* 8:121–125.
59. Horton, J.R., Liebert, K., Hattman, S., Jeltsch, A., and Cheng, X. (2005). Transition from nonspecific to specific DNA interactions along the substrate-recognition pathway of dam methyltransferase. *Cell* 121:349–361.
60. Malone, T., Blumenthal, R.M., and Cheng, X. (1995). Structure-guided analysis reveals nine sequence motifs conserved among DNA amino-methyltransferases, and suggests a catalytic mechanism for these enzymes. *J Mol Biol* 253:618–632.
61. Newby, Z.E., Lau, E.Y., and Bruice, T.C. (2002). A theoretical examination of the factors controlling the catalytic efficiency of the DNA-(adenine-N6)-methyltransferase from Thermus aquaticus. *Proc Natl Acad Sci USA* 99:7922–7927.
62. Martin, J.L., Begun, J., McLeish, M.J., Caine, J.M., and Grunewald, G.L. (2001). Getting the adrenaline going: crystal structure of the adrenaline-synthesizing enzyme PNMT. *Structure* 9:977–985.
63. Zubieta, C., He, X.Z., Dixon, R.A., and Noel, J.P. (2001). Structures of two natural product methyltransferases reveal the basis for substrate specificity in plant O-methyltransferases. *Nat Struct Biol* 8:271–279.
64. Djordjevic, S., and Stock, A.M. (1997). Crystal structure of the chemotaxis receptor methyl-transferase CheR suggests a conserved structural motif for binding S-adenosylmethionine. *Structure* 5:545–558.
65. Thakker, D.R., Boehlert, C., Kirk, K.L., Antkowiak, R., and Creveling, C.R. (1986). Regioselectivity of catechol O-methyltransferase. The effect of pH on the site of O-methylation of fluorinated norepinephrines. *J Biol Chem* 261:178–184.
66. Zheng, Y.-J., and Bruice, T.C. (1997). A theoretical examination of the factors controlling the catalytic efficiency of a transmethylation enzyme: catechol O-methyltransferase. *J Am Chem Soc* 119:8137–8145.
67. Selmer, T., Kahnt, J., Goubeaud, M., Shima, S., Grabarse, W., Ermler, U., and Thauer, R.K. (2000). The biosynthesis of methylated amino acids in the active site region of methyl-coenzyme M reductase. *J Biol Chem* 275:3755–3760.
68. Kumar, S., Horton, J.R., Jones, G.D., Walker, R.T., Roberts, R.J., and Cheng, X. (1997). DNA containing 4'-thio-2'-deoxycytidine inhibits methylation by HhaI methyltransferase. *Nucleic Acids Res* 25:2773–2783.
69. Wu, J.C., and Santi, D.V. (1987). Kinetic and catalytic mechanism of HhaI methyltransferase. *J Biol Chem* 262:4778–4786.
70. Reid, R., Greene, P.J., and Santi, D.V. (1999). Exposition of a family of RNA m(5)C methyltrans-ferases from searching genomic and proteomic sequences. *Nucleic Acids Res* 27:3138–3145.
71. Liu, Y., and Santi, D.V. (2000). m5C RNA and m5C DNA methyl transferases use different cysteine residues as catalysts. *Proc Natl Acad Sci USA* 97:8263–8265.

72. Foster, P.G., Nunes, C.R., Greene, P., Moustakas, D., and Stroud, R.M. (2003). The first structure of an RNA m5C methyltransferase, Fmu, provides insight into catalytic mechanism and specific binding of RNA substrate. *Structure* 11:1609–1620.
73. Lee, T.T., Agarwalla, S., and Stroud, R.M. (2004). Crystal structure of RumA, an iron-sulfur cluster containing E. coli ribosomal RNA 5-methyluridine methyltransferase. *Structure* 12:397–407.
74. Lee, T.T., Agarwalla, S., and Stroud, R.M. (2005). A unique RNA Fold in the RumA-RNA-cofactor ternary complex contributes to substrate selectivity and enzymatic function. *Cell* 120:599–611.
75. Keller, J.P., Smith, P.M., Benach, J., Christendat, D., deTitta, G.T., and Hunt, J.F. (2002). The crystal structure of MT0146/CbiT suggests that the putative precorrin-8w decarboxylase is a methyltransferase. *Structure* 10:1475–1487.
76. Shepherd, M., Reid, J.D., and Hunter, C.N. (2003). Purification and kinetic characterization of the magnesium protoporphyrin IX methyltransferase from Synechocystis PCC6803. *Biochem J* 371:351–360.
77. Min, J., Feng, Q., Li, Z., Zhang, Y., and Xu, R.M. (2003). Structure of the catalytic domain of human DOT1L, a non-SET domain nucleosomal histone methyltransferase. *Cell* 112:711–723.
78. Sawada, K., Yang, Z., Horton, J.R., Collins, R.E., Zhang, X., and Cheng, X. (2004). Structure of the conserved core of the yeast Dot1p, a nucleosomal histone H3 lysine 79 methyltransferase. *J Biol Chem* 279:43296–43306.
79. Katz, J.E., Dlakic, M., and Clarke, S. (2003). Automated identification of putative methyltransferases from genomic open reading frames. Mol Cell Proteomics 2:525–540.
80. Ludwig, M.L., and Matthews, R.G. (1997). Structure-based perspectives on B12-dependent enzymes. *Annu Rev Biochem* 66:269–313.
81. Evans, J.C., Huddler, D.P., Hilgers, M.T., Romanchuk, G., Matthews, R.G., and Ludwig, M.L. (2004). Structures of the N-terminal modules imply large domain motions during catalysis by methionine synthase. *Proc Natl Acad Sci USA* 101:3729–3736.

Part II

Modification of Lysine and Arginine Residues in Signal Transduction, Transcription, Translation, and Other Functions

2

The Family of Protein Arginine Methyltransferases *

MARK T. BEDFORD

The University of Texas M.D. Anderson Cancer Center
Science Park, Research Division
P.O. Box 389
Smithville, TX 78957, USA

I.   Abstract

Arginine methylation is a widespread posttranslational modification found on both nuclear and cytoplasmic proteins. The methylation of arginine residues is catalyzed by the protein arginine *N*-methyltransferase (PRMT) family of enzymes, of which there are at least nine members in mammals. PRMTs are evolutionarily conserved and are found in organisms from yeast to man, but not in bacteria. Proteins that are arginine methylated are involved in a number of different cellular processes, including transcriptional regulation, RNA metabolism, and DNA damage repair. How arginine methylation impacts these cellular actions is unclear, although it is likely through the regulation of protein-protein and protein-DNA/RNA interactions.

The different PRMTs display varying degrees of substrate specificity, and a certain amount of redundancy is likely to exist between different PRMT family members. Most PRMTs methylate glycine- and arginine-rich patches within their substrates. These regions have been termed GAR motifs. The complexity of the

*This work was supported by a grant from the Welch Foundation (G-1495).

THE ENZYMES, Vol. XXIV

methylarginine mark is enhanced by the ability of this residue to be methylated in three different fashions on the guanidino group (with different functional consequences for each methylated state): monomethylated, symmetrically dimethylated, and asymmetrically dimethylated. This chapter outlines the biochemistry of arginine methylation, including a detailed description of the enzymes involved, the motifs methylated, and the prospects of inhibiting these enzymes with small molecules.

II. Introduction

The amount of information encoded by the 20 amino acids incorporated into proteins by the ribosomes is enormous. After synthesis, many proteins are given a further level of complexity by posttranslational modifications on some of the incorporated amino acids. These modifications include phosphorylation, acetylation, and methylation. The latter modification occurs predominantly on arginine, lysine, and histidine residues, and is catalyzed by S-adenosyl-L-methionine (AdoMet)-dependent enzymes that donate a methyl group to the side-chain nitrogen atoms of these residues. The metabolic price of methylation is high. In the case of a reaction that is catalyzed by a kinase, the amount of metabolic energy expended is one ATP equivalent. In reactions where AdoMet serves as a methyl donor, the metabolic cost of the reaction is 12 ATP equivalents, which makes active methyl the most expensive metabolic compound on a per-carbon basis in the cell [1]. Thus, if a particular methylation event is not important to the cell it will very likely not survive evolutionary pressure. Studies on small-molecule RNA and DNA methylation have led the way to our understanding of AdoMet-dependent reactions, and we are only now starting to understand the biological significance of protein methylation.

Amino acid side-chain methylation of proteins was first reported in the mid 1960s [2-4]. With extraordinary foresight, it was proposed that protein methylation and acetylation may regulate transcription [5]. In these early studies, methylated derivatives of arginine residues were identified by incubating calf thymus nuclei with S-adenosyl-[^{14}C-*methyl*]-L-methionine, followed by acid hydrolysis of the labeled proteins and elution from a cation-exchange column [2, 6]. Two peaks eluted close to free arginine and they were termed *unknown I* and *unknown II*. Further analysis of proteins in urine and *in vitro* methylated brain and liver proteins clearly identified ω-N^G-monomethylarginine (MMA), ω-N^G,N^G-asymmetric dimethylarginine (aDMA), and ω-N^G,N'^G-symmetric dimethylarginine (sDMA) as methylarginine species [7, 8]. In hindsight, it has become clear that the peak originally identified as *unknown I* was a mixture of aDMA and sDMA, and that *unknown II* was MMA [9].

Arginine methylation is a relatively abundant posttranslational modification with about 2% of arginine residues asymmetrically dimethylated in rat liver nuclei [10].

Within the nuclear compartment, aDMA residues are further enriched in the heterogeneous nuclear ribonucleoprotein (hnRNP) fraction. About 12% of the arginine residues isolated from hnRNPs are asymmetrically dimethylated [10, 11]. The tissue concentration of aDMA is always greater than sDMA and MMA [12, 13].

Shortly after the initial identification of methylated arginine residues, attempts were made at isolating the enzyme responsible, which was at first called "protein methylase I" [6]. In the following 27 years, at least 15 reports have been published on the purification of protein arginine N-methyltransferases (reviewed in [9]). It was discovered that histones and myelin basic protein were methylated by different ammonium sulfate fractions, which implied the existence of at least two different arginine methyltransferases [14]. Although many of these studies were able to substantially purify the enzymatic activity up to 550-fold [11], they all failed to identify a specific band(s) that represented the enzyme activity. Thus, despite the number of purification attempts to isolate protein N-methyl-transferases using traditional biochemical approaches no peptide sequence was ever obtained that would allow the identification of these enzymes. Ultimately, proof for the identification of protein arginine methyltransferases had to wait for candidate proteins to be tested in their recombinant forms against a panel of putative substrates for specific enzymatic activity.

III. The Protein Arginine Methyltransferases Family of Enzymes

It has been predicted that over 1% of genes in the mammalian genome encode methyltransferases [15]. The methyltransferase family includes the enzymes that catalyze the protein arginine N-methylation reactions. The PRMT family all harbor the set of four conserved amino acid sequence motifs (I, post-I, II, and III) common to the seven beta-strand methyltransferases [15, 16], as well as motifs specific for this class of enzymes [17]. These motifs have facilitated the classi-fication of the PRMT family members. Nine PRMTs have been identified to date (Figure 2.1). Seven have been shown to catalyze the transfer of a methyl group from S-adenosyl-L-methionine (AdoMet) to a guanidino nitrogen of arginine resulting in S-adenosyl-L-homocysteine (AdoHcy) and methylarginine (Figure 2.2).

No activity has been demonstrated for PRMT2 and PRMT9. Arginine methyl-transferases have been identified in yeast [9, 18], *Drosophila melanogaster* [19], plants, *Caenorhabditis elegans*, and fish [20], but not in bacteria. PRMTs are classified as either type I, type II, type III or type IV enzymes (Figure 2.2). Types I, II, and III enzymes methylate the terminal (or ω) guanidino nitrogen atoms. Type I and type II enzymes catalyze the formation of an MMA intermediate, and then type I PRMTs (PRMT1, 3, 4, 6, and 8) further catalyze the production of aDMA. Type II PRMTs (PRMT5) catalyze the formation of sDMA. PRMT7 exhibits

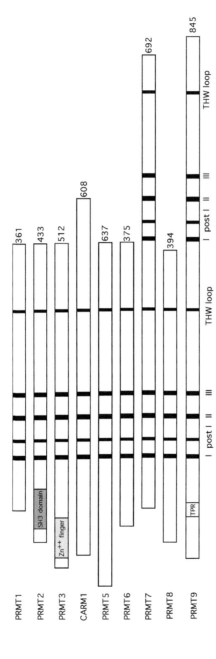

Fig. 2.1. The protein arginine methyltransferase family. There are currently nine mammalian members of the PRMT family, all of which harbor signature motifs I, post-I, II, and III, and the conserved THW loop (in black). PRMT7 and PRMT9 have duplications of these motifs. PRMT2, PRMT3, and PRMT9 have a SH3 domain, a Zn^{2+} finger, and a TPR repeat (respectively), which likely facilitate substrate recognition. The accession numbers for the PRMTs are as follows: AAF62893 for hPRMT1, AAH00727 for hPRMT2, AAC39837 for hPRMT3, NP_954592 for CARM1, AAF04502 for hPRMT5, Q96LA8 for hPRMT6, NP_061896 for hPRMT7, DAA01382 for mPRMT8, and AAH64403 for hPRMT9. The number of residues is indicated at the C-terminus of the PRMTs.

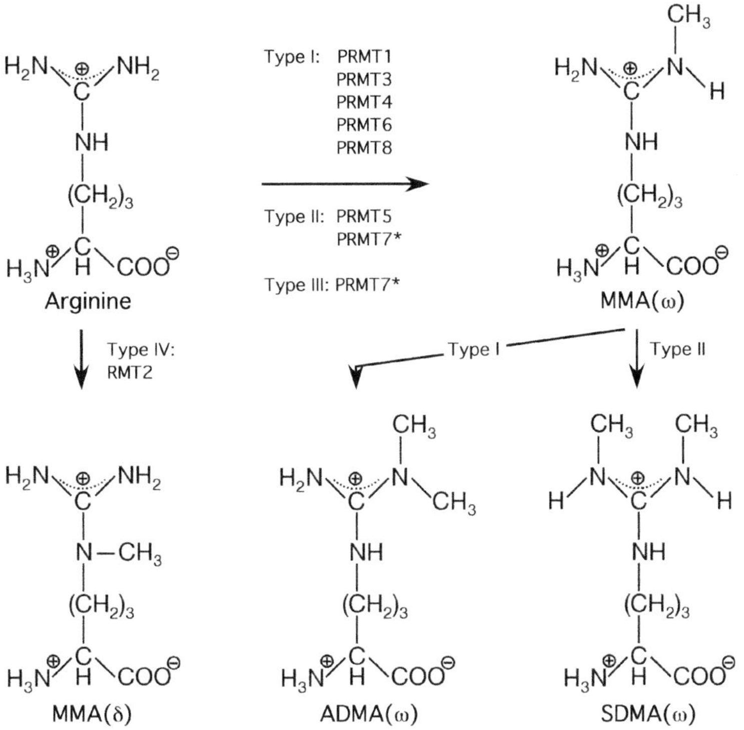

FIG. 2.2. The types of methylation on arginine residues. Types I, II, and III PRMTs generate monomethylarginine on one of the terminal (ω) guanidino nitrogen atoms. These nitrogen atoms are equivalent. The subsequent generation of asymmetric dimethylarginine is catalyzed by type I enzymes, and the production of symmetric dimethylarginine is catalyzed by the type II enzymes. The type III PRMT only monomethylates. The asterisk (*) denotes that PRMT7 has type II activity on certain peptide substrates, and type III activity on others. The type IV arginine methyltransferase monomethylates the internal (δ) nitrogen atom.

type III enzymatic activity with its propensity to catalyze the formation of MMA (it may also have some type II activity on selected substrates). Type IV enzymes catalyze the monomethylation of the internal (or δ) guanidino nitrogen atom.

IV. Mammalian Protein Arginine Methyltransferases

A. PRMT1

PRMT1, the predominant mammalian type I enzyme, was identified by homology to the yeast arginine methyltransferase Hmt1/Rmt1 and as an interacting protein of the immediate early genes TIS21/BTG1 [21]. The interaction of PRMT1 with these binding partners can regulate its substrate specificity and activity [21], and may have a role in erythroid differentiation [22]. PRMT1 has also been found to interact with the cytoplasmic portion of the type I interferon receptor [23], which implicates arginine methylation in signal transduction processes. PRMT1 is broadly expressed, and at least three splice variants have been described for this gene [24].

PRMT1 localizes to both the cytoplasm and the nucleus and has substrates in both these cellular compartments [17]. It primarily methylates GAR domain-containing RNA-binding proteins [11]. These hnRNP molecules often shuttle between the cytoplasm and the nucleus, and arginine methylation plays a role in regulating this process [25, 26]. Studies centered on the arginine methylation of one of these shuttling molecules, Sam68, revealed the negative effects methylation has on the binding of proline-rich ligands to Src homology 3 (SH3), but not to WW domain protein modules [27]. Apart from hnRNP molecules, PRMT1 also methylates histone H4 at arginine 3 [28, 29], thus contributing to the histone code [32].This modification on histone H4 functions as a transcriptional activation mark, which could either result in the recruitment of methyl-binding proteins or influence the deposition of other posttranslational marks in the vicinity.

Indeed, PRMT1 can function synergistically with other PRMTs as a nuclear receptor coactivator [30, 31]. However, PRMT1 is not a dedicated coactivator of nuclear receptors. It is also recruited to promoters by numerous other transcription factors, including p53 [31] and YY1 [33]. The central role PRMT1 plays as a regulator of protein function is revealed by the disruption of this enzyme in mice. PRMT1 knockout mice die shortly after implantation, although embryonic stem cells severely deficient in PRMT1 levels (not null) do survive [34]. These ES cells provide a vital tool for the analysis of PRMT1 function in many studies [25, 35]. The crystal structure of PRMT1 in complex with the reaction product (AdoHcy) and a GAR motif has been described [36], which reveals three different peptide binding channels, possibly reflecting alternate docking orientations for different GAR motif-containing substrates.

B. PRMT2

The first homologue of PRMT1 to be identified was PRMT2 (also called HRMT1L1) [37]. Like PRMT1, PRMT2 is ubiquitously expressed [38]. A novel feature of PRMT2 is that it harbors a SH3 domain at its N-terminus [37, 38]. This SH3 domain likely mediates interactions with proline-rich PRMT2 binding proteins [39]. Indeed, the SH3 domain of PRMT2 is required for its reported interaction with the proline-rich hnRNP, E1B-AP5 [40]. PRMT2 has coactivator activity for the estrogen receptor and facilitates transactivation in ERE-luciferase reporter assays [41]. Although PRMT2 has been shown to bind AdoMet [41], it has not yet been demonstrated to be an active enzyme. It is thus unclear whether PRMT2 displays type I or type II activity, or if it is an enzyme-dead PRMT that possibly carries out scaffolding or dominant-negative functions. PRMT2 is evolutionarily conserved and is found in flies and fish [19, 20], suggesting an as yet unidentified central function for this enzyme.

C. PRMT3

PRMT3 was identified as a PRMT1 binding protein in a yeast two-hybrid screen [42]. However, gel filtration analysis of RAT1 cells demonstrated that PRMT3 occurs as a monomer and is not complexed *in vivo* with PRMT1 [42]. In addition, it is unlikely that PRMT1 and PRMT3 occur in a complex due to the fact that PRMT1 is a predominantly nuclear protein, whereas PRMT3 is cytoplasmic. Indeed, PRMT3 is the only type I arginine methyltransferase that does not display a nuclear localization [42, 43]. Another unique property of PRMT3 is that it harbors a zinc-finger domain at its N-terminus. It has been proposed that this domain may play a role in the regulation of PRMT3 activity or in the recognition of PRMT3 substrates [42, 44].

Deletion analysis studies demonstrated that PRMT3 lacking the zinc-finger domain is still active *in vitro*. However, the zinc-finger minus enzyme loses its ability to methylate substrates when presented with a complex mix of hypomethylated proteins isolated from RAT1 cells [44]. This suggests that zinc-finger is the substrate recognition module of PRMT3. It was recently found that the 40S ribosomal protein S2 (rpS2) is a zinc-finger dependent substrate of PRMT3 [45]. Importantly, in fission yeast this same enzyme/substrate pair (PRMT3/rpS2) exists [46], and the disruption of the *prmt3* gene in this organism results in an imbalance in the 40S:60S free subunit ratio. Taken together, these findings demonstrate a highly conserved role for arginine methylation in ribosomal assembly and protein biosynthesis. Finally, binding of the tumor suppressor DAL-1 to PRMT3 acts as an inhibitor of enzyme activity, both in *in vitro* reactions and in cell lines [47].

D. PRMT4/CARM1

CARM1 (also referred to as PRMT4) was identified in the yeast two-hybrid to associate with GRIP1, the p160 steroid receptor coactivator [48]. The recruitment of CARM1 to transcriptional promoters results in the methylation of both histone H3 at arginine 17 [48] and the histone acetyltransferases (HATs), p300/CBP [49–51]. This methylation has a positive effect on transcription, and therefore CARM1 is considered a coactivator [52]. As a coactivator, CARM1 functions synergistically with PRMT1 [30] and HATs [53]. Indeed, this synergy can in part be explained by the fact that prior acetylation of histone H3 (by p300) makes it a better substrate for CARM1 methylation [54]. Additional substrates for CARM1 have been identified by performing large-scale enzyme reactions on protein arrays [55]. Substrates identified in this manner include the poly-A binding protein 1 (PABP1) and the T-cell specific factor, TARPP.

CARM1 is not only a steroid receptor coactivator but an enhancer of transcription/translation rates that are initiated by other transcription factor pathways, such as the myocyte enhancer factor-2C (MEF2C) [56], β-catenin [57], the tumor suppressor p53 [31], and NF-κB (which is a master regulator of genes involved in inflammation and cell survival) [58]. As a coactivator, CARM1 likely functions with the help of other factors and is found in a complex of at least 10 proteins, called the nucleosomal methylation activator complex (NUMAC) [59]. CARM1 within NUMAC acquires the ability to methylate nucleosomal histone H3, whereas recombinant CARM1 preferentially methylates free histone H3. The relative importance of the scaffolding versus the enzymatic functions of CARM1 have yet to be determined.

Gene targeting of CARM1 in mice has been performed. These knockout mice die just after birth and are smaller than their wild-type littermates [60]. CARM1 knockout mouse embryonic fibroblasts (MEFs) were established and have been used to confirm CARM1s role as a coactivator for the estrogen receptor [60] and NF-κB [58]. These MEFs have also been used to genetically verify that PABP1 and TARPP are indeed specific substrates for CARM1, and that CARM1 lies in a nonredundant pathway [60, 61]. Furthermore, knockout embryos have a partial T-cell development block that may be due to the hypomethylated status of TARPP [61].

CARM1 orthologues have been identified in fish and flies [19, 20]. *Drosophila* CARM1 (CARMER) modulates the steroid hormone function in this organism, where it is a coactivator for the ecdysone receptor and is required for ecdysone-induced cell death [62]. Thus, the role of CARM1 as a steroid receptor coactivator has been evolutionarily conserved. Finally, the fact that CARM1 is a coactivator for nuclear receptors makes it a likely candidate for overexpression in prostate and breast cancers. Indeed, it has been found that increased expression of CARM1 correlates with androgen independence in human prostate carcinoma [63].

E.   PRMT5

PRMT5 was cloned as Jak2-binding protein and shown to have type II activity [64, 65]. It methylates histones (H2A and H4) and myelin basic protein *in vitro* [64]. In the cell, PRMT5 is a homo-oligomer [66] or is associated with one of at least three different protein complexes. Two complexes are nuclear, and the third is cytoplasmic. In the cytoplasm, PRMT5 is found in the "methylosome," where it is involved in the methylation of Sm proteins, thus implicating PRMT5 in snRNP biogenesis [67]. Nuclear PRMT5 associates with the regulator of transcriptional elongation, SPT4 and SPT5, and with pICln (which is also a component of the methylosome) [68]. Nuclear PRMT5 also complexes with the hSWI/SNF chromatin remodeling proteins BRG and BRM, and this association enhances PRMT5 methyltransferase activity. In this case, the target for methylation is arginine 8 on histone H3, and acetylation of lysine 9 prevents methylation [69].

For certain substrates, PRMR5 has the ability to methylate the same arginine residues as are recognized by type I PRMTs. The ability for the same site to carry an aDMA or sDMA modification may have opposing biological consequences. The snRNP core proteins, SmB /B′, are methylated by PRMT5 in an sDMA fashion [67], which facilitates an interaction with the tudor domain of SMN [70]. In addition, these splicing factors can be asymmetrically methylated *in vitro* by CARM1 [19, 71]. Recently it was shown that SmB, known to shuttle between the cytosol and nucleus, harbors sDMA in the cytosol and both sDMA and aDMA in the nucleus [72]. It has yet to be determined if common arginine residues in the SmB C-terminal region are targeted by PRMT5 and CARM1. Along the same lines, histone H4 is asymmetrically methylated at arginine 3 by PRMT1 [29] and symmetrically methylated at the same residue by PRMT5 [69]. PRMT1 is a transcriptional coactivator and PRMT5 is generally regarded as a corepressor [73]. Thus, the type of methylation at arginine 3 of histone H4 may dictate the accessibility of a locus to transcription factors. Furthermore, a single arginine residue (R698) in RNA polymerase II binding domain of the transcriptional elongation factor, SPT5, can also be methylated by both PRMT1 and PRMT5 [68]. This again raises the possibility that aDMA and sDMA may antagonize each other and differentially regulate protein function.

F.   PRMT6

PRMT6 is another arginine methyltransferase with a nuclear restricted pattern of localization. It is a type I PRMT that is ubiquitously expressed and has the ability to methylate itself [17]. A PRMT6 orthologue has been identified in fish, where expression is seen predominantly in the brain [20]. Like PRMT1, PRMT6 methylates a GAR motif. However, it displays unique substrate specificity; that is, it methylates histones H3 and H4 *in vitro*, whereas PRMT1 only methylates histone H4 [74]. Another specific substrate for PRMT6 is the HIV transactivator protein, Tat [75]. Arginine methylation of Tat negatively regulates its transactivation activity and a knockdown of PRMT6 increases HIV-1 production in cells.

G.   PRMT7

This arginine methyltransferase is one of two PRMTs that harbor two putative AdoMet-binding motifs [76]. It has a strong propensity to catalyze the formation of MMA but not DMA when using the R1 peptide, GGFGGRGGFG, derived from fibrillarin as a substrate [76]. This study thus classified PRMT7 as a type III enzyme. Using a different peptide, SGRGKGGKGLGKGGAKRHRK, it was shown that PRMT7 catalyzes the formation of sDMA, which classifies it as a type II enzyme [77]. It is possible that distinct substrates are methylated in different fashions by this enzyme. PRMT7 orthologues have been identified in fish and flies [19, 20]. PRMT7 was first identified in a genetic screen for susceptibility to chemotherapeutic cytotoxicity [78], which implicated PRMTs in the DNA damage response pathway for the first time. In certain mouse strains, doxorubicin (an anthracycline antibiotic) induces nephropathy. In a study that focused on identifying loci that impart susceptibility to this trait, PRMT7 was identified [79]. High levels of PRMT7 have a protective effect on end-stage renal failure induced by xenobiotics.

H. PRMT8

PRMT8 was identified due to its high degree of homology with the predominant arginine methyltransferase, PRMT1 [80]. This PRMT has been annotated in the National Center for Biotechnology Information (NCBI) database as hnRNP methyltransferase-like 4 (Hrmt1l4) (NM_019854) and Hrmt1l3 (BK001349). This PRMT1-like enzyme has recently been termed *PRMT8* [81, 82], and we will use this nomenclature. The existence of this PRMT1-like molecule has been noted [20, 36], although the unique N-terminal region of 76 amino acids was only recently recognized. This N-terminal region is conserved between mouse and man, and an orthologue also exists in fish (fuguL3) [20]. The unique N-terminal end harbors a functional myristoylation motif that facilitates its association with the plasma membrane. The second singular property of PRMT8 is its tissue-specific expression pattern (it is largely expressed in the brain). A glutathione *S*-transferase fusion protein of PRMT8 has type I PRMT activity, catalyzing the formation of ω-N^G-monomethylated and asymetrically ω-N^G,N^G-dimethylated arginine residues on a recombinant glycine- and arginine-rich substrate [80]. PRMT8 is thus an active arginine methyltransferase that is membrane associated and tissue specific.

I. PRMT9

The PRMT family is still expanding, albeit slowly. Database searches reveal evidence for at least one more PRMT, which we designate PRMT9 (NP_612373), and it may prove to be the last member of this family. The human PRMT9 maps to 4q31.2 and its open reading frame is 845 amino acids in size. PRMT9 encodes a protein that is similar to PRMT7, in that it harbors two putative AdoMet-binding motifs. In addition, at its N-terminal end PRMT9 has a TPR repeat. TPR repeats form antiparallel α-helices, which are protein-protein interaction modules [83].

V.   Yeast Protein Arginine Methyltransferases

A.   RMT1/HMT1

The predominant type I protein arginine methyltransferase in *Saccharomyces cerevisiae* was identified at almost the same time by two different groups [84, 85]. Using mammalian PRMT1, the Clarke laboratory performed an *in silico* search of the yeast DNA sequence database to identify its protein orthologue, which they called Rmt1p (protein-arginine methyltransferase 1) [84]. Silver and co-workers identified the same enzyme, which they called Hmt1p (hnRNP methyltransferase), based on its ability to interact with the poly(A)-binding protein Npl3p [85]. *rmt1/hmt1* null cells are viable and display normal growth rates at 14° C and 36° C, but this allele is synthetically lethal with the temperature-sensitive *npl3-1* allele [85]. Using an *in vivo* nuclear export assay, it was shown that *rmt1/hmt1* null cells fail to export the RNA-binding proteins Npl3p, Hrp1p, and Nab2p from the nucleus [86, 87]. These three GAR-motif-containing proteins are substrates for Rmt1p/Hmt1p, the enzymatic activity of which is required for these proteins' correct subcellular localization [88]. These studies identified nuclear export as a biological function for arginine methylation. The structure of Rmt1p/Hmt1p has been solved and it is clear that this protein forms dimers and hexamers [89].

Dimerization is required for enzymatic activity. A yeast two-hybrid screen has identified Air1p (arginine methyltransferase-interacting RING finger protein) and a homologue (Air2p) as Rmt1p/Hmt1p-binding proteins that can regulate the methylation of Npl3p [90]. Npl3p is also a substrate for the kinase Sky1p, which phosphorylates it at a single site in the C-terminus. This phosphorylation site is relatively close to the arginine methylated GAR motifs of Npl3p, and it was found that methylation prevents the subsequent phosphorylation of Npl3p [91]. Phosphorylation of Npl3p is required for its interaction with the nuclear import receptor, Mtr10p. Thus, the subcellular localization of Npl3p is regulated by both methylation and phosphorylation. Npl3p contains 15 arginine residues, within RGG motifs, that are methylated. To further investigate the role of arginine methylation in nuclear export, all of these RGG motifs were replaced with KGG motifs [92, 93]. This mutant form of Npl3p could exit the nucleus independently of Rmt1p/Hmt1p activity, indicating a direct requirement of methylation for Npl3p transport by weakening contacts with nuclear proteins.

Further, it was demonstrated that methylation resulted in the weakening of interactions with the transcription elongation factor Tho2p [93, 94]. Thus, it is proposed that arginine methylation of Npl3p relaxes contacts with nuclear proteins, thereby facilitating export. Rmt1p/Hmt1p has additional functions in the nucleus apart from its clear role in regulating the export of proteins to the cytoplasm. This became clear with the identification of a group (Gar1p, Nop1p, and Nsr1p) of proteins that are methylated and are not subjected to export out of the nucleus [95]. These proteins display a nucleolar localization and it is not apparent what the function of arginine methylation is within this cellular compartment. In addition, Rmt1p/Hmt1p is recruited to a specific functional class of actively transcribed genes, where it is thought to play a role in transcriptional elongation and the recruitment of RNA-processing factors [94].

B. RMT2

Until recently, arginine methylation was thought to occur only on the terminal (or ω) guanidino nitrogen atoms, and the existence of δ-N-monomethylarginine was ruled out early on in the development of this field [12, 96]. However, it was found that when yeast cells lacking the predominant protein arginine methyltransferase Rmt1p/Hmt1p (displaying no ω-N^G,N^G-asymmetric dimethylarginine peak of their cellular proteins) were subjected to acid hydrolysis and fractionation over a cation exchange column, a previously obscured methylated arginine species was unmasked [97].

Base treatment of this minor peak did not result in a volatile [^{3}H]methyl group, but did generate δ-N-methylornithine, thus identifying this unique methylated residue as δ-N-monomethylarginine. The enzyme responsible for this activity was identified using an *in silico*-based screen and gene disruption, and designated Rmt2p [16]. The Rmt2p enzyme thus catalyzes type IV methylation (Figure 2.2). Recently, the yeast ribosomal protein L12 was identified as a specific substrate for Rmt2p [98]. A single δ-N-monomethylarginine was identified in L12 in the sequence, IQNRQAA. Therefore, Rmt2p does not methylate a GAR motif as Rmt1p/Hmt1p does. It is unclear if there is a mammalian orthologue of Rmt2p, and δ-N-monomethylarginine has not yet been detected in higher eukaryotes.

C. HSL7

The predominant type II protein arginine methyltransferase in *Saccharomyces cerevisiae*, Hsl7p (Histone synthetic-lethal 7), was identified because of its homology to PRMT5 [64, 99]. The orthologue in *Saccharomyces pombe* is Skb1p [100]. Hsl7p was identified in a second-site mutation screen for mutations that were fatal in combination with a deletion of the amino terminus of histone H3 [101]. This was one of the first studies to genetically implicate arginine methylation in transcriptional regulation and possibly the histone code. Hsl7p can regulate the Wee1 family kinase, Swe1p [101], by targeting this kinase for degradation [102]. The ability of Hsl7p to promote the degradation of Swe1p requires the kinase, Hsl1p [102, 103]. These three proteins (Hsl1p-Hsl7p-Swe1p) form a complex that localizes to the bud neck, where it takes part in the yeast morphogenesis checkpoint. Although Hsl7p is an active arginine methyltransferase [99] that can methylate histones (H2A and H4) *in vitro*, the *in vivo* substrates of this enzyme have yet to be identified.

VI. PRMT Substrate Specificity

It has long been known that proteins that harbor glycine- and arginine-rich (GAR) motifs are often targets for PRMTs [104]. Specific PRMT substrates have been identified through candidate approaches [48, 105], through serendipitous discovery [49, 106], and more recently through focused *in vitro* substrate screens [55, 61, 107] and by using proteomic-based mass spectrometry approaches [108, 109]. The type I enzymes (PRMT1, 3 and 6) generally recognize GAR-motif-containing substrates, whereas CARM1 displays a higher degree of specificity and does not methylate GAR motifs. There is no obvious motif that is recognized by CARM1, making it difficult to predict potential substrates using *in silico* searches of primary protein sequences. The type II enzymes, PRMT5 and 7, methylate isolated arginine residues as well as arginines within GAR motifs. An extensive list of PRMT substrates can be found in the following chapter (Chapter 3, Table 4.1) and more substrates are likely to be identified, both with and without GAR motifs. Three enzymes that methylate GAR motifs have been crystallized, and their core structures have proven very similar [36, 89, 110] (see Chapter 4 for details).

VII. Small-Molecule Inhibitors of Protein Arginine Methyltransferases

Two types of compounds are used to disrupt methyltransferase activity. First, small molecules that inhibit AdoHcy-hydrolase, such as AdOx, result in a substantial intracellular accumulation of AdoHcy. Most methylation reactions are affected through feedback inhibition by elevated levels of AdoHcy. Second, analogues of AdoMet—such as sinefungin and MTA (methylthioadenosine)— also function as inhibitors of methylation. The inhibition of AdoHcy hydrolase by small molecules and the use of AdoMet analogues can negatively influence cellular methylation of phospholipids, proteins, DNA, and RNA. Thus, current methyltransferase inhibitors display limited specificity, indiscriminately targeting all enzymes that use AdoMet. A "bump-and-hole" approach has been used to generate analogues of AdoMet that selectively inhibit a mutant form of the yeast protein methyltransferase, Hmt1p [111].

This approach demonstrates the feasibility of generating specific methyltransferase inhibitors but does not provide reagents for the inhibition of endogenous PRMTs. Small-molecule inhibitors of PRMT were recently identified [112].

AMIs (arginine methyltransferase inhibitors) selectively inhibit PRMTs, not lysine methyltransferases. The AMIs display no specificity for individual PRMTs, demonstrating that further primary or analogue screens are required to identify PRMT-specific inhibitors. Importantly, one of these compounds, AMI1, is able to inhibit the coactivator function of PRMTs in a luciferase reporter assay. These studies provide encouraging evidence that specific PRMT inhibitors may be useful to treat hormone-responsive tumors, or hormone-independent tumors that resulted from elevated activities of coactivators.

REFERENCES

1. Atkinson, D. E. (1977). Cellular energy metabolism and its regulation, Academic Press, New York.
2. Paik, W. K., and Kim, S. (1967). Enzymatic methylation of protein fractions from calf thymus nuclei. *Biochem Biophys Res Commun* 29:14–20.
3. Murray, K. (1964). The occurrence of epsilon-N-methyl lysine in histones. *Biochemistry* 127:10–15.
4. Asatoor, A. M., and Armstrong, M. D. (1967). 3-methylhistidine, a component of actin. *Biochem Biophys Res Commun* 26:168–174.
5. Allfrey, V. G., Faulkner, R., and Mirsky, A. E. (1964). Acetylation and methylation of histones and their possible role in the regulation of RNA synthesis. *Proc Natl Acad Sci USA* 51:786–794.
6. Paik, W. K., and Kim, S. (1968). Protein methylase I. Purification and properties of the enzyme. *J Biol Chem* 243:2108–2114.
7. Kakimoto, Y. (1971). Methylation of arginine and lysine residues of cerebral proteins. *Biochim Biophys Acta* 243:31–37.
8. Kakimoto, Y., and Akazawa, S. (1970). Isolation and identification of N-G,N-G- and N-G, N'-G-dimethyl-arginine, N-epsilon-mono-, di-, and trimethyllysine, and glucosylgalactosyl- and galactosyl-delta-hydroxylysine from human urine. *J Biol Chem* 245:5751–5758.
9. Gary, J. D., and Clarke, S. (1998). RNA and protein interactions modulated by protein arginine methylation. *Prog Nucleic Acid Res Mol Biol* 61:65–131.
10. Boffa, L. C., Karn, J., Vidali, G., and Allfrey, V. G. (1977). Distribution of NG, NG,-dimethyl-larginine in nuclear protein fractions. *Biochem Biophys Res Commun* 74:969–976.
11. Liu, Q., and Dreyfuss, G. (1995). In vivo and in vitro arginine methylation of RNA-binding proteins. *Mol Cell Biol* 15:2800–2808.
12. Nakajima, T., Matsuoka, Y., and Kakimoto, Y. (1971). Isolation and identification of N-G-monomethyl, N-G, N-G-dimethyl- and N-G,N' G-dimethylarginine from the hydrolysate of proteins of bovine brain. *Biochim Biophys Acta* 230:212–222.
13. Kakimoto, Y., Matsuoka, Y., Miyake, M., and Konishi, H. (1975). Methylated amino acid residues of proteins of brain and other organs. *J Neurochem* 24:893–902.
14. Miyake, M., and Kakimoto, Y. (1973). Protein methylation by cerebral tissue. *J Neurochem* 20:859–871.
15. Katz, J. E., Dlakic, M., and Clarke, S. (2003). Automated identification of putative methyltransferases from genomic open reading frames. *Mol Cell Proteomics* 2:525–540.
16. Niewmierzycka, A., and Clarke, S. (1999). S-Adenosylmethionine-dependent methylation in Saccharomyces cerevisiae. Identification of a novel protein arginine methyltransferase. *J Biol Chem* 274:814–824.

17. Frankel, A., Yadav, N., Lee, J., Branscombe, T. L., Clarke, S., and Bedford, M. T. (2002). The novel human protein arginine N-methyltransferase PRMT6 is a nuclear enzyme displaying unique substrate specificity. *J Biol Chem* 277:3537–3543.
18. McBride, A. E., and Silver, P. A. (2001). State of the arg: protein methylation at arginine comes of age. *Cell* 106:5–8.
19. Boulanger, M. C., Miranda, T. B., Clarke, S., Di Fruscio, M., Suter, B., Lasko, P., and Richard, S. (2004). Characterization of the Drosophila protein arginine methyltransferases DART1 and DART4. *Biochem J* 379:283–289.
20. Hung, C. M., and Li, C. (2004). Identification and phylogenetic analyses of the protein arginine methyltransferase gene family in fish and ascidians. *Gene* 340:179–187.
21. Lin, W. J., Gary, J. D., Yang, M. C., Clarke, S., and Herschman, H. R. (1996). The mammalian immediate-early TIS21 protein and the leukemia-associated BTG1 protein interact with a protein-arginine N-methyltransferase. *J Biol Chem* 271:15034–15044.
22. Bakker, W. J., Blazquez-Domingo, M., Kolbus, A., Besooyen, J., Steinlein, P., Beug, H., Coffer, P. J., Lowenberg, B., von Lindern, M., and van Dijk, T. B. (2004). FoxO3a regulates erythroid differentiation and induces BTG1, an activator of protein arginine methyl transferase 1. *J Cell Biol* 164:175–184.
23. Abramovich, C., Yakobson, B., Chebath, J., and Revel, M. (1997). A protein-arginine methyltransferase binds to the intracytoplasmic domain of the IFNAR1 chain in the type I interferon receptor. *Embo J* 16:260–266.
24. Scorilas, A., Black, M. H., Talieri, M., and Diamandis, E. P. (2000). Genomic organization, physical mapping, and expression analysis of the human protein arginine methyltransferase 1 gene [in-process citation]. *Biochem Biophys Res Commun* 278:349–359.
25. Cote, J., Boisvert, F. M., Boulanger, M. C., Bedford, M. T., and Richard, S. (2003). Sam68 RNA binding protein is an in vivo substrate for protein arginine N-methyltransferase 1. *Mol Biol Cell* 14:274–287.
26. Herrmann, F., Bossert, M., Schwander, A., Akgun, E., and Fackelmayer, F. O. (2004). Arginine methylation of scaffold attachment factor A by heterogeneous nuclear ribonucleoprotein particle-associated PRMT1. *J Biol Chem* 279:48774–48779.
27. Bedford, M. T., Frankel, A., Yaffe, M. B., Clarke, S., Leder, P., and Richard, S. (2000). Arginine methylation inhibits the binding of proline-rich ligands to Src homology 3, but not WW, domains. *J Biol Chem* 275:16030–16036.
28. Strahl, B. D., Briggs, S. D., Brame, C. J., Caldwell, J. A., Koh, S. S., Ma, H., Cook, R. G., Shabanowitz, J., Hunt, D. F., Stallcup, M. R., and Allis, C. D. (2001). Methylation of histone H4 at arginine 3 occurs in vivo and is mediated by the nuclear receptor coactivator PRMT1. *Curr Biol* 11:996–1000.
29. Wang, H., Huang, Z. Q., Xia, L., Feng, Q., Erdjument-Bromage, H., Strahl, B. D., Briggs, S. D., Allis, C. D., Wong, J., Tempst, P., and Zhang, Y. (2001). Methylation of histone H4 at arginine 3 facilitating transcriptional activation by nuclear hormone receptor. *Science* 293:853–857.
30. Koh, S. S., Chen, D., Lee, Y. H., and Stallcup, M. R. (2001). Synergistic enhancement of nuclear receptor function by p160 coactivators and two coactivators with protein methyltransferase activities. *J Biol Chem* 276:1089–1098.
31. An, W., Kim, J., and Roeder, R. G. (2004). Ordered cooperative functions of PRMT1, p300, and CARM1 in transcriptional activation by p53. *Cell* 117:735–748.
32. Jenuwein, T., and Allis, C. D. (2001). Translating the histone code. *Science* 293: 1074–1080.
33. Rezai-Zadeh, N., Zhang, X., Namour, F., Fejer, G., Wen, Y. D., Yao, Y. L., Gyory, I., Wright, K., and Seto, E. (2003). Targeted recruitment of a histone H4-specific methyltransferase by the transcription factor YY1. *Genes Dev* 17:1019–1029.

34. Pawlak, M. R., Scherer, C. A., Chen, J., Roshon, M. J., and Ruley, H. E. (2000). Arginine N-methyltransferase 1 is required for early postimplantation mouse development, but cells deficient in the enzyme are viable. *Mol Cell Biol* 20:4859–4869.

35. Pawlak, M. R., Banik-Maiti, S., Pietenpol, J. A., and Ruley, H. E. (2002). Protein arginine methyltransferase I: substrate specificity and role in hnRNP assembly. *J Cell Biochem* 87: 394–407.

36. Zhang, X., and Cheng, X. (2003). Structure of the predominant protein arginine methyltransferase PRMT1 and analysis of its binding to substrate peptides. *Structure (Camb)* 11:509–520.

37. Katsanis, N., Yaspo, M. L., and Fisher, E. M. (1997). Identification and mapping of a novel human gene, HRMT1L1, homologous to the rat protein arginine N-methyltransferase 1 (PRMT1) gene. *Mamm Genome* 8:526–529.

38. Scott, H. S., Antonarakis, S. E., Lalioti, M. D., Rossier, C., Silver, P. A., and Henry, M. F. (1998). Identification and characterization of two putative human arginine methyltransferases (HRMT1L1 and HRMT1L2). *Genomics* 48:330–340.

39. Alexandropoulos, K., Cheng, G., and Baltimore, D. (1995). Proline-rich sequences that bind to Src homology 3 domains with individual specificities. *Proc Natl Acad Sci USA* 92:3110–3114.

40. Kzhyshkowska, J., Schutt, H., Liss, M., Kremmer, E., Stauber, R., Wolf, H., and Dobner, T. (2001). Heterogeneous nuclear ribonucleoprotein E1B-AP5 is methylated in its Arg-Gly-Gly (RGG) box and interacts with human arginine methyltransferase HRMT1L1. *Biochem J* 358:305–314.

41. Qi, C., Chang, J., Zhu, Y., Yeldandi, A. V., Rao, S. M., and Zhu, Y. J. (2002). Identification of protein arginine methyltransferase 2 as a coactivator for estrogen receptor alpha. *J Biol Chem* 277:28624–28630.

42. Tang, J., Gary, J. D., Clarke, S., and Herschman, H. R. (1998). PRMT 3, a type I protein arginine N-methyltransferase that differs from PRMT1 in its oligomerization, subcellular localization, substrate specificity, and regulation. *J Biol Chem* 273:16935–16945.

43. Frankel, A., Yadav, N., Lee, J., Branscombe, T. L., Clarke, S., and Bedford, M. T. (2002). The novel human protein arginine N-methyltransferase PRMT6 is a nuclear enzyme displaying unique substrate specificity. *J Biol Chem* 277:3537–3543.

44. Frankel, A., and Clarke, S. (2000). PRMT3 is a distinct member of the protein arginine N-methyltransferase family. Conferral of substrate specificity by a zinc-finger domain. *J Biol Chem* 275:32974–32982.

45. Swiercz, R., Person, M. D., and Bedford, M. T. (2005). Ribosomal protein S2 is a substrate for mammalian PRMT3 (protein arginine methyltransferase 3). *Biochem J* 386:85–91.

46. Bachand, F., and Silver, P. A. (2004). PRMT3 is a ribosomal protein methyltransferase that affects the cellular levels of ribosomal subunits. *Embo J* 23:2641–2650.

47. Singh, V., Miranda, T. B., Jiang, W., Frankel, A., Roemer, M. E., Robb, V. A., Gutmann, D. H., Herschman, H. R., Clarke, S., and Newsham, I. F. (2004). DAL-1/4.1B tumor suppressor interacts with protein arginine N-methyltransferase 3 (PRMT3) and inhibits its ability to methylate substrates in vitro and in vivo. *Oncogene* 23:7761–7771.

48. Chen, D., Ma, H., Hong, H., Koh, S. S., Huang, S. M., Schurter, B. T., Aswad, D. W., and Stallcup, M. R. (1999). Regulation of transcription by a protein methyltransferase. *Science* 284:2174–2177.

49. Xu, W., Chen, H., Du, K., Asahara, H., Tini, M., Emerson, B. M., Montminy, M., and Evans, R. M. (2001). A transcriptional switch mediated by cofactor methylation. *Science* 294:2507–2511.

50. Chevillard-Briet, M., Trouche, D., and Vandel, L. (2002). Control of CBP co-activating activity by arginine methylation. *Embo J* 21:5457–5466.

51. Lee, Y. H., Coonrod, S. A., Kraus, W. L., Jelinek, M. A., and Stallcup, M. R. (2005). Regulation of coactivator complex assembly and function by protein arginine methylation and demethylimination. *Proc Natl Acad Sci USA* 102:3611–3616.

52. Bauer, U. M., Daujat, S., Nielsen, S. J., Nightingale, K., and Kouzarides, T. (2002). Methylation at arginine 17 of histone H3 is linked to gene activation. *EMBO Rep* 3:39–44.
53. Lee, Y. H., Koh, S. S., Zhang, X., Cheng, X., and Stallcup, M. R. (2002). Synergy among nuclear receptor coactivators: selective requirement for protein methyltransferase and acetyltransferase activities. *Mol Cell Biol* 22:3621–3632.
54. Daujat, S., Bauer, U. M., Shah, V., Turner, B., Berger, S., and Kouzarides, T. (2002). Crosstalk between CARM1 methylation and CBP acetylation on histone H3. *Curr Biol* 12:2090–2097.
55. Lee, J., and Bedford, M. T. (2002). PABP1 identified as an arginine methyltransferase substrate using high-density protein arrays. *EMBO Rep* 3:268–273.
56. Chen, S. L., Loffler, K. A., Chen, D., Stallcup, M. R., and Muscat, G. E. (2002). The coactivator-associated arginine methyltransferase is necessary for muscle differentiation: CARM1 coactivates myocyte enhancer factor-2. *J Biol Chem* 277:4324–4333.
57. Koh, S. S., Li, H., Lee, Y. H., Widelitz, R. B., Chuong, C. M., and Stallcup, M. R. (2002). Synergistic coactivator function by coactivator-associated arginine methyltransferase (CARM) 1 and beta-catenin with two different classes of DNA-binding transcriptional activators. *J Biol Chem* 277:26031–26035.
58. Covic, M., Hassa, P. O., Saccani, S., Buerki, C., Meier, N. I., Lombardi, C., Imhof, R., Bedford, M. T., Natoli, G., and Hottiger, M. O. (2005). Arginine methyltransferase CARM1 is a promoter-specific regulator of NF-kappaB-dependent gene expression. *Embo J* 24:85–96.
59. Xu, W., Cho, H., Kadam, S., Banayo, E. M., Anderson, S., Yates, J. R. III, Emerson, B. M., and Evans, R. M. (2004). A methylation-mediator complex in hormone signaling. *Genes Dev* 18:144–156.
60. Yadav, N., Lee, J., Kim, J., Shen, J., Hu, M. C., Aldaz, C. M., and Bedford, M. T. (2003). Specific protein methylation defects and gene expression perturbations in coactivator-associated arginine methyltransferase 1-deficient mice. *Proc Natl Acad Sci USA* 100:6464–6468.
61. Kim, J., Lee, J., Yadav, N., Wu, Q., Carter, C., Richard, S., Richie, E., and Bedford, M. T. (2004). Loss of CARM1 results in hypomethylation of TARPP and deregulated early T cell development. *J Biol Chem* 279:25339–25344.
62. Cakouros, D., Daish, T. J., Mills, K., and Kumar, S. (2004). An arginine-histone methyltransferase, CARMER, coordinates ecdysone-mediated apoptosis in Drosophila cells. *J Biol Chem* 279:18467–18471.
63. Hong, H., Kao, C., Jeng, M. H., Eble, J. N., Koch, M. O., Gardner, T. A., Zhang, S., Li, L., Pan, C. X., Hu, Z., MacLennan, G. T., and Cheng, L. (2004). Aberrant expression of CARM1, a transcriptional coactivator of androgen receptor, in the development of prostate carcinoma and androgen-independent status. *Cancer* 101:83–89.
64. Pollack, B. P., Kotenko, S. V., He, W., Izotova, L. S., Barnoski, B. L., and Pestka, S. (1999). The human homologue of the yeast proteins Skb1 and Hsl7p interacts with Jak kinases and contains protein methyltransferase activity. *J Biol Chem* 274:31531–31542.
65. Branscombe, T. L., Frankel, A., Lee, J. H., Cook, J. R., Yang, Z., Pestka, S., and Clarke, S. (2001). Prmt5 (janus kinase-binding protein 1) catalyzes the formation of symmetric dimethylarginine residues in proteins. *J Biol Chem* 276:32971–32976.
66. Rho, J., Choi, S., Seong, Y. R., Cho, W. K., Kim, S. H., and Im, D. S. (2001). Prmt5, which forms distinct homo-oligomers, is a member of the protein-arginine methyltransferase family. *J Biol Chem* 276:11393–11401.
67. Friesen, W. J., Paushkin, S., Wyce, A., Massenet, S., Pesiridis, G. S., Van Duyne, G., Rappsilber, J., Mann, M., and Dreyfuss, G. (2001). The methylosome, a 20S complex containing JBP1 and pICln, produces dimethylarginine-modified Sm proteins. *Mol Cell Biol* 21:8289–8300.
68. Kwak, Y. T., Guo, J., Prajapati, S., Park, K. J., Surabhi, R. M., Miller, B., Gehrig, P., and Gaynor, R. B. (2003). Methylation of SPT5 regulates its interaction with RNA polymerase II and transcriptional elongation properties. *Mol Cell* 11:1055–1066.

69. Pal, S., Vishwanath, S. N., Erdjument-Bromage, H., Tempst, P., and Sif, S. (2004). Human SWI/SNF-associated PRMT5 methylates histone H3 arginine 8 and negatively regulates expression of ST7 and NM23 tumor suppressor genes. *Mol Cell Biol* 24:9630–9645.

70. Friesen, W. J., Massenet, S., Paushkin, S., Wyce, A. and Dreyfuss, G. (2001). SMN, the product of the spinal muscular atrophy gene, binds preferentially to dimethylarginine-containing protein targets. *Mol Cell* 7:1111–1117.

71. Cheng, D., Yadav, N., King, R. W., Swanson, M. S., Weinstein, E. J., and Bedford, M. T. (2004). Small molecule regulators of protein arginine methyltransferases. *J Biol Chem* 279:23892–23899.

72. Miranda, T. B., Khusial, P., Cook, J. R., Lee, J. H., Gunderson, S. I., Pestka, S., Zieve, G. W., and Clarke, S. (2004). Spliceosome Sm proteins D1, D3, and B/B′ are asymmetrically dimethylated at arginine residues in the nucleus. *Biochem Biophys Res Commun* 323: 382–387.

73. Fabbrizio, E., El Messaoudi, S., Polanowska, J., Paul, C., Cook, J. R., Lee, J. H., Negre, V., Rousset, M., Pestka, S., Le Cam, A., and Sardet, C. (2002). Negative regulation of transcription by the type II arginine methyltransferase PRMT5. *EMBO Rep* 3:641–645

74. Lee, J., Cheng, D., and Bedford, M. T. (2004). Techniques in protein methylation. *Methods Mol Biol* 284:195–208.

75. Boulanger, M. C., Liang, C., Russell, R. S., Lin, R., Bedford, M. T., Wainberg, M. A., and Richard, S. (2005). Methylation of Tat by PRMT6 regulates human immunodeficiency virus type 1 gene expression. *J Virol* 79:124–131.

76. Miranda, T. B., Miranda, M., Frankel, A., and Clarke, S. (2004). PRMT7 is a member of the protein arginine methyltransferase family with a distinct substrate specificity. *J Biol Chem.*

77. Lee, J. H., Cook, J. R., Yang, Z. H., Mirochnitchenko, O., Gunderson, S. I., Felix, A. M., Herth, N., Hoffmann, R., and Pestka, S. (2005). PRMT7, a new protein arginine methyltransferase that synthesizes symmetric dimethylarginine. *J Biol Chem* 280:3656–3664.

78. Gros, L., Delaporte, C., Frey, S., Decesse, J., de Saint-Vincent, B. R., Cavarec, L., Dubart, A., Gudkov, A. V., and Jacquemin-Sablon, A. (2003). Identification of new drug sensitivity genes using genetic suppressor elements: protein arginine N-methyltransferase mediates cell sensitivity to DNA-damaging agents. *Cancer Res* 63:164–171.

79. Zheng, Z., Schmidt-Ott, K. M., Chua, S., Foster, K. A., Frankel, R. Z., Pavlidis, P., Barasch, J., D'Agati, V. D., and Gharavi, A. G. (2005). A Mendelian locus on chromosome 16 determines susceptibility to doxorubicin nephropathy in the mouse. *Proc Natl Acad Sci USA* 102: 2502–2507.

80. Lee, J., Sayegh, J., Daniel, J., Clarke, S., and Bedford, M. T. (2005). PRMT8, a new membrane-bound tissue-specific member of the protein arginine methyltransferase family. *J Biol Chem* (in press).

81. Bedford, M. T., and Richard, S. (2005). Arginine methylation an emerging regulator of protein function. *Mol Cell* 18:263–272.

82. Boisvert, F. M., Chenard, C. A., and Richard, S. (2005). Protein interfaces in signaling regulated by arginine methylation. *Sci STKE* 271(2):1–10.

83. Blatch, G. L., and Lassle, M. (1999). The tetratricopeptide repeat: a structural motif mediating protein-protein interactions. *Bioessays* 21:932–939.

84. Gary, J. D., Lin, W. J., Yang, M. C., Herschman, H. R., and Clarke, S. (1996). The predominant protein-arginine methyltransferase from Saccharomyces cerevisiae. *J Biol Chem* 271: 12585–12594.

85. Henry, M. F., and Silver, P. A. (1996). A novel methyltransferase (Hmt1p) modifies poly(A)+-RNA-binding proteins. *Mol Cell Biol* 16:3668–3678.

86. Shen, E. C., Henry, M. F., Weiss, V. H., Valentini, S. R., Silver, P. A., and Lee, M. S. (1998). Arginine methylation facilitates the nuclear export of hnRNP proteins. *Genes Dev* 12:679–691.

87. Green, D. M., Marfatia, K. A., Crafton, E. B., Zhang, X., Cheng, X., and Corbett, A. H. (2002). Nab2p is required for poly(A) RNA export in Saccharomyces cerevisiae and is regulated by arginine methylation via Hmt1p. *J Biol Chem* 277:7752–7760.

88. McBride, A. E., Weiss, V. H., Kim, H. K., Hogle, J. M., and Silver, P. A. (2000). Analysis of the yeast arginine methyltransferase Hmt1p/Rmt1p and its in vivo function. Cofactor binding and substrate interactions. *J Biol Chem* 275:3128–3136.

89. Weiss, V. H., McBride, A. E., Soriano, M. A., Filman, D. J., Silver, P. A., and Hogle, J. M. (2000). The structure and oligomerization of the yeast arginine methyltransferase, Hmt1. *Nat Struct Biol* 7:1165–1171.

90. Inoue, K., Mizuno, T., Wada, K., and Hagiwara, M. (2000). Novel RING finger proteins, Air1p and Air2p, interact with Hmt1p and inhibit the arginine methylation of Npl3p. *J Biol Chem* 275:32793–32799.

91. Yun, C. Y., and Fu, X. D. (2000). Conserved SR protein kinase functions in nuclear import and its action is counteracted by arginine methylation in Saccharomyces cerevisiae. *J Cell Biol* 150:707–718

92. Xu, C., and Henry, M. F. (2004). Nuclear export of hnRNP Hrp1p and nuclear export of hnRNP Npl3p are linked and influenced by the methylation state of Npl3p. *Mol Cell Biol* 24:10742–10756

93. McBride, A. E., Cook, J. T., Stemmler, E. A., Rutledge, K. L., McGrath, K. A., and Rubens, J. A. (2005). Arginine methylation of yeast mRNA-binding protein Npl3 directly affects its function, nuclear export and intranuclear protein interactions. *J Biol Chem* (in press).

94. Yu, M. C., Bachand, F., McBride, A. E., Komili, S., Casolari, J. M., and Silver, P. A. (2004). Arginine methyltransferase affects interactions and recruitment of mRNA processing and export factors. *Genes Dev* 18:2024–2035.

95. Xu, C., Henry, P. A., Setya, A., and Henry, M. F. (2003). In vivo analysis of nucleolar proteins modified by the yeast arginine methyltransferase Hmt1/Rmt1p. RNA 9:746–759.

96. Paik, W. K., and Kim, S. (1970). Omega-N-methylarginine in protein. *J Biol Chem* 245:88–92.

97. Zobel-Thropp, P., Gary, J. D., and Clarke, S. (1998). delta-N-methylarginine is a novel posttranslational modification of arginine residues in yeast proteins. *J Biol Chem* 273: 29283–29286.

98. Chern, M. K., Chang, K. N., Liu, L. F., Tam, T. C., Liu, Y. C., Liang, Y. L., and Tam, M. F. (2002). Yeast ribosomal protein L12 is a substrate of protein-arginine methyltransferase 2. *J Biol Chem* 277:15345–15353.

99. Lee, J. H., Cook, J. R., Pollack, B. P., Kinzy, T. G., Norris, D., and Pestka, S. (2000). Hsl7p, the yeast homologue of human JBP1, is a protein methyltransferase. *Biochem Biophys Res Commun* 274:105–111.

100. Gilbreth, M., Yang, P., Wang, D., Frost, J., Polverino, A., Cobb, M. H., and Marcus, S. (1996). The highly conserved skb1 gene encodes a protein that interacts with Shk1, a fission yeast Ste20/PAK homolog. *Proc Natl Acad Sci USA* 93:13802–13807.

101. Ma, X. J., Lu, Q., and Grunstein, M. (1996). A search for proteins that interact genetically with histone H3 and H4 amino termini uncovers novel regulators of the Swe1 kinase in Saccharomyces cerevisiae. *Genes Dev* 10:1327–1340.

102. McMillan, J. N., Longtine, M. S., Sia, R. A., Theesfeld, C. L., Bardes, E. S., Pringle, J. R., and Lew, D. J. (1999). The morphogenesis checkpoint in Saccharomyces cerevisiae: cell cycle control of Swe1p degradation by Hsl1p and Hsl7p. *Mol Cell Biol* 19:6929–6939.

103. Cid, V. J., Shulewitz, M. J., McDonald, K. L., and Thorner, J. (2001). Dynamic localization of the Swe1 regulator Hsl7 during the Saccharomyces cerevisiae cell cycle. *Mol Biol Cell* 12:1645–1669.

104. Najbauer, J., Johnson, B. A., Young, A. L., and Aswad, D. W. (1993). Peptides with sequences similar to glycine, arginine-rich motifs in proteins interacting with RNA are efficiently recognized

by methyltransferase(s) modifying arginine in numerous proteins. *J Biol Chem* 268: 10501–10509.

105. Li, H., Park, S., Kilburn, B., Jelinek, M. A., Henschen-Edman, A., Aswad, D. W., Stallcup, M. R., and Laird-Offringa, I. A. (2002). Lipopolysaccharide-induced methylation of HuR, an mRNA-stabilizing protein, by CARM1. Coactivator-associated arginine methyltransferase. *J Biol Chem* 277:44623–44630.

106. Wong, G., Muller, O., Clark, R., Conroy, L., Moran, M. F., Polakis, P., and McCormick, F. (1992). Molecular cloning and nucleic acid binding properties of the GAP-associated tyrosine phosphoprotein p62. *Cell* 69:551–558.

107. Wada, K., Inoue, K., and Hagiwara, M. (2002). Identification of methylated proteins by protein arginine N-methyltransferase 1, PRMT1, with a new expression cloning strategy. *Biochim Biophys Acta* 1591:1–10.

108. Boisvert, F. M., Cote, J., Boulanger, M. C., and Richard, S. (2003). A proteomic analysis of arginine-methylated protein complexes. *Mol Cell Proteomics* 2:1319–1330.

109. Ong, S.-E., Mittler, G., and Mann, M. (2004). Identifying and quantifying in vivo methylation sites by heavy methyl SILAC. *Nature Methods* 1:119–126.

110. Zhang, X., Zhou, L., and Cheng, X. (2000). Crystal structure of the conserved core of protein arginine methyltransferase PRMT3. *Embo J* 19:3509–3519.

111. Lin, Q., Jiang, F., Schultz, P. G., and Gray, N. S. (2001). Design of allele-specific protein methyltransferase inhibitors. *J Am Chem Soc* 123:11608–11613.

112. Cheng, D., Yadav, N., King, R. W., Swanson, M. S., Weinstein, E. J., and Bedford, M. T. (2004). Small molecule regulators of protein arginine methyltransferases. *J Biol Chem* 279:23892–23899.

3

Diverse Roles of Protein Arginine Methyltransferases

ANNE E. McBRIDE

Department of Biology
Bowdoin College
85 Union Street
Brunswick, ME 04011, USA

I. Abstract

Methylation of arginine residues within proteins results in a more subtle change than phosphorylation, but one that also has profound impacts on eukaryotic organisms. The rapidly expanding list of protein arginine methyltransferases (PRMTs) and PRMT substrates has implicated arginine methylation in a wide variety of cellular processes. This chapter explores the diverse functions of PRMTs by examining evidence for effects of arginine methylation and PRMTs on specific cellular processes, and connecting the data to molecular mechanisms that have been proposed to explain these effects. First, effects of arginine methylation on intermolecular interactions are addressed, focusing on how methylation affects protein-nucleic acid and protein-protein interactions. Next, the numerous links among PRMTs, arginine methylation, and cellular processes are described, including roles of PRMTs and methylation in transcription, cell signaling, protein transport, pre-mRNA splicing, ribosome assembly, and DNA damage repair. Finally, the broader implications of arginine methylation are considered by examining the

THE ENZYMES, Vol. XXIV

connections between PRMTs and their substrates and development, differentiation, and disease.

II. Introduction

Eukaryotic organisms use a wide range of strategies at all levels of gene expression to control their molecular repertoire. Transcriptional regulation and posttranscriptional mechanisms including alternative splicing, translational control, regulation of protein and RNA stability, and posttranslational protein modification all influence the capabilities of a cell at a given time. Recently, protein arginine methyltransferases have emerged as significant players in numerous cellular processes [1-3].

The discovery of modified arginine residues in histones [4] and myelin basic protein (MBP) [5, 6] led to the identification of two different enzymatic activities in mammalian cell extracts. Type I arginine methyltransferases transfer one or two methyl groups from S-adenosyl-L-methionine (AdoMet) to one of the guanidino nitrogens on arginine residues in substrates including histones, forming monomethylarginine (MMA), and asymmetric dimethylarginine (aDMA; Figure 3.1) [7]. Type II arginine methyltransferases also catalyze monomethylation but then add a second methyl group to the other guanidino nitrogen, forming symmetric dimethylarginine (sDMA) in substrates including MBP (Figure 3.1) [7]. Although specific sites of

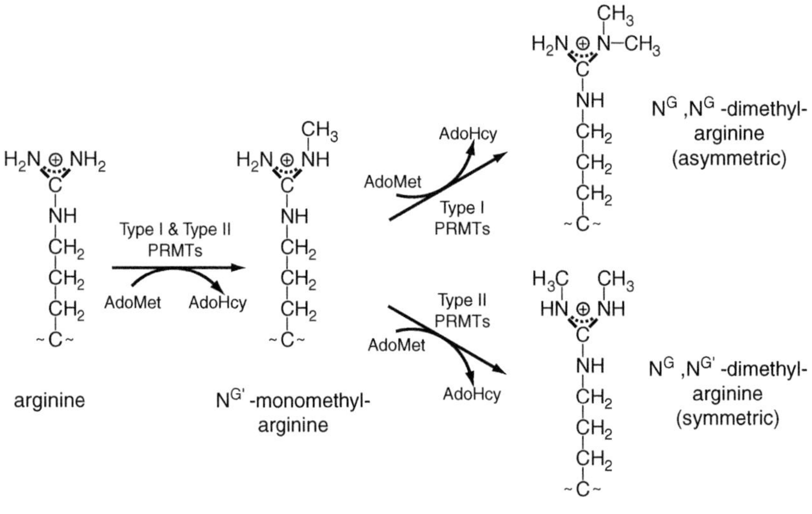

FIG. 3.1. Arginine methylation. Protein arginine methyltransferases catalyze the transfer of methyl groups from S-adenosyl-L-methionine (AdoMet) to arginine residues (~C~ denotes the α carbon), also generating S-adenosyl-L-homocysteine (AdoHcy). A yeast PRMT that monomethylates the internal δ nitrogen has also been identified (see the section IV.E).

arginine dimethylation were discovered in RNA-binding proteins—including heterogeneous nuclear ribonucleoprotein A1 (hnRNP A1), nucleolin, and fibrillarin—in the 1980s [8, 9], the cloning of the first protein arginine methyltransferase genes in the mid 1990s revolutionized the study of protein arginine methylation [10-12].

In the past decade, genomic and protein-interaction studies have allowed the identification of eight protein arginine methyltransferases (PRMTs) in mammalian cells as well as numerous homologs in other eukaryotic systems (see Chapter 2). These studies, in addition to the identification of arginine-glycine (RG)-rich domains as substrates for several methyltransferases [13], have facilitated the identification of a plethora of substrates (Table 3.1). Proteomic approaches have also

TABLE 3.1

PROTEIN ARGININE METHYLTRANSFERASE SUBSTRATES

Chromatin-Associated Proteins, Transcription Factors, and Coactivators				
Substrate	*In Vitro* PRMT	*In Vivo* PRMT[a]	Organism[b]	Notes[c]
Histone H2A[d]	PRMT1 (41)			
	CARM1 (41)			
	PRMT5 (84)			
	HSL7 (95)			
Histone H3[d]	CARM1 (41)	type I (44)	calf	amino acid analysis
	PRMT6 (228)	nd (45)	calf	α-mR17-H3
	CARMER (57)	nd (46)	human	α-mR17-H3
	DART4 (157)	PRMT5 (70)	mouse	α-mR8-H3 (PRMT5 I-RNA)
Histone H4[d]	PRMT1 (41)	PRMT1 (59)	human	MS (+PRMT1 wt,mut)
	PRMT5 (84)	PRMT1 (60)	mouse	α-mR3-H4 (PRMT1–/–)
	PRMT6 (228)			
	DART4 (157)			
	Hsl7 (95)			
CBP	CARM1 (35)	CARM1 (35)	human	AdoMet* (+CARM1)
	CARM1 (49)	CARM1 (49)	human	α-mR742-CBP (CARM1 I-
	DART4 (157)			RNA)
p300	CARM1 (35)	CARM (50)	human	α-R2141-p300 (+ CARM1)
	DART4 (157)			
PGC-1α	PRMT1 (61)		human	
SPT5	PRMT1,	nd (66)	human	AdoMet*
	PRMT5 (66)			
ZF5	PRMT1 (229)			
TAFII 68	PRMT1 (229)			
NF-AT 90		nd (16)	human	MS
TBP-associated factor 2N		nd (16)	human	MS
FCP1	PRMT5 (133)	type II (133)	human	IB α-sDMA (IP α-FCP1)
HMGA1a		nd (230)	human, rat	MS
		types I & II (231)	human	MS

Continued

TABLE 3.1

PROTEIN ARGININE METHYLTRANSFERASE SUBSTRATES—cont'd

Signaling Proteins				
Substrate	*In Vitro* PRMT	*In Vivo* PRMT[a]	Organism[b]	Notes[c]
TIS21	type I (232)			
ILF3	PRMT1 (64)		rat	
	(224)		human	
NF90	PRMT1 (64)		rat	
TARPP	CARM1 (224)	nd (162)	mouse	AdoMet*
STAT1[e]	PRMT1 yes: (77)	yes: nd (77,217)	human	yes: MS, IP: α–DMA
	PRMT1 no: (79)	no: (79,80)		(IB: α–STAT1)
				no: MS, AdoMet*, IP:
				α–DMA (IB: α–STAT1)
STAT3		nd (27)	human	IP: α–DMA (IB: α–Flag)
STAT6		nd (78)	human	IP: α–DMA (IB: α–Flag)
NIP45	PRMT1 (82)	PRMT1 (82)	mouse	IP: α–DMA (IB: α–Flag,
				PRMT1–/–)
HMW FGF-2	PRMT1 Hmt1/	nd (122)	mouse	MS
	Rmt1 (122)			
Ras-GAP- binding protein 1		nd (16)	human	MS
Ras-GAP- binding protein 2		nd (16)	human	MS
FKBP12		nd (225)		AdoMet*
Vav-1		nd (233)		
hnRNPs and hnRNP-like Proteins				
hnRNP A0		nd (16)	human	MS
hnRNP A1	PRMT1 (12) HMT1/RMT1 (10,11)	nd (8,16,234)	calf	peptide seq., MS
hnRNP A2	PRMT1 (114)	nd (16,101)	human	AdoMet*
hnRNP A/B		nd (16)	human	MS
hnRNP B1, B2		nd (16,101)	human	AdoMet*
hnRNP D		nd (101)	human	AdoMet*
hnRNP D0		nd (16)	human	MS
hnRNP E		nd (101)	human	AdoMet*
hnRNP G		nd (16,101)	human	AdoMet*, MS
hnRNP H		nd (101)	human	AdoMet*
hnRNP J		nd (101)	human	AdoMet*
hnRNP JKTBP		nd (16)	human	MS
hnRNP K	PRMT1 (229)	nd (16,101,113)	human	AdoMet*, MS
hnRNP P		nd (101)	human	AdoMet*
hnRNP Q		nd (101)	human	AdoMet*
hnRNP R	PRMT1 (229)	nd (101)	human	AdoMet*
hnRNP U/SAF-A	PRMT1 (104,224)	nd (16,101)	human	AdoMet*, MS

Continued

TABLE 3.1

PROTEIN ARGININE METHYLTRANSFERASE SUBSTRATES—cont'd

hnRNPs and hnRNP-like Proteins—cont'd				
Substrate	*In Vitro* PRMT	*In Vivo* PRMT[a]	Organism[b]	Notes[c]
similar to hnRNP U		nd (16)	human	MS
E1B-AP5		nd (103)	human	AdoMet*
FUS/TLS	PRMT1 (224,229)	type I (220)	human	MS
Npl3	Hmt1/Rmt1 (11) PRMT1 (235) PRMT3 (219) PRMT6 (236)	Hmt1/Rmt1 (11,237)	*Saccharomyces*	AdoMet* (*hmt1Δ*)
Hrp1	Hmt1/Rmt1 (109)	Hmt1/Rmt1 (26,106)	*Saccharomyces*	AdoMet* (*hmt1Δ*)
Hrb1	Hmt1/Rmt1 (109)		*Saccharomyces*	
Nab2	Hmt1/Rmt1 (107)	Hmt1/Rmt1 (107)	*Saccharomyces*	AdoMet* (*hmt1Δ*)
P38		nd (238)	*Artemia*	peptide seq.
Squid	DART1 (157)		*Drosophila*	
Other mRNA-binding Proteins				
CIRP	PRMT1 (224)		human	
CIRP2	PRMT1 (118)		*Xenopus*	
RNA helicase A	PRMT1 (115,224)		Human	
PABP1	CARM1 (224) DART4 (157)	CARM1 (48)	Mouse	AdoMet* (CARM1 −/−)
PABP2	PRMT1, PRMT3 (121)	nd (121)	Calf	MS
PABP4		nd (16)	Human	MS
HuR	CARM1 (120)	CARM1 (120)	mouse	α–mR217-HuR (+CARM1)
Sam68	PRMT1 (34,113)	PRMT1 (16,113)	mouse	AdoMet* (PRMT1−/−)
SLM-1		nd (113)	human[f]	AdoMet*
SLM-2		nd (113)	human[f]	AdoMet*
GRP33		nd (113)	human[f]	AdoMet*
QKI-5		nd (113)	human[f]	AdoMet*
Yra1		Hmt1/Rmt1 (106)	*Saccharomyces*	AdoMet* (*hmt1Δ*)
Aly/REF		nd (16)	human	MS
FMR-related protein 1		nd (16)	human	MS
Vasa	DART1 (157)		*Drosophila*	
Splicing Proteins				
SmB, SmB′	PRMT5 (126,127) CARM1, DART4 (157)	type I (36) types I&II (135)	human mouse	MS, protein seq. amino acid analysis
SmD1	PRMT5 (126,127)	type I (124) types I&II (135)	human mouse	MS, protein seq. amino acid analysis
SmD3	PRMT5 (126,127)	type I (124) types I&II (135)	human mouse	MS, protein seq. amino acid analysis

Continued

TABLE 3.1

PROTEIN ARGININE METHYLTRANSFERASE SUBSTRATES—cont'd

Splicing Proteins—cont'd				
Substrate	*In Vitro* PRMT	*In Vivo* PRMT[a]	Organism[b]	Notes[c]
LSm4		type I (36)	human	MS, protein seq.
Coilin		type II (128)	human	MS (PRMT5 I-RNA)
splicing factor, P,Q-rich		nd (16)	human	MS
splicing factor SF2, P33		nd (16)	human	MS
R/S-rich splicing factor 10		nd (16)	human	MS

Nucleolar and Ribosomal Proteins				
nucleolin		nd (9)	rat	peptide seq.
		nd (239)	CHO	peptide seq.
fibrillarin	PRMT1, HMT1/RMT1 (240)	nd (9)	rat	peptide seq.
		nd (138)	*Physarum*	peptide seq.
Nop1	Hmt1/Rmt1 (108)	Hmt1/Rmt1 (108)	*Saccharomyces*	AdoMet* (*hmt1Δ*)
Gar1	Hmt1/Rmt1 (108)	Hmt1/Rmt1 (108)	*Saccharomyces*	AdoMet* (*hmt1Δ*)
Nsr1	Hmt1/Rmt1 (108)	Hmt1/Rmt1 (108)	*Saccharomyces*	AdoMet* (*hmt1Δ*)
rpS2	PRMT3 (141) PRMT1, PRMT3, PRMT6 (142)	PRMT3 (141) nd (142)	*Schizosaccha- romyces* mouse	AdoMet* (*prmt3Δ*) AdoMet*
rpL12		Rmt2 (144)	*Saccharomyces*	AdoMet* (*rmt2Δ*)

Other Nucleic-Acid-Binding Proteins				
EWS	PRMT1 (117)	nd (16,241)	human	MS
MRE11	PRMT1 (146)	PRMT1 (146)	human mouse	MS AdoMet* (PRMT1–/–)
RBP1	PRMT1	nd (242)	trypanosome	MS
GRY-RBP	PRMT1 (229)		human	
RBP58	PRMT1 (229)		human	
putative RNA-binding protein 3		nd (16)	human	MS
DEAD-box protein p68		nd (16)	human	MS
Ataxin-2-related domain protein		nd (16)	human	MS

Viral Proteins				
Tat	PRMT6 (211)	PRMT6 (211)	human immun- odeficiency virus-1	AdoMet* (+PRMT6 wt,mut)

Continued

TABLE 3.1

PROTEIN ARGININE METHYLTRANSFERASE SUBSTRATES—cont'd

Substrate	*In Vitro* PRMT	*In Vivo* PRMT[a]	Organism[b]	Notes[c]
Viral Proteins—cont'd				
ICP27		nd (209)	herpes simplex virus	AdoMet*
NS3	PRMT1 (27)	nd (27)	hepatitis C virus	IP:α–DMA (IB: α–Flag)
Small hepatitis delta antigen	PRMT1 (212)		hepatitis delta virus	
L4-100 kDa	PRMT1 (210)	nd (210)	Adenovirus	AdoMet*
EBNA2	PRMT5 (213)	nd (213)	Epstein-Barr virus	AdoMet*
Golgi-Associated Proteins[g]				
putative transmembrane methyltransferase		nd (15)	rat	MS
Integral membrane protein Tmp21-I		nd (15)	rat	MS
p24B (*cis*-Golgi)		nd (15)	rat	MS
TGN38, *trans*-Golgi network protein 1		nd (15)	rat	MS
Golgin-84		nd (15)	rat	MS
cis-Golgi SNARE (p28)		nd (15)	rat	MS
mannoside acetylglu-cosaminyl-transferase 1		nd (15)	rat	MS
mannoayl (α–1,3)-glycoprotein β-1,4-N-acetylglucoasminyl-transferase 2		nd (15)	rat	MS
mannosidase 1, α		nd (15)	rat	MS
GRASP55		nd (15)	rat	MS
cytochrome P450, 2d2		nd (15)	rat	MS
cytochrome P450, 2c29		nd (15)	rat	MS
epoxide hydrolase 1		nd (15)	rat	MS
protein disulfide isomerase A3		nd (15)	rat	MS
flavin-containing monooxygenase 5		nd (15)	rat	MS
transferrin		nd (15)	rat	MS
EMP70, member 2		nd (15)	rat	MS
EMP70, member 3		nd (15)	rat	MS
Other Proteins				
myelin basic protein	PRMT5 (84) Hsl7 (95) CARM1, DART4 (157)	nd (5)	bovine	amino acid analysis
myosin		nd (243)		amino acid analysis

Continued

TABLE 3.1

PROTEIN ARGININE METHYLTRANSFERASE SUBSTRATES—cont'd

Other Proteins—cont'd				
Substrate	In Vitro PRMT	In Vivo PRMT[a]	Organism[b]	Notes[c]
p137-GPI	PRMT1 (229)		human	
PRMT6	PRMT6 (236)		human	
keratin, type II cytoskeletal 7		nd (16)	human	MS
protein C9 or f10		nd (16)	human	MS

[a]nd, not determined.

[b]Organism for which in vivo arginine methylation has been demonstrated.

[c]Abbreviations: AdoMet*, transfer of methyl group(s) from radiolabeled AdoMet; +PRMT, expression of exogenous PRMT (wt = wild type, mut = AdoMet-binding mutant); PRMT –/– or prmtΔ, cells with homozygous PRMT disruption; α-m = anti-methylarginine antibody; sDMA, symmetric dimethylarginine; I-RNA, inhibitory RNA (RNAi, small inhibitory RNA, or "antisense" RNA); IP, immunoprecipitation; IB, immunoblot; MS, mass spectrometry; seq., sequencing.

[d]General histone arginine methylation was first shown by Paik and Kim [4].

[e]There are data that support both methylation and non-methylation of STAT1 (see the section IV.B.1).

[f]These nonhuman proteins were expressed in human cells to test for methylation.

[g]Although these proteins co-purify with Golgi fractions, several endoplasmic reticulum, secretory, and endosomal proteins were also detected in this study.

indicated that arginine methylated proteins play roles in multiple cellular processes [14-16]. In spite of the many known PRMTs and PRMT substrates, the evidence for in vivo methylation of individual substrates by specific enzymes is limited (Table 3.1). Understanding the diverse roles of PRMTs requires relating the effects of methylation on protein structure and molecular interactions to the in vivo function of targeted proteins.

Numerous in vitro studies have probed the molecular mechanisms that underlie the effects of arginine methylation on cellular processes. These studies have focused on defining biochemical activities and requirements for protein complex formation through the addition of purified enzymes, substrates, and other interacting molecules. The results using simplified systems, however, do not always correspond to studies of complexes found in vivo. To determine the roles of PRMTs in vivo, studies have focused on altering PRMT expression through addition of exogenous PRMTs, disruption of endogenous PRMT genes, or addition of inhibitory RNAs. In addition, mutagenesis of targeted arginines in substrate proteins has suggested roles for arginine methylation of specific substrates in cellular processes. Finally, use of methyltransferase inhibitors such as adenosine dialdehyde (AdOx) and 5′-methyl-thioadenosine (MTA) has helped to define the importance of arginine methylation.

Whereas most methyltransferase inhibitors are relatively nonspecific, inhibiting the activity of many enzymes that use AdoMet as a substrate, disruption of PRMT expression is expected primarily to affect the modification of its own substrates. The presence of numerous substrates for a single PRMT, however, presents a challenge to understanding the mechanism for effects of PRMT expression on cellular processes. On the other hand, mutagenesis of PRMT substrates changes not only their ability to be methylated but the functionality of the residue, making it difficult to distinguish between effects caused by the lack of methylation and those caused by the absence of arginine. Given these limitations, understanding the diverse functions of protein arginine methyltransferases requires a combination of *in vitro* and *in vivo* techniques that address PRMT function by investigating both the enzymes and their substrates.

To address the question of what molecular mechanisms underlie the function of protein arginine methyltransferases, this chapter first explores the effects of arginine methylation on intermolecular interactions. It then addresses the proposed roles of PRMTs in cellular processes, from transcription and cell signaling to protein transport and splicing, and the data supporting these roles. Finally, it describes connections between PRMT expression and activity and more complex processes including differentiation, development, and disease progression.

III. Arginine Methylation and Intermolecular Interactions

One striking feature of proteins known to be arginine methylated is that the majority are nucleic-acid-binding proteins and proteins involved in transcriptional regulation (Table 3.1). The positive charge and hydrogen-bonding properties of arginine facilitate interactions with the phosphodiester backbone of nucleic acids [17, 18]. Molecular modeling suggests that methylation of arginine restricts its hydrogen-bonding capabilities as well as increasing the bulkiness and hydrophobicity of the residue (Figure 3.1) [18]. Although arginine should maintain a net positive charge following methylation, *in vitro* methylation of hnRNP A1 lowers its pI, indicating a reduction in basicity of the protein [19]. Each of these changes might affect the ability of arginine residues to interact with nucleic acids. The possibility that arginine methylation affects nucleic-acid binding activity of its substrates is supported by the finding that deletion of the RG motifs targeted for methylation in hnRNP U and nucleolin disrupts their interaction with RNA *in vitro* [20, 21]. However, mutagenesis of the RG-rich domain in hnRNP A1 and yeast hnRNP-like protein Npl3 also points toward roles of RG-rich regions in protein-protein interactions [22, 23]. Work over the past decade has addressed the role of arginine methylation in both protein-nucleic acid and protein-protein interactions.

A.   PROTEIN-NUCLEIC ACID INTERACTIONS

Studies with mammalian and yeast RNA-binding proteins have found relatively subtle effects of arginine methylation on the affinity of PRMT1 targets for nucleic acids. Methylated hnRNP A1 elutes from single-stranded (ss) DNA columns at a slightly lower salt concentration than unmethylated hnRNP A1 (0.59 M versus 0.63 M), and 35 mM less NaCl is required to decrease binding to ssRNA by 50% [19]. In addition, methylation does not affect the binding of a nucleolin peptide to ssDNA or ssRNA [18]. The affinity of yeast 3′ end processing factor Hrp1 for specific RNA ligands is not significantly changed by methylation [24]. Together these results suggest that methylation might lower nonspecific nucleic acid binding without affecting binding to specific ligands, consistent with the proposal that arginine methylation might act as a switch between nonspecific and specific RNA-binding modes of a protein [17]. This possibility is not supported by the RNA-binding activity of the fragile-X mental retardation protein FMRP following *in vitro* translation in the presence or absence of the methyltransferase inhibitor AdOx [25]. Although the binding of FMRP from AdOx-treated lysates to three of four homoribopolymers was increased over that from untreated lysates, its binding for several individual mRNAs was slightly reduced by the presence of AdOx during translation [25]. For all but poly(rA), the effect of AdOx treatment was twofold or less [25]. Deleting the predominant yeast arginine methyltransferase gene also does not affect the amount of its substrate Npl3 that can be cross-linked to bulk poly(A) RNA *in vivo* [26].

Although the RNA-binding affinity of PRMT substrates is not dramatically affected by methylation, a number of experiments point to structural implications of methylation and RNA binding. Circular dichroism studies indicate that the conformation of RNAs bound by a methylated and unmethylated nucleolin peptide differs, whereas DNA melting curves are similar in the presence of these peptides [18]. The binding of RNA to proteins can also influence protein methylation. *In vitro* methylation of Hrp1 [24] and the hepatitis C NS3 protein [27] are decreased in the presence of RNA, and RNAase treatment of yeast or RAT1 cell lysates alters the ability of lysate proteins to be methylated by recombinant PRMTs [28]. In capillary electrophoresis experiments, the migration pattern of a human immunodeficiency virus (HIV) Tat peptide-RNA complex changes upon methylation of a single arginine within the peptide, suggesting structural differences in the two complexes [29]. Therefore, although RNA-protein interactions can affect protein methylation and arginine methylation can influence nucleic acid structure results to date do not suggest specific roles for methylation in cellular processes through the modulation RNA-protein interactions.

B. PROTEIN-PROTEIN INTERACTIONS

RG domains are frequently found in proteins that contain other RNA-binding domains such as RNA-recognition motifs (RRMs) [30] or hnRNP K-homology (KH) domains [31] and RNA binding is not always dependent on the presence of the RG domain [32]. Although RG domains can mediate RNA interactions [20], a plethora of evidence points to roles for RG domains in protein-protein interactions. For example, mutations in the RG domain of hnRNP A1 decrease its ability to self-associate *in vitro* [20]. A two-hybrid screen for proteins that interact with the RG domain of yeast Npl3 revealed numerous RG-containing proteins [33], and mutations in this domain affect the protein-protein interactions of yeast Npl3 *in vivo* [23].

Bedford and colleagues presented the first clear evidence for a role of methylation in controlling protein complex formation using peptides from the mammalian RNA-binding protein Sam68 [34]. *In vitro* Sam68 binds to a number of proteins with either Src-homology-3 (SH3) or WW domains. Interestingly, methylation of Sam68 peptides containing four or more arginines inhibited binding to SH3-domain proteins but not to WW-domain-containing proteins [34]. Methylation of HIV NEF, which only has a single arginine, also inhibits its binding to SH3-domain proteins [34]. Similarly, methylation serves as a binding switch for the CREB-binding protein (CBP)/p300 transcriptional coactivator proteins [35]. The CARM1 protein arginine methyltransferase and CBP/p300 synergistically coactivate nuclear hormone receptors (see section IV.A.1.b), and CARM1 methylation of p300 decreases its affinity for CREB, a competing transcription factor [35]. Methylation of snRNP proteins SmB/B′, D1, and D3 also affects their binding to the survival of motor neurons (SMN) protein [36, 37]. Structural studies revealed that aromatic residues in the SMN tudor domain interact with symmetrically dimethylated arginines in Sm proteins [38].

These results, and other examples (discussed in material following), present overwhelming evidence for roles of PRMTs in modulating protein-protein interactions, although these effects do not rule out subtle primary or secondary effects on protein-nucleic acid interactions. The following sections address the question of how these effects of methylation on molecular interactions relate to experiments that investigate the function of PRMTs in cellular processes. For clarity, all mammalian PRMTs other than CARM1 are referred to by their PRMT number rather than by alternative names (e.g., HRMT1L1=PRMT2, HRMT1L2=PRMT1, JBP1=PRMT5).

IV.   Roles of PRMTs in Cellular Processes

A. TRANSCRIPTION

Protein arginine methyltransferases have been implicated in transcription through their genetic and biochemical interactions with numerous transcription factors and transcriptional coactivators, as well as through their methylation of histones. Many studies in the 1990s focused on lysine acetylation and deacetylation in N-terminal histone tails and the importance of these modifications for chromatin remodeling [39]. The identification of both arginine and lysine methyltransferases that target histones and experiments that have addressed the roles of these enzymes in transcriptional control have enhanced our understanding of the complexity of the "histone code" [40]. A role for PRMTs in transcription was first suggested by the cloning of a coactivator-associated arginine methyltransferase, CARM1 (PRMT4) [41], and subsequent studies have revealed intriguing roles for a number of PRMTs in different aspects of transcription.

1.   CARM1

a.   Transcriptional Coactivation of Nuclear Hormone Receptors

Prior to the cloning of CARM1, the p160 protein family (including GRIP-1 and SRC-1) and the p300/CREB binding protein (CBP) family were both known to act as coactivators for nuclear hormone receptors (NHRs): co-transfection of a member of each family along with a nuclear hormone receptor gene allows enhanced transcriptional activation on hormone stimulation [42]. A two-hybrid screen to identify GRIP-1 interacting proteins revealed a novel protein that is similar to the predominant mammalian arginine methyltransferase PRMT1 and that methylates histone H3 *in vitro*, leading it to be named CARM1 [41]. CARM1 transfection results in stimulation of GRIP-1-dependent coactivation of estrogen-, thyroid-, and androgen-inducible reporter genes in the presence of NHR and hormone [41] (Figure 3.2). In addition, CARM1 and CBP act as synergistic coactivators, although a p160 family member is required for each protein to coactivate transcription [43]. CARM1 coactivation is abrogated by mutating a conserved VLD within the AdoMet-binding domain to AAA, consistent with methyltransferase activity being crucial for its role in transcription [41].

In vitro CARM1 methylates histone H3 at R2, R17, and R26 [44], and many studies have focused on methylation of R17, using antibodies raised against a peptide in which R17 is asymmetrically dimethylated (mR17-H3) [45, 46]. Chromatin immunoprecipitation (ChIP) experiments have found promoter regions that can be precipitated with both CARM1 antibodies and the mR17-H3 antibodies specifically under inducing conditions [45–47]. The importance of CARM1 in transcription of NHR-dependent genes is underscored by studies of murine CARM1−/− embryonic fibroblasts: estrogen-receptor-dependent transactivation of a reporter gene is severely compromised and after estrogen exposure endogenous estrogen-responsive genes are expressed at lower levels than in CARM+/+ cells [48]. Remarkably, a comparison of proteins from CARM1−/− and CARM1+/+ cells by immunoblotting with mR17-H3 antibodies reveals a decrease in the signal for several high molecular weight bands in CARM1−/− cells but no

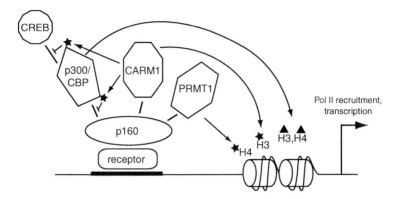

FIG. 3.2. Transcriptional coactivation by CARM1 and PRMT1. Arginine methyltransferases CARM1 and PRMT1 and histone acetyltransferase p300/CREB-binding protein (CBP) can coactivate transcription of nuclear hormone receptor-targeted genes by interacting with p160 family transcriptional coactivators [41, 43, 45, 46, 55, 58-60, 62]. Dashed arrows indicate physical interactions, solid arrows indicate enzymatic modifications (star = methylation; triangle = acetylation), and flat arrowheads indicate inhibition of interactions through methylation of CBP [35, 49, 50]. Modification of histones H3 and H4 by PRMTs and CBP can influence subsequent modifications (see the section IV.A.5). p160 = GRIP-1, ACT-R, SRC-1; receptor = estrogen, androgen, thyroid, glucocorticoid, farnesoid X, or orphan receptors; and Pol II = RNA polymerase II.

decrease in H3 signal [48]. These results suggest that ChIP experiments using mR17-H3 antibodies may reflect DNA binding of other CARM1 substrates in addition to histone H3.

b.   Methylation of p300/CREB Binding Protein

The influence of histone methylation on chromatin structure is likely to be critical for CARM1 transcriptional coactivation, but CARM1 also affects transcription through modification of other substrates. CARM1 methylates the p300/CBP family of coactivators in at least three different regions *in vitro*: within the CREB-interacting KIX domain [35], just C-terminal to the KIX domain [49], and in the C-terminus [50]. Methylation of arginines in different regions of p300/CBP appears to have different effects on transcription (Figure 3.2) [35, 49, 50].

p300/CBP methylation has been linked to a switch between transcription mediated by the cyclic AMP-response-element binding protein (CREB) and CARM1-coactivated transcription [35]. Transfection of increasing amounts of a CARM1 expression vector results in a dose-dependent reduction of expression of cAMP-responsive endogenous or reporter genes. This effect is lost if an inactive mutant CARM1 is used [35]. Treatment of NIH3T3 cells with nerve growth factor (NGF) normally induces CREB-dependent Bcl-2 expression and cell survival, but expression of wild-type and not mutant CARM1 induces apoptosis in NGF-treated cells [35]. The antagonistic effect of endogenous CARM1 on

CREB-dependent transcription is supported by a dose-dependent reduction in endogenous CARM1 expression as cells are treated with higher concentrations of NGF [35]. KIX-domain methylation by CARM1 *in vitro* is decreased by a lysine substitution at an arginine that helps mediate the CREB interaction, suggesting that methylation of this residue may also directly inhibit p300/CBP-CREB binding (Figure 3.2) [35, 51].

Conversely, substitution of alanine for three CARM1-targeted arginines in a region immediately C-terminal to the KIX domain of CBP leads to a loss of GRIP-1-dependent transcriptional activation, whereas CREB-dependent transcription is not affected [49]. Expression of the triple R-to-A mutant CBP also leads to a dose-dependent inhibition of estrogen-dependent transcription in MCF-7 cells, supporting the importance of these arginines in NHR-mediated transcription [49]. CARM1 modification of R2142 in the C-terminal domain of p300 inhibits binding to GRIP-1, and a R2142K substitution inhibits p300 coactivation [50]. The p300-GRIP-1 interaction is increased by the expression of protein arginine deiminase 4 (PAD4), which catalyzes the formation of citrulline at mono- or unmethylated arginines [50, 52, 53]. Together, these data suggest that CARM1 methylation or lysine substitution at R2142 of p300/CBP can negatively regulate hormone-dependent transcriptional activation by decreasing the p300-GRIP-1 interaction [50], whereas citrulline does not inhibit this interaction and blocks methylation. Therefore, methylation of p300/CBP both within and outside the KIX domain may switch a cell between transcriptional programs by influencing binding to coactivating partners (Figure 3.2).

c.   Interactions with Other Transcription Factors and Coactivators

Proposed roles for CARM1 in nuclear factor kappa B (NF-KB)- and farnesoid X receptor (FXR)-mediated transcription show similarities to the nuclear hormone receptor system. When the NF-KB subunit p65 is limiting, CARM1, p300, and GRIP-1 synergistically coactivate transcription from a NF-KB-dependent reporter gene [54]. Lipopolysaccharide (LPS) treatment of CARM1−/− cells results in the expression of only a subset of NF-KB-regulated genes (including IL-6), whereas expression of other genes (such as IP-10 and MIP-2) is severely reduced [54]. After LPS treatment of CARM1 +/+ cells, p65 localizes to IP-10, MIP-2, and IL-6 promoters. In LPS-treated CARM1 −/− cells, however, p65 is recruited to the IL-6 promoter, but not the IP-10 and MIP-2 promoters [54]. These results suggest a role for CARM1 in recruitment of p65 to specific genes [54]. Although mutations in the active sites of p300 or CARM1 do not block complex formation, they partially inhibit cooperative transactivation of an NF-KB-responsive HIV long-terminal-repeat (LTR) reporter gene [54].

FXR co-purifies with CARM1 from cell extracts [55]. In the presence of its ligand, levels of FXR, CARM1, and mR17-H3 are increased at the FXR-controlled

endogenous bile salt export bump (BSEP) promoter [55]. The increase in endogenous BSEP levels upon transfection of a CARM1 expression plasmid suggests
that CARM1 acts as a transcriptional coactivator for FXR [55]. Whereas p160
protein SRC-1a is required for coactivation of BSEP transcription by CARM1,
effects of CBP/p300 on FXR-mediated transcription have not yet been addressed.

In a search for CARM1-interacting proteins, Xu, Evans, and co-workers
isolated a complex they termed the nucleosomal methylation activator complex
(NUMAC), which contains many members of the SWI/SNF transcriptional system
(including ATPase BRG1) in addition to CARM1 [56]. Whereas recombinant
CARM1 preferentially modifies free histones independent of ATP hydrolysis, purified NUMAC methylates nucleosomal histones to a greater extent than free histone
H3 and its activity is inhibited by a non-hydrolyzable ATP analog [56]. CARM1
both binds to BRG1 and can stimulate its ATPase activity [56]. Although CARM1
and BRG1 both assemble on and cooperatively coactivate an estrogen-responsive
target gene, these activities are not affected by mutations that reduce methyltransferase activity of CARM1 [56]. One possible explanation for this result is
that the role of CARM1 in the NUMAC complex may be in promoting complex
assembly rather than in methylation of specific targets. Further exploration of the
role of this complex *in vivo* remains.

Work in *Drosophila* indicates that roles of CARM1 in transcription coactivation are conserved. The *Drosophila* CARM1 homolog, Carmer, co-precipitates
with the ecdysone receptor and can methylate H3 *in vitro*, unless VLD-to-AAA
mutations are introduced in the Carmer AdoMet-binding domain [57]. Carmer
expression results in hormone-dependent transcription of a luciferase reporter
gene. This activity is also reduced by the VLD-to-AAA mutations [57]. Ecdysone
induces apoptosis in cultured *Drosophila* cells, but Carmer inhibitory RNA
(RNAi) can rescue cell death [57]. The decrease in caspase activity and inhibition
of induction of several apoptotic genes in the presence of Carmer RNAi suggests
the importance of this PRMT for coactivation of ecdysone-induced genes.

2.   PRMT1

Although roles of PRMTs in transcription activation were first elucidated in
CARM1 studies, PRMT1 (the predominant mammalian arginine methyltransferase)
has also been implicated in transcriptional control. Like CARM1, PRMT1 binds
GRIP-1 *in vitro* and PRMT1 expression can coactivate transcription from nuclear
hormone-dependent promoters, an activity that requires the presence of the p160
protein (Figure 3.2) [58]. At low levels of androgen or orphan receptors, PRMT1
and CARM1 synergistically coactivate transcription of reporter genes [58]. Unlike
CARM1, however, PRMT1 methylates histone H4 *in vitro* on R3 [59] and an
mR3-H4-specific antibody does not bind proteins in immunoblots of PRMT1−/−
cell extracts [60].

PRMT1 also affects transcription from endogenous genes controlled by PGC-1α (peroxisome-proliferator-activated receptor γ coactivator) or the farnesoid-X receptor. PRMT1 synergistically coactivates transcription from estrogen-responsive promoters with its substrate PGC-1α [61]. Substitution of lysine for arginine at three sites eliminates *in vitro* methylation of PGC-1α by PRMT1 [61]. Induction of endogenous genes by PGC-1α is decreased either by introduction of the R-to-K mutations in PGC-1α or by PRMT1 siRNA treatment [61]. Transfection of PRMT1 both coactivates an FXR-dependent reporter gene and affects the expression from endogenous genes regulated by FXR [62]. Whereas BSEP and SHP are upregulated on PRMT1 expression, and PRMT1 and mR3-H4 are cross-linked to the promoters of these genes, an inhibitory effect is seen on the expression of two other FXR-regulated genes [62]. Thus, although PRMT1 has a positive effect on transcription of NHR- and PGC-1α-regulated genes it has both positive and negative effects on FXR-regulated genes.

PRMT1 has also been implicated in transcription driven by the YY1 transcription factor. Both PRMT1 and a double-stranded RNA-binding protein DRBP76 co-purify with YY1 from HeLa cell extracts [63]. Because DRBP76 is an alternative splice product of the interleukin-enhancer-binding factor 3 (ILF3), which is a known PRMT1-binding protein [64], the YY1-PRMT1 interaction may be mediated by DRBP76/ILF3 [63]. YY1 transfection has a dose-dependent effect on the presence of PRMT1 at the c-*myc* promoter and PRMT1 RNAi decreases expression from an YY1-dependent reporter gene [63]. These results, combined with the ability of DRBP76 and ILF3 to enhance PRMT1 activity *in vitro*, suggest that PRMT1, YY1, and DRBP76/ILF3 cooperate in transcription of YY1-target genes [63, 64]. Activation of the Grp78/BiP chaperone gene upon endoplasmic reticulum (ER) stress is also mediated by YY1 [65]. Although YY1 and PRMT1 co-purify in the presence or absence of ER stress, coactivation of the ATF6 transcription factor by PRMT1 and p300 requires both YY1 and inducing conditions. Similarly, PRMT1 and mR3-H4 localization at the promoter is also increased by ER stress [65]. Thus, although PRMT1 and CARM1 target different histones these PRMTs are both able to coactivate transcription in combination with a variety of other coactivators and transcription factors.

3. PRMT5

Whereas CARM1 and PRMT1 are implicated primarily in positive regulation of transcription, studies of the type II arginine methyltransferase PRMT5 suggest that it may have both positive and negative effects on transcription involving the SPT5 transcription elongation factor [66]. SPT5 binds to and acts as a substrate for both PRMT1 and PRMT5 *in vitro*, but the presence of SPT5 and the PRMTs at a cytokine-inducible promoter are inversely correlated: after tumor necrosis factor α (TNFα) treatment, levels of SPT5 rise and levels of PRMT1 and PRMT5 decrease at the I-KB and IL-8 promoters [66]. SPT5 also enhances Tat-dependent HIV-1 transcription elongation [67], and AdOx treatment increases Tat-dependent transcription *in vitro*, suggesting that methylation down-regulates elongation by SPT5 [66]. Similarly, addition of mutant SPT5 protein with lysine substitutions at sites that are normally methylated results in less inhibition of transcription than addition of wild-type protein [66]. Conversely, transfection of PRMT1 or PRMT5 leads to a decrease in Tat activation of a reporter gene [66]. Co-precipitation of SPT5 and RNA polymerase II, as well as the presence of these proteins at the IL-8 promoter, is increased by either AdOx treatment or expression of the R-to-K mutant SPT5 proteins [66]. These results suggest that both PRMT1 and PRMT5 negatively regulate transcription elongation at cytokine and Tat-activated genes.

A negative role for PRMT5 is also implied by its presence in a cyclin E1 repressor complex [68]. Coexpression of wild-type but not mutant PRMT5 with cyclin E1 activators decreases promoter activity [68]. In addition, transfection of PRMT5 leads to lower levels of cyclin E and cell cycle arrest at the G0-G1 boundary [68]. Although levels of mR3-H4 at the CycE transcriptional start site increase upon PRMT5 expression [68], it is not clear whether this increase reflects PRMT5 methylation of H4 because anti-mR3-H4 antibodies have been raised against an aDMA-containing peptide [59, 60]. Further studies may clarify the mechanism for PRMT5 negative regulation of cyclin E transcription.

PRMT5 co-purifies with both BRG1 and hBRM, key members of SWI/SNF activator complexes, and has been linked to both negative and positive regulation of SWI/SNF activity [69, 70]. Although the complex containing PRMT5 and SWI/SNF components can methylate histones H3 and H4 *in vitro* [69], the inability of the complex to methylate peptides with R8A-H3 or R3A-H4 mutations suggests that these are the sites of PRMT5 methylation [70]. A comparison of NIH3T3 cells transfected with flag-PRMT5, vector, or antisense PRMT5 constructs revealed that with decreasing levels of PRMT5 expression of NM23 and ST7 increase and levels of MYT11 mRNA decrease [70]. The increases correlated with lower levels of PRMT5 and mR8-H3 at the NM23 and ST7 promoters, whereas PRMT5 and mR8-H3 were not present at the MYT11 promoter [70]. Interestingly, the likely site of H3 methylation by PRMT5 [70] does not overlap with the major CARM1 sites [44] but both PRMT1 and PRMT5 appear to modify R3 in H4 [59, 60, 70]. Future studies may reveal whether symmetric and asymmetric dimethylation have different effects on histone function or chromatin remodeling.

4. PRMT2

The majority of PRMT studies have focused on PRMT1, CARM1, and PRMT5, but recent work has pointed to a possible role of PRMT2 in transcription regulation [71]. Although no *in vitro* methyltransferase activity has been demonstrated for PRMT2, one study suggests that this PRMT, like PRMT1 and CARM1, may act as a coactivator for nuclear hormone receptors [71]. The lone PRMT with an SH3-domain, PRMT2 binds p160 member SRC-1 and coactivates transcription of a reporter gene in the presence of the α1 estrogen receptor, although coactivation is not synergistic with SRC-1 [71]. Wheras AdoMet binding domain mutations disrupt coactivation function, removal of the SH3 domain has no effect [71]. The importance of endogenous PRMT2 and its SH3 domain therefore remain to be elucidated.

5. Interplay of Histone Modifying Enzymes

The interaction of PRMTs with histone acetyltransferase p300/CBP and the proximity of modified arginine and lysine residues within histone N-terminal tails have sparked interesting studies investigating possible interplay between arginine methylation and lysine acetylation [60, 72]. After estrogen treatment, CBP and acetylated K18-H3 localize to the endogenous pS2 promoter in MCF-7 cells more quickly than CARM1 and mR17-H3, suggesting that acetylation of a neighboring lysine may promote arginine methylation [72]. Indeed, when CBP (but not an inactive mutant) is overexpressed, CBP, AcK18-H3, CARM1, and mR17-H3 all co-precipitate pS2 promoter regions even in the absence of estrogen [72]. *In vitro* p300 increases CARM1 methylation of H3 and acetylated peptides are better CARM1 substrates than unmodified peptides [72]. The positive influence of acetylation on methylation of H3 may be due in part to increased substrate-enzyme binding [72].

Conversely, H4 methylated at R3 is a better substrate for p300 than an unmodified R3-H4 peptide. PRMT1 stimulates p300 acetylation of H4 [60], and this effect is abrogated by an R3K-H4 mutation [60]. In addition, acetylation decreases H4 methylation [60]. Thus, in the case of H4 evidence points to methylation-enhancing acetylation and acetylation-inhibiting methylation. Similarly, pure BRG1 and hBRM complexes that contain PRMT5 can methylate hypoacetylated HeLa core histones *in vitro* [60], but hyperacetylation of core histones or acetylation at K9 or 14 of H3 inhibits their methylation by the complex [70].

In an extensive study of coactivation of p53-dependent transcription, An, Kim, and Roeder explored the interplay of histone acetylation and methylation using an *in vitro* transcription system reconstituted with recombinant p300, CARM1, PRMT1 and histones [73]. Consistent with earlier studies, addition of p300 before CARM1 enhances H3 methylation and addition of PRMT1 before p300 increases H4 acetylation [73]. These orders of coactivator addition also increase transcription in comparison to simultaneous enzyme addition [73]. When the system is reconstituted with H3 containing K-to-R mutations in CARM1-target sites, however, CARM1 coactivation is lost [73]. This result is the first demonstration of a direct connection between histone H3 methylation and CARM1 coactivator function. These results therefore link the interactions among histone modifying enzymes to transcriptional effects.

Although preincubation with p300 does not inhibit PRMT1 methylation of H4 in this system [73], p300 preincubation does reduce PRMT1 coactivation of transcription, suggesting inhibition of PRMT1 activity by acetylation. Surprisingly, an R3K mutation in H4 does not dramatically reduce transcription in the presence of PRMT1 [73]. One possible explanation of this effect is that PRMT1 methylation of H2A in the presence of the R3K-H4 protein may partially compensate for unmethylated H4 in this system [73].

For almost a decade, models of chromatin remodeling have invoked reversible lysine acetylation in transcription [39], and the recent discovery of PAD4-catalyzed deimination of arginine residues to citrulline within histones [52, 53] has sparked further interest in understanding the interplay of histone-modifying enzymes. Reversibility of arginine methylation and its implications are explored in depth in Chapters 5 and 6. The complexity of the "histone code" is underscored by the ability of either asymmetric or symmetric dimethylation to block PAD4 deimination [53]. Although monomethylarginine is often considered a short-lived precursor to dimethylarginine, PAD4 specificity indicates the importance of considering roles for monomethylarginine in histone function. Future studies on the effect of citrulline on lysine modifications, as well as the effects of lysine modifications on PAD4 activity, should help further elucidate the "histone code" and its role in transcription.

B. CELL SIGNALING

1. STAT

The cloning of PRMT1 as a two-hybrid interactor of the interferon α/β (IFN α/β) receptor [74] led to experiments addressing roles of PRMTs in cell signaling. Initial experiments demonstrated that antisense PRMT1 oligoribonucleotides reduce interferon-induced growth inhibition of a human cell line [74]. Stable expression of antisense PRMT1 RNAs also reduces IFNβ inhibition of viral replication and cell proliferation [75]. Neither of these studies, however, identified any substrates that might be involved in mediating this effect. The signal transducer and activation of transcription (STAT) family of proteins act in combination with Janus kinases (Jaks) downstream of the interferon receptor to transduce cytokine signals to the nucleus [76]. The discovery that STAT1 both co-precipitates with PRMT1 and is an *in vitro* PRMT1 substrate suggested that methylation of STAT1 is involved in interferon signaling [77]. Subsequent studies have offered both complementary and contradictory results.

In vivo methylation of STATs has been tested with various techniques (Table 3.1): immunoprecipitation with an anti-dimethylarginine (DMA) antibody, followed by anti-STAT immunoblotting; labeling of cells in the presence of translation inhibitors and radiolabeled AdoMet, followed by STAT immunoprecipitation; and mass spectrometric analysis of STAT1 from a human cell line. Antibodies raised against STAT1, STAT3, and STAT6 have all been shown to recognize proteins precipitated by an anti-DMA antibody [27, 77, 78] and anti-DMA-precipitated STAT1 levels increased following interferon treatment, particularly in nuclear fractions [77]. The presence of bands recognized by a monoclonal STAT1 antibody in anti-DMA precipitates from STAT1–/– and STAT3–/– cells, however, indicates possible cross-reactivity of this antibody [79]. In addition, *in vivo* labeling followed by immunoprecipitation of STAT1 with a non cross-reacting polyclonal antiserum did not result in a detectable band [80]. However, it is not clear whether the conditions were significantly sensitive enough to detect a single site of methylation, particularly if STAT1 were not fully methylated. Mass spectrometric analyses have also shown conflicting data on the *in vivo* methylation of STAT1 [80], and in some experiments PRMT1 does not methylate STAT1 *in vitro* [79].

Substitution of alanine for R31 in STAT1 decreased its *in vitro* methylation by PRMT1, suggesting methylation of this residue [77]. Whereas R31A and R31E mutations in STAT1 increased interferon-dependent transcription of a reporter gene in one study [77], others found that the R31A mutation led to a loss of interferon-induced transcription [80]. STAT1 proteins with mutations in residues that surround R31 in the crystal structure are less stable than wild-type STAT1 [80]. Similarly, STAT1, STAT3, and STAT6 with mutations in R31 or its analogous residue have shorter half-lives than wild-type STATs [78, 79], consistent with a loss of function.

Two primary techniques have been used to block PRMT activity *in vivo* to look at effects on STAT function: treatment with exogenous methyltransferase inhibitors such as MTA and disruption of the MTA phosphorylase (MTAP) gene, which allows accumulation of endogenous MTA. In both systems, decreases in transcription of STAT-responsive genes as well as decreased inhibition of MTAP–/– cell proliferation after interferon treatment support the importance of methylation for interferon signaling [77]. Methyltransferase inhibitors both increase co-precipitation of protein inhibitor of activated STATs (PIAS1) with STAT1 [77] and decrease STAT1 interactions with the T-cell protein tyrosine phosphatase (TcPTP) [81]. These data led to the model that a PRMT facilitates interferon signaling by decreasing interaction of STAT1 with the PIAS1 inhibitor and allowing binding to TcPTP [81]. Evidence exists, however, for methyltransferase inhibitors having nonspecific effects on phosphorylation of multiple signaling proteins outside the Jak-STAT pathway [79].

Therefore, whether STATs are targets for methylation *in vivo* and the possible roles of STAT methylation remain controversial. These studies indicate the interpretative challenges posed by the use of general methyltransferase inhibitors and mutated substrates when trying to understand the role of arginine methylation of a specific protein in cellular processes.

2. *NFAT*

Both PRMT1 and PRMT5 have been implicated in T-cell signaling through the nuclear factor of activated T-cells (NFAT) transcription factor. The 45 kDa

NFAT-interacting protein NIP45 has an extensive RG-rich N-terminal domain, is a PRMT1 substrate both *in vitro* and *in vivo* (Table 3.1), and the NIP45-PRMT1 interaction requires the N-terminus of NIP45 [82]. Co-precipitation of NIP45 with NFAT also requires the RG-rich region and is reduced by methyltransferase inhibitors [82]. Whereas PRMT1 and NFAT co-precipitate from cell extracts, expression of NIP45 lacking the N-terminus inhibits this interaction [82]. Together, these results suggest the presence of a complex in which NIP45 binds to both PRMT1 and NFAT through its N-terminal RG-rich domain, with methylation stimulating the NIP45-NFAT interaction. Both NIP45 and PRMT1 can coactivate transcription by cMaf/NFAT at the IL-4 promoter, but expression of NIP45 lacking the RG-rich region inhibits PRMT1 coactivation [82]. These results are consistent with a PRMT1-NIP45-NFAT complex playing a role in cytokine transcription. Stimulation of helper T-cells increases PRMT1 levels, but it does not dramatically affect levels of NIP45 *in vivo* methylation [82]. Therefore, PRMT1 may also coactivate cytokine transcription through methylation of substrates other than NIP45.

A role for methyltransferases in T-cell signaling is also implied by methyltransferase inhibitor treatment decreasing NFAT reporter activation on anti-CD3 ligation of Jurkat cells [83]. Although neither stimulation nor PRMT overexpression changed the overall pattern of proteins recognized by antibodies specific for aDMA (ASYM24) or sDMA (SYM11), expression of PRMT5 siRNA decreased transcription from an NFAT-dependent reporter gene under a number of inducing conditions [83]. PRMT5 siRNA also inhibited IL-2 secretion [83]. Interestingly, ChIP experiments revealed that although PRMT5 is not located at the IL-2 promoter, symmetrically dimethylated proteins recognized by the SYM11 antibody were present at this promoter [83]. Taken together, these results suggest a positive role for PRMT5 in induction of NFAT-dependent genes through the methylation and/or recruitment of substrates.

3. *PRMT5: Cell Cycle and Morphological Control*

Human PRMT5 was first identified as a binding partner of Janus-kinase-2 [84] and kinase interactions have also implicated PRMT5 homologs from other organisms in cell cycle and morphological control. The *Saccharomyces cerevisiae HSL7* (histone synthetic lethal 7) gene, a PRMT5 homolog, was first identified in a screen for mutations that are lethal in combination with an N-terminal deletion of histone H3 [85]. Hsl7 plays a role in the morphogenesis checkpoint at the G2-M boundary through its interaction with two kinases: Hsl1 and Swe1. The cell cycle regulatory kinase Swe1 phosphorylates the cyclin-dependent kinase Cdc28 to prevent entry into mitosis [86] and Hsl1 and Hsl7 target Swe1 for degradation [87]. Like the *S. cerevisiae* protein, *Xenopus* Hsl7 promotes entry into mitosis: addition of excess xHsl7 to cycling oocyte extracts results in faster mitotic entry, whereas anti-Hsl7 antibodies slow mitotic entry [88]. The ability of xHsl7 to bind wee1, a Swe1 homolog, and promote its degradation in oocyte nuclei suggests the conservation of this cell cycle control mechanism [88]. In contrast, the *Schizosaccharomyces pombe* PRMT5 homolog, Skb1, inhibits mitotic entry and expression of human PRMT5 in *S. pombe* has the same effect [89]. Skb1 binds to Shk1, a p21$^{Cdc42/rac}$-activated kinase (PAK), and both proteins co-precipitate with the cyclin-dependent kinase Cdc2 (Cdc28 homolog) *in vivo* [89, 90]. Inhibition of mitotic entry by Skb1 is at least partially dependent on wee1 [89]. The mechanism for Skb1 inhibition of mitotic entry in *S. pombe* has not yet been elucidated.

Both Hsl7 and Skb1 also play a role in cell morphology. *S. cerevisiae* cells that lack Hsl7 display filamentous growth on rich medium [91], and this pheno-type is dependent on the PAK Ste20, suggesting a role of Hsl7 in controlling a downstream mitogen-activated protein (MAP) kinase cascade implicated in filamentation [91]. Although *S. pombe* cells lacking Skb1 (*skb1Δ*) are slightly less elongated than wild-type cells under normal growth conditions, this pheno-type is exacerbated by hyperosmotic shock [90, 92]. Skb1 normally localizes to cell tips, septation sites, and nuclei, but becomes delocalized from cell tips and nuclei after shock [92]. Localization at cell tips of Orb6, a second Skb1-interacting kinase important for cell polarity, is decreased in an *skb1Δ* strain and hyperosmotic shock also causes delocalization of this protein [93]. Deletion of *SKB1* also hampers redistribution of F-actin to cell tips following hyperosmotic shock [92]. These results suggest that Skb1 may be involved in subcellular local-ization of these proteins involved in cell morphology. Human PRMT5 has also been implicated in localization of the somatostatin receptor subtype 1 at the cell surface [94].

Few experiments have addressed the importance of methyltransferase activity for the function of these PRMT5 homologs. Delocalization of Skb1 upon hyper-osmotic shock correlates with an increase in its methyltransferase activity, but no Skb1 substrates have been identified in *S. pombe* [92]. The only known *in vitro* substrates for *S. cerevisiae* Hsl7 are histones H2A and H4 [95]. Because the N-terminal tails of H3 and H4 are redundant in yeast, the lethality of *hsl7* muta-tions in combination with an H3 N-terminal deletion [85] suggests that methyla-tion of the H4 N-terminal tail may be critical for viability of cells lacking the H3 N-terminal tail. The isolation of *hsl1* and *cdc28* mutations in the same synthetic lethal screen [85] and the localization of Hsl7 at the bud neck [96], however, point to decreased Cdc28 activity as a probable cause for lethality of the H3 dele-tion strain. Surprisingly, a mutation that abrogates *in vitro* methyltransferase activ-ity of Hsl7 does not affect its ability to regulate Swe1 [97]. Therefore, Hsl7 may function in cell cycle regulation of *S. cerevisiae* through its protein-protein inter-actions rather than through the methylation of specific substrates. In contrast, PRMT5 inhibition of mitogenic signaling through oncogenic ras-p21 in *Xenopus* oocytes is partially relieved by introduction of an AdoMet-binding mutation in

PRMT5 [98]. Further experiments that specifically address the importance of the methyltransferase activity of Hsl7 and Skb1 *in vivo* will help define roles for this type II arginine methyltransferase in cell cycle and morphogenesis.

C. MRNA-BINDING PROTEINS AND INTRACELLULAR TRANSPORT

Following transcription, pre-mRNAs are processed in the nucleus and transported to the cytoplasm for translation. Each of these steps requires multiple RNA-binding proteins, many of which are PRMT1 substrates (Table 3.1). Early studies of arginine methylation revealed that heterogeneous nuclear ribonucleoproteins (hnRNPs), which play key roles in pre-mRNA processing and transport [99], contain the majority of asymmetric dimethylarginine in HeLa cell nuclei [100]. Although numerous hnRNPs are methylated *in vivo* and *in vitro* (Table 3.1; [101]), the composition of hnRNP complexes does not differ significantly between PRMT1+/+ and PRMT1−/− cells [102]. Treatment with methyltransferase inhibitors, however, increases co-purification of PRMTs with hnRNPs [103, 104]. Functional studies of the effect of arginine methylation on hnRNPs and a number of other RNA-binding proteins have revealed roles for PRMTs in intracellular transport.

1.   Yeast mRNA-Binding Protein Methylation and Nuclear Export

Many studies probing the role of methylation on RNA-binding protein transport have focused on hnRNP-like proteins in *S. cerevisiae*. A number of yeast RNA-binding proteins have been shown to be methylated by the predominant yeast type I arginine methyltransferase, termed Hmt1 (hnRNP methyltransferase) or Rmt1 (arginine methyltransferase) [10, 11]. Hmt1 *in vivo* substrates include Npl3 (11,105), Hrp1(26,106), Nab2 (107), and Yra1(106)—which are implicated in mRNA export—and nucleolar proteins Nop1, Gar1, and Nsr1 [108]. Although deletion of the *HMT1* gene is not lethal in an otherwise wild-type background [11], Hmt1 becomes essential in cells containing specific mutations in Npl3 [11] or lacking the 80 kDa subunit of the nuclear mRNA cap binding complex, Cbp80 [109]. A cold-sensitive mutant Hmt1 protein isolated in a strain lacking Cbp80 shows decreased *in vitro* methylation of Npl3 but not Hrp1 at low temperature [110]. This specificity suggests that Npl3 methylation may be important for growth in the absence of Cbp80, perhaps by facilitating mRNA export. Introduction of increasing numbers of R-to-K mutations within the RG domain of Npl3 inhibits the growth of cells lacking Cbp80 but not of wild-type cells, suggesting that overall methylation of the extensive RG domain is important for full function of Npl3 [23].

Although Npl3 and Hrp1 are predominantly nuclear at steady state, they shuttle between the nucleus and cytoplasm and their nuclear export can be tested by blocking import [111]. The deletion of *HMT1* inhibits export of both Npl3 and Hrp1 in this assay, indicating a role for arginine methylation in nuclear export [109].

To test whether Hmt1 has a direct effect on nuclear export of Npl3 and Hrp1, mutations were introduced in the RG domains of each protein and export tested in cells with and without the methyltransferase [23, 26]. The export of Hrp1 with each of its two methylated arginines changed to lysine was still dependent on Hmt1, suggesting that methylation indirectly affects Hrp1 export [26]. On the other hand, Npl3 with 15 R-to-K mutations (all within RGG tripeptides shown to be methylated in wild-type cells [23]) was not detectably methylated *in vivo* [26] and was able to exit the nucleus in the presence or absence of Hmt1 [23, 26].

The interaction of Npl3 with two nuclear proteins, Npl3 and transcription elongation factor Tho2, is increased by deletion of *HMT1* [106]. These interactions are also reduced by the introduction of the 15 R-to-K mutations, which could disrupt arginine-specific contacts [23]. In combination, these results suggest that methylation directly affects Npl3 export by loosening contacts with nuclear proteins. Interestingly, the expression of the Npl3-RK mutant promotes the export of Hrp1 in the absence of Hmt1, suggesting that Hrp1 export is dependent on Npl3 export [26]. *In vitro* experiments exploring the interaction of Npl3 with its import receptor have also implicated Hmt1 in Npl3 import through effects of Hmt1 on Npl3 phosphorylation [112].

Although no bulk mRNA export defect is seen in cells that lack Hmt1, they show slowed kinetics of production of a heat shock protein mRNA [106]. This link between Hmt1 and transcription is substantiated by its localization at active promoters [106], as found for PRMT1 and CARM1 in mammalian systems. Interestingly, Hmt1 has varying effects on the localization of its substrates on chromatin. In genome-wide localization experiments, *hmt1* deletion has a large effect on which genes are bound by Hrp1 and a slight effect on the subset of genes bound by Yra1 and Nab2, but no significant effect on genes bound by Npl3 or Tho2 [106]. The molecular mechanisms underlying these differences remain to be elucidated.

2. Intracellular Transport and Mammalian mRNA-Binding Proteins

The intracellular localization of a number of metazoan, primarily mammalian, mRNA-binding proteins has also been linked to arginine methylation. In PRMT1−/− cells or AdOx-treated PRMT1+/+ cells, Sam68 is more cytoplasmic than in untreated PRMT1+/+ cells [113]. Similarly, AdOx treatment causes hnRNP A2 but not hnRNP A1 or hnRNP L to become more cytoplasmic [114]. Deleting or mutating the RG domain of each protein also increases its cytoplasmic localization [113, 114]. In contrast to the yeast hnRNP proteins, these results are consistent with arginine methylation facilitating nuclear import or slowing nuclear export. Studies of the nuclear transport domain (NTD) of RNA helicase A showed that this RG-rich domain is methylated by PRMT1 [115]. After injection of methylated and unmethylated NTD proteins into the cytoplasm of cells pre-treated with methyltransferase inhibitors, only the methylated form entered the nucleus, demonstrating the importance of methylation specifically in nuclear import [115]. In the same assay, a mutant helicase lacking the RG domain also entered the nucleus, supporting a model in which arginine methylation releases this protein from a cytoplasmic retention factor [115].

In the case of both Sam68 and hnRNPA2, AdOx treatment has also been linked to defects in mRNA trafficking. Sam68 can substitute for HIV rev in facil-itating export of rev-responsive-element-containing mRNAs. After AdOx treat-ment, rev (but not Sam68) can still increase expression of an RRE-CAT reporter gene [113]. HnRNP A2 is implicated in intracellular transport of myelin basic protein (MBP) mRNA and a change in the number of cytoplasmic hnRNP A2 granules on AdOx treatment correlates with the presence of MBP mRNA granules closer to the nucleus [116].

PRMT1 overexpression causes two other substrates, the Ewing sarcoma RNA-binding protein EWS and a *Xenopus* cold-inducible RNA-binding protein CIRP2, to become more cytoplasmic [117, 118]. Whereas the *in vivo* function of CIRP2 is not known, EWS can coactivate transcription with hepatocyte nuclear factor 4 [119]. In contrast to the positive role of PRMT1 in transcription coacti-vation described in section IV.A.2, coexpression of PRMT1 with EWS decreases EWS coactivation [117]. Because deletion of two RG domains in EWS also decreases coactivation without affecting binding to HNF-4 and results in cyto-plasmic EWS localization, methylation of the RG domain may down-regulate EWS coactivation through nuclear exclusion [117]. CARM1 also methylates the HuR RNA-binding protein in a region that is required for shuttling [120], and stimulation of Jurkat cells with LPS increases HuR methylation [120]. HuR is involved in mRNA stabilization, but it is not known if its methylation affects its activity or localization.

Thus, although arginine methylation has been implicated in intracellular localization of many eukaryotic RNA-binding proteins the effects vary among proteins. When nuclear export and nuclear import have been tested directly, data are consistent with methylation-disrupting interactions: either interactions with nuclear factors in facilitating the export of a yeast hnRNP [23] or interactions with cytoplasmic factors in allowing import of RNA helicase A [115]. Although poly(A) binding proteins (PABP) 1 and 2 have been shown to be methylated *in vivo* [48, 121] (Table 3.1), the functional consequences of these modifications are unknown. Methylation has also been linked to nucleocytoplasmic distribution of another *in vitro* PRMT1 substrate, high-molecular-weight fibroblast growth factor (HMW FGF). HMW FGF purified from cultured cells contains eight dimethylarginine residues [122], and methyltransferase inhibition results in greater cytoplasmic localization [123].

D.   PRMTs AND PRE-MRNA SPLICING

1.   PRMT5 and snRNP Complex Formation

The discovery of symmetrically dimethylated arginines in spliceosomal proteins SmD1, SmD3, and SmB as major epitopes for autoantibodies in patients with lupus erythematosus [124] coincided with studies showing the importance of an RG-rich region in the same proteins for their binding to the survival of motor neurons (SMN) protein [125]. Subsequent studies revealed that symmetric dimethylation of these Sm proteins and Sm-like protein LSm4 is required for SMN binding via the SMN tudor domain [36, 37], one of the first examples of methylation-dependent protein-protein interactions. The co-purification of SmD1 and D3 in a 20S complex that contained Janus-kinase-binding protein 1 (JBP1) indicates that this type II arginine methyltransferase, also called PRMT5, is a prime candidate for the Sm PRMT [126, 127]. Either PRMT5 purified from extracts or the 20S complex, termed the methylosome, methylates recombinant SmD1, D3, and B [126]. PRMT5 siRNA expression lowers the recognition of SmB by an sDMA-specific antibody, indicating that PRMT5 modifies SmB *in vivo* [128].

The methylosome also contains the pICln protein and a WD-repeat-containing 45- to 50-kDa protein, called WD45 or MEP50, and can be purified using anti-pICln or anti-PRMT5 antibodies [127, 129]. Whereas pICln binds Sm proteins, it does not bind SMN [127, 130]. In addition, recombinant SmD1 and D3 proteins bind to PRMT5 and pICln from HeLa extracts, but symmetric dimethylation of these proteins abrogates these interactions and increases binding to SMN [126]. The assembly of methylated Sm proteins on U1 snRNA is decreased if the reaction is incubated with a pICln complex that lacks Sm proteins, suggesting that the role of the pICln complex in spliceosome assembly is not solely enzymatic [127]. Reconstitution of U snRNPs in a HeLa cell extract reveals that either an SMN complex or a complex containing PRMT5 pICln and relatively lower amounts of SMN complex proteins allows formation of a U1 snRNP complex *in vitro* [131]. The PRMT5-containing complex, however, shows increased efficiency of assembly and release of the snRNP from the complex [131].

Taken together, these data have led to a model for the role of PRMT5 in promoting assembly of snRNPs, shown in Figure 3.3. Although anti-aDMA antibodies or a methyltransferase inhibitor reduce *in vitro* splicing twofold, effects of PRMT5 on splicing *in vivo* have yet to be tested [128]. The ability of the RNA polymerase II phosphatase FCP1 to co-precipitate PRMT5, MEP50, SmB, and U1 snRNA from nuclear extracts [132] and its *in vitro* methylation by PRMT5 [133] reflect an intriguing connection between transcription and splicing proteins modified by PRMT5 that awaits further study.

Type II PRMT modification of nuclear proteins has also been linked to subnuclear localization. Treatment with a methyltransferase inhibitor disrupts the colocalization of sDMA-containing proteins and SMN [128]. Coilin is a major

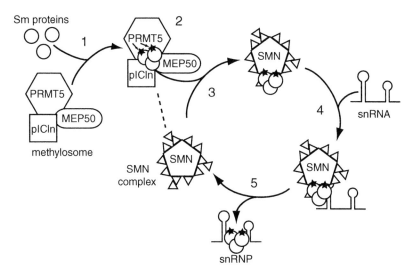

FIG. 3.3. Roles of PRMT5- and SMN-complexes in spliceosome assembly. (1) A complex containing PRMT5, pICln, and methylosome protein MEP50 recruits newly synthesized Sm proteins. (2) SmB/B′, SmD1, and SmD3 are methylated by PRMT5. (3) The PRMT5 complex transiently associates with a complex containing SMN and gemin proteins (triangles) and transfers methylated Sm proteins to this complex. (4) The SMN complex binds snRNA and helps coordinate assembly of snRNP. (5) The properly assembled snRNP is released [36, 37, 126, 127, 129-131, 226, 227].

sDMA-containing nuclear protein that interacts with SMN, and the RG domain of coilin is necessary for its colocalization with SMN [134]. The SMN-coilin interaction is disrupted by treatment with methyltransferase inhibitors, as reflected both by co-immunoprecipitation and colocalization studies [134].

Type I PRMTs may also play a role in splicing or spliceosome assembly. Nuclear forms of SmB, D1/D2, and D3 contain asymmetric dimethylarginine residues [135], although the functional significance of this modification has not yet been elucidated. Interestingly, the asymmetrically dimethylated nucleolar protein GAR1 binds the SMN tudor domain, but neither the methyltransferase inhibitor S-adenosyl-L-homocysteine (AdoHcy) nor PRMT1 expression affects the GAR1-SMN interaction [136]. This result points to the specificity of the regulation seen in SMN binding to sDMA-containing proteins.

2. Carm1 and Alternative Splicing

Recent data have revealed that a splice variant of CARM1 also plays a role in splicing [137]. CARM1 splice variant 3, which has a C-terminal sequence that is unique among known CARM isoforms, was identified through an interaction with the U1C snRNP protein [137]. Expression of CARM1 v.3, but not of other

CARM1 isoforms, leads to a biased alternative splicing of an E1A minigene, and this splicing pattern is dependent on the unique C-terminus [137]. CARM1 v.3 can methylate H3 *in vitro* and activate estrogen-receptor-dependent transcription, although the GRIP-1 p160 family member is not necessary for this activity [137]. Notably, AdoMet-binding domain mutations (VLD-to-AAA) inhibit H3 methylation but not the alternative splicing and coactivation activities of CARM1 v.3, implying that methyltransferase activity is not critical for this variant's role in these processes [137]. A distinct tissue expression pattern compared to other CARM1 isoforms suggests that this variant may function in the spleen and kidney [137]. The specific substrates that are critical for CARM1 v.3 function *in vivo* or whether CARM1 displays a non-enzymatic function remain to be determined.

E.   RIBOSOME BIOGENESIS

Two of the major highly conserved nucleolar proteins involved in ribosome assembly, called fibrillarin and nucleolin in mammals and Nop1 and Gar1 in *S. cerevisiae*, have extensive RG domains that are heavily methylated *in vivo* [9, 108, 138]. PRMT1 and perhaps PRMT5 co-purify with fibrillarin from human cell extracts and these three proteins comigrate in sucrose gradients [139]. Yeast Hmt1 methylates nucleolar proteins Nop1, Gar1, and Nsr1, but it is not required for their nucleolar localization [108]. R-to-K mutations within nucleolin still allow localization to the nucleolus [140]. Deletion of the RG-rich domain in nucleolin, however, causes this protein to distribute throughout the nucleus and also affects the localization of another methylated nucleolar protein, B23, without altering the localization of Nopp140 or fibrillarin [140]. Thus, although the RG domain is important for nucleolar localization methylation does not seem to influence this function. Methylation of these nucleolar proteins may modulate complex formation in ribosomal processing or assembly, but no experiments have yet addressed this possibility.

The small ribosomal subunit protein rpS2 co-purifies with the zinc-finger-containing type I methyltransferase PRMT3 from *S. pombe* and mammalian cells [141, 142]. PRMT3 can methylate rpS2 both *in vitro* and *in vivo* and the zinc finger is required for *in vitro* activity [141, 142]. In both systems, a fraction of PRMT3 comigrates with the 40S small ribosomal subunit peak in sucrose gradients [141, 142]. Although there is no defect in pre-rRNA processing/maturation when *PRMT3* is deleted in *S. pombe*, the ratio between large and small ribosomal subunits increases [141]. This result indicates a role for PRMT3 in a late stage of ribosome assembly [141].

A unique protein arginine methyltransferase, called Rmt2, was uncovered in an extensive search of the *S. cerevisiae* genome for Ado-Met binding domains [143]. This enzyme is not a homolog of mammalian PRMT2 and putative Rmt2 homologs are found predominantly in other fungi. Rmt2 monomethylates the internal (δ) nitrogen in arginine, consistent with its sequence similarity to the small molecule guanidinoacetate N-methyltransferase [143]. The only known Rmt2 substrate is the large ribosomal subunit protein L12 (144), but the functional importance of this modification is unknown.

F.   DNA DAMAGE REPAIR

Studies with topoisomerase II inhibitors—including adriamycin/doxorubicin, etoposide, and 9-OH-ellipticine (9-OH-E)—have revealed connections between DNA damage repair and protein arginine methyltransferases [145-148]. Induction of DNA damage through adriamycin treatment of cultured cells increases expression of PRMT1-interacting proteins BTG1 and BTG2/TIS21 [145]. Addition of recombinant BTG1 or BTG2 increases methylation of proteins in RAT1 cell extracts [12], and histone methyltransferase activity increases upon adriamycin treatment of a human cell line [145]. Notably, this increase in activity is dependent on wild-type p53 and correlates with BTG2 expression [145], suggesting a role for BTG2 in positive regulation of PRMT activity.

The DNA repair protein MRE11 was identified in a proteomic screen for asymmetrically dimethylated proteins [14]. Methylated by PRMT1 both *in vivo* and *in vitro*, MRE11 isolated from HeLa cells contains nine DMA residues [146]. When eight of these arginines are mutated to either alanine or lysine, MRE11 still interacts with other DNA repair factors NBS1 and RAD50 [146]. The exonuclease activity of MRE11, however, is completely abrogated by the R-to-A mutations and significantly reduced by the R-to-K mutations [146]. Methyltransferase inhibitor or PRMT1 RNAi treatment of cells leads to checkpoint failure after stimulation of DNA damage with etoposide [146]. Transfection of a purified MRE11/NBS1/RAD50 complex can facilitate checkpoint arrest unless the MRE11 R-to-A mutant protein is used [146]. In addition, the G2-M arrest induced by DNA damage is reduced in PRMT1−/− cells in comparison to PRMT1+/+ cells [146]. Recent work has also revealed colocalization of PRMT1 with MRE11 in PML nuclear bodies and the failure of MRE11 to localize to sites of DNA damage on methyltransferase inhibition [149]. In combination, these results indicate a role for PRMT1 in DNA damage repair through its modification of MRE11.

Two studies have linked PRMT7, a type II arginine methyltransferase [150], to DNA damage. A portion of the PRMT7 gene was found in a screen for genetic suppressor elements (GSEs) that increase resistance to 9-OH-E [147]. This GSE has an antisense effect, decreasing levels of two isoforms of PRMT7, p77 and p82. Conversely, overexpression of p82 increases 9-OH-E sensitivity [147]. These results are consistent with PRMT7 mediating sensitivity to DNA-damaging drugs. The original DNA fragment that reduced PRMT7 levels also increased resistance to etoposide while decreasing resistance to UV-irradiation and bleomycin, indicating that PRMT7 does not influence cellular responses to all DNA damage [147].

Doxorubicin can induce nephropathy in a subset of mouse strains. Sensitivity to doxorubicin maps to Chromosome 16 [148], as does PRMT7 [151]. In contrast to the 9-OE-E results, sensitivity to doxorubicin correlates with lower levels of PRMT7 [148]. Differences in PRMT7 levels are seen before damage, implying that the effects on PRMT7 expression do not result from damage [148]. Symmetric arginine dimethylation may therefore have a role in damage-induced pathways, but the mechanisms at work remain obscure and may be specific for particular modes of DNA damage.

V.   Roles of Protein Arginine Methyltransferases in Development and Differentiation

Cellular differentiation and organismal development require complex coordinated changes at the molecular level, including altered patterns of protein modification and gene expression. *In vitro* models for differentiation and *in vivo* developmental experiments have implicated arginine methylation in these processes. Some studies have drawn direct connections between differentiation and cellular roles of arginine methylation (discussed previously). At this early stage, however, most studies offer tantalizing data that point to the importance of arginine methylation in development and differentiation, whereas the molecular mechanisms underlying the role of PRMTs in these processes remain to be elucidated.

A.   DEVELOPMENT

PRMT disruption phenotypes and expression patterns of PRMTs in mice support the importance of these enzymes for early development. Although embryonic stem (ES) cells with homozygous disruptions of either PRMT1 or CARM1 genes are viable, both disruptions are lethal: PRMT1−/− embryos do not develop past E6.5 [152] and CARM1−/− mice die perinatally [48]. PRMT1 is expressed along the neural plate midline and forming headfold between E7.5 and E8.5 and in the developing central nervous system from E8.5 to E13.5 [152]. CARM1 is expressed in the neural fold, somites, posterior lateral plate, and fore- and hindbrain at ~8.25 d post-coitus, and in the neural tube and somites at 8.75 d [153]. PRMT1 protein levels in rat liver decrease slightly postnatally and increase after hepatectomy. These effects are stronger in comparisons of PRMT enzyme activity using hnRNP A1 as an *in vitro* substrate [154]. These results suggest that post-transcriptional regulation of PRMT1 or other PRMTs may act in liver development. A temporary increase in type II PRMT activity, but not PRMT5 levels, is also detected 20 days after birth [154]. Small increases in PRMT5 are detected with *in vitro* models for renal differentiation and injury [155]. Comprehensive analysis of all PRMTs, as well as identification of substrates critical for differentiation, will help clarify the connections between PRMT expression and developmental processes.

In both mouse and *Drosophila*, arginine methylation has been associated with germ cell formation. Localization of arginine-methylated histones varies during mouse oogenesis [156]. Although mR3-H4 and mR17-H3 epitopes significantly colocalize with DNA in immature oocytes, after ovulation this colocalization is temporarily disrupted [156]. Methyl-R3-H4 and acetyl-H4 patterns are similar, with signals disappearing after ovulation, whereas the mR17-H3 pattern becomes more dispersed. The reduction in signal may be due in part to deimination of histones [52, 156]. Both mR3-H4 and mR17-H3 signals are nuclear following fertilization, but do not always correlate with DNA localization in early embryonic stages [156]. The consequences of these changes in histone modification and localization for murine development have not yet been determined.

Genetic experiments have implicated PRMTs and their interacting partners and substrates in *Drosophila* pole cell formation. Vasa (a DEAD-box RNA helicase) and Squid (an hnRNP) are methylated *in vitro* by DART1, a PRMT1 homolog [157]. Both of these DART1 substrates play posttranscriptional regulatory roles in pole cell development [158]. Mutations in either Capsuleen, a PRMT5 homolog, or the MEP50-like protein Valois also lead to defects in oogenesis [159, 160]. These two proteins co-precipitate from ovarian extracts. Valois can also bind to the Tudor protein, its WD domain mediating interactions with tudor domains [159]. These interactions reflect similarities with the PRMT5-MEP50 complex, which binds transiently to the SMN complex in mammalian cells [131]. Although SMN contains tudor domains that bind to methylated Sm proteins, Tudor is not the *Drosophila* SMN ortholog [161]. Capsuleen-Valois complexes may therefore perform a function in pole cell formation other than a role in spliceosome assembly. Further investigation of Capsuleen and DART1 will be necessary to elucidate possible roles of these methyltransferases in pole cell formation.

A non-histone substrate of CARM1, thymocyte cAMP-regulated phosphoprotein TARPP, is expressed in an immature but not a mature T-cell line, suggesting that CARM1 might play a role in T-cell development [162]. In comparison to wild-type littermates, CARM1−/− embryos have fewer total cells in the thymus [162]. The numbers of both CD4−CD8− and CD4+CD8+ cells are reduced, but the percentage of CD4−CD8− cells increase slightly while that of CD4+CD8+ cells slightly decrease [162]. Although the mechanism underlying these differences has not yet been addressed, the presence of an immature T-cell-specific CARM1 substrate suggests that CARM1 may act in T-cell development outside its involvement in histone modification.

Arginine methylation of myelin basic protein has been implicated in the formation of myelin sheaths around neurons [163]. PRMT activity increases during myelination [164] and inhibition of methyltransferases results in less compact forms of myelin in a cultured cell system [165]. Although MBP methylation increases association of lipid vesicles [166], suggesting a role for methylation in myelin compaction, levels of MBP methylation are higher in less compact myelin [167]. Thus, the role of PRMTs in compact myelin assembly may not be due to methylated MBP tightly linking lipid layers within compact myelin. Although MBP is a prototypical type II PRMT substrate [7], MBP in developing mouse brain contains asymmetric as well as symmetric dimethylarginine [167]. Over the course of mouse development, the amounts of aDMA decrease and those of sDMA and MMA increase. This pattern is also seen in other mammalian systems [167]. Mice with mutations that cause demyelination show increased levels of aDMA [167], but the functional significance of this correlation is unclear.

B.   DIFFERENTIATION

In the case of muscle differentiation and retinoid-induced differentiation of myeloid cells, *in vitro* studies have linked differentiation to known transcriptional roles of PRMTs. Transcriptional coactivation by CARM1 has been implicated in muscle differentiation. CARM1 and GRIP-1 synergistically coactivate transcription controlled by the myocyte-enhancing factor MEF-2C and CARM1 enhances a mammalian two-hybrid interaction between MEF-2C and GRIP-1 [153]. Association of CARM1 and MEF2 with MEF2-binding sites at the muscle creatine kinase enhancer is increased in differentiated myotubes in comparison to proliferating myoblasts [153]. These data, combined with AdOx inhibition of muscle differentiation *in vitro*, indicate that CARM1 coactivation of transcription may be involved in muscle differentiation [153]. CARM1 can also cooperatively coactivate transcription with GRIP-1 and Flightless-I [168], an actin-binding protein that is essential for *Drosophila* and mouse development [169, 170], although this effect has not been connected with differentiation of a specific cell type.

Retinoid-induced differentiation of the HL-60 cell line involves two steps: first cells are "primed" to enter a precommitted intermediate state and then retinoid treatment induces later events that lead directly to myeloid-specific differentiation [171]. PRMT1 and R3-H4 methylation have been linked to the priming step in this process [172]. Priming increases methylated R3-H4 occupancy at the enhancer for tissue transglutaminase type 2 (TGM2), a marker for the retinoid response [172], and mR3-H4 occupancy positively correlates with levels of TGM2 mRNA [172]. Methylated R3-H4 levels at the TGM2 enhancer as well as levels of three retinoid-responsive mRNAs are all reduced by AdOx treatment, whereas wild-type but not mutant PRMT1 overexpression increases TGM2 mRNA levels [172]. Conversely, inhibiting histone deacetylases or activating protein arginine deiminases also reduces mR3-H4 occupancy and TGM2 mRNA levels after priming and retinoid treatment [172]. Because acetylation of H4 [60] or deimination of R3 should inhibit methylation at R3, these results are consistent with

H4 methylation by PRMT1 functioning in the priming step [172]. Curiously, PAD4 expression increases upon priming and PAD4 overexpression does not repress PRMT1-stimulated TGM2 expression, but overexpression of an inactive PAD4 with PRMT1 synergistically enhances TGM2 expression [172]. These results indicate the complexity of the interplay of histone modifying enzymes in retinoid-induced differentiation.

Connections between neuronal differentiation and protein methylation were first noted by changes in the patterns of protein methylation after nerve growth factor (NGF) treatment of PC12 cells [173]. PRMT activity in PC12 extracts increases with time of NGF treatment, although PRMT1 levels do not change significantly [174]. Inhibiting S-adenosyl-L-homocysteine hydrolase (SAHH), which leads to an accumulation of the methyltransferase inhibitor AdoHcy, does not affect the number of cells in a culture but does inhibit neurite outgrowth [174]. Similarly, treatment with the copper chelator TEPA results in shorter neurites after NGF treatment and TEPA inhibits SAHH activity *in vitro* [175]. Therefore, the decrease in PRMT activity in TEPA-NGF treated cell extracts, combined with steady PRMT1 levels, suggests that TEPA inhibition of SAHH leads to methyltransferase inhibition and neuronal differentiation defects [175].

NGF treatment of PC12 cells also induces expression of the BTG2 protein [176]. The BTG2 gene, also called TIS21, is closely related to BTG1, a gene located close to a chromosomal translocation breakpoint correlated with chronic lymphocytic leukemia [177]. Both BTG1 and BTG2 have been implicated in negative regulation of cell proliferation, interact with PRMT1, and can increase methylation of cell extract proteins by PRMT1 [12]. BTG1 and BTG2 share a region (Box C) that is necessary but not sufficient for its interaction with PRMT1, and a Box C fusion protein can inhibit PRMT activity in PC12 extracts [178]. The Box C fusion protein also inhibits the differentiation of embryonic stem cells to neuronal cell types and increases apoptosis of terminally differentiated neuronal cells [178].

These correlations between effects of BTG proteins on PRMT activity and on neuronal differentiation support a possible role for protein arginine methylation in this process. Although specific PRMT targets involved in differentiation have not yet been identified, BTG and Box C fusion proteins affect methylation of 20-kDa, 55-kDa, and 65-kDa proteins in cell extracts [12, 178]. BTG1 is also upregulated during erythroid development and a time course of erythroid development shows fluctuation in levels of arginine-methylated proteins [179]. The inhibition of hemoglobin accumulation on MTA treatment also points to the importance of methylation in erythroid development [179]. Thus, PRMTs and potential regulators are likely to play ubiquitous roles in development and differentiation.

VI. Roles of Protein Arginine Methyltransferases in Disease

A. CANCER

Work on protein methylation in the 1970s noted connections between arginine methylation and cancer. Histones isolated from bone marrow erythroid precursors from patients with chronic erythremic myelosis show higher methylarginine content than histones from patients with other forms of anemia [180]. In rats, rates of hepatoma growth correlate with PRMT enzymatic activity [181]. Subsequent work has supported the possible involvement of arginine methylation in tumor growth.

Like BTG1 and BTG2, the DAL1/4.1 protein is a tumor suppressor that has been proposed as a possible PRMT regulator. DAL1/4.1 was identified as being differentially expressed in adenocarcinoma of the lung [182] and has been linked to growth suppression of meningioma and breast tumors [183, 184]. DAL1/4.1 binds to PRMT3, PRMT5, and PRMT6 upon coexpression in human cells, as well as co-precipitating PRMT5 and pICln [185]. Although DAL1/4.1 is not a PRMT3 substrate, addition of this protein decreases PRMT3-catalyzed methylation of a glycine-arginine-rich domain [185]. In PRMT5-catalyzed reactions, DAL1/4.1 increases MBP methylation while decreasing Sm protein methylation [186]. Induction of DAL1/4.1 *in vivo* results in increased modification of extract proteins by PRMT3 *in vitro*, but both increases and decreases methylation of proteins by PRMT5, indicating that DAL1/4.1 might be both a positive and negative regulator of PRMT function [185, 186]. It is not clear either whether this regulatory function of DAL1/4.1 is associated with its effects on cell proliferation or whether DAL1/4.1 can regulate other PRMTs.

The importance of androgen for progression of prostate carcinoma [187] and the ability of CARM1 to act as a coactivator for the androgen receptor [41] suggest that CARM1 may play a role in this form of cancer. A comparison of numbers of CARM1+ cells and intensity of CARM1 staining in prostate specimens reveals increases in CARM1 levels over tumor progression, with highest levels in prostate carcinomas [188]. In addition, samples from androgen-independent tumors, which can no longer be controlled by androgen-ablation therapy, contain slightly fewer CARM1+ cells but show more intense CARM1 staining [188]. These results are consistent with CARM1 involvement in prostate carcinoma progression.

B. CARDIOVASCULAR DISEASES

Elevated plasma levels of aDMA in cardiovascular risk states such as hypertension [189], hypercholesterolemia [190], and atherosclerosis [191] indicate the possible involvement of PRMTs in cardiovascular diseases. This potent inhibitor of nitric oxide synthase results from the degradation of asymmetrically dimethy-lated proteins, suggesting that PRMT activity might correlate with aDMA produc-tion associated with disease [192]. LDL cholesterol treatment both increases aDMA release from endothelial cells and causes a slight upregulation of PRMT1, PRMT2, and PRMT3, connecting PRMTs with hypercholesterolemia [193]. High salt intake in human hypertension increases sheer stress, which both increases aDMA release from endothelial cells in culture and upregulates PRMT1 expres-sion about twofold [194]. Inhibition of NF-KB signaling, however, blocks aDMA release and PRMT1 upregulation by stress, pointing to the importance of this pathway for the sheer stress response [194]. Antibodies specific for aDMA- and sDMA-containing proteins stain tissue sections from patients with pulmonary hypertension more strongly than those from normal human lung tissue, correlat-ing with plasma levels of free aDMA and sDMA [195]. Although sDMA does not inhibit nitric oxide synthase directly, it may have an indirect effect through inhi-bition of a cationic amino acid transporter [196]. The recent identification of a large family of PRMTs will facilitate the investigation of whether type I PRMTs other than PRMT1 show correlations with cardiovascular diseases.

C.   MULTIPLE SCLEROSIS

Arginine modification of MBP has implications for the demyelinating autoim-mune disease multiple sclerosis. Citrulline levels within MBP correlate with dis-ease severity [197, 198], and PAD deimination of MBP increases its susceptibility to cathepsin D digestion [199]. Together, these results point to the possibility that deimination of MBP leads to increased fragmentation of MBP and an MBP-specific immunological response that results in demyelination [199]. Because arginine dimethylation of MBP inhibits PAD activity [200], PRMT activity may help protect MBP from degradation and therefore limit the autoimmune response. Cationicity of MBP has also been postulated to play a role in stabilizing myelin [201], and unlike deimination methylation affects hydrogen bonding without dra-matically altering charge. Levels of both methylarginine and citrulline residues in two forms of MBP, however, are increased in brains from multiple sclerosis patients compared to controls [202]. Although the relevance of this higher methy-larginine content in MBP from multiple sclerosis patients has not yet been fully addressed, it is likely that arginine methylation and deimination do not have iden-tical effects on myelin dynamics.

D.   SPINAL MUSCULAR ATROPHY

Spinal muscular atrophy (SMA) is a neurodegenerative disease that affects motor neurons, leading to paralysis and muscular atrophy. This recessive genetic disorder results from lesions in the SMN gene [203], and although SMA severity correlates with SMN protein levels in many patients [204] missense mutations within the SMN tudor domain have also been found in SMA patients [205, 206]. The E134K mutation does not disrupt the overall structure of the SMN tudor domain but alters the charge distribution on a negatively charged surface that interacts with the RG domains of Sm proteins [207]. Although this point mutation disrupts an *in vitro* interaction between SMN and Sm proteins [205], the tudor domain was not essential in an avian cell genetic assay for SMN function [208]. In cells from an SMA patient with low levels of SMN, however, sDMA-containing proteins are mislocalized [128]. Therefore, although altered interactions between SMN and sDMA-containing proteins may be connected to SMA pathogenesis, mechanisms that may link spliceosome assembly to motor neuron function are not clear.

E.   VIRAL REPLICATION AND HOST INTERACTIONS

Connections between viral processes and arginine methylation have been highlighted both through direct methylation of viral proteins and through viral interactions with cellular machinery. The hepatitis C NS3 helicase-like protein [27], the herpes simplex virus RNA-binding protein ICP27 [209], and the adenovirus L4-100 nonstructural protein [210] are all methylated *in vivo*, but the functional consequences of arginine methylation of these proteins have not been probed. For HIV-1 and hepatitis delta virus (HDV), however, arginine methylation of viral proteins has been linked to viral replication.

Transfection of wild-type PRMT6, but not an AdoMet-binding mutant, allows *in vivo* methylation of the HIV Tat protein and decreases Tat activation of the HIV LTR [211]. Conversely, PRMT6 small inhibitory RNA (siRNA) expression increases activation of an LTR reporter [211]. When cells are transfected with proviral HIV-1 DNA, levels of the p24 capsid protein serve as a marker for virus production. Whereas PRMT6 transfection decreases p24 levels, PRMT6 siRNA increases these levels [211]. These results point to a negative role for PRMT6 in HIV replication through the methylation of the HIV Tat protein. Future experiments to investigate the effects of Tat methylation on its molecular interactions will help address the mechanism responsible for PRMT6 negative regulation of viral transcription.

PRMT1 methylates the small hepatitis delta antigen *in vitro* (sHDAg) and methylation is blocked by an R13A mutation in sHDAg [212]. When methylated or unmethylated sHDAg is cotransfected with genomic or antigenomic RNA, the methylated form enhances first-round RNA synthesis, particularly from the antigenomic template. Wild-type sHDAg is normally nuclear, but AdoHcy treatment or introduction of the R13A mutation increases cytoplasmic localization of sHDAg. Although methylation does not affect *in vitro* RNA binding of sHDAg, unmethylated sHDAg allows transport of genomic but not antigenomic RNA into the nucleus [212]. Thus, methylation of sHDAg appears to facilitate antigenomic replication by enhancing its nuclear transport.

Methylation of Epstein-Barr virus nuclear antigen 2 (EBNA2) has been linked to its interaction with host protein SMN [213]. Whereas EBNA1 variably co-purifies with PRMT5 (214), EBNA2 is methylated in HeLa cell extracts and PRMT5 depletion reduces EBNA2 methylation [213]. As with Sm proteins, EBNA2 binding to SMN (which is mediated by the SMN tudor domains) is reduced in the absence of methylation or by deletion of the RG domain in EBNA2 [213]. The transcriptional coactivation of a viral reporter gene by EBNA2 and SMN suggests that the EBNA2-SMN interaction plays a role in viral infection [213]. Surprisingly, whereas deletion of the tudor domain in SMN eliminates coactivation, deletion of the RG domain in EBNA2 does not affect coactivation by either SMN or EBNA2 [213]. On its own, EBNA2ΔRG shows greater coactivation than wild-type EBNA2 and reduces but does not eliminate B-cell transformation [215]. Thus, specific roles of EBNA2-SMN interactions and of EBNA2 methylation in viral transcription and B-cell transformation are not yet clearly defined.

Viral success depends in large part on effective evasion of host defenses. One proposed mechanism for immune evasion by hepatitis C virus (HCV) is inhibition of interferon signaling through the Jak-STAT pathway [216]. One study using transgenic mice expressing HCV proteins, a cell line overexpressing the catalytic subunit of protein phosphatase 2A (PP2A), and samples from chronic hepatitis C patients suggests that HCV protein expression inhibits interferon signaling through PP2A upregulation and decreased STAT1 methylation [217]. Whereas there is no clear mechanistic link between PP2A and STAT1 methylation, another study indicates STAT1 degradation as a likely mechanism for inhibition of interferon signaling by HCV protein expression [218]. Because the role of STAT1 methylation in interferon signaling remains controversial (see section IV.B.1), further work is required to determine whether protein methylation may play a role in HCV inhibition of interferon signaling.

VII.   Conclusions and Future Directions

PRMTs have been implicated in a vast array of cellular processes, yet results to date suggest molecular themes that unify the roles of arginine methylation. Modification of numerous PRMT substrates affects their interactions with different binding partners, in some cases enhancing interactions and in others inhibiting them. These positive and negative effects of methylation can influence assembly of macromolecular complexes that are important either for general "housekeeping" cellular processes or for specific responses to intra- and extracellular signals. Trying to link a certain effect to methylation of an individual substrate by a particular PRMT, however, poses technical challenges that have hindered the elucidation of mechanisms underlying cellular processes. A number of experimental approaches described recently should advance our understanding of PRMT roles *in vivo*.

The use of specific PRMT inhibitors and novel mass spectrometric approaches should help address questions of PRMT function through the study of both enzymes and substrates. Whereas the majority of work to date has relied on general methyltransferase inhibitors, screens for PRMT modulators have revealed molecules that inhibit or enhance PRMT activity in both enzyme- and substrate-specific fashions [219]. Inhibitory RNA approaches also allow the targeting of a specific PRMT *in vivo* [49, 61, 83, 128, 211] and are likely to be used more widely to probe PRMT function. Given the connections between arginine methylation and diseases described previously, specific PRMT inhibitors or activators may also prove useful for future therapeutic strategies.

Proteomic mass spectrometric techniques have allowed the identification of numerous arginine-methylated proteins [14–16]. In particular, the use of stable isotope labeling by amino acids in cell culture (SILAC) in combination with aDMA- and sDMA-specific antibodies facilitates both identification and quantitation of PRMT substrates by mass spectrometry [16]. The observation that aDMA and sDMA residues produce different fragmentation patterns also helps distinguish type I and type II arginine methylation by mass spectrometry [220–222]. The design of algorithms for predicting peptide peaks and matching predictions to experimental data is particularly challenging for arginine methylation studies. In addition to the presence of multiple target arginines, there are numerous possible *m/z* predictions for a particular peptide, given mono- or dimethylation of each arginine as well as type-I- and type-II-specific fragmentation patterns. Partial purification of methylated proteins, either by cell fractionation [15] or immunoprecipitation [16], has helped overcome these issues.

Two outstanding questions about the roles of PRMTs in metazoans are: How are PRMT expression and activity controlled and what is the significance of the temporal and spatial control of PRMTs? The construction of conditional PRMT alleles and the use of model systems may help elucidate roles of specific PRMTs in differentiation and development. To date, PRMT studies in non-mammalian model systems other than *S. cerevisiae* have been quite limited. The conservation of PRMTs in *Drosophila melanogaster*, *Caenorhabditis elegans*, and *Danio rerio* suggests that genetic approaches in fly, worm, and zebrafish might be useful for understanding the roles of PRMTs. In addition, the quantitation offered by the SILAC mass spectrometric approach could facilitate studies of how arginine methylation changes under different growth conditions or during development. A mass spectrometric approach has already been used to investigate changes in histone modifications over the course of *Drosophila* development [223]. Quantitative mass spectrometric techniques should also allow putative PRMT regulators such as Dal1/4.1 and BTG proteins to be tested for *in vivo* regulatory activity.

To date, the vast majority of PRMT research has focused on proteins that are involved in nuclear processes or cytoplasmic RNA metabolism. This bias is likely to be due in part to the presence of RG-rich proteins involved in these processes. These proteins can be multiply modified and RG-domains are easily recognizable as probable PRMT substrates [13]. Neither CARM1 nor PRMT7, however, methylates RG-rich domains [151, 224]. In addition, the discovery of a number of arginine methylated proteins associated with the Golgi complex [15], as well as PRMT1 activation of ryanodine receptor-calcium release channels [225], suggests that PRMTs are likely to play even more diverse roles in eukaryotic cells than are now known. Some results have also suggested that PRMTs and PRMT-containing complexes may play non-enzymatic roles in cellular processes [56, 97, 127, 137]. Future research should therefore expand our understanding of the diverse functional roles of protein arginine methyltransferases in these and other unanticipated directions.

ACKNOWLEDGMENTS

I am grateful to Mark Bedford, Sara Nakielny, Michael Stallcup, and Michael Yu for critical reading of the manuscript and the National Science Foundation for support (MCB-0235590).

REFERENCES

1. McBride, A. E., and Silver, P. A. (2001). State of the arg: protein methylation at arginine comes of age. *Cell* 106:5–8.
2. Bedford, M. T., and Richard, S. (2005). Arginine methylation an emerging regulator of protein function. *Mol Cell* 18:263–272.
3. Boisvert, F. M., Chenard, C. A., and Richard, S. (2005). Protein interfaces in signaling regulated by arginine methylation. *Sci STKE* 2005:re2.
4. Paik, W. K., and Kim, S. (1967). Enzymatic methylation of protein fractions from calf thymus nuclei. *Biochem Biophys Res Commun* 29:14–20.
5. Baldwin, G. S., and Carnegie, P. R. (1971). Isolation and partial characterization of methylated arginines from the encephalitogenic basic protein of myelin. *Biochem J* 123:69–74.
6. Eylar, E. H., Brostoff, S., Hashim, G., Caccam, J., and Burnett, P. (1971). Basic A1 protein of the myelin membrane. The complete amino acid sequence. *J Biol Chem* 246:5770–5784.
7. Ghosh, S. K., Paik, W. K., and Kim, S. (1988). Purification and molecular identification of two protein methylases I from calf brain. Myelin basic protein- and histone-specific enzyme. *J Biol Chem* 263:19024–19033.
8. Williams, K. R., Stone, K. L., LoPresti, M. B., Merrill, B. M., and Planck, S. R. (1985). Amino acid sequence of the UP1 calf thymus helix-destabilizing protein and its homology to an analogous protein from mouse myeloma. *Proc Natl Acad Sci USA* 82:5666–5670.
9. Lischwe, M. A., Cook, R. G., Ahn, Y. S., Yeoman, L. C., and Busch, H. (1985). Clustering of glycine and NG, NG-dimethylarginine in nucleolar protein C23. *Biochemistry* 24:6025–6028.
10. Gary, J. D., Lin, W. J., Yang, M. C., Herschman, H. R., and Clarke, S. (1996). The predominant protein-arginine methyltransferase from *Saccharomyces cerevisiae*. *J Biol Chem* 271:12585–12594.
11. Henry, M. F., and Silver, P. A. (1996). A novel methyltransferase (Hmt1p) modifies poly(A)+-RNA-binding proteins. *Mol Cell Biol* 16:3668–3678.

12. Lin, W. J., Gary, J. D., Yang, M. C., Clarke, S., and Herschman, H. R. (1996). The mammalian immediate-early TIS21 protein and the leukemia-associated BTG1 protein interact with a protein-arginine N-methyltransferase. *J Biol Chem* 271:15034–15044.
13. Gary, J. D., and Clarke, S. (1998). RNA and protein interactions modulated by protein arginine methylation. *Prog Nucleic Acid Res Mol Biol* 61:65–131.
14. Boisvert, F. M., Cote, J., Boulanger, M. C., and Richard, S. (2003). A proteomic analysis of arginine-methylated protein complexes. *Mol Cell Proteomics* 2:1319–1330.
15. Wu, C. C., MacCoss, M. J., Mardones, G., Finnigan, C., Mogelsvang, S., Yates, J. R., 3rd, and Howell, K. E. (2004). Organellar proteomics reveals Golgi arginine dimethylation. *Mol Biol Cell* 15:2907–2919.
16. Ong, S. E., Mittler, G., and Mann, M. (2004). Identifying and quantifying *in vivo* methylation sites by heavy methyl SILAC. *Nat Methods* 1:119–126.
17. Calnan, B. J., Tidor, B., Biancalana, S., Hudson, D., and Frankel, A. D. (1991). Arginine-mediated RNA recognition: the arginine fork. *Science* 252:1167–1171.
18. Raman, B., Guarnaccia, C., Nadassy, K., Zakhariev, S., Pintar, A., Zanuttin, F., Frigyes, D., Acatrinei, C., Vindigni, A., Pongor, G., and Pongor, S. (2001). N(omega)-arginine dimethylation modulates the interaction between a Gly/Arg-rich peptide from human nucleolin and nucleic acids. *Nucleic Acids Res* 29:3377–3384.
19. Rajpurohit, R., Paik, W. K., and Kim, S. (1994). Effect of enzymic methylation of heterogeneous ribonucleoprotein particle A1 on its nucleic-acid binding and controlled proteolysis. *Biochem J* 304:903–909.
20. Kiledjian, M., and Dreyfuss, G. (1992). Primary structure and binding activity of the hnRNP U protein: binding RNA through RGG box. *Embo J* 11:2655–2664.
21. Heine, M. A., Rankin, M. L., and DiMario, P. J. (1993). The Gly/Arg-rich (GAR) domain of *Xenopus* nucleolin facilitates *in vitro* nucleic acid binding and *in vivo* nucleolar localization. *Mol Biol Cell* 4:1189–1204.
22. Cartegni, L., Maconi, M., Morandi, E., Cobianchi, F., Riva, S., and Biamonti, G. (1996). hnRNP A1 selectively interacts through its Gly-rich domain with different RNA-binding proteins. *J Mol Biol* 259:337–348.
23. McBride, A. E., Cook, J. T., Stemmler, E. A., Rutledge, K. L., McGrath, K. A., and Rubens, J. A. (2005). Arginine methylation of yeast mRNA-binding protein Npl3 directly affects its function, nuclear export and intranuclear protein interactions. *J Biol Chem* 280:30888–30898.
24. Valentini, S. R., Weiss, V. H., and Silver, P. A. (1999). Arginine methylation and binding of Hrp1p to the efficiency element for mRNA 3′-end formation. *RNA* 5:272–280.
25. Denman, R. B. (2002). Methylation of the arginine-glycine-rich region in the fragile X mental retardation protein FMRP differentially affects RNA binding. *Cell Mol Biol Lett* 7:877–883.
26. Xu, C., and Henry, M. F. (2004). Nuclear export of hnRNP Hrp1p and nuclear export of hnRNP Npl3p are linked and influenced by the methylation state of Npl3p. *Mol Cell Biol* 24:10742–10756.
27. Rho, J., Choi, S., Seong, Y. R., Choi, J., and Im, D. S. (2001). The arginine-1493 residue in QRRGRTGR1493G motif IV of the hepatitis C virus NS3 helicase domain is essential for NS3 protein methylation by the protein arginine methyltransferase 1. *J Virol* 75:8031–8044.
28. Frankel, A., and Clarke, S. (1999). RNase treatment of yeast and mammalian cell extracts affects *in vitro* substrate methylation by type I protein arginine N-methyltransferases. *Biochem Biophys Res Commun* 259:391–400.
29. Mucha, P., Szyk, A., Rekowski, P., and Agris, P. F. (2003). Using capillary electrophoresis to study methylation effect on RNA-peptide interaction. *Acta Biochim Pol* 50:857–864.

30. Birney, E., Kumar, S., and Krainer, A. R. (1993). Analysis of the RNA-recognition motif and RS and RGG domains: conservation in metazoan pre-mRNA splicing factors. *Nucleic Acids Res* 21:5803–5816.

31. Lukong, K. E., and Richard, S. (2003). Sam68, the KH domain-containing superSTAR. *Biochim Biophys Acta* 1653:73–86.

32. Bagni, C., and Lapeyre, B. (1998). Gar1p binds to the small nucleolar RNAs snR10 and snR30 *in vitro* through a nontypical RNA binding element. *J Biol Chem* 273:10868–10873.

33. Inoue, K., Mizuno, T., Wada, K., and Hagiwara, M. (2000). Novel RING finger proteins, Air1p and Air2p interact with Hmt1p and inhibit the arginine methylation of Npl3p. *J Biol Chem* 275:32793–32798.

34. Bedford, M. T., Frankel, A., Yaffe, M. B., Clarke, S., Leder, P., and Richard, S. (2000). Arginine methylation inhibits the binding of proline-rich ligands to Src homology 3, but not WW, domains. *J Biol Chem* 275:16030–16036.

35. Xu, W., Chen, H., Du, K., Asahara, H., Tini, M., Emerson, B. M., Montminy, M., and Evans, R. M. (2001). A transcriptional switch mediated by cofactor methylation. *Science* 294:2507–2511.

36. Brahms, H., Meheus, L., de Brabandere, V., Fischer, U., and Luhrmann, R. (2001). Symmetrical dimethylation of arginine residues in spliceosomal Sm protein B/B′ and the Sm-like protein LSm4, and their interaction with the SMN protein. *RNA* 7:1531–1542.

37. Friesen, W. J., Massenet, S., Paushkin, S., Wyce, A., and Dreyfuss, G. (2001). SMN, the product of the spinal muscular atrophy gene, binds preferentially to dimethylarginine-containing protein targets. *Molecular Cell* 7:1111–1117.

38. Sprangers, R., Groves, M. R., Sinning, I., and Sattler, M. (2003). High-resolution X-ray and NMR structures of the SMN Tudor domain: conformational variation in the binding site for symmetrically dimethylated arginine residues. *J Mol Biol* 327:507–520.

39. Tsukiyama, T., and Wu, C. (1997). Chromatin remodeling and transcription. *Curr Opin Genet Dev* 7:182–191.

40. Jenuwein, T., and Allis, C. D. (2001). Translating the histone code. *Science* 293:1074-1080.

41. Chen, D., Ma, H., Hong, H., Koh, S. S., Huang, S. M., Schurter, B. T., Aswad, D. W., and Stallcup, M. R. (1999). Regulation of transcription by a protein methyltransferase. *Science* 284:2174–2177.

42. McKenna, N. J., Xu, J., Nawaz, Z., Tsai, S. Y., Tsai, M. J., and O'Malley, B. W. (1999). Nuclear receptor coactivators: multiple enzymes, multiple complexes, multiple functions. *J Steroid Biochem Mol Biol* 69:3–12.

43. Chen, D., Huang, S. M., and Stallcup, M. R. (2000). Synergistic, p160 coactivator-dependent enhancement of estrogen receptor function by CARM1 and p300. *J Biol Chem* 275:40810–40816.

44. Schurter, B. T., Koh, S. S., Chen, D., Bunick, G. J., Harp, J. M., Hanson, B. L., Henschen-Edman, A., Mackay, D. R., Stallcup, M. R., and Aswad, D. W. (2001). Methylation of histone H3 by coactivator-associated arginine methyltransferase 1. *Biochemistry* 40:5747–5756.

45. Ma, H., Baumann, C. T., Li, H., Strahl, B. D., Rice, R., Jelinek, M. A., Aswad, D. W., Allis, C. D., Hager, G. L., and Stallcup, M. R. (2001). Hormone-dependent, CARM1-directed, arginine-specific methylation of histone H3 on a steroid-regulated promoter. *Curr Biol* 11:1981–1985.

46. Bauer, U. M., Daujat, S., Nielsen, S. J., Nightingale, K., and Kouzarides, T. (2002). Methylation at arginine 17 of histone H3 is linked to gene activation. *EMBO Rep* 3:39–44.

47. Kang, Z., Janne, O. A., and Palvimo, J. J. (2004). Coregulator recruitment and histone modifications in transcriptional regulation by the androgen receptor. *Mol Endocrinol* 18:2633–2648.

48. Yadav, N., Lee, J., Kim, J., Shen, J., Hu, M. C., Aldaz, C. M., and Bedford, M. T. (2003). Specific protein methylation defects and gene expression perturbations in coactivator-associated arginine methyltransferase 1-deficient mice. *Proc Natl Acad Sci USA* 100:6464–6468.

49. Chevillard-Briet, M., Trouche, D., and Vandel, L. (2002). Control of CBP co-activating activity by arginine methylation. *EMBO J* 21:5457–5466.

50. Lee, Y. H., Coonrod, S. A., Kraus, W. L., Jelinek, M. A., and Stallcup, M. R. (2005). Regulation of coactivator complex assembly and function by protein arginine methylation and demethylimination. *Proc Natl Acad Sci USA* 102:3611–3616.

51. Wei, Y., Horng, J. C., Vendel, A. C., Raleigh, D. P., and Lumb, K. J. (2003). Contribution to stability and folding of a buried polar residue at the CARM1 methylation site of the KIX domain of CBP. *Biochemistry* 42:7044–7049.

52. Wang, Y., Wysocka, J., Sayegh, J., Lee, Y. H., Perlin, J. R., Leonelli, L., Sonbuchner, L. S., McDonald, C. H., Cook, R. G., Dou, Y., Roeder, R. G., Clarke, S., Stallcup, M. R., Allis, C. D., and Coonrod, S. A. (2004). Human PAD4 regulates histone arginine methylation levels via demethylimination. *Science* 306:279–283.

53. Cuthbert, G. L., Daujat, S., Snowden, A. W., Erdjument-Bromage, H., Hagiwara, T., Yamada, M., Schneider, R., Gregory, P. D., Tempst, P., Bannister, A. J., and Kouzarides, T. (2004). Histone deimination antagonizes arginine methylation. *Cell* 118:545–553.

54. Covic, M., Hassa, P. O., Saccani, S., Buerki, C., Meier, N. I., Lombardi, C., Imhof, R., Bedford, M. T., Natoli, G., and Hottiger, M. O. (2005). Arginine methyltransferase CARM1 is a promoter-specific regulator of NF-kappaB-dependent gene expression. *EMBO J* 24:85–96.

55. Ananthanarayanan, M., Li, S., Balasubramaniyan, N., Suchy, F. J., and Walsh, M. J. (2004). Ligand-dependent activation of the farnesoid X-receptor directs arginine methylation of histone H3 by CARM1. *J Biol Chem* 279:54348–54357.

56. Xu, W., Cho, H., Kadam, S., Banayo, E. M., Anderson, S., Yates, J. R., 3rd, Emerson, B. M., and Evans, R. M. (2004). A methylation-mediator complex in hormone signaling. *Genes Dev* 18:144–156.

57. Cakouros, D., Daish, T. J., Mills, K., and Kumar, S. (2004). An arginine-histone methyltransferase, CARMER, coordinates ecdysone-mediated apoptosis in *Drosophila* cells. *J Biol Chem* 279:18467–18471.

58. Koh, S. S., Chen, D., Lee, Y. H., and Stallcup, M. R. (2001). Synergistic enhancement of nuclear receptor function by p160 coactivators and two coactivators with protein methyltransferase activities. *J Biol Chem* 276:1089–1098.

59. Strahl, B. D., Briggs, S. D., Brame, C. J., Caldwell, J. A., Koh, S. S., Ma, H., Cook, R. G., Shabanowitz, J., Hunt, D. F., Stallcup, M. R., and Allis, C. D. (2001). Methylation of histone H4 at arginine 3 occurs *in vivo* and is mediated by the nuclear receptor coactivator PRMT1. *Curr Biol* 11:996–1000.

60. Wang, H., Huang, Z. Q., Xia, L., Feng, Q., Erdjument-Bromage, H., Strahl, B. D., Briggs, S. D., Allis, C. D., Wong, J., Tempst, P., and Zhang, Y. (2001). Methylation of histone H4 at arginine 3 facilitating transcriptional activation by nuclear hormone receptor. *Science* 293:853–857.

61. Teyssier, C., Ma, H., Emter, R., Kralli, A., and Stallcup, M. R. (2005). Activation of nuclear receptor coactivator PGC-1alpha by arginine methylation. *Genes Dev* 19:1466–1473.

62. Rizzo G, Renga B, Antonelli E, Passeri D, Pellicciari R, Fiorucci S. (2005) The Methyl Transferase PRMT1 Functions as Co-Activator of Farnesoid X Receptor (FXR)/9-cis Retinoid X Receptor and Regulates Transcription of FXR Responsive Genes. *Mol Pharmacol* 68:551–8.

63. Rezai-Zadeh, N., Zhang, X., Namour, F., Fejer, G., Wen, Y. D., Yao, Y. L., Gyory, I., Wright, K., and Seto, E. (2003). Targeted recruitment of a histone H4-specific methyltransferase by the transcription factor YY1. *Genes Dev* 17:1019–1029.

64. Tang, J., Kao, P. N., and Herschman, H. R. (2000). Protein-arginine methyltransferase I, the predominant protein-arginine methyltransferase in cells, interacts with and is regulated by interleukin enhancer-binding factor 3. *J Biol Chem* 275:19866–19876.

65. Baumeister, P., Luo, S., Skarnes, W. C., Sui, G., Seto, E., Shi, Y., and Lee, A. S. (2005). Endoplasmic reticulum stress induction of the Grp78/BiP promoter: activating mechanisms mediated by YY1 and its interactive chromatin modifiers. *Mol Cell Biol* 25:4529–4540.

66. Kwak, Y. T., Guo, J., Prajapati, S., Park, K. J., Surabhi, R. M., Miller, B., Gehrig, P., and Gaynor, R. B. (2003). Methylation of SPT5 regulates its interaction with RNA polymerase II and transcriptional elongation properties. *Mol Cell* 11:1055–1066.

67. Wu-Baer, F., Lane, W. S., and Gaynor, R. B. (1998). Role of the human homolog of the yeast transcription factor SPT5 in HIV-1 Tat-activation. *J Mol Biol* 277:179–197.

68. Fabbrizio, E., El Messaoudi, S., Polanowska, J., Paul, C., Cook, J. R., Lee, J. H., Negre, V., Rousset, M., Pestka, S., Le Cam, A., and Sardet, C. (2002). Negative regulation of transcription by the type II arginine methyltransferase PRMT5. *EMBO Rep* 3:641–645.

69. Pal, S., Yun, R., Datta, A., Lacomis, L., Erdjument-Bromage, H., Kumar, J., Tempst, P., and Sif, S. (2003). mSin3A/histone deacetylase 2- and PRMT5-containing Brg1 complex is involved in transcriptional repression of the Myc target gene cad. *Mol Cell Biol* 23:7475–7487.

70. Pal, S., Vishwanath, S. N., Erdjument-Bromage, H., Tempst, P., and Sif, S. (2004). Human SWI/SNF-associated PRMT5 methylates histone H3 arginine 8 and negatively regulates expression of ST7 and NM23 tumor suppressor genes. *Mol Cell Biol* 24:9630–9645.

71. Qi, C., Chang, J., Zhu, Y., Yeldandi, A. V., Rao, S. M., and Zhu, Y. J. (2002). Identification of protein arginine methyltransferase 2 as a coactivator for estrogen receptor alpha. *J Biol Chem* 277:28624–28630.

72. Daujat, S., Bauer, U. M., Shah, V., Turner, B., Berger, S., and Kouzarides, T. (2002). Crosstalk between CARM1 methylation and CBP acetylation on histone H3. *Curr Biol* 12:2090–2097.

73. An, W., Kim, J., and Roeder, R. G. (2004). Ordered cooperative functions of PRMT1, p300, and CARM1 in transcriptional activation by p53. *Cell* 117:735–748.

74. Abramovich, C., Yakobson, B., Chebath, J., and Revel, M. (1997). A protein-arginine methyl-transferase binds to the intracytoplasmic domain of the IFNAR1 chain in the type I interferon receptor. *EMBO J* 16:260–266.

75. Altschuler, L., Wook, J. O., Gurari, D., Chebath, J., and Revel, M. (1999). Involvement of receptor-bound protein methyltransferase PRMT1 in antiviral and antiproliferative effects of type I interferons. *J Interferon Cytokine Res* 19:189–195.

76. Darnell, J. E., Jr., Kerr, I. M., and Stark, G. R. (1994). Jak-STAT pathways and transcriptional activation in response to IFNs and other extracellular signaling proteins. *Science* 264:1415–1421.

77. Mowen, K. A., Tang, J., Zhu, W., Schurter, B. T., Shuai, K., Herschman, H. R., and David, M. (2001). Arginine methylation of STAT1 modulates IFNalpha/beta-induced transcription. *Cell* 104:731–741.

78. Chen, W., Daines, M. O., and Hershey, G. K. (2004). Methylation of STAT6 modulates STAT6 phosphorylation, nuclear translocation, and DNA-binding activity. *J Immunol* 172:6744–6750.

79. Komyod, W., Bauer, U. M., Heinrich, P. C., Haan, S., and Behrmann, I. (2005). Are STATs arginine-methylated? *J Biol Chem* 280:21700–21705.

80. Meissner, T., Krause, E., Lodige, I., and Vinkemeier, U. (2004). Arginine methylation of STAT1: a reassessment. *Cell* 119:587-589; discussion 589–590.

81. Zhu, W., Mustelin, T., and David, M. (2002). Arginine methylation of STAT1 regulates its dephosphorylation by T cell protein tyrosine phosphatase. *J Biol Chem* 277:35787–35790.

82. Mowen, K. A., Schurter, B. T., Fathman, J. W., David, M., and Glimcher, L. H. (2004). Arginine methylation of NIP45 modulates cytokine gene expression in effector T lymphocytes. *Mol Cell* 15:559–571.

83. Richard, S., Morel, M., and Cleroux, P. (2005). Arginine methylation regulates IL-2 gene expression: a role for PRMT5. *Biochem J* 388:379–386.
84. Pollack, B. P., Kotenko, S. V., He, W., Izotova, L. S., Barnoski, B. L., and Pestka, S. (1999). The human homologue of the yeast proteins Skb1 and Hsl7p interacts with Jak kinases and contains protein methyltransferase activity. *J Biol Chem* 274:31531–31542.
85. Ma, X. J., Lu, Q., and Grunstein, M. (1996). A search for proteins that interact genetically with histone H3 and H4 amino termini uncovers novel regulators of the Swe1 kinase in *Saccharomyces cerevisiae*. *Genes Dev* 10:1327–1340.
86. Leu, J. Y., and Roeder, G. S. (1999). The pachytene checkpoint in *S. cerevisiae* depends on Swe1-mediated phosphorylation of the cyclin-dependent kinase Cdc28. *Mol Cell* 4:805–814.
87. McMillan, J. N., Longtine, M. S., Sia, R. A., Theesfeld, C. L., Bardes, E. S., Pringle, J. R., and Lew, D. J. (1999). The morphogenesis checkpoint in *Saccharomyces cerevisiae:* cell cycle control of Swe1p degradation by Hsl1p and Hsl7p. *Mol Cell Biol* 19:6929–6939.
88. Yamada, A., Duffy, B., Perry, J. A., and Kornbluth, S. (2004). DNA replication checkpoint control of Wee1 stability by vertebrate Hsl7. *J Cell Biol* 167:841–849.
89. Gilbreth, M., Yang, P., Bartholomeusz, G., Pimental, R. A., Kansra, S., Gadiraju, R., and Marcus, S. (1998). Negative regulation of mitosis in fission yeast by the shk1 interacting protein skb1 and its human homolog, Skb1Hs. *Proc Natl Acad Sci USA* 95:14781–14786.
90. Gilbreth, M., Yang, P., Wang, D., Frost, J., Polverino, A., Cobb, M. H., and Marcus, S. (1996). The highly conserved skb1 gene encodes a protein that interacts with Shk1, a fission yeast Ste20/PAK homolog. *Proc Natl Acad Sci USA* 93:13802–13807.
91. Fujita, A., Tonouchi, A., Hiroko, T., Inose, F., Nagashima, T., Satoh, R., and Tanaka, S. (1999). Hsl7p, a negative regulator of Ste20p protein kinase in the *Saccharomyces cerevisiae* filamentous growth-signaling pathway. *Proc Natl Acad Sci USA* 96:8522–8527.
92. Bao, S., Qyang, Y., Yang, P., Kim, H., Du, H., Bartholomeusz, G., Henkel, J., Pimental, R., Verde, F., and Marcus, S. (2001). The highly conserved protein methyltransferase, Skb1, is a mediator of hyperosmotic stress response in the fission yeast *Schizosaccharomyces pombe*. *J Biol Chem* 276:14549–14552.
93. Wiley, D. J., Marcus, S., D'Urso, G., and Verde, F. (2003). Control of cell polarity in fission yeast by association of Orb6p kinase with the highly conserved protein methyltransferase Skb1p. *J Biol Chem* 278:25256–25263.
94. Schwarzler, A., Kreienkamp, H. J., and Richter, D. (2000). Interaction of the somatostatin receptor subtype 1 with the human homolog of the Shk1 kinase-binding protein from yeast. *J Biol Chem* 275:9557–9562.
95. Lee, J. H., Cook, J. R., Pollack, B. P., Kinzy, T. G., Norris, D., and Pestka, S. (2000). Hsl7p, the yeast homologue of human JBP1, is a protein methyltransferase. *Biochem Biophys Res Commun* 274:105–111.
96. Shulewitz, M. J., Inouye, C. J., and Thorner, J. (1999). Hsl7 localizes to a septin ring and serves as an adapter in a regulatory pathway that relieves tyrosine phosphorylation of Cdc28 protein kinase in *Saccharomyces cerevisiae*. *Mol Cell Biol* 19:7123–7137.
97. Theesfeld, C. L., Zyla, T. R., Bardes, E. G., and Lew, D. J. (2003). A monitor for bud emergence in the yeast morphogenesis checkpoint. *Mol Biol Cell* 14:3280-3291.
98. Chie, L., Cook, J. R., Chung, D., Hoffmann, R., Yang, Z., Kim, Y., Pestka, S., and Pincus, M. R. (2003). A protein methyl transferase, PRMT5, selectively blocks oncogenic ras-p21 mitogenic signal transduction. *Ann Clin Lab Sci* 33:200–207.
99. Dreyfuss, G., Kim, V. N., and Kataoka, N. (2002). Messenger-RNA-binding proteins and the messages they carry. *Nat Rev Mol Cell Biol* 3:195–205.
100. Boffa, L. C., Karn, J., Vidali, G., and Allfrey, V. G. (1977). Distribution of NG, NG,-dimethylarginine in nuclear protein fractions. *Biochem Biophys Res Commun* 74:969–976.
101. Liu, Q., and Dreyfuss, G. (1995). *In vivo* and *in vitro* arginine methylation of RNA-binding proteins. *Mol Cell Biol* 15:2800–2808.

102. Pawlak, M. R., Banik-Maiti, S., Pietenpol, J. A., and Ruley, H. E. (2002). Protein arginine methyltransferase I: substrate specificity and role in hnRNP assembly. *J Cell Biochem* 87:394–407.
103. Kzhyshkowska, J., Schutt, H., Liss, M., Kremmer, E., Stauber, R., Wolf, H., and Dobner, T. (2001). Heterogeneous nuclear ribonucleoprotein E1B-AP5 is methylated in its Arg-Gly-Gly (RGG) box and interacts with human arginine methyltransferase HRMT1L1. *Biochem J* 358:305–314.
104. Herrmann, F., Bossert, M., Schwander, A., Akgun, E., and Fackelmayer, F. O. (2004). Arginine methylation of scaffold attachment factor A by heterogeneous nuclear ribonucleoprotein particle-associated PRMT1. *J Biol Chem* 279:48774–48779.
105. Siebel, C. W., Feng, L., Guthrie, C., and Fu, X. D. (1999). Conservation in budding yeast of a kinase specific for SR splicing factors. *Proc Natl Acad Sci USA* 96:5440–5445.
106. Yu, M. C., Bachand, F., McBride, A. E., Komili, S., Casolari, J. M., and Silver, P. A. (2004). Arginine methyltransferase affects interactions and recruitment of mRNA processing and export factors. *Genes Dev* 18:2024–2035.
107. Green, D. M., Marfatia, K. A., Crafton, E. B., Zhang, X., Cheng, X., and Corbett, A. H. (2002). Nab2p is required for poly(A) RNA export in *Saccharomyces cerevisiae* and is regulated by arginine methylation via Hmt1p. *J Biol Chem* 277:7752–7760.
108. Xu, C., Henry, P. A., Setya, A., and Henry, M. F. (2003). *In vivo* analysis of nucleolar proteins modified by the yeast arginine methyltransferase Hmt1/Rmt1p. *RNA* 9:746–759.
109. Shen, E. C., Henry, M. F., Weiss, V. H., Valentini, S. R., Silver, P. A., and Lee, M. S. (1998). Arginine methylation facilitates the nuclear export of hnRNP proteins. *Genes Dev* 12:679–691.
110. McBride, A. E., Weiss, V. H., Kim, H. K., Hogle, J. M., and Silver, P. A. (2000). Analysis of the yeast arginine methyltransferase Hmt1p/Rmt1p and its *in vivo* function. Cofactor binding and substrate interactions. *J Biol Chem* 275:3128–3136.
111. Lee, M. S., Henry, M., and Silver, P. A. (1996). A protein that shuttles between the nucleus and the cytoplasm is an important mediator of RNA export. *Genes Dev* 10:1233–1246.
112. Yun, C. Y., and Fu, X. D. (2000). Conserved SR protein kinase functions in nuclear import and its action is counteracted by arginine methylation in *Saccharomyces cerevisiae*. *J Cell Biol* 150:707–718.
113. Cote, J., Boisvert, F. M., Boulanger, M. C., Bedford, M. T., and Richard, S. (2003). Sam68 RNA binding protein is an *in vivo* substrate for protein arginine N-methyltransferase 1. *Mol Biol Cell* 14:274–287.
114. Nichols, R. C., Wang, X. W., Tang, J., Hamilton, B. J., High, F. A., Herschman, H. R., and Rigby, W. F. (2000). The RGG domain in hnRNP A2 affects subcellular localization. *Exp Cell Res* 256:522–532.
115. Smith, W. A., Schurter, B. T., Wong-Staal, F., and David, M. (2004). Arginine methylation of RNA helicase a determines its subcellular localization. *J Biol Chem* 279:22795–22798.
116. Maggipinto, M., Rabiner, C., Kidd, G. J., Hawkins, A. J., Smith, R., and Barbarese, E. (2004). Increased expression of the MBP mRNA binding protein HnRNP A2 during oligodendrocyte differentiation. *J Neurosci Res* 75:614–623.
117. Araya, N., Hiraga, H., Kako, K., Arao, Y., Kato, S., and Fukamizu, A. (2005). Transcriptional down-regulation through nuclear exclusion of EWS methylated by PRMT1. *Biochem Biophys Res Commun* 329:653–660.
118. Aoki, K., Ishii, Y., Matsumoto, K., and Tsujimoto, M. (2002). Methylation of Xenopus CIRP2 regulates its arginine- and glycine-rich region-mediated nucleocytoplasmic distribution. *Nucleic Acids Res* 30:5182–5192.
119. Araya, N., Hirota, K., Shimamoto, Y., Miyagishi, M., Yoshida, E., Ishida, J., Kaneko, S., Kaneko, M., Nakajima, T., and Fukamizu, A. (2003). Cooperative interaction of EWS with CREB-binding protein selectively activates hepatocyte nuclear factor 4-mediated transcription. *J Biol Chem* 278:5427–5432.

120. Li, H., Park, S., Kilburn, B., Jelinek, M. A., Henschen-Edman, A., Aswad, D. W., Stallcup, M. R., and Laird-Offringa, I. A. (2002). Lipopolysaccharide-induced methylation of HuR, an mRNA-stabilizing protein, by CARM1. Coactivator-associated arginine methyltransferase. *J Biol Chem* 277:44623–44630.

121. Smith, J. J., Rucknagel, K. P., Schierhorn, A., Tang, J., Nemeth, A., Linder, M., Herschman, H. R., and Wahle, E. (1999). Unusual sites of arginine methylation in Poly(A)-binding protein II and *in vitro* methylation by protein arginine methyltransferases PRMT1 and PRMT3. *J Biol Chem* 274:13229–13234.

122. Klein, S., Carroll, J. A., Chen, Y., Henry, M. F., Henry, P. A., Ortonowski, I. E., Pintucci, G., Beavis, R. C., Burgess, W. H., and Rifkin, D. B. (2000). Biochemical analysis of the arginine methylation of high molecular weight fibroblast growth factor-2. *J Biol Chem* 275:3150–3157.

123. Pintucci, G., Quarto, N., and Rifkin, D. B. (1996). Methylation of high molecular weight fibroblast, growth factor-2 determines post-translational increases in molecular weight and affects its intracellular distribution. *Mol Biol Cell* 7:1249–1258.

124. Brahms, H., Raymackers, J., Union, A., de Keyser, F., Meheus, L., and Luhrmann, R. (2000). The C-terminal RG dipeptide repeats of the spliceosomal Sm proteins D1 and D3 contain symmetrical dimethylarginines, which form a major B-cell epitope for anti-Sm autoantibodies. *J Biol Chem* 275:17122–17129.

125. Friesen, W. J., and Dreyfuss, G. (2000). Specific sequences of the Sm and Sm-like (Lsm) proteins mediate their interaction with the spinal muscular atrophy disease gene product (SMN). *J Biol Chem* 275:26370–26375.

126. Friesen, W. J., Paushkin, S., Wyce, A., Massenet, S., Pesiridis, G. S., Van Duyne, G., Rappsilber, J., Mann, M., and Dreyfuss, G. (2001). The methylosome, a 20S complex containing JBP1 and pICln, produces dimethylarginine-modified Sm proteins. *Mol Cell Biol* 21:8289–8300.

127. Meister, G., Eggert, C., Buhler, D., Brahms, H., Kambach, C., and Fischer, U. (2001). Methylation of Sm proteins by a complex containing PRMT5 and the putative U snRNP assembly factor pICln. *Curr Biol* 11:1990–1994.

128. Boisvert, F. M., Cote, J., Boulanger, M. C., Cleroux, P., Bachand, F., Autexier, C., and Richard, S. (2002). Symmetrical dimethylarginine methylation is required for the localization of SMN in Cajal bodies and pre-mRNA splicing. *J Cell Biol* 159:957–969.

129. Friesen, W. J., Wyce, A., Paushkin, S., Abel, L., Rappsilber, J., Mann, M., and Dreyfuss, G. (2002). A novel WD repeat protein component of the methylosome binds Sm proteins. *J Biol Chem* 277:8243–8247.

130. Pu, W. T., Krapivinsky, G. B., Krapivinsky, L., and Clapham, D. E. (1999). pICln inhibits snRNP biogenesis by binding core spliceosomal proteins. *Mol Cell Biol* 19:4113–4120.

131. Meister, G., and Fischer, U. (2002). Assisted RNP assembly: SMN and PRMT5 complexes cooperate in the formation of spliceosomal UsnRNPs. *Embo J* 21:5853–5863.

132. Licciardo, P., Amente, S., Ruggiero, L., Monti, M., Pucci, P., Lania, L., and Majello, B. (2003). The FCP1 phosphatase interacts with RNA polymerase II and with MEP50 a component of the methylosome complex involved in the assembly of snRNP. *Nucleic Acids Res* 31:999–1005.

133. Amente, S., Napolitano, G., Licciardo, P., Monti, M., Pucci, P., Lania, L., and Majello, B. (2005). Identification of proteins interacting with the RNAPII FCP1 phosphatase: FCP1 forms a complex with arginine methyltransferase PRMT5 and it is a substrate for PRMT5-mediated methylation. *FEBS Lett* 579:683–689.

134. Hebert, M. D., Shpargel, K. B., Ospina, J. K., Tucker, K. E., and Matera, A. G. (2002). Coilin methylation regulates nuclear body formation. *Dev Cell* 3:329–337.

135. Miranda, T. B., Khusial, P., Cook, J. R., Lee, J. H., Gunderson, S. I., Pestka, S., Zieve, G. W., and Clarke, S. (2004). Spliceosome Sm proteins D1, D3, and B/B′ are asymmetrically dimethylated at arginine residues in the nucleus. *Biochem Biophys Res Commun* 323:382–387.

136. Whitehead, S. E., Jones, K. W., Zhang, X., Cheng, X., Terns, R. M., and Terns, M. P. (2002). Determinants of the interaction of the spinal muscular atrophy disease protein SMN with the dimethylarginine-modified box H/ACA small nucleolar ribonucleoprotein GAR1. *J Biol Chem* 277:48087–48093.

137. Ohkura, N., Takahashi, M., Yaguchi, H., Nagamura, Y., and Tsukada, T. (2005). Coactivator-associated arginine methyltransferase 1, CARM1, affects pre-mRNA splicing in an isoform-specific manner. *J Biol Chem* 280:28927–28935.

138. Christensen, M. E., and Fuxa, K. P. (1988). The nucleolar protein, B-36, contains a glycine and dimethylarginine-rich sequence conserved in several other nuclear RNA-binding proteins. *Biochem Biophys Res Commun* 155:1278–1283.

139. Yanagida, M., Hayano, T., Yamauchi, Y., Shinkawa, T., Natsume, T., Isobe, T., and Takahashi, N. (2004). Human fibrillarin forms a sub-complex with splicing factor 2-associated p32, protein arginine methyltransferases, and tubulins alpha 3 and beta 1 that is independent of its association with preribosomal ribonucleoprotein complexes. *J Biol Chem* 279:1607–1614.

140. Pellar, G. J., and DiMario, P. J. (2003). Deletion and site-specific mutagenesis of nucleolin's carboxy GAR domain. *Chromosoma* 111:461–469.

141. Bachand, F., and Silver, P. A. (2004). PRMT3 is a ribosomal protein methyltransferase that affects the cellular levels of ribosomal subunits. *EMBO J* 23:2641–2650.

142. Swiercz, R., Person, M. D., and Bedford, M. T. (2005). Ribosomal protein S2 is a substrate for mammalian PRMT3 (protein arginine methyltransferase 3). *Biochem J* 386:85–91.

143. Niewmierzycka, A., and Clarke, S. (1999). S-Adenosylmethionine-dependent methylation in *Saccharomyces cerevisiae*. Identification of a novel protein arginine methyltransferase. *J Biol Chem* 274:814–824.

144. Chern, M. K., Chang, K. N., Liu, L. F., Tam, T. C., Liu, Y. C., Liang, Y. L., and Tam, M. F. (2002). Yeast ribosomal protein L12 is a substrate of protein-arginine methyltransferase 2. *J Biol Chem* 277:15345–15353.

145. Cortes, U., Moyret-Lalle, C., Falette, N., Duriez, C., Ghissassi, F. E., Barnas, C., Morel, A. P., Hainaut, P., Magaud, J. P., and Puisieux, A. (2000). BTG gene expression in the p53-dependent and -independent cellular response to DNA damage. *Mol Carcinog* 27:57–64.

146. Boisvert, F. M., Dery, U., Masson, J. Y., and Richard, S. (2005). Arginine methylation of MRE11 by PRMT1 is required for DNA damage checkpoint control. *Genes Dev* 19:671–676.

147. Gros, L., Delaporte, C., Frey, S., Decesse, J., de Saint-Vincent, B. R., Cavarec, L., Dubart, A., Gudkov, A. V., and Jacquemin-Sablon, A. (2003). Identification of new drug sensitivity genes using genetic suppressor elements: protein arginine N-methyltransferase mediates cell sensitivity to DNA-damaging agents. *Cancer Res* 63:164–171.

148. Zheng, Z., Schmidt-Ott, K. M., Chua, S., Foster, K. A., Frankel, R. Z., Pavlidis, P., Barasch, J., D'Agati, V. D., and Gharavi, A. G. (2005). A Mendelian locus on chromosome 16 determines susceptibility to doxorubicin nephropathy in the mouse. *Proc Natl Acad Sci USA* 102:2502–2507.

149. Boisvert, F. M., Hendzel, M. J., Masson, J. Y., and Richard, S. (2005). Methylation of MRE11 regulates its nuclear compartmentalization. *Cell Cycle* 4:981–989.

150. Lee, J. H., Cook, J. R., Yang, Z. H., Mirochnitchenko, O., Gunderson, S. I., Felix, A. M., Herth, N., Hoffmann, R., and Pestka, S. (2005). PRMT7, a new protein arginine methyltransferase that synthesizes symmetric dimethylarginine. *J Biol Chem* 280:3656–3664.

151. Miranda, T. B., Miranda, M., Frankel, A., and Clarke, S. (2004). PRMT7 is a member of the protein arginine methyltransferase family with a distinct substrate specificity. *J Biol Chem* 279:22902–22907.

152. Pawlak, M. R., Scherer, C. A., Chen, J., Roshon, M. J., and Ruley, H. E. (2000). Arginine N-methyltransferase 1 is required for early postimplantation mouse development, but cells deficient in the enzyme are viable. *Mol Cell Biol* 20:4859–4869.

153. Chen, S. L., Loffler, K. A., Chen, D., Stallcup, M. R., and Muscat, G. E. (2002). The coactivator-associated arginine methyltransferase is necessary for muscle differentiation: CARM1 coactivates myocyte enhancer factor-2. *J Biol Chem* 277:4324–4333.

154. Lim, Y., Kwon, Y. H., Won, N. H., Min, B. H., Park, I. S., Paik, W. K., and Kim, S. (2005). Multimerization of expressed protein-arginine methyltransferases during the growth and differentiation of rat liver. *Biochim Biophys Acta* 1723:240–247.

155. Braun, M. C., Kelly, C. N., Prada, A. E., Mishra, J., Chand, D., Devarajan, P., and Zahedi, K. (2004). Human PRMT5 expression is enhanced during in vitro tubule formation and after in vivo ischemic injury in renal epithelial cells. *Am J Nephrol* 24:250–257.

156. Sarmento, O. F., Digilio, L. C., Wang, Y., Perlin, J., Herr, J. C., Allis, C. D., and Coonrod, S. A. (2004). Dynamic alterations of specific histone modifications during early murine development. *J Cell Sci* 117:4449–4459.

157. Boulanger, M. C., Miranda, T. B., Clarke, S., Di Fruscio, M., Suter, B., Lasko, P., and Richard, S. (2004). Characterization of the *Drosophila* protein arginine methyltransferases DART1 and DART4. *Biochem J* 379:283–289.

158. Johnstone, O., and Lasko, P. (2001). Translational regulation and RNA localization in *Drosophila* oocytes and embryos. *Annu Rev Genet* 35:365–406.

159. Anne, J., and Mechler, B. M. (2005). Valois, a component of the nuage and pole plasm, is involved in assembly of these structures, and binds to Tudor and the methyltransferase Capsuleen. *Development* 132:2167–2177.

160. Schupbach, T., and Wieschaus, E. (1986). Germline autonomy of maternal-effect mutations altering the embryonic body pattern of *Drosophila*. *Dev Biol* 113:443–448.

161. Chan, Y. B., Miguel-Aliaga, I., Franks, C., Thomas, N., Trulzsch, B., Sattelle, D. B., Davies, K. E., and van den Heuvel, M. (2003). Neuromuscular defects in a *Drosophila* survival motor neuron gene mutant. *Hum Mol Genet* 12:1367–1376.

162. Kim, J., Lee, J., Yadav, N., Wu, Q., Carter, C., Richard, S., Richie, E., and Bedford, M. T. (2004). Loss of CARM1 results in hypomethylation of thymocyte cyclic AMP-regulated phosphoprotein and deregulated early T cell development. *J Biol Chem* 279:25339–25344.

163. Kim, S., Lim, I. K., Park, G. H., and Paik, W. K. (1997). Biological methylation of myelin basic protein: enzymology and biological significance. *Int J Biochem Cell Biol* 29:743–751.

164. Crang, A. J., and Jacobson, W. (1982). The relationship of myelin basic protein (arginine) methyltransferase to myelination in mouse spinal cord. *J Neurochem* 39:244–247.

165. Amur, S. G., Shanker, G., Cochran, J. M., Ved, H. S., and Pieringer, R. A. (1986). Correlation between inhibition of myelin basic protein (arginine) methyltransferase by sinefungin and lack of compact myelin formation in cultures of cerebral cells from embryonic mice. *J Neurosci Res* 16:367–376.

166. Young, P. R., Vacante, D. A., and Waickus, C. M. (1987). Mechanism of the interaction between myelin basic protein and the myelin membrane; the role of arginine methylation. *Biochem Biophys Res Commun* 145:1112–1118.

167. Rawal, N., Lee, Y. J., Paik, W. K., and Kim, S. (1992). Studies on NG-methylarginine derivatives in myelin basic protein from developing and mutant mouse brain. *Biochem J* 287:929–935.

168. Lee, Y. H., Campbell, H. D., and Stallcup, M. R. (2004). Developmentally essential protein flightless I is a nuclear receptor coactivator with actin binding activity. *Mol Cell Biol* 24:2103–2117.

169. Campbell, H. D., Schimansky, T., Claudianos, C., Ozsarac, N., Kasprzak, A. B., Cotsell, J. N., Young, I. G., de Couet, H. G., and Miklos, G. L. (1993). The *Drosophila melanogaster* flightless-I gene involved in gastrulation and muscle degeneration encodes gelsolin-like and leucine-rich repeat domains and is conserved in *Caenorhabditis elegans* and humans. *Proc Natl Acad Sci USA* 90:11386–11390.

170. Campbell, H. D., Fountain, S., McLennan, I. S., Berven, L. A., Crouch, M. F., Davy, D. A., Hooper, J. A., Waterford, K., Chen, K. S., Lupski, J. R., Ledermann, B., Young, I. G., and Matthaei, K. I. (2002). Fliih, a gelsolin-related cytoskeletal regulator essential for early mammalian embryonic development. *Mol Cell Biol* 22:3518–3526.

171. Yen, A. (1985). Control of HL-60 myeloid differentiation. Evidence of uncoupled growth and differentiation control, S-phase specificity, and two-step regulation. *Exp Cell Res* 156:198–212.

172. Balint, B. L., Szanto, A., Madi, A., Bauer, U. M., Gabor, P., Benko, S., Puskas, L. G., Davies, P. J., and Nagy, L. (2005). Arginine methylation provides epigenetic transcription memory for retinoid-induced differentiation in myeloid cells. *Mol Cell Biol* 25:5648–5663.

173. Cimato, T. R., Ettinger, M. J., Zhou, X., and Aletta, J. M. (1997). Nerve growth factor-specific regulation of protein methylation during neuronal differentiation of PC12 cells. *J Cell Biol* 138:1089–1103.

174. Cimato, T. R., Tang, J., Xu, Y., Guarnaccia, C., Herschman, H. R., Pongor, S., and Aletta, J. M. (2002). Nerve growth factor-mediated increases in protein methylation occur predominantly at type I arginine methylation sites and involve protein arginine methyltransferase 1. *J Neurosci Res* 67:435–442.

175. Birkaya, B., and Aletta, J. M. (2005). NGF promotes copper accumulation required for optimum neurite outgrowth and protein methylation. *J Neurobiol* 63:49–61.

176. Bradbury, A., Possenti, R., Shooter, E. M., and Tirone, F. (1991). Molecular cloning of PC3, a putatively secreted protein whose mRNA is induced by nerve growth factor and depolarization. *Proc Natl Acad Sci USA* 88:3353–3357.

177. Rimokh, R., Rouault, J. P., Wahbi, K., Gadoux, M., Lafage, M., Archimbaud, E., Charrin, C., Gentilhomme, O., Germain, D., Samarut, J., and et al. (1991). A chromosome 12 coding region is juxtaposed to the MYC protooncogene locus in a t(8;12)(q24;q22) translocation in a case of B-cell chronic lymphocytic leukemia. *Genes Chromosomes Cancer* 3:24–36.

178. Berthet, C., Guehenneux, F., Revol, V., Samarut, C., Lukaszewicz, A., Dehay, C., Dumontet, C., Magaud, J. P., and Rouault, J. P. (2002). Interaction of PRMT1 with BTG/TOB proteins in cell signalling: molecular analysis and functional aspects. *Genes Cells* 7:29–39.

179. Bakker, W. J., Blazquez-Domingo, M., Kolbus, A., Besooyen, J., Steinlein, P., Beug, H., Coffer, P. J., Lowenberg, B., von Lindern, M., and van Dijk, T. B. (2004). FoxO3a regulates erythroid differentiation and induces BTG1, an activator of protein arginine methyl transferase 1. *J Cell Biol* 164:175–184.

180. Kass, L., and Zarafonetis, C. J. (1974). Methylated arginines in chronic erythemic myelosis. *Proc Soc Exp Biol Med* 145:944-947.

181. Paik, W. K., Kim, S., Ezirike, J., and Morris, H. P. (1975). S-adenosylmethionine:protein methyltransferases in hepatomas. *Cancer Res* 35:1159–1163.

182. Tran, Y. K., Bogler, O., Gorse, K. M., Wieland, I., Green, M. R., and Newsham, I. F. (1999). A novel member of the NF2/ERM/4.1 superfamily with growth suppressing properties in lung cancer. *Cancer Res* 59:35–43.

183. Gutmann, D. H., Donahoe, J., Perry, A., Lemke, N., Gorse, K., Kittiniyom, K., Rempel, S. A., Gutierrez, J. A., and Newsham, I. F. (2000). Loss of DAL-1, a protein 4.1-related tumor suppressor, is an important early event in the pathogenesis of meningiomas. *Hum Mol Genet* 9:1495–1500.

184. Charboneau, A. L., Singh, V., Yu, T., and Newsham, I. F. (2002). Suppression of growth and increased cellular attachment after expression of DAL-1 in MCF-7 breast cancer cells. *Int J Cancer* 100:181–188.

185. Singh, V., Miranda, T. B., Jiang, W., Frankel, A., Roemer, M. E., Robb, V. A., Gutmann, D. H., Herschman, H. R., Clarke, S., and Newsham, I. F. (2004). DAL-1/4.1B tumor suppressor interacts with protein arginine N-methyltransferase 3 (PRMT3) and inhibits its ability to methylate substrates *in vitro* and *in vivo*. *Oncogene* 23:7761–7771.

186. Jiang, W., Roemer, M. E., and Newsham, I. F. (2005). The tumor suppressor DAL-1/4.1B modulates protein arginine N-methyltransferase 5 activity in a substrate-specific manner. *Biochem Biophys Res Commun* 329:522–530.

187. Jenster, G. (1999). The role of the androgen receptor in the development and progression of prostate cancer. *Semin Oncol* 26:407–421.

188. Hong, H., Kao, C., Jeng, M. H., Eble, J. N., Koch, M. O., Gardner, T. A., Zhang, S., Li, L., Pan, C. X., Hu, Z., MacLennan, G. T., and Cheng, L. (2004). Aberrant expression of CARM1, a transcriptional coactivator of androgen receptor, in the development of prostate carcinoma and androgen-independent status. *Cancer* 101:83–89.

189. Goonasekera, C. D., Rees, D. D., Woolard, P., Frend, A., Shah, V., and Dillon, M. J. (1997). Nitric oxide synthase inhibitors and hypertension in children and adolescents. *J Hypertens* 15:901–909.

190. Boger, R. H., Bode-Boger, S. M., Szuba, A., Tsao, P. S., Chan, J. R., Tangphao, O., Blaschke, T. F., and Cooke, J. P. (1998). Asymmetric dimethylarginine (ADMA): a novel risk factor for endothelial dysfunction: its role in hypercholesterolemia. *Circulation* 98:1842–1847.

191. Miyazaki, H., Matsuoka, H., Cooke, J. P., Usui, M., Ueda, S., Okuda, S., and Imaizumi, T. (1999). Endogenous nitric oxide synthase inhibitor: a novel marker of atherosclerosis. *Circulation* 99:1141–1146.

192. Leiper, J., and Vallance, P. (1999). Biological significance of endogenous methylarginines that inhibit nitric oxide synthases. *Cardiovasc Res* 43:542–548.

193. Boger, R. H., Sydow, K., Borlak, J., Thum, T., Lenzen, H., Schubert, B., Tsikas, D., and Bode-Boger, S. M. (2000). LDL cholesterol upregulates synthesis of asymmetrical dimethylarginine in human endothelial cells: involvement of S-adenosylmethionine-dependent methyltransferases. *Circ Res* 87:99–105.

194. Osanai, T., Saitoh, M., Sasaki, S., Tomita, H., Matsunaga, T., and Okumura, K. (2003). Effect of shear stress on asymmetric dimethylarginine release from vascular endothelial cells. *Hypertension* 42:985–990.

195. Pullamsetti, S., Kiss, L., Ghofrani, H. A., Voswinckel, R., Haredza, P., Klepetko, W., Aigner, C., Fink, L., Muyal, J. P., Weissmann, N., Grimminger, F., Seeger, W., and Schermuly, R. T. (2005). Increased levels and reduced catabolism of asymmetric and symmetric dimethylarginines in pulmonary hypertension. *FASEBJ* 19:1175–1177.

196. Closs, E. I., Basha, F. Z., Habermeier, A., and Forstermann, U. (1997). Interference of L-arginine analogues with L-arginine transport mediated by the y+ carrier hCAT-2B. *Nitric Oxide* 1:65–73.

197. Moscarello, M. A., Wood, D. D., Ackerley, C., and Boulias, C. (1994). Myelin in multiple sclerosis is developmentally immature. *J Clin Invest* 94:146–154.

198. Wood, D. D., Bilbao, J. M., O'Connors, P., and Moscarello, M. A. (1996). Acute multiple sclerosis (Marburg type) is associated with developmentally immature myelin basic protein. *Ann Neurol* 40:18–24.

199. Pritzker, L. B., Joshi, S., Gowan, J. J., Harauz, G., and Moscarello, M. A. (2000). Deimination of myelin basic protein. 1. Effect of deimination of arginyl residues of myelin basic protein on its structure and susceptibility to digestion by cathepsin D. *Biochemistry* 39:5374–5381.

200. Pritzker, L. B., Joshi, S., Harauz, G., and Moscarello, M. A. (2000). Deimination of myelin basic protein. 2. Effect of methylation of MBP on its deimination by peptidylarginine deiminase. *Biochemistry* 39:5382–5388.

201. Mastronardi, F. G., and Moscarello, M. A. (2005). Molecules affecting myelin stability: a novel hypothesis regarding the pathogenesis of multiple sclerosis. *J Neurosci Res* 80:301–308.

202. Kim, J. K., Mastronardi, F. G., Wood, D. D., Lubman, D. M., Zand, R., and Moscarello, M. A. (2003). Multiple sclerosis: an important role for post-translational modifications of myelin basic protein in pathogenesis. *Mol Cell Proteomics* 2:453–462.

203. Lefebvre, S., Burglen, L., Reboullet, S., Clermont, O., Burlet, P., Viollet, L., Benichou, B., Cruaud, C., Millasseau, P., Zeviani, M., and et al. (1995). Identification and characterization of a spinal muscular atrophy-determining gene. *Cell* 80:155–165.

204. Lefebvre, S., Burlet, P., Liu, Q., Bertrandy, S., Clermont, O., Munnich, A., Dreyfuss, G., and Melki, J. (1997). Correlation between severity and SMN protein level in spinal muscular atrophy. *Nat Genet* 16:265–269.

205. Buhler, D., Raker, V., Luhrmann, R., and Fischer, U. (1999). Essential role for the tudor domain of SMN in spliceosomal U snRNP assembly: implications for spinal muscular atrophy. *Hum Mol Genet* 8:2351–2357.

206. Cusco, I., Barcelo, M. J., del Rio, E., Baiget, M., and Tizzano, E. F. (2004). Detection of novel mutations in the SMN Tudor domain in type I SMA patients. *Neurology* 63:146–149.

207. Selenko, P., Sprangers, R., Stier, G., Buhler, D., Fischer, U., and Sattler, M. (2001). SMN tudor domain structure and its interaction with the Sm proteins. *Nat Struct Biol* 8:27-31.

208. Wang, J., and Dreyfuss, G. (2001). Characterization of functional domains of the SMN protein *in vivo*. *J Biol Chem* 276:45387–45393.

209. Mears, W. E., and Rice, S. A. (1996). The RGG box motif of the herpes simplex virus ICP27 protein mediates an RNA-binding activity and determines in vivo methylation. *J Virol* 70:7445–7453.

210. Kzhyshkowska, J., Kremmer, E., Hofmann, M., Wolf, H., and Dobner, T. (2004). Protein arginine methylation during lytic adenovirus infection. *Biochem J* 383:259–265.

211. Boulanger, M. C., Liang, C., Russell, R. S., Lin, R., Bedford, M. T., Wainberg, M. A., and Richard, S. (2005). Methylation of Tat by PRMT6 regulates human immunodeficiency virus type 1 gene expression. *J Virol* 79:124–131.

212. Li, Y. J., Stallcup, M. R., and Lai, M. M. (2004). Hepatitis delta virus antigen is methylated at arginine residues, and methylation regulates subcellular localization and RNA replication. *J Virol* 78:13325–13334.

213. Barth, S., Liss, M., Voss, M. D., Dobner, T., Fischer, U., Meister, G., and Grasser, F. A. (2003). Epstein-Barr virus nuclear antigen 2 binds via its methylated arginine-glycine repeat to the survival motor neuron protein. *J Virol* 77:5008–5013.

214. Holowaty, M. N., Zeghouf, M., Wu, H., Tellam, J., Athanasopoulos, V., Greenblatt, J., and Frappier, L. (2003). Protein profiling with Epstein-Barr nuclear antigen-1 reveals an interaction with the herpesvirus-associated ubiquitin-specific protease HAUSP/USP7. *J Biol Chem* 278:29987–29994.

215. Tong, X., Yalamanchili, R., Harada, S., and Kieff, E. (1994). The EBNA-2 arginine-glycine domain is critical but not essential for B-lymphocyte growth transformation; the rest of region 3 lacks essential interactive domains. *J Virol* 68:6188–6197.

216. Heim, M. H., Moradpour, D., and Blum, H. E. (1999). Expression of hepatitis C virus proteins inhibits signal transduction through the Jak-STAT pathway. *J Virol* 73:8469–8475.

217. Duong, F. H., Filipowicz, M., Tripodi, M., La Monica, N., and Heim, M. H. (2004). Hepatitis C virus inhibits interferon signaling through up-regulation of protein phosphatase 2A. *Gastroenterology* 126:263–277.

218. Lin, W., Choe, W. H., Hiasa, Y., Kamegaya, Y., Blackard, J. T., Schmidt, E. V., and Chung, R. T. (2005). Hepatitis C virus expression suppresses interferon signaling by degrading STAT1. *Gastroenterology* 128:1034–1041.
219. Cheng, D., Yadav, N., King, R. W., Swanson, M. S., Weinstein, E. J., and Bedford, M. T. (2004). Small molecule regulators of protein arginine methyltransferases. *J Biol Chem* 279:23892–23899.
220. Rappsilber, J., Friesen, W. J., Paushkin, S., Dreyfuss, G., and Mann, M. (2003). Detection of arginine dimethylated peptides by parallel precursor ion scanning mass spectrometry in positive ion mode. *Anal Chem* 75:3107–3114.
221. Gehrig, P. M., Hunziker, P. E., Zahariev, S., and Pongor, S. (2004). Fragmentation pathways of N(G)-methylated and unmodified arginine residues in peptides studied by ESI-MS/MS and MALDI-MS. *J Am Soc Mass Spectrom* 15:142–149.
222. Brame, C. J., Moran, M. F., and McBroom-Cerajewski, L. D. (2004). A mass spectrometry based method for distinguishing between symmetrically and asymmetrically dimethylated arginine residues. *Rapid Commun Mass Spectrom* 18:877–881.
223. Bonaldi, T., Imhof, A., and Regula, J. T. (2004). A combination of different mass spectroscopic techniques for the analysis of dynamic changes of histone modifications. *Proteomics* 4:1382–1396.
224. Lee, J., and Bedford, M. T. (2002). PABP1 identified as an arginine methyltransferase substrate using high-density protein arrays. *EMBO Rep* 3:268–273.
225. Chen, Y. F., Zhang, A. Y., Zou, A. P., Campbell, W. B., and Li, P. L. (2004). Protein methylation activates reconstituted ryanodine receptor-ca release channels from coronary artery myocytes. *J Vasc Res* 41:229–240.
226. Meister, G., Eggert, C., and Fischer, U. (2002). SMN-mediated assembly of RNPs: a complex story. *Trends Cell Biol* 12:472–478.
227. Yong, J., Wan, L., and Dreyfuss, G. (2004). Why do cells need an assembly machine for RNA-protein complexes? *Trends Cell Biol* 14:226–232.
228. Lee, J., Cheng, D., and Bedford, M. T. (2004). Techniques in protein methylation. *Methods Mol Biol* 284:195-208.
229. Wada, K., Inoue, K., and Hagiwara, M. (2002). Identification of methylated proteins by protein arginine N-methyltransferase 1, PRMT1, with a new expression cloning strategy. *Biochim Biophys Acta* 1591:1–10.
230. Sgarra, R., Diana, F., Bellarosa, C., Dekleva, V., Rustighi, A., Toller, M., Manfioletti, G., and Giancotti, V. (2003). During apoptosis of tumor cells HMGA1a protein undergoes methylation: identification of the modification site by mass spectrometry. *Biochemistry* 42:3575–3585.
231. Zou, Y., and Wang, Y. (2005). Tandem mass spectrometry for the examination of the posttranslational modifications of high-mobility group A1 proteins: symmetric and asymmetric dimethylation of Arg25 in HMGA1a protein. *Biochemistry* 44:6293–6301.
232. Lim, I. K., Park, T. J., Kim, S., Lee, H. W., and Paik, W. K. (1998). Enzymatic methylation of recombinant TIS21 protein-arginine residues. *Biochem Mol Biol Int* 45:871–878.
233. Blanchet, F., Cardona, A., Letimier, F. A., Herschfield, M. S., and Acuto, O. (2005). CD28 costimulating signal induces protein arginine methylation in T cells. *J Exp Med* 202:371–377.
234. Kim, S., Merrill, B. M., Rajpurohit, R., Kumar, A., Stone, K. L., Papov, V. V., Schneiders, J. M., Szer, W., Wilson, S. H., Paik, W. K., and Williams, K. R. (1997). Identification of N(G)-methylarginine residues in human heterogeneous RNP protein A1: Phe/Gly-Gly-Gly-Arg-Gly-Gly-Gly/Phe is a preferred recognition motif. *Biochemistry* 36:5185–5192.
235. Scott, H. S., Antonarakis, S. E., Lalioti, M. D., Rossier, C., Silver, P. A., and Henry, M. F. (1998). Identification and characterization of two putative human arginine methyltransferases (HRMT1L1 and HRMT1L2). *Genomics* 48:330–340.

236. Frankel, A., Yadav, N., Lee, J., Branscombe, T. L., Clarke, S., and Bedford, M. T. (2002). The novel human protein arginine N-methyltransferase PRMT6 is a nuclear enzyme displaying unique substrate specificity. *J Biol Chem* 277:3537–3543.

237. Siebel, C. W., and Guthrie, C. (1996). The essential yeast RNA binding protein Np13p is methylated. *Proc Natl Acad Sci USA* 93:13641–13646.

238. Pype, S., Slegers, H., Moens, L., Merlevede, W., and Goris, J. (1994). Tyrosine phosphorylation of a M(r) 38,000 A/B-type hnRNP protein selectively modulates its RNA binding. *J Biol Chem* 269:31457–31465.

239. Lapeyre, B., Amalric, F., Ghaffari, S. H., Rao, S. V., Dumbar, T. S., and Olson, M. O. (1986). Protein and cDNA sequence of a glycine-rich, dimethylarginine-containing region located near the carboxyl-terminal end of nucleolin (C23 and 100 kDa). *J Biol Chem* 261:9167–9173.

240. Lin, C. H., Huang, H. M., Hsieh, M., Pollard, K. M., and Li, C. (2002). Arginine methylation of recombinant murine fibrillarin by protein arginine methyltransferase. *J Protein Chem* 21:447–453.

241. Belyanskaya, L. L., Gehrig, P. M., and Gehring, H. (2001). Exposure on cell surface and extensive arginine methylation of ewing sarcoma (EWS) protein. *J Biol Chem* 276:18681–18687.

242. Pelletier, M., Xu, Y., Wang, X., Zahariev, S., Pongor, S., Aletta, J. M., and Read, L. K. (2001). Arginine methylation of a mitochondrial guide RNA binding protein from *Trypanosoma brucei*. *Mol Biochem Parasitol* 118:49–59.

243. Reporter, M., and Corbin, J. L. (1971). N G,N G,-dimethylarginine in myosin during muscle development. *Biochem Biophys Res Commun* 43:644–650.

4

Structure of Protein Arginine Methyltransferases[1]

XING ZHANG • XIAODONG CHENG

Department of Biochemistry
Emory University School of Medicine
1510 Clifton Road
Atlanta, GA 30322, USA

I. Abstract

With genome sequencing nearing completion for the model organisms used in biomedical research, there is a rapidly growing appreciation that proteomics (including the study of covalent modification to proteins) and transcriptional regulation will likely dominate the research headlines in the next decade. Protein methylation plays a central role in both of these fields, as several different residues (Arg, Lys, Gln) are methylated in cells and methylation plays a central role in regulating chromatin structure and impacts transcription. In some cases, a single arginine can be mono-, symmetrically di-, or asymmetrically di-methylated, with different

---

[1]This chapter has been modified and updated from the article "Structural and Sequence Motifs of Protein (Histone) Methylation Enzymes" (originally published in *Annu. Rev. Biophys. Biomol. Struct.* 2005, vol. 34, pp. 267–294). It is reproduced by permission of the *Annual Review of Biophysics and Biomolecular Structure*, copyright 2005 by Annual Reviews (*www.annualreviews.org*).

functional consequences for each of the three forms. This review summarizes the progress that has been made in structural studies of protein arginine methyltransferases and their related sequence conservations; it also discusses, somewhat speculatively, their mechanisms.

II. Protein Arginine Methylation

Protein N(itrogen)-methylation occurs on residues of arginine, lysine, glutamine, asparagines, histidine, and the amino group at the N-terminus (Figure 4.1). Protein arginine methylation is a common posttranslational modification in eukaryotes. There are many recent reviews [1] in this fast-moving field, as outlined in the

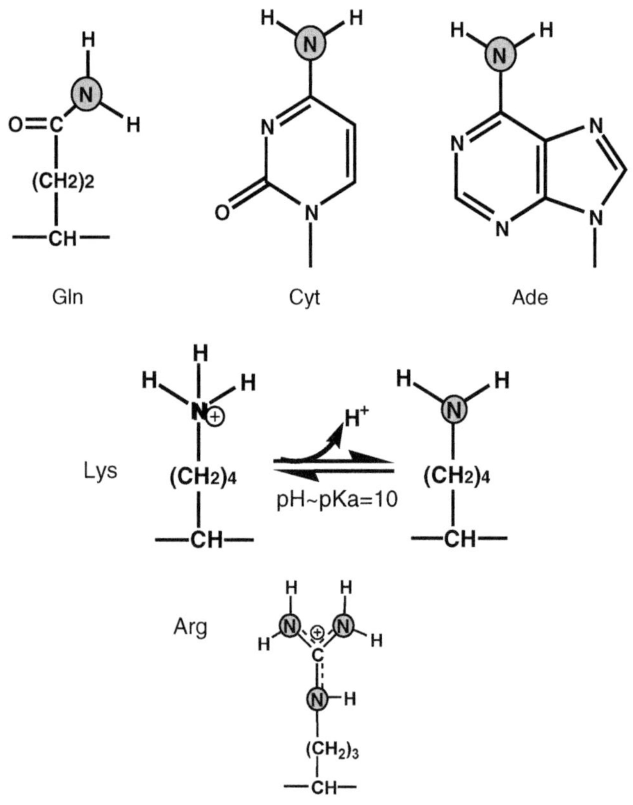

FIG. 4.1. Examples of known targets of amino methylation. Only the deprotonated amino group (NH$_2$) has a free lone pair of electrons capable of nucleophilic attack on the AdoMet methyl group.

other chapters of this volume. Two major types of protein arginine (R) methyl-transferases (PRMTs) transfer the methyl group from AdoMet to the guanidino group of arginines in protein substrates [2]. Both catalyze the formation of monomethylarginine, but type I PRMTs also form asymmetric dimethylarginine and type II PRMTs form symmetric dimethylarginine [3] (Figure 4.2). Among the known PRMTs, only PRMT5/JBP1 is a type II PRMT [3], which symmetri-cally dimethylates specific arginines in a few proteins. Fewer than 20 proteins have been identified in the last 40 years as containing dimethylated arginine(s) [4], including myelin basic protein [5], spliceosomal Sm proteins [6], and his-tones H3 and H4 [7]. PRMT5 was initially identified as a Jak kinase-binding protein (JBP1) [8, 9] and has been found to coexist with substrates in multipro-tein complexes [6, 10–13].

Multiple PRMT genes are present in eukaryotes from fungi to plants and animals (Figure 4.3 and Table 4.1). For example, nine very similar paralogous mammalian PRMT genes have been reported so far: PRMT1 [14–16], PRMT1′ [17] (also called PRMT8 in [1]), PRMT2 [18], PRMT3 [19], CARM1/PRMT4 [20], JBP1/PRMT5 [8, 21], PRMT6 [22], PRMT7 [23, 24], and PRMT8 [25].

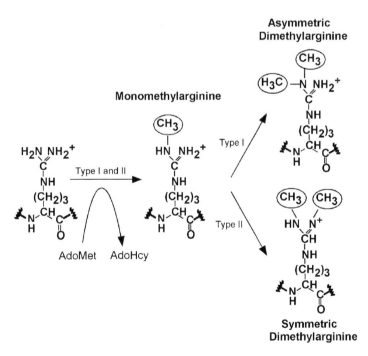

FIG. 4.2. Two major types of protein arginine methylation.

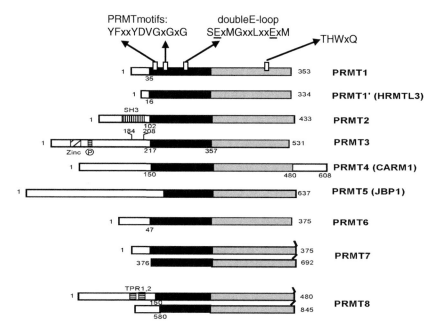

FIG. 4.3. Members of PRMT family. The conserved MTase domain is in black and the unique β-barrel domain to the PRMT family is in gray. The N- and C-termini of the proteins and the first invariant residue are labeled.

Nine PRMTs are present in the completed *Drosophila melanogaster* genome [26], and complete PRMT genes from *S. pombe*, *Arabdopsis*, and *C. elegans* have also been identified through genome sequencing projects. In addition, ESTs with strong homology to PRMT1 can also be found in *Xenopus*, zebrafish, sea urchin, rice, and tomato, indicating that PRMT is a highly conserved family of proteins in eukaryotes. The presence of multiple PRMTs may signify the diverse roles they can play.

Two well-studied enzymes, PRMT1 and PRMT4/CARM1, methylate histones H3 [20, 27–29], H4 [30, 31], and H2B [32] in addition to many other substrates. Histone arginine methylation is a component of the "histone code" that directs a variety of processes involving chromatin [33, 34]. For example, methylation of Arg-3 of histone H4 by PRMT1 facilitates H4 acetylation and enhances transcriptional activation by nuclear hormone receptors synergistically with CARM1 [30, 31, 35, 36], in that CARM1 prefers acetylated histone tails in generating H3 Arg-17 methylation [31, 37]. *In vitro*, p53-mediated transcription was stimulated the greatest when all three coactivators (PRMT1, CARM1, and p300) were present, whether added sequentially or at the same time [32]. Preincubation of a chromatin template with p53 and PRMT1 significantly stimulated the histone

TABLE 4.1
MEMBERS OF PRMT FAMILY (AS OF JULY 2005)

	Human PRMT Genes							Presence of PRMT Homologous in Other Organisms				
Enzyme	Activity	Chromosome	EST	Coding Exon	Genomic Size (kb)	Protein Accession Number	Protein Size (Residues)	Arabidopsis	Drosophila	C. elegans	S. pombe	S. cerevisiae
PRMT1	+++	19q13	+++	9-10	10	CAA71764	361	CAB79709	AAF54556	CAB54335	CAB63498	P38074
PRMT1′ (HRMTL3)	+	12p13	+	9	52	AAF91390	334	AAC62148	-	-	-	-
PRMT2	-	21q22	++	10	30	P55345	433	-	-	-	-	-
PRMT3	+	11p15	+	13	50	AAH64831	531	AAG51062	AAF55147	-	CAA17825	-
PRMT4 (CARM1)	+	19p13	++	16	50	NP_954592	608	NP_199713	AAF54471	-	-	-
PRMT5 (JBP1)	+ (type II)	14q11	++	17	8.5	AAF04502	637	NP_194841	AAF04504	AAK95874	P78963	P38274
PRMT6	+	1p13	+/-	1	2.5	AAK85733	375	BAB01859	-	-	-	-
PRMT7	+	16q22	++	17	41	Q9NVM4	692	NP_567508	AAM29595	CAA22252 (*)	-	-
PRMT8	?	4q31	+	10	40	AAH64403	845	-	-	CAB07676 (?)	-	-
Additional members								AAF40450	AAF51032 (DART2) AAF55002 (DART6) NP_609478 (DART8) NP_650321 (DART9)			

*The entry CAA22252 misses 17 conserved residues. W06D4.gc4 of the WormBase contains the complete gene.

** ? indicates *unknown* or *uncertain*.

acetyltransferase activity of p300, and similarly preincubation of the template with p53 and p300 stimulated H3 arginine methylation by CARM1.

III.   PRMT1

PRMT1 is the predominant type I PRMT in mammalian cells, accounting for 85% of cellular PRMT activity [38]. It is essential for early post-implantation development, as shown by the embryonic lethality of mouse Prmt1$^{-/-}$ mutants [39]. Although PRMT1 is expressed at detectable levels in all tissues examined [14, 18, 19, 39], the expression is highest in developing neural structures in embryos [39], and PRMT1 has been implicated in neuronal differentiation [40]. PRMT1 has at least six alternatively spliced transcripts that would produce proteins with an N-terminus of 20 to 40 amino acids [17, 18], and these proteins may have different substrate specificities [41].

PRMT1 gene is found in all eukaryotes examined and is highly conserved (Table 4.1). The sequence identity is over 90% among mammals, zebrafish, and *Xenopus*, and about 50% even between human and *S. cerevisiae*. There appears to be another gene (PRMT1′) closely related to PRMT1 genes both in *A. thaliana* and in humans (HRMT1L3, AAF91390, on chromosome 12p13), which in each case share 80% amino acid identity with PRMT1. Except for their N-termini, the two genes have identical genomic structure: each pair has eight introns inserted at identical positions, and the locations of seven of those introns are also shared between humans and *A. thaliana*. No functionality regarding this PRMT1-like gene (PRMT1′) has been reported. Like *S. cerevisiae*, *C. elegans* and *S. pombe* have only one copy of PRMT1 and share some of the splicing sites used by humans and *A. thaliana*. *D. melanogaster* encodes four to six PRMT proteins of similar size, but of these only DmPRMT1 (AAF54556) has a high-percentage identity with the mammalian PRMT1 (65% versus 15 to 35% for the others).

The best-known substrates for PRMT1 are RNA-binding proteins involved in various aspects of RNA processing and/or transport such as hnRNPs, fibrillarin, nucleolin [42], and poly(A)-binding protein II [43]. A growing number of other proteins were found to be substrates of PRMT1, including high-molecular-weight fibroblast growth factor 2 (HMW FGF-2), a nuclear growth factor [44]; interleukin enhancer-binding factor 3 (ILF3) [38]; SPT5, a regulator of transcriptional elongation [45]; and histones H4 [30, 31] and H2B [32]. Arginine methylation at residue 31 of STAT1, a transcription factor activated by extracellular signals, was proposed to enhance its DNA binding by reducing association with the specific inhibitor PIAS1, thus intensifying the growth-restraining activities of the interferons [46]. This observation was reassessed by Meissner et al. [47], who provided evidence that contradicts previous results that stated methylation of arginine 31. Thus, alternative explanations to STAT1 methylation (perhaps at a different residue) need to be explored in order to understand the molecular mechanisms that underlie the reduced interferon sensitivity of many tumor cells.

IV. CARM1/PRMT4

PRMT4 was discovered as a transcriptional coactivator-associated arginine (R) MTase (CARM1) [20]. CARM1 enhances gene activation by nuclear receptors in a synergistic collaboration with two other classes of coactivators: the p160 coactivators and the protein acetyltransferases p300/CBP [20, 36]. CARM1 can methylate specific arginine residues in the N-terminal tail of histone H3 [20, 27–29].

Both CARM1 and PRMT1 act in concert with the acetyltransferase CBP/ p300, along with the p160 coactivator family, to enhance transcription from hormone-responsive promoters [48, 49]. Similarly, CARM1 and PRMT1 act as coactivators in the tumor suppressor protein p53-mediated transcription, via direct interactions with p53 and its associated coactivator partner p300 [32]. These results provide compelling evidence that the histones are relevant targets for CARM1, PRMT1, and p300, and that the resulting histone modifications are directly important for transcription.

V. A Conserved PRMT Core

The PRMT proteins vary in length from 348 amino acids in *S. cerevisiae* RMT to 608 in CARM1, but they all contain a conserved core region of approximately 310 amino acids (Figure 4.3). The sequences beyond the conserved PRMT core region are all N-terminal additions. However, CARM1 also has a C-terminal addition. The size of the N-terminal addition varies from ~20 amino acids in *S. cerevisiae* RMT1 to 200 in PRMT3. The varied N-termini could subject each PRMT to a different regulation. For example, PRMT2 includes an SH3 domain that mediates protein-protein interaction in the Src family signaling pathways.

However, the recombinant PRMT2 with or without the N-terminal SH3 domain is inactive using either myelin basic protein or purified proteins containing GAR domain as substrate [18]. PRMT3 contains a C2H2 zinc finger, whose structure is recently resolved by NMR (PDB code 1WIR) (Figure 4.4), and a tyrosine phosphorylation site, suggesting a link to other signaling pathways. The 40S ribosomal protein S2 is the only physiological substrate of PRMT3 identified so far [50]. *In vitro* studies showed that the zinc-finger domain of PRMT3 is necessary and sufficient for binding to substrate ribosomal protein S2 [51]. On the other hand, DAL-1/4.1B tumor suppressor, which is not a substrate for PRMT3, interacts with PRMT3 via its C-terminal catalytic core domain and inhibits its ability to methylate substrates [52].

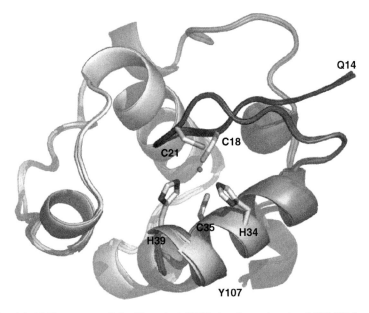

Fig. 4.4. NMR structure of the N-terminal C2H2 zinc-finger domain of PRMT3 from mus musculus (PDB code 1WIR). Residues 14 (N-terminus) to 107 (C-terminus) are shown. The zinc atom is shown as a ball, and its ligands as sticks. (See color plate.)

VI. Structure of the Conserved PRMT Core

Three crystal structures of PRMTs are currently available: rat PRMT1 (amino acids 41 through 353) [17], rat PRMT3 catalytic core (amino acids 208 through 528) [53], and yeast RMT1/Hmt1 (amino acids 30 through 348) [54]. These structures reflect a striking structural conservation of the PRMT catalytic core (Figure 4.5). The overall monomeric structure of the PRMT core can be divided into three parts: an MTase domain, a β barrel, and a dimerization arm. The MTase domain has the consensus fold conserved in Class-I AdoMet-dependent MTases that harbor an AdoMet-binding site [55, 56]. The β-barrel domain is unique to the PRMT family [53].

VII. PRMT Dimerization is Essential for AdoMet Binding and Enzymatic Activity

An identical hydrophobic dimer interface is observed in PRMT1 [17], PRMT3 core [53], and yeast RMT1/Hmt1 [54] (Figure 4.5), despite very different crystallization conditions, space groups, and cell dimensions. This observation supports the notion that dimer formation is a conserved feature in the PRMT family [53]. A mutant of yeast RMT1/Hmt1 that replaces the dimerization arm

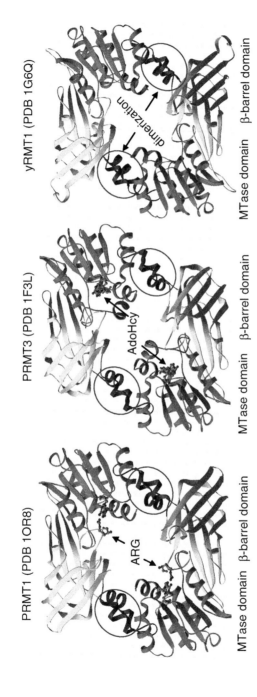

FIG. 4.5. Dimer structures of PRMT cores: (*left-hand panel*) rat PRMT1, (*middle panel*) rat PRMT3, and (*right-hand panel*) yeast RMT1/Hmt1. (See color plate.)

with alanines results in the loss of dimer formation and methylation activity [54]. The mutant PRMT1 ΔARM that lacks the entire dimerization arm (residues 188 through 222) elutes as a monomer on a gel filtration column, and completely lacks enzymatic activity—most likely because it is unable to bind the AdoMet cofactor, as determined by UV cross-linking experiments [17]. In the crystal structure, the dimer interface forms between the arm and the outer surface of the AdoMet binding site (Figure 4.5). It is conceivable that dimerization is required to engage the residues in the AdoMet-binding site in a manner in which they can interact with AdoMet properly. Interestingly, the higher-order oligomerization of PRMT1 [14, 18, 19, 39] does not occur in the absence of dimerization (i.e., in the case of ΔARM) [17].

Another potential function of the conserved PRMT dimer might be to allow processive production of the final methylation product, asymmemetric dimethylarginine. PRMT substrates isolated *in vivo* are usually completely or nearly completely dimethylated [43, 44, 57–59]. *In vitro*, PRMT6 forms dimethylarginine in a processive manner [22]. It is conceivable that a ring-like dimer could allow the product of the first methylation reaction, monomethylarginine, to enter the active site of the second molecule of the dimer without releasing the substrate from the ring or replenishing the methyl donor. An interesting feature of PRMT7 [23], and PRMT8 (Figure 4.3 and Table 4.1) is that they seem to have arisen from a gene duplication event and contain two conserved core regions, each with a putative AdoMet-binding motif, although the C-terminal copy is much more divergent in both cases. The N-terminal and C-terminal halves of PRMT7 and PRMT8 might mimic a homodimer.

VIII. Multiple Substrate Binding Grooves

Most PRMT1 substrates contain glycine- and arginine-rich sequences that include multiple arginines in RGG context [42–44]. Peptides that contain three copies of the consensus RGG repeat sequence (R3) were co-crystallized with PRMT1 [17]. Three peptide binding grooves were identified (Figure 4.6), which probably represent a mixture of binding modes of the R3 peptide, which contains three potential methylation targets (at positions 3, 9, and 15). Additional acidic grooves running parallel to site P3 were identified (Figure 4.6). These grooves could form additional binding sites for protein substrates with more RGG repeats.

IX. Asymmetric and Symmetric Dimethylarginines

The target arginine is situated in a deep acidic pocket between the MTase domain and the β-barrel domain (Figure 4.6). The residues that make up the active site are conserved across the PRMT family, and a "double-E" hairpin loop (Figure 4.3) contributes most of the residues in the active site pocket. Two invariant

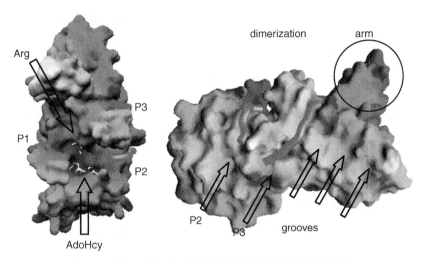

R3 peptide (19): GGRGGFGGRGGFGGRGGFG

FIG. 4.6. Peptide binding grooves (P1, P2, and P3) in the structure of ternary complex of PRMT1-AdoHcy-R3 peptide (sequence shown at the bottom): solvent-accessible molecular surface with bound AdoHcy and Arg shown as stick models and indicated by the arrows (*left-hand panel*). If the central Arg9 were the target bound in the active site, connecting peptide-binding sites P1 and P2 would cover the active site and the entire length of the peptide. When the end arginine (either Arg3 or Arg15) is bound in the active site, connection of peptide-binding sites P2 and P3 would account for the length of the entire peptide (*right-hand panel*). Site P3 corresponds to one of the grooves perpendicular to the strands of the β-barrel domain. (See color plate.)

glutamates (E144 and E153 of PRMT1, and E326 and E335 of PRMT3) are used to neutralize the positive charge on the substrate guanidino group (Figure 4.7). The interaction with E153 of PRMT1 (or E335 of PRMT3) redistributes the positive charge on the guanidino group toward one amino group, while leaving a lone pair of electrons on the other amino group to attack the cationic methylsulfonium moiety of AdoMet [53]. The corresponding mutant in CARM1/PRMT4 (E267Q) has been used to demonstrate that the MTase activity of CARM1 was required for synergy among nuclear receptor coactivators [36].

The three solved PRMT structures (rat PRMT1 and PRMT3 core, and yeast RMT1) are all type I enzymes. Interestingly, all of the PRMTs except mammalian PRMT5 (or yeast Hsl7) contain an active site methionine. This is the last residue of the double-E loop (amino acid 155 for PRMT1, amino acid 337 for PRMT3, and amino acid 143 for yeast RMT1; see Figure 4.7), which has been proposed to exclude binding of monomethylated Arg in a conformation that would allow its symmetric methylation [53]. However, in both PRMT5 (amino acid 446) and Hsl7 (amino acid 474) the residue corresponding to M155 of PRMT1 is serine. The smaller bulk of the side chain of this residue may allow for symmetric di-methyl arginine formation by the type II enzymes PRMT5 and possibly Hsl7 [3].

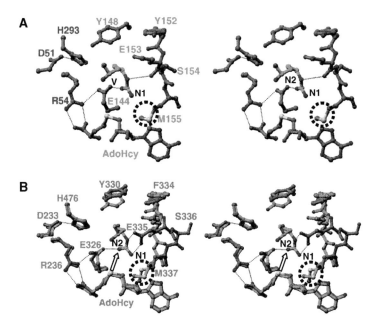

FIG. 4.7. Active sites of PRMT1 and PRMT3 (A) PRMT1 active site with bound Arg in stereo. (B) Superimposition of PRMT1 and PRMT3 active sites in stereo. Only the PRMT3 residues are labeled. The arrow indicates transfer of the methyl group (attached to the sulfur atom of AdoHcy) to the N2 nitrogen atom of bound Arg. The methionine (circled by dashed line) would exclude a mono-methylated amino group from occupying N1 site. (See color plate.)

X.   Antagonize Arginine Methylation

Despite recent advances in identifying MTases, we still know very little about what regulates their activities or determines their specificity. Methylation can function as a reversible signal, as in the case of O(xygen)-methylation, in which the side-chain carboxyl groups of glutamate residues or the C-terminal carboxyl groups are reversibly methylated. However, it is unclear until recently whether the N(itrogen)-methylations (of arginine, lysine, glutamine, asparagine, histidine residues, and the amino group at the N-terminus) are reversible in the cell or the N-methylation function is a more permanent modification that affects the activity or surface hydrophobicity of a substrate. A human nuclear amine oxidase, LSD1, functions as a histone H3 Lys4 di/momomethyl-lysine demethylase via an oxidation reaction [60, 61]. Interestingly, it was proposed recently that Epe1, a hydroxylase-like protein, might be able to demethylate mono-, di-, or trimethylated lysines [62]. Indeed the human and S. cerevisiae homologs of Epe have been confirmed to be able to demethylate histone H3 methyl-K36 [63].

Regarding the protein arginine demethylation, a human nuclear peptidyl arginine deiminase (PAD4) has been shown to antagonize methylation on the arginine residues by converting arginine to citrulline [64, 65]. One crucial question on the role of PAD4 in "histone code" is whether it can act on methylated arginine. *In vitro* and *in vivo* biochemistry data showed that di-methylated arginine is not a substrate for PAD4. Although antibody data suggested that mono-methyl-arginine can be substrate for PAD4, this has not been clearly demonstrated biochemically, perhaps due to the relatively low reactivity of methyl arginine versus arginine. Further study on protein arginine demethylation is warranted. It is worth noting that although specific modification and demodification of histones contribute to gene expression in important ways, recent study on histone H4 acetylation suggests a simple, rather than a combinational histone code for gene expression is in doubt [66, 67].

ACKNOWLEDGMENTS

Work in our laboratories was supported in part by a grant from the National Institute of Health (GM61355). The authors are currently supported by a grant from NIH (GM068680), and Xiaodong Cheng is a Georgia Research Alliance Eminent Scholar.

REFERENCES

1. Bedford, M.T., and Richard, S. (2005). Arginine methylation an emerging regulator of protein function. *Mol Cell* 18:263–272.
2. Lee, H.W., Kim, S., and Paik, W.K. (1977). S-adenosylmethionine: protein-arginine methyltransferase. Purification and mechanism of the enzyme. *Biochemistry* 16:78–85.
3. Branscombe, T.L., Frankel, A., Lee, J.H., Cook, J.R., Yang, Z., Pestka, S., and Clarke, S. (2001). PRMT5 (Janus kinase-binding protein 1) catalyzes the formation of symmetric dimethylarginine residues in proteins. *J Biol Chem* 276:32971–32976.
4. Boisvert, F.M., Cote, J., Boulanger, M.C., and Richard, S. (2003). A proteomic analysis of arginine-methylated protein complexes. *Mol Cell Proteomics* 2:1319–1330.
5. Kim, S., Lim, I.K., Park, G.H., and Paik, W.K. (1997). Biological methylation of myelin basic protein: enzymology and biological significance. *Int J Biochem Cell Biol* 29:743–751.
6. Friesen, W.J., Paushkin, S., Wyce, A., Massenet, S., Pesiridis, G.S., Van Duyne, G., Rappsilber, J., Mann, M., and Dreyfuss, G. (2001). The methylosome, a 20S complex containing JBP1 and pICln, produces dimethylarginine-modified Sm proteins. *Mol Cell Biol* 21:8289–8300.
7. Pal, S., Vishwanath, S.N., Erdjument-Bromage, H., Tempst, P., and Sif, S. (2004). Human SWI/SNF-associated PRMT5 methylates histone H3 arginine 8 and negatively regulates expression of ST7 and NM23 tumor suppressor genes. *Mol Cell Biol* 24:9630–9645.
8. Pollack, B.P., Kotenko, S.V., He, W., Izotova, L.S., Barnoski, B.L., and Pestka, S. (1999). The human homologue of the yeast proteins Skb1 and Hsl7p interacts with Jak kinases and contains protein methyltransferase activity. *J Biol Chem* 274:31531–31542.
9. Rho, J., Choi, S., Seong, Y.R., Cho, W.K., Kim, S.H., and Im, D.S. (2001). Prmt5, which forms distinct homo-oligomers, is a member of the protein-arginine methyltransferase family. *J Biol Chem* 276:11393–11401.

10. Yanagida, M., Hayano, T., Yamauchi, Y., Shinkawa, T., Natsume, T., Isobe, T., and Takahashi, N. (2004). Human fibrillarin forms a sub-complex with splicing factor 2-associated p32, protein arginine methyltransferases, and tubulins alpha 3 and beta 1 that is independent of its association with preribosomal ribonucleoprotein complexes. *J Biol Chem* 279:1607–1614.

11. Pal, S., Yun, R., Datta, A., Lacomis, L., Erdjument-Bromage, H., Kumar, J., Tempst, P., and Sif, S. (2003). mSin3A/histone deacetylase 2- and PRMT5-containing Brg1 complex is involved in transcriptional repression of the Myc target gene cad. *Mol Cell Biol* 23:7475–7487.

12. Meister, G., and Fischer, U. (2002). Assisted RNP assembly: SMN and PRMT5 complexes cooperate in the formation of spliceosomal UsnRNPs. *Embo J* 21:5853–5863.

13. Meister, G., Eggert, C., Buhler, D., Brahms, H., Kambach, C., and Fischer, U. (2001). Methylation of Sm proteins by a complex containing PRMT5 and the putative U snRNP assembly factor pICln. *Curr Biol* 11:1990–1994.

14. Lin, W.J., Gary, J.D., Yang, M.C., Clarke, S., and Herschman, H.R. (1996). The mammalian immediate-early TIS21 protein and the leukemia-associated BTG1 protein interact with a protein-arginine N-methyltransferase. *J Biol Chem* 271:15034–15044.

15. Abramovich, C., Yakobson, B., Chebath, J., and Revel, M. (1997). A protein-arginine methyltransferase binds to the intracytoplasmic domain of the IFNAR1 chain in the type I interferon receptor. *Embo J* 16:260–266.

16. Katsanis, N., Yaspo, M.L., and Fisher, E.M. (1997). Identification and mapping of a novel human gene, HRMT1L1, homologous to the rat protein arginine N-methyltransferase 1 (PRMT1) gene. *Mamm Genome* 8:526–529.

17. Zhang, X., and Cheng, X. (2003). Structure of the predominant protein arginine methyltransferase PRMT1 and analysis of its binding to substrate peptides. *Structure* 11:509–520.

18. Scott, H.S., Antonarakis, S.E., Lalioti, M.D., Rossier, C., Silver, P.A., and Henry, M.F. (1998). Identification and characterization of two putative human arginine methyltransferases (HRMT1L1 and HRMT1L2). *Genomics* 48:330–340.

19. Tang, J., Gary, J.D., Clarke, S., and Herschman, H.R. (1998). PRMT 3, a type I protein arginine N-methyltransferase that differs from PRMT1 in its oligomerization, subcellular localization, substrate specificity, and regulation. *J Biol Chem* 273:16935–16945.

20. Chen, D., Ma, H., Hong, H., Koh, S.S., Huang, S.M., Schurter, B.T., Aswad, D.W., and Stallcup, M.R. (1999). Regulation of transcription by a protein methyltransferase. *Science* 284:2174–2177.

21. Lee, J.H., Cook, J.R., Pollack, B.P., Kinzy, T.G., Norris, D., and Pestka, S. (2000). Hsl7p, the yeast homologue of human JBP1, is a protein methyltransferase. *Biochem Biophys Res Commun* 274:105–111.

22. Frankel, A., Yadav, N., Lee, J., Branscombe, T.L., Clarke, S., and Bedford, M.T. (2002). The novel human protein arginine N-methyltransferase PRMT6 is a nuclear enzyme displaying unique substrate specificity. *J Biol Chem* 277:3537–3543.

23. Miranda, T.B., Miranda, M., Frankel, A., and Clarke, S. (2004). PRMT7 is a member of the protein arginine methyltransferase family with a distinct substrate specificity. *J Biol Chem* 279:22902–22907.

24. Lee, J.H., Cook, J.R., Yang, Z.H., Mirochnitchenko, O., Gunderson, S.I., Felix, A.M., Herth, N., Hoffmann, R., and Pestka, S. (2005). PRMT7, a new protein arginine methyltransferase that synthesizes symmetric dimethylarginine. *J Biol Chem* 280:3656–3664.

25. Cheng, X., Collins, R.E., and Zhang, X. (2005). Structural and sequence motifs of protein (histone) methylation enzymes. *Annu Rev Biophys Biomol Struct* 34:267–294.

26. Boulanger, M.C., Miranda, T.B., Clarke, S., Di Fruscio, M., Suter, B., Lasko, P., and Richard, S. (2004). Characterization of the Drosophila protein arginine methyltransferases DART1 and DART4. *Biochem J* 379:283–289.

27. Schurter, B.T., Koh, S.S., Chen, D., Bunick, G.J., Harp, J.M., Hanson, B.L., Henschen-Edman, A., Mackay, D.R., Stallcup, M.R., and Aswad, D.W. (2001). Methylation of histone H3 by coactivator-associated arginine methyltransferase 1. *Biochemistry* 40:5747–5756.

28. Ma, H., Baumann, C.T., Li, H., Strahl, B.D., Rice, R., Jelinek, M.A., Aswad, D.W., Allis, C.D., Hager, G.L., and Stallcup, M.R. (2001). Hormone-dependent, CARM1-directed, arginine-specific methylation of histone H3 on a steroid-regulated promoter. *Curr Biol* 11:1981–1985.

29. Bauer, U.M., Daujat, S., Nielsen, S.J., Nightingale, K., and Kouzarides, T. (2002). Methylation at arginine 17 of histone H3 is linked to gene activation. *EMBO Rep* 3:39–44.

30. Strahl, B.D., Briggs, S.D., Brame, C.J., Caldwell, J.A., Koh, S.S., Ma, H., Cook, R.G., Shabanowitz, J., Hunt, D.F., Stallcup, M.R., and Allis, C.D. (2001). Methylation of histone H4 at arginine 3 occurs in vivo and is mediated by the nuclear receptor coactivator PRMT1. *Curr Biol* 11:996–1000.

31. Wang, H., Huang, Z.Q., Xia, L., Feng, Q., Erdjument-Bromage, H., Strahl, B.D., Briggs, S.D., Allis, C.D., Wong, J., Tempst, P., and Zhang, Y. (2001). Methylation of histone H4 at arginine 3 facilitating transcriptional activation by nuclear hormone receptor. *Science* 293:853–857.

32. An, W., Kim, J., and Roeder, R.G. (2004). Ordered cooperative functions of PRMT1, p300, and CARM1 in transcriptional activation by p53. *Cell* 117:735–748.

33. Strahl, B.D., and Allis, C.D. (2000). The language of covalent histone modifications. *Nature* 403:41–45.

34. Kouzarides, T. (2002). Histone methylation in transcriptional control. *Curr Opin Genet Dev* 12:198–209.

35. Xu, W., Chen, H., Du, K., Asahara, H., Tini, M., Emerson, B.M., Montminy, M., and Evans, R.M. (2001). A transcriptional switch mediated by cofactor methylation. *Science* 294:2507–2511.

36. Lee, Y.H., Koh, S.S., Zhang, X., Cheng, X., and Stallcup, M.R. (2002). Synergy among nuclear receptor coactivators: selective requirement for protein methyltransferase and acetyltransferase activities. *Mol Cell Biol* 22:3621–3632.

37. Daujat, S., Bauer, U.M., Shah, V., Turner, B., Berger, S., and Kouzarides, T. (2002). Crosstalk between CARM1 methylation and CBP acetylation on histone H3. *Curr Biol* 12:2090–2097.

38. Tang, J., Kao, P.N., and Herschman, H.R. (2000). Protein-arginine methyltransferase I, the predominant protein-arginine methyltransferase in cells, interacts with and is regulated by interleukin enhancer-binding factor 3. *J Biol Chem* 275:19866–19876.

39. Pawlak, M.R., Scherer, C.A., Chen, J., Roshon, M.J., and Ruley, H.E. (2000). Arginine N-methyltransferase 1 is required for early postimplantation mouse development, but cells deficient in the enzyme are viable. *Mol Cell Biol* 20:4859–4869.

40. Cimato, T.R., Tang, J., Xu, Y., Guarnaccia, C., Herschman, H.R., Pongor, S., and Aletta, J.M. (2002). Nerve growth factor-mediated increases in protein methylation occur predominantly at type I arginine methylation sites and involve protein arginine methyltransferase 1. *J Neurosci Res* 67:435–442.

41. Pawlak, M.R., Banik-Maiti, S., Pietenpol, J.A., and Ruley, H.E. (2002). Protein arginine methyltransferase I: substrate specificity and role in hnRNP assembly. *J Cell Biochem* 87:394–407.

42. Gary, J.D., and Clarke, S. (1998). RNA and protein interactions modulated by protein arginine methylation. *Prog Nucleic Acid Res Mol Biol* 61:65–131.

43. Smith, J.J., Rucknagel, K.P., Schierhorn, A., Tang, J., Nemeth, A., Linder, M., Herschman, H.R., and Wahle, E. (1999). Unusual sites of arginine methylation in Poly(A)-binding protein II and in vitro methylation by protein arginine methyltransferases PRMT1 and PRMT3. *J Biol Chem* 274:13229–13234.

44. Klein, S., Carroll, J.A., Chen, Y., Henry, M.F., Henry, P.A., Ortonowski, I.E., Pintucci, G., Beavis, R.C., Burgess, W.H., and Rifkin, D.B. (2000). Biochemical analysis of the arginine methylation of high molecular weight fibroblast growth factor-2. *J Biol Chem* 275:3150–3157.
45. Kwak, Y.T., Guo, J., Prajapati, S., Park, K.J., Surabhi, R.M., Miller, B., Gehrig, P., and Gaynor, R.B. (2003). Methylation of SPT5 regulates its interaction with RNA polymerase II and transcriptional elongation properties. *Mol Cell* 11:1055–1066.
46. Mowen, K.A., Tang, J., Zhu, W., Schurter, B.T., Shuai, K., Herschman, H.R., and David, M. (2001). Arginine methylation of STAT1 modulates IFNalpha/beta-induced transcription. *Cell* 104:731–741.
47. Meissner, T., Krause, E., Lodige, I., and Vinkemeier, U. (2004). Arginine methylation of STAT1: a reassessment. *Cell* 119:587–589; discussion 589–590.
48. Stallcup, M.R., Kim, J.H., Teyssier, C., Lee, Y.H., Ma, H., and Chen, D. (2003). The roles of protein-protein interactions and protein methylation in transcriptional activation by nuclear receptors and their coactivators. *J Steroid Biochem Mol Biol* 85:139–145.
49. Stallcup, M.R. (2001). Role of protein methylation in chromatin remodeling and transcriptional regulation. *Oncogene* 20:3014–3020.
50. Bachand, F., and Silver, P.A. (2004). PRMT3 is a ribosomal protein methyltransferase that affects the cellular levels of ribosomal subunits. *Embo J* 23:2641–2650.
51. Swiercz, R., Person, M.D., and Bedford, M.T. (2005). Ribosomal protein S2 is a substrate for mammalian PRMT3 (protein arginine methyltransferase 3). *Biochem J* 386:85–91.
52. Singh, V., Miranda, T.B., Jiang, W., Frankel, A., Roemer, M.E., Robb, V.A., Gutmann, D.H., Herschman, H.R., Clarke, S., and Newsham, I.F. (2004). DAL-1/4.1B tumor suppressor interacts with protein arginine N-methyltransferase 3 (PRMT3) and inhibits its ability to methylate substrates in vitro and in vivo. *Oncogene* 23:7761–7771.
53. Zhang, X., Zhou, L., and Cheng, X. (2000). Crystal structure of the conserved core of protein arginine methyltransferase PRMT3. *Embo J* 19:3509–3519.
54. Weiss, V.H., McBride, A.E., Soriano, M.A., Filman, D.J., Silver, P.A., and Hogle, J.M. (2000). The structure and oligomerization of the yeast arginine methyltransferase, Hmt1. *Nat Struct Biol* 7:1165–1171.
55. Cheng, X., and Roberts, R.J. (2001). AdoMet-dependent methylation, DNA methyltransferases and base flipping. *Nucleic Acids Res* 29:3784–3795.
56. Schubert, H.L., Blumenthal, R.M., and Cheng, X. (2003). Many paths to methyltransfer: a chronicle of convergence. *Trends Biochem Sci* 28:329–335.
57. Kim, S., Merrill, B.M., Rajpurohit, R., Kumar, A., Stone, K.L., Papov, V.V., Schneiders, J.M., Szer, W., Wilson, S.H., Paik, W.K., and Williams, K.R. (1997). Identification of N(G)-methylarginine residues in human heterogeneous RNP protein A1: Phe/Gly-Gly-Gly-Arg-Gly-Gly-Gly/Phe is a preferred recognition motif. *Biochemistry* 36:5185–5192.
58. Lischwe, M.A., Cook, R.G., Ahn, Y.S., Yeoman, L.C., and Busch, H. (1985). Clustering of glycine and NG,NG-dimethylarginine in nucleolar protein C23. *Biochemistry* 24:6025–6028.
59. Lischwe, M.A., Ochs, R.L., Reddy, R., Cook, R.G., Yeoman, L.C., Tan, E.M., Reichlin, M., and Busch, H. (1985). Purification and partial characterization of a nucleolar scleroderma antigen (Mr = 34,000; pI, 8.5) rich in NG,NG-dimethylarginine. *J Biol Chem* 260:14304–14310.
60. Shi, Y., Lan, F., Matson, C., Mulligan, P., Whetstine, J.R., Cole, P.A., and Casero, R.A. (2004). Histone demethylation mediated by the nuclear amine oxidase homolog LSD1. *Cell* 119:941–953.
61. Forneris, F., Binda, C., Vanoni, M.A., Mattevi, A., and Battaglioli, E. (2005). Histone demethylation catalysed by LSD1 is a flavin-dependent oxidative process. *FEBS Lett* 579:2203–2207.
62. Trewick, S.C., McLaughlin, P.J., and Allshire, R.C. (2005). Methylation: lost in hydroxylation? *EMBO Rep* 6:315–320.

63. Tsukada, Y., Fang, J., Erdjument-Bromage, H., Warren, M. E., Borchers, C. H., Tempst, P., and Zhang, Y. (2006) Histone demethylation by a family of JmjC domain-containing proteins. *Nature* 439:811–816.

64. Cuthbert, G.L., Daujat, S., Snowden, A.W., Erdjument-Bromage, H., Hagiwara, T., Yamada, M., Schneider, R., Gregory, P.D., Tempst, P., Bannister, A.J., and Kouzarides, T. (2004). Histone deimination antagonizes arginine methylation. *Cell* 118:545–553.

65. Wang, Y., Wysocka, J., Sayegh, J., Lee, Y.H., Perlin, J.R., Leonelli, L., Sonbuchner, L.S., McDonald, C.H., Cook, R.G., Dou, Y., Roeder, R.G., Clarke, S., Stallcup, M.R., Allis, C.D., and Coonrod, S.A. (2004). Human PAD4 regulates histone arginine methylation levels via demethylimination. *Science* 306:279–283.

66. Dion, M.F., Altschuler, S.J., Wu, L.F., and Rando, O.J. (2005). Genomic characterization reveals a simple histone H4 acetylation code. *Proc Natl Acad Sci USA* 102:5501–5506.

67. Henikoff, S. (2005). Histone modifications: combinatorial complexity or cumulative simplicity? *Proc Natl Acad Sci USA* 102:5308–5309.

5

Methylation and Demethylation of Histone Arg and Lys Residues in Chromatin Structure and Function

YANMING WANG

Department of Biochemistry and Molecular Biology
Pennsylvania State University
108 Althouse Lab
University Park, PA 16802, USA

I. Abstract

Chromatin is the physiological template of all eukaryotic genomic activities. Histone proteins are the fundamental building elements of chromatin, which are the subject of various posttranslational modifications, including methylation. Adding and removing the methyl moieties from histones plays an important epigenetic role to ensure the release of the appropriate genetic information. Both Lys and Arg residues in histones can be dynamically methylated and demethylated by different enzymes. The processes of adding and removing methyl groups on histone Lys residues are catalyzed by histone Lys methyltransferases (HKMTs) and histone-Lys-specific demethylase (LSD), respectively. Protein Arg methyltransferases (PRMTs) add methyl groups to histone Arg residues. On the other hand, peptidyl-larginine deiminases remove the methyl groups in conjunction with the amine group, leaving the citrulline amino acid in histones. The fate of citrulline residues in histone is currently unknown. Importantly, methylation has been implicated as playing a major role in regulating gene expression to control normal cell growth,

THE ENZYMES, Vol. XXIV

proliferation, and differentiation. The steady-state balance of histone methylation is important for the normal development and the health of an organism.

II. Introduction

DNA in eukaryotic cells is organized with histones to form chromatin. The regulation of chromatin structure plays a pivotal role in many nuclear events utilizing DNA as a template. Recently, the role of posttranslational histone modifications (such as methylation, acetylation, phosphorylation, and so on) has been extensively studied in various nuclear events that include chromosome condensation/decondensation, transcription, DNA damage and repair, and chromosome rearrangement. An emerging theme is that histone modifications mediate these nuclear events in coordination with DNA methylation and ATP-dependent chromatin remodeling machineries. It has been proposed that these covalent histone modifications work as a "histone code" to incorporate upstream signaling inputs as well as to dictate downstream chromatin outputs [1–4]. Because histones are separated together with DNA into daughter cells, the information imprinted on histones by covalent modifications is heritable [5]. It becomes evident that histone modifications serve as important "epigenetic information" to ensure that genetic information is appropriately released during cell differentiation and development [6]. Recently, epigenetics is often used to refer the study of heritable changes of chromatin function without altering the DNA sequence per se.

In keeping with the theme of methyltransferase in this book, this chapter focuses on enzymes involved in histone methylation/demethylation and their roles in chromatin biology. Many Lys and Arg residues on histones are target sites of methylation, and most of these sites are clustered on the tails of histones. The first enzyme responsible for histone Lys methylation was reported in 2000. Since then, dozens of histone lysine methyltransferases (HKMTs) have been discovered [7]. Many of the identified HKMTs play a role in transcriptional regulation and/or in the organization of higher-order chromatin structure. Moreover, recent evidence suggests that histone Lys methylation may be involved in DNA damage and repair. In contrast to the diverse functions of histone Lys methylation, histone Arg methylation catalyzed by protein Arg methyltransferases (PRMTs) mainly regulates transcription. Because the abnormal gene expression patterns can lead to tumorigenesis and because of the role of histone methylation as a part of epigenetic information in cell differentiation, this field is attracting more and more scientists from fields of cell and developmental biology as well from cancer and stem cell biology.

III. Chromatin Structure and Function

Each diploid human cell contains over 2×10^9 bp of DNA, which is about 2 meters in length. To fit this genetic material into a nucleus with a diameter of

5 to 10μm, DNA is highly folded and organized to form chromatin. At the basic level of chromatin structure, about 146 bp of DNA is first wrapped around a core histone octamer (including two each of histones H3, H4, H2A, and H2B) to form a nucleosome core particle, the basic structure unit of all eukaryotic chromatin [8]. Upon the association of linker histone H1 and additional histone/DNA binding factors, the string of nucleosomes is further organized to form higher-order chromatin structures [9]. Recent evidence suggests a two-start model for the organization of nucleosomes to form the 30-nm chromatin filament [10]. However, it is not clear how chromatin is organized beyond the level of nucleosomal core particles under the physiological conditions in a living cell. Nevertheless, the organization of chromatin has to allow the dynamic change of chromatin conformation during cell cycles.

Chromatin is loosely diffused throughout the nucleus in an interphase cell. Depending on the level of compaction, chromatin is classified as euchromatin or heterochromatin. Euchromatin is composed of highly decondensed chromatin fibers, which is permissible to transcription. Heterochromatin is highly condensed and repressive for transcription [11]. Levels of DNA condensation can be easily visualized by electron or fluorescence microscopes. As shown in Figure 5.1, heterochromatin regions are dark in color in the electron microscopic picture of a nucleus in a terminal-differentiated human granulocyte, whereas euchromatin is light in color in this electron microscope picture. The formation of euchromatin and heterochromatin is regulated by molecular mechanisms. In *Drosophila* Schneider 2 cells, heterochromatin is organized to form a distinct structure in the nucleus, which is enriched in heterochromatin protein 1 (HP1). Heterochromatin is composed of highly repetitive DNA sequences (e.g., the satellite DNA) with no or very few genes. However, several *Drosophila* genes are localized in the heterochromatin region of the genome (e.g., light). The heterochromatin environment is important for the proper expression of these heterochromatic genes. In contrast, euchromatin is composed of nonrepetitive DNA sequences with high gene density. How the underlying DNA sequence directly impacts on the organization of the higher-order chromatin structure remains an interesting question for future exploration. Currently, multiple mechanisms have been implicated in this process, which include binding proteins for specific DNA sequences, histone-modifying enzymes and binding proteins, ATP-dependent chromatin remodeling complexes, and others.

The strong interaction between DNA and histones restricts the access of DNA-binding proteins to their cognate binding sites. For transcription factors to gain access to DNA, local chromatin structures containing the binding sites are dramatically remodeled during the transcription processes [12]. The interaction of histones with DNA also imposes a barrier when RNA polymerase II (Pol II) passes along the DNA template to synthesize RNAs. It has been observed that the chromatin structure of heat shock genes and nuclear hormone inducible genes

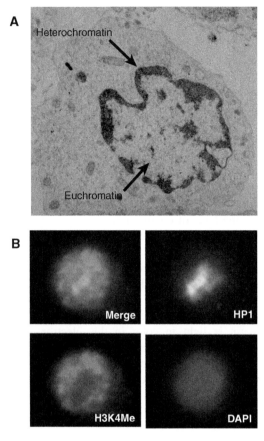

FIG. 5.1. Chromatin in the nucleus of a cell is organized into euchromatin and heterochromatin. (A) In the electron microscopy picture of a human granulocyte, heterochromatin is evident as dark regions (denoted by the arrow), whereas euchromatic regions are light in color, suggesting that less DNA and histones are compacted at the euchromatin regions compared with that of the heterochromatin. (B) Immunofluorescence staining of a *Drosophila* tissue culture S2 cell. Heterochromatin is associated with HP1 (green color), whereas euchromatin is strongly stained with a histone H3 Lys4 methyl antibody (red color). DAPI staining was applied to show the entire nucleus. Note that in this type of *Drosophila* cell heterochromatin is organized together to form a unique compartment in the nucleus, as compared with the organization of heterochromatin underneath the nuclear envelope in the human granulocyte. (See color plate.)

is dramatically decondensed in a time course of several minutes after stimulation by activation signals. The fast dynamic incorporation and replacement of H2A and H2B from chromatin in *Physarum* suggested that the H2A.H2B dimer is replaced from the nucleosome during the elongation phase of transcription [13]. Although the replacement of the $(H3.H4)_2$ tetramer is theoretically possible, the slow dynamic of the exchange of H3 and H4 from chromatin suggests that the $(H3.H4)_2$ tetramer is likely associated with DNA during the elongation process [13–15].

In addition to transcription, chromatin is also the physiological template for DNA replication, chromatin rearrangement, and DNA fragmentation during apoptosis. Moreover, during the M phase of the cell cycle, chromatin is highly compacted to form chromosomes to facilitate the faithful segregation of genetic materials to daughter cells. Histone methylation is associated with the processes mentioned previously, including DNA damage and repair [16, 17]. One challenge facing the field is to analyze the precise role of histone modifications. That is, to discover whether histone methylation is a causal event or only a downstream consequence of these chromatin events. Toward this direction, it is both attractive and challenging to apply genetic methods to study histone modifications in multicellular organisms such as fruit flies and mice.

IV. Histone Lys Methylation

A. THE DISCOVERY OF HISTONE METHYLATION

The presence of methylation on histone Lys residues was first demonstrated over three decades ago [18]. Histone Lys methylation is a very abundant modification in bulk histones purified from various tissues, such as calf thymus, human spleen, and chicken erythrocytes. Because of their abundance, the protein microsequencing method was able to detect the particular site in histones containing methylated Lys residues. For example, among the five Lys residues (Lys4, 8, 12, 16, and 20) on the histone H4 N-terminal tail, Lys20 was the only residue found to be methylated. Moreover, the E-amino group of the Lys residues can be mono-, di-, or tri-methylated. Each of these modification states seems to play a particular role in chromatin biology. Recently, an increasing number of histone methylation sites have been discovered due to the advancement of mass spectrometry techniques as well as the development of site- and methyl-specific antibodies. Figure 5.2 summarizes the known histone Lys methylation sites.

B. IDENTIFICATION OF HISTONE LYS METHYLTRANSFERASES (HKMTS)

The function of Lys methylation in histones was unknown for a long time in part because the identity of the methyltransferase had remained mysterious. The first histone Lys methyltransferase was reported by the Jenuwein group in 2000 [19]. In this paper, mammalian Suv39h1 was first speculated to work as a histone methyltransferase because of its sequence similarity with other methyltransferases in

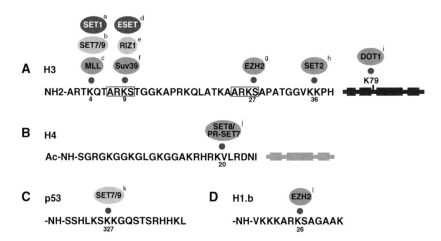

FIG. 5.2. A diagram of known Lys methylation sites in H3, H4, linker histone H1.b, and p53. The known enzymes that target these particular sites are shown. References for each enzyme are denoted by letters: a [88–91], b [92, 93], c [94], d [25, 95], e [97, 105], f [19, 96], g [98], h [99], i [100–102], j [92, 103], k [106], and l [26].

plants. It was shown that Suv39h1 can methylate histone H3 at the Lys9 site in *in vitro* biochemical assays using *S*-adenosyl-L-methione (SAM) as a cofactor.

The initial discovery of Suv39h1 as a histone Lys methyltransferase spurred very strong interest in identifying additional HKMTs as well as to explore the role of these enzymes in chromatin structure and function. Many HKMTs have been identified in the recent years by biochemistry and genetic approaches (Figure 5.2). Most of these proteins contain the SET domain as their catalytic motif, which was named after three founding members of this protein family in *Drosophila*, including Su(var)3–9, Enhance of zeste [E(z)], and trithorax (Trx). Further, these three *Drosophila* proteins were found to methylate histone H3 at Lys9, Lys27, and Lys4, respectively. The crystal structures of several HKMTs have been dissolved. (See the chapter by Dr. Gamblin for further details.)

In *Drosophila*, Su(var)3–9 (the homologue of mammalian Suv39h1 protein) is a well-characterized regulatory protein for the formation of heterochromatin [11]. Su(var)3–9 was originally identified in genetic screens searching for chromatin modifiers that regulate position effect variegation (PEV), a phenomenon where reporter genes (e.g., the *white* eye-color gene) juxtaposed to heterochromatin demonstrate variegated expression levels in genetically identical cells in *Drosophila* eyes. The loss-of-function mutations of proteins that facilitate the formation of heterochromatin increase the white gene expression. Therefore, these genes are called Su(var) genes. Many chromatin and histone modifiers have been identified in these genetic screens, including HP1 (heterochromatin protein 1, also called Su(var)2-5). In contrast, the loss-of-function mutations of proteins that play negative roles in heterochromatin formation decrease the white gene expression. Therefore, these genes are called E(var) genes.

In immunofluorescence studies, high levels of methylation of histone H3 Lys9 were found to be associated with heterochromatin [20]. Moreover, the chromo-domain of heterochromatin protein 1 (HP1) directly interacts with the methylation mark on histone H3 Lys9, which is also enriched on the heterochromatin [21]. Together with the finding that Su(var)3-9 methylates the H3 Lys9 residue, these results offered a biochemical pathway wherein HP1 and Su(var)3-9 work through the H3 Lys9 methylation epigenetic mark for heterochromatin formation. This pathway has established a paradigm that histone methylation catalyzed by the HKMTs exert their biological functions via recruiting additional histone-binding proteins and chromatin modifiers.

C. THE COMPLEXITY OF HISTONE LYS METHYLATION AND ITS IMPLICATION IN CHROMATIN BIOLOGY

Multiple Lys residues on histones H3, H4, and linker histone H1.b can be methylated. In fact, each HKMT has its preferred target site in histones (Figure 5.2). For example, although Lys9 and Lys27 are embedded in very similar sequence context on the H3 N-terminal tail, including four identical residues ARKS (Figure 5.2), Su(var)3-9 prefers the Lys9 for methylation, whereas E(z) prefers the Lys27 site for methylation. These results suggest that different Lys methylation sites and their corresponding HKMTs have been evolved for separate regulatory pathways. In the *Drosophila* system, it is clear that H3 Lys9 di- and tri-methylation is involved in the regulation of pericentric heterochromatin or the chromocenter on polytene chromosomes [22]. In contrast, H3 K27 tri-methylation functions together with Polycomb (Pc) to regulate homeotic gene silencing in order to establish the correct body segmentation during early embryonic development.

Moreover, each Lys residue can exist in three methyl states: mono-, di-, or tri-methylation. Recent evidence suggests that different methylation states of a singular Lys site are catalyzed by distinct enzymes. For example, the HKMT G9a can generate mono-methylation and di-methylation on H3 Lys9, which is correlated with the repression of euchromatic genes in mammalian cells [23]. In contrast, tri-methylation of Lys9 on H3 produced by Suv39h1 regulates the formation of constitutive heterochromatin [23].

At the global level, the methylation state of a given chromatin region also demonstrates certain plasticity. In wild-type mouse embryonic stem cells, peri-centric heterochromatin is enriched with H3 Lys9 tri-methylation and Lys27 mono-methylation [24]. However, in Suv39h1 and Suv39h2 double-null cells H3 Lys9 mono-methylation and H3 Lys27 tri-methylation become enriched on the pericentric heterochromatin, indicating that alternative silencing machineries are recruited to the pericentric heterochromatin in the absence of the Suv39h1 and Suv39h2 functions. However, the surveillance mechanism(s) that ensures the recruitment of the appropriate silencing machinery is not very clear.

Theoretically, the combination of different methylation sites and methylation states can confer a tremendous amount of epigenetic information on chromatin. One view of this complexity is that different modifications can appear on the same histone molecule at a given time point during the cell cycle. Alternatively, only a particular combination of modifications can occur simultaneously at the same histone molecule or at different histones of the same nucleosome. Existing evidence in the literature seems to favor the second view. Chromatin in an interphase cell nucleus does not exist as a random mixture but is organized into functional compartments (e.g., euchromatin and heterochromatin). Histone modifications associated with heterochromatin (e.g., tri-methylation at Lys9 of H3) unlikely coexist with tri-methylation at Lys4 of H3 that is enriched on euchromatin and active genes. However, during the transition from an active gene state to a repressed gene state it is conceivable that all active marks on histone proteins need to be removed, which is followed by the addition of repressive histone marks. Limited evidence is available about the chromatin and histone modification change during such a transition process.

D.   THE PROTEIN COMPLEXES FORMED BY HISTONE LYS METHYLTRANSFERASES

Histone Lys methylation plays a variety of roles in regulating chromatin structure and function. Five Lys residues on histone H3 are methylation sites, including Lys4, 9, 27, 36, and 79 (Figure 5.2). Histone H3 methylation regulates distinct chromatin functions, such as transcriptional activation (Lys4 and 36), pericentric heterochromatin formation (Lys9), homeotic gene silencing (Lys27), and the maintenance of chromatin boundaries (e.g., euchromatin/heterochromatin boundary by Lys79) (Figure 5.3).

FIG. 5.3. The action modes of histone Lys methylation. (A). The HKMT and the binding protein are organized in the same protein complex (e.g., Suv39 and HP1). The interaction between the enzyme and the binding protein allows retaining the HKMT at the vicinity where the methyl-Lys marks were first generated. In turn, the retained enzyme can produce additional methyl-Lys marks, which may help to spread the methyl marks along a big block of chromatin. This mode can facilitate the formation of condensed chromatin over the pericentric heterochromatin. (B) Some histone Lys methyltransferases (e.g., EZH2) are organized into different complexes with the methyl-Lys binding protein (e.g., Polycomb). Therefore, after the HKMT is recruited to a chromatin locus to produce methyl-Lys marks the enzyme is dissociated. Then, chromo-domain-containing binding proteins are recruited to these sites to exert downstream functions, such as chromatin structure organization and transcription. (C) Alternatively, the methyl marks produced may not directly recruit binding proteins but prevent the spreading of other proteins (protein X). This may help to establish a boundary, such as that observed between yeast telomeres and silenced mating locus with the adjacent euchromatin. (See color plate.)

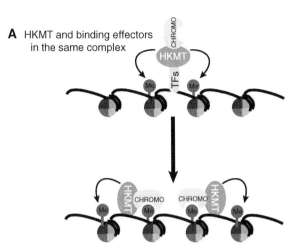

A HKMT and binding effectors
   in the same complex

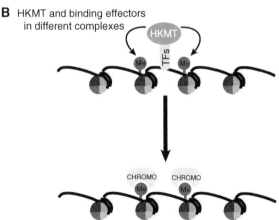

B HKMT and binding effectors
   in different complexes

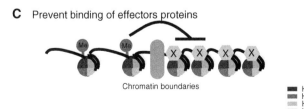

C Prevent binding of effectors proteins

Chromatin boundaries

H3
H2B
H2A
H4

Histone methyltransferases are often incorporated into large protein complexes to perform their functions. As such, histone modification activities are specifically recruited to particular chromatin regions/compartments. It was shown that associating proteins can alter the methyl state produced on a singular Lys residue [25] or the substrate preference of the methyltransferase [26]. For example, in the absence of interacting proteins ESET can produce di-methyl-Lys9 on histone H3 [25]. The interaction between mAM and ESET confers the ability of ESET to generate tri-methyl Lys9 at H3, which enhances the ability of ESET to repress gene expression [25]. Moreover, due to the presence of different EED isoform in the EZH2 complex the substrate preference of EZH2 can be switched from H3 Lys27 to H1.b Lys26 [26]. Hence, protein-protein interaction is important for regulating the function of HKMTs.

The methyltransferases of histone H3 Lys4 in humans (MLL) and yeast (ySet1) are evolutionarily conserved. Protein components of the ySet1 complex in yeast are similar to those of the MLL complex in humans (Table 5.1). Importantly, chromosome translocations that fuse MLL with over 50 other genes have been linked with the development of leukemia [27]. This suggests that normal MLL function and histone H3 Lys4 methylation are important in maintaining the homeostasis of blood cell proliferation and differentiation, likely through its role in regulating homeotic gene expression.

MLL complexes composed of different protein components have been described [28, 29, 29b]. Using immunoaffinity purification procedures, a large MLL protein complex containing MLL together with more than 29 other proteins was identified, including factors of the ATP-dependent chromatin-remodeling complexes (Mi2 and Brm, hSNF2H), histone deacetylases (HDAC1 and 2), and histone acetyltransferase (TAF250) [29]. Accordingly, this complex demonstrated ATP-dependent chromatin remodeling, histone acetylation, histone deacetylation, and histone methylation activities [29]. In contrast, another MLL complex characterized by the Cleary group identified a smaller (1MDa) protein complex with fewer protein components, including ASH2L, MLL[N], MLL[C], HCF-2, RbBP5, Menin, and WDR5 [28]. The second protein complex has very similar components to that of the yeast Set1 protein complex. Most recently, a third MLL containing protein complex was purified using a Hela S3 cell line stably expressing Flag-tagged WDR5 fusion protein (29b). This protein complex contains human MOF in addition to the MLL[N] and MLL[C] proteins, suggesting that MLL and MOF are coordinated to regulate homeotic gene expression.

The histone H3 Lys9 methyltransferase Suv39h1 forms a complex with HP1, which contains a chromodomain binding to H3 methyl-Lys9 [30]. Suv39h1 and HP1 have been found to interact with several transcriptional factors, such as Rb and AML1 [31, 32]. Therefore, in addition to their function with pericentric heterochromatin H3 Lys9 methylation and Suv39 can facilitate the silencing

TABLE 5.1

COMPLEXES OF HISTONE METHYLATION AND DEMETHYLATION ENZYMES

Enzyme	ySET1[1]	hSET1[2]	MLL[3]	ESET[4]	RIZ1[5]	Suv39[6]	Su(var)3-9[7]	EZH2[8]	CARM1[9]	LSD1[10]
Components identified by biochemical purification	Bre2 Swd1 Spp1 Swd2 Swd3 Sdc1[a]	Ash2 WDR5 Sin3A HDAC1/2 RbAp48/46 Sds3[b]	MLL[N,C] HCF-2 Menin RBBP5 ASH2L WDR5				HP1 Su(var)3-7	SUZ12 AEBP2 EED RbAp48	BRG1 BAF250 BAF170 BAF155 BAF60a Ini-1[c]	G9a HDAC1/2 CDYL HPC2 CtBP RREB-1[d]
Identified by other methods				ERG	Rb	Rb HP1				

(1) See reference [91]. (2) See reference [104]. (3) See reference [28]. (4) See reference [95]. (5) See reference [105]. (6) See reference [31]. (7) See reference [96]. (8) See reference [98]. (9) See reference [60]. (10) See reference [45].

(a) Additional components include Shg1. (b) Additional components include HCF-1c, HSP70, HSP90, and so on. (c) Additional components include BAF57, p105, HSP70, HSP73, HSP90, Keratin 8, α-actin, and β-actin. (d) Additional components include ZEB1 and 2, ZNF217, CoRest, CtBP2, p40, p80, and p90.

of Rb or AML1 target genes. Further, Suv39h1 interacts with other chromatin factors required for the establishment of repressive chromatin structure, such as HDAC1 and HDAC2, as well as MBD1 (methyl CpG binding protein 1) [33, 34]. In mouse cells double null for Suv39h1 and h2, telomere and chromosome stability was compromised. Mutant mice display a higher risk of B-cell lymphoma [24]. These results suggest a role of the H3 Lys9 HKMTs in gene silencing, chromatin structure, and genome stability.

The function of H3 Lys27 methylation is mediated by two protein complexes: PRC1 and PRC2. The *Drosophila* PRC1 complex contains Polycomb (Pc), Polyhomeotic (Ph), Posterior sex combs (Psc), sex combs on midleg (Scm), and dRING/sex combs extra [35]. The Pc protein contains a chromodomain, which can specifically recognize methylation of histone H3 at Lys27 [22]. The mammalian or *Drosophila* PRC2 complex contains EED (embryonic ectoderm development), EZH2 (enhancer of zeste 2), Suz12 (suppressor of zeste 12), RbAp48, and AEBP2 [36]. Three components of the complex (EED, EZH2, and Suz12) define the minimal components of the complex required for the enzymatic activity of EZH2 to nucleosomal histone substrates, as well as for the repression of hoxa9 gene in Hela cells [36]. In addition to its role in silencing of homeotic genes, the Polycomb pathway is implicated in the inactivation of one of the X chromosomes in female cells to reach dosage compensation in mammals [37, 38].

It is not clear whether the histone H3 Lys79 methyltransferase, Dot1, is stably incorporated into a large protein complex in a manner similar to that of the histone H3 K4 and K9 methyltransferases MLL and E(z). Recently, Okada et al. (2005) showed that human hDOT1L interacts with the transcription factor AP1 in yeast two-hybrid screens (38b). In addition, it was found that the HKMT activity of hDOT1L contributes to the leukemogenic activity of the MLL-AF10 fusion protein. One of the downstream target genes of MLL-AF10 and hDOT1L is hoxa9. The MLL-AF10 protein can transform wild-type primary bone marrow cells but not the hoxa9$^{-/-}$ cells, suggesting hoxa9 gene is an important downstream target of the MLL-AP10 fusion protein during leukemogenesis. Interestingly, Lys79 of histone H3 is the only known methylation site in the core domain of histone proteins, whereas other methylation sites are located at the N-termini of histones H3 and H4.

In general, HKMTs as well as many other chromatin- and histone-modifying activities are incorporated into big protein complexes in order to perform their functions properly. As mentioned previously, the methyltransferase activity of HKMTs is often modulated by interacting protein partners, which either increase the efficacy of the methyltransferase activity or alter the substrate preference of the enzyme. The association with other protein factors may target HKMTs to particular loci of the chromatin. Therefore, the pattern of histone Lys methylation can be deliberately generated along the entire genome in a living cell.

E. Modes of Action for Histone Methylation in Chromatin Biology

Depending on the components of the HKMT complex or the way histone methylation affects the chromatin function, different modes of histone methylation action have evolved in chromatin biology. HKMTs are sometimes incorporated into the same protein complex with the methyl-Lys binding proteins. For example, Suv39h1 is in the same protein complex with HP1. Trx/MLL is associated with WDR5, a protein that was recently found to interact with histone H3 Lys4 methylation [38c]. The advantage for the incorporation of the enzyme and the methyl-Lys binding protein in the same complex is not very clear. However, one may postulate that the interaction of HKMTs and methyl-Lys binding protein (or effector proteins) may help to facilitate the spread of a certain histone Lys methylation patterns along a large domain of chromatin (Figure 5.3a). This mechanism is probably very helpful in the case of the pericentric heterochromatin, which consists of a large block of DNA. However, because histone Lys4 methylation is mainly localized to the promoter region of genes the extent of H3 Lys4 methylation may be relatively restricted [39]. Therefore, how the spread of H3 Lys4 methylation is finely tuned during transcription is an interesting question for future studies.

As discussed previously, the methyl-Lys27 binding protein Pc and the methyltransferase E(z) are incorporated into the PRC1 and PRC2 complexes, respectively. Histone H3 Lys27 methylation or the silencing of homeotic genes during development or hematopoiesis has to be restricted to a particular promoter. Otherwise, this silencing mark may spread along the chromatin, leading to the unexpected silencing of the adjacent genes. To avoid the wide spreading of histone H3 Lys27 methylation, the silencing of homeotic genes by histone Lys methylation may occur through a two-step process (Figure 5.3b). The methyltransferase complex is first recruited to produce the histone H3 Lys27 methylation mark. The methyl-Lys27 binding protein is then recruited to facilitate the formation of repressive chromatin conformation.

In budding yeast, the methylation of H3 Lys79 catalyzed by Dot1 was found to play a role in the maintenance of the boundary between euchromatin and heterochromatin [40]. H3 Lys79 methylation is very abundant and is enriched on euchromatic regions in budding yeast. Interestingly, loss of Lys79 methylation on H3 leads to the de-silencing of genes in the vicinity of telomeres [40]. It was proposed that telomere-associated proteins spread along the chromosome arms in the absence of the Dot1 function, thereby leading to the compromised function of heterochromatin (Figure 5.3c).

F. The Role of Histone Lys Methylation in DNA Damage and Repair

Recently, a role of histone Lys methylation in damaged-DNA repair was reported. It was found that the methylation of histone H3 at Lys79 directly recruits 53BP1 to the sites of DNA double-strand break in mammalian cells [17]. Mutations of the HKMTs for H3 Lys79, Dot1L, demonstrate defects in recruiting 53BP1 to the double-strand breaks [17]. That the two tandem Tudor domains of 53BP1 recognize the methyl-Lys79 H3 site was shown by crystal studies [17]. In contrast, the 53BP1 homolog in fission yeast, the Crb2 protein, was found to mediate the DNA damage response pathway through recognizing histone H4 at Lys20 methylation [16]. It will be interesting to dissect whether these two pathways complement each other during DNA damage response in each organism. Alternatively, these two Lys methylation pathways were evolved separately in yeast and humans to perform the same function.

V. Histone Lys Demethylase

Unlike other histone modifications (e.g., phosphorylation and acetylation), which are dynamically added and removed, histone methylation seems to be relatively stable. A number of studies have found that the turnover rate of methylation in histones is very slow [41, 42]. In contrast, Hempel et al. (1979) reported that the half-life of histone methylation is only a few days in cat kidneys [43]. Given that cells in this tissue rarely divide, a demethylation mechanism was implicated. In agreement with this later observation, an enzymatic activity that can remove the methyl group from methyl-Lys residues was reported by Paik and Kim in 1961 [18]. It was proposed that the process is an oxidative reaction with formaldehyde as a product of the reaction. In fact, their proposal is consistent with the recent discovery of LSD1 (see below).

With these intriguing results in the early literature, the identity of enzymes that can demethylate histone Lys had remained mysterious for a long time. Recently, a long-awaited histone-Lys-specific demethylase (LSD1) was identified [44]. LSD1 has sequence similarity to amine oxidase (Figure 5.4). *In vitro*, this enzyme can demethylate mono- or di-methyl H3 Lys4 using FAD as a cofactor, producing hydrogen peroxide (H_2O_2) and formaldehyde (CHO) as side products (Figure 5.4). When levels of LSD1 protein were decreased by siRNA treatment in Hela cells, genes repressed by LSD1 (e.g., M4 AchR and SCN1A) were de-silenced [44]. In agreement with a role of LSD1 as a histone Lys4 demethylase, levels of histone Lys4 methylation were increased on these de-silenced promoters [44].

The LSD1 family of proteins is evolutionarily conserved from fission yeast to humans, but is not identified in budding yeast based on sequence similarity search [44]. However, because LSD1 is specific to mono- and di-methyl Lys4 on H3, several questions remained. Is there another enzyme(s) that is capable of demethylating tri-methyl Lys4? Are there additional enzymes that demethylate other histone methylation sites, such as H3 Lys9, Lys27, Lys36, and Lys79, or H4 Lys20? If the specificity toward mono-, di-, or tri-methyl Lys exists for

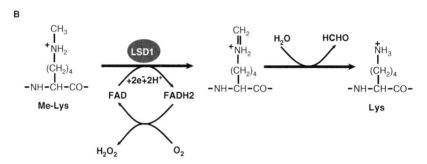

FIG. 5.4. Histone Lys demethylation. (A). Conserved features of LSD1 protein family. Shown here are two human proteins, LSD1 and AOF1, and the LSD1 homolog in *Drosophila* (dLSD1). The catalytic domain shows homology with amine oxidase (denoted by yellow boxes). In addition, the catalytic domain is often juxtaposed with a SWIRM domain, the function of which is unclear. (B) The proposed mechanism of the LSD1 Lys demethylation reaction. FAD is used as a cofactor, and formaldehyde is produced in this reaction.

other Lys demethylase as well, there will be a large number of Lys demethylases. Further, there are only two LSD1 homologous proteins in the human genome. What are those other enzymes, if they exist? It can be speculated that more LSD (Lys-specific demethylase) proteins will be discovered in the future, and their enzymatic mechanisms might be different from that of LSD1.[1]

Consistent with the recurring theme that histone-modifying activities are often organized into big protein complexes, LSD1 is a component of the CtBP (C-terminal binding protein) protein complex, which is involved in transcriptional repression [45]. CtBP directly interacts with the adenovirus protein E1A and likely plays a role during oncogenic transformation by E1A [45]. In *Drosophila*, dCtBP interacts with Knirps to regulate gene expression during the embryonic pattern formation [46]. This complex also contains histone deacetylases HDAC1 and HDAC2 and histone Lys methyltransferases G9a and EuHMT [45]. Hence, this multiprotein complex possesses activities sufficient to modify histones in

---

Recently, a JmjC domain containing protein was identified by the Zhang laboratory as no other family of proteins that demethylate histone Lys in addition to the LSD1 protein family [107].

three ways: histone deacetylation by HDAC1 and 2, Lys methylation to produce repressive methyl-Lys marks on H3 Lys9 by G9a, and Lys demethylation by LSD1 to remove the active methyl marks on H3 Lys4. By combining these three histone-modifying activities, the CtBP complex is very efficient in repressing gene expression.

VI.   Histone Arg Methylation and Transcription

In addition to Lys residues, histone Arg residues can be modified by another family of enzymes: protein Arg methyltransferases (PRMTs) (Figure 5.5). So far, there are seven PRMTs identified in the human genome [47]. (For more information on the PRMT family of proteins, see the chapter by Mark Bedford.) According to the type of dimethyl Arg produced, PRMTs can be classified into two types. Type I PRMTs (including PRMT1, 2, 3, 4, and 6) produce monomethyl Arg and asymmetric dimethyl Arg, whereas type II PRMT (PRMT5) produces monomethyl Arg and symmetric dimethyl Arg. The recently characterized PRMT7 can generate monomethyl Arg and to a lesser extent symmetric dimethyl Arg. Therefore, PRMT7 is a new member of the type II PRMTs. Subcellular localization studies

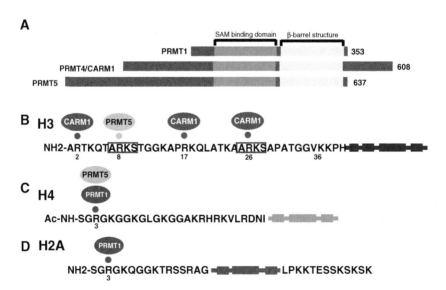

FIG. 5.5. Histone Arg methylation. (A) Three members of the human protein Arg methyltransferases (PRMT1, 4, and 5), which target histone Arg residues, are shown. They have conserved domain structures, including the SAM binding domain and the β-barrel structure C-terminal to the SAM binding. (B–D) Known methylation sites by these three enzymes on histones H3, H4, and H2A are shown.

of GFP-PRMT fusion proteins in Hela cells found that PRMT1, 2, 4, and 6 are localized mainly in the nucleus, whereas PRMT3 and 5 are enriched in the cytoplasm [48]. Noteworthy is that PRMTs target many cytoplasmic and nuclear proteins in addition to histones. Protein Arg methylation has been correlated with various cellular processes, such as transcription, RNA processing, myelination of axons, and cell signaling [49–52]. Arg methylation of non-histone targets in the context of cell signaling has recently been reviewed by Boisvert et al. (2005) [53]. Here, we focus on histone Arg methylation.

PRMTs demonstrate site preference to histone proteins (Figure 5.5). On histone H3, Arg17 and 26 are the preferred sites for PRMT4/CARM1 [54, 55]. On histone H4, Arg3 is the preferred target site by PRMT1 [56, 57]. Recently, it was demonstrated that PRMT5 targets histone H3 Arg8 and H4 Arg3 for methylation [58]. However, most of the enzymatic specificity of PRMTs came from *in vitro* biochemical assays, and little is known about how other interacting proteins may affect the substrate specificity or the degree of methylation to produce monomethyl- or dimethyl-Arg.

Histone Arg methylation has mainly been correlated with transcriptional regulation. PRMT4/CARM1 was first found to interact with the p160 family of coactivators (SRC-1, GRIP1/TIF2, and p/CIP) in yeast two-hybrid screens. CARM1 can preferentially methylate histone H3 in *in vitro* biochemical assays [54]. Moreover, CARM1 enhances the nuclear hormone response by a direct interaction with the p160 coactivators [54]. Although Arg8 and Arg26 on H3 are embedded into the similar sequence context with four conserved residues ARKS at each site, H3 Arg8 is not a target site of CARM1. Interestingly, H3 Arg8 was shown to be the favored methylation site by PRMT5 [58].

CARM1 is an essential gene in mice. CARM1$^{-/-}$ mice embryos are smaller and die perinatally [59]. The methylation of PABP1 [poly(A) binding protein] and p300/CBP by CARM1 in mutant mice cells is abolished. However, histone H3 Arg17 methylation was largely unaffected, which suggests that other PRMTs may target histone H3 Arg17 when CARM1 function is disrupted. Although the global level of H3 Arg17 was largely unaltered, cells lost their ability to transcribe estrogen-dependent genes. It is possible that the CARM1 activity is responsible for H3 methylation at particular promoters, such as those of the estrogen-responsive genes.

Recent studies suggest that CARM1 also associates with other proteins to form a multiprotein complex [60]. When CARM1 and its associated proteins were purified from MCF-7 cells stably expressing epitope-tagged CARM1 protein, a huge protein complex containing at least eight components of the SWI/SNF complex (including the ATPase BRG1) was found. The methyltransferase activity of CARM1 toward nucleosomal histones is greatly increased by the associated proteins. Reciprocally, CARM1 stimulates the ATPase activity of BRG1 in nucleosomal remodeling. *In vivo*, CARM1 and Brg1 cooperatively increase the estrogen-dependent gene expression. Therefore, a mechanistic linkage between histone Arg methylation and ATP-dependent chromatin remodeling has been implicated in nuclear receptor-dependent transcription.

Another extensively studied protein Arg methyltransferase is PRMT1, which has also been found to facilitate the transcription of nuclear hormone response genes [57]. PRMT1 preferentially methylates histone H4 at the Arg3 site. In PRMT1$^{-/-}$ mouse ES cells, the level of histone H4 Arg3 methylation was dramatically decreased, suggesting that PRMT1 is the major enzyme responsible for H4 Arg3 methylation in mice [57]. Further, in enzymatic assays H4 Arg3 was found to be the major site of methylation in the H4 N-terminal tail (residues 1 through 25) [56]. It has been shown that histone Arg methylation functions synergistically with histone Lys acetylation in response to nuclear hormone [57, 61]. Moreover, methylation of H4 Arg3 seems to increase the p300/CPB histone acetyltransferase activity. In contrast, acetylated H4 N-terminal peptide is a poor substrate of PRMT1 [57]. These results suggest interesting regulation of Lys acetylation and Arg methylation of histone proteins through the cross-talk of these two histone marks.

PRMT1 is important for early development of mouse embryos [62]. Using the promoter trap method, the expression of the PRMT1 was found to be highest along the midline of the neural plate, in the forming headfold in E7.5-(embryonic day 7.5)-to-E8.5 embryos, and in the developing central nervous system in E8.5-to-E13.5 embryos. PRMT1$^{-/-}$ embryos die before the E6.5 stage, which suggests that PRMT1 plays a fundamental role during embryonic development. However, PRMT1$^{-/-}$ ES cell lines were established from PRTM1$^{-/-}$ blastocysts. Since these PRMT1$^{-/-}$ ES cells still have about 1% Prmt1 transcription of that of the wild-type cells, it complicated the determination whether PRMT1 is essential for ES cell survival. Total amount of protein Arg methyltransferase activity in PRMT1$^{-/-}$ cells was reduced by 85% of that of wild-type cells, suggesting that PRMT1 is the major methyltransferase responsible for protein Arg methylation in mice.

Recently, PRMT1, CARM1, and histone acetyltransferase p300 were shown to participate in the p53-mediated transcriptional activation [63]. Using reconstituted recombinant chromatin templates and purified cofactors, An et al. found that PRMT1 and CARM1 cooperate with p300 in p53 function. The ability of PRMT1 and CARM1 to activate transcription depends on their activity to modify the chromatin template. An ordered accumulation of PRMTs and p300 and their corresponding histone modifications on the p53 regulated GADD45 promoter was observed upon UV irradiation. Together, these studies suggest that Arg methylation of histones by PRMT1 and CARM1 is involved in gene activation from chromatin templates *in vitro* and *in vivo*. Because p53 is involved in the stability of the genome, protein Arg methylation very likely plays a role in cancers caused by abnormal p53 functions.

PRMT5/JBP1 was first identified in a yeast two-hybrid screen searching for proteins that interact with Jak kinases [64]. In contrast to PRMT1 and

CARM1/PRMT4, PRMT5 produces symmetric dimethyl-Arg in proteins [65]. Although PRMT5 was shown to be mainly cytoplasmic by transient transfection assays, several nuclear proteins have been identified as the targets of PRMT5, including the transcriptional elongation factor SPT5, as well as a phosphatase FCP1 targeting RNA Pol II [66, 67]. Recently, PRMT5 was shown to be associated with SWI/SNF proteins in Hela cells [58]. It was shown that PRMT5/SWI/SNF complex preferential methylates H3. Moreover, the methylation sites of PRMT5 were shown as the Arg8 residue at H3 and the Arg3 residue at H4. In contrast to the function of PRMT1 and CARM1 in transcriptional activation, the activity of PRMT5 seems to repress target genes, including two tumor suppressor genes *ST7* and *NM23* [58]. In that H4 Arg3 is also a major target site by PRMT1, it seems that the methylation of H4 Arg3 by type I PRMT or type II PRMT, respectively, may have opposite functions.

It is an attractive possibility that methyl-Arg residues are recognized by other protein domains in a manner similar to that of methyl-Lys recognition by chromodomain containing proteins. It is known that the Tudor domain of SMN (surviving of motor neurons) interacts with the methyl-Arg containing Sm proteins [68, 69]. This interaction was suggested to play a role in RNA processing. However, the identity of proteins that can recognize methyl-Arg in histones is currently unknown. Interestingly, Tudor, Chromo, and MBT (malignant brain tumor) domains were proposed to form similar 3D structures [70]. Therefore, these domains might be structurally and functionally related during evolution.

VII. Histone Arg Demethylating Enzymes

A. IS HISTONE ARG METHYLATION DYNAMIC?

Like histone Lys demethylase, enzymes that demethylate methyl-Arg in proteins had remained unknown for decades. The paradigm existing for the importance of reversible histone phosphorylation and acetylation in gene regulation suggests that identifying enzymes that govern the dynamic balance of Arg methylation (i.e., discovery of the enzymes that function in opposition to PRMTs) will be greatly important.

The first line of evidence that histone Arg methylation is dynamic came from the analyses of the global change of histone methylation by immunofluorescence in early mouse eggs and embryos [71]. During the transition of eggs from the germinal vesicle stage to the metaphase II stage, global levels of histone Arg methylation (e.g., H3 Arg17 and H4 Arg3) are dramatically decreased [71]. In contrast, the global levels of histone H3 Lys4 methylation are unaltered in the early mouse eggs and embryos. This observation suggests that factors may exist in early mouse eggs for histone Arg demethylation. However, other possibilities may exist, such as histone replacement by the unmodified histones or the clipping of histone tails [72]. This global-level change of histone Arg methylation in early mouse eggs and embryos may have an important role in early development. As witnessed by animal cloning experiments, early mammalian eggs can reprogram a somatic cell nucleus to restore its totipotency for early development. Therefore, it is an attractive idea that the global change of histone Arg methylation may play a role in restoring the totipotency of somatic nuclei for early embryonic development.

It is well established that histone Arg methylation plays a role in transcription mediated by p53 or nuclear hormone receptors [54, 57, 63]. In response to extracellular signals, cells need to alter their gene expression program to adjust to environmental or physiological changes. Therefore, histone modifications on the active promoters need to be erased to acquire repressive histone modifications. As histone Arg methylation is correlated with active transcription, mechanisms that dynamically regulate this modification should exist. The question is whether an enzymatic mechanism is operating on histone Arg residues.

B.   The PAD Protein Family and Human Diseases

To study the reversibility of histone Arg methylation, we have started searching for the putative enzymes with Arg-specific demethylating activities in the Allis and the Coonrod laboratories in 2002. The family of enzymes of particular interest to us at the moment was the peptidylarginine deiminases (PADs).

PADs are a family of enzymes previously well known for their ability to convert protein Arg to Citrulline (Cit) in a calcium- and DTT-dependent manner [73]. Citrullination neutralizes the positive charges carried by protein Arg residues, which likely affects the charge property and the 3D structure of a protein. PAD enzymes have substrates in various tissues, including skin keratin, myelin basic protein in the nervous system, and the recently reported histones [74–76]. PAD4, the only nuclear isoform of PADs, was first identified in HL-60 granulocytes. Citrullination of histones H3, H2A, and H4 was observed after treatment of the HL-60 granulocytes with calcium ionophore, suggesting that calcium signaling is required for the activation of PAD4 [76, 77]. Structural studies find that PAD4 has five calcium binding sites. Comparing the calcium-bound and calcium-free PAD4 crystal structures, the association of calcium with PAD4 induces a conformational change of PAD4, which facilitates the substrate binding and the catalytic activity of the enzyme [78]. Moreover, Cys645 of human PAD4 is directly involved in forming the active site of the enzyme [78].

By sequence similarity searches, PADs seem to be restricted only to vertebrates. There is one PAD protein in the genome of zebrafish or *Xenopus*. In the mouse, rat, and human genomes, there are five PADs, including PAD1, 2, 3, 4, and 6

(PAD6 was also called ePAD) (Figure 5.6). PADs of the same organism share above 40% sequence identity among one another. In addition, the five PAD genes are organized into a cluster on chromosome 1 in humans or on chromosome 4 in mice, suggesting that all PADs may have recently evolved from a single gene by duplication [73]. Strikingly, human and mouse PAD4 share about 73% identities, suggesting that these two enzymes play a similar role in the two organisms.

PAD proteins have been correlated with various human diseases. For example, PAD1 is expressed mainly in the epidermis and uterus, and keratin is one of the PAD1 substrates [75]. In psoriasis patients, keratin is not properly citrullinated, suggesting that PAD1 may be deregulated in this type of disease [79]. Further, when a higher percentage of myelin basic protein (MBP) is citrullinated, multiple sclerosis—a chronic inflammatory disorder of the CNS (central nervous system)—occurs [80, 81]. PAD2 is proposed to mediate the citrullination of MBP in the nervous system. The catalytic activity of PAD2 to MBP is increased by the addition of phosphatidylserine to the reaction. Finally, PAD4 has been genetically correlated with rheumatoid arthritis [82]. Increased levels of PAD4 protein have been found in the joints of patients with rheumatoid arthritis (RA) [81]. Citrullination of joint proteins is likely associated with the pathological progress

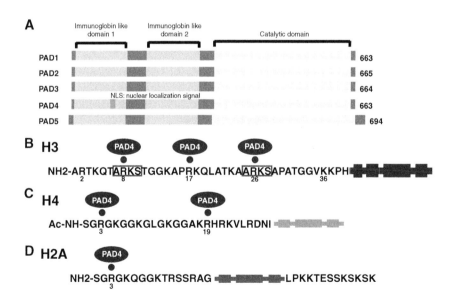

FIG. 5.6. Peptidylarginine deiminases (PADs). (A) The five members of the PAD protein family in the human genome. The N-terminal half of each protein can form two immunoglobin-like domains. The C-terminal of about 300 amino acids of the protein forms the catalytic domain. (B–D) Known target sites of PAD4 on histones H3, H4, and H2A are shown.

of this disease. Sera of RA patients contain a large number of autoimmune antibodies, with significant amount of antibodies against citrullinated proteins. Because PAD4 is a very abundant protein in human polymorphonuclear granulocytes, it can be directly recruited to the inflammatory spots under inflammatory conditions.

Currently, it is not very clear how PADs affect the pathological processes of these several autoimmune diseases. It is possible that the structure of protein targeted by PADs is altered after citrullination, which affects the normal function of these proteins. For example, citrullination of keratin may affect how keratin form bundles during the keratinization process in the skin. Similarly, citrullination of MBP may alter its biophysical properties to interact with phospholipids to form the myelin sheath surrounding axons of oligodendrocytes and neurons. In RA patients, citrulline containing proteins (such as fibrin) may be recognized as foreign antigens by the surveillance mechanism of the immune system.

C.   PAD4 FUNCTIONS AS AN ARG DEMETHYLATING ENZYME

PADs are attractive enzymes for histone Arg demethylases for several reasons. First, as mentioned previously PADs are a family of enzymes previously known to convert protein Arg to citrulline (Cit). Second, PADs display primary sequence and structural similarities with dimethylarginine dimehylaminohydrolase (DDAH), a protein that converts methyl-Arg amino acids to citrulline [83]. These observations led us to hypothesize that PAD may function as an Arg-specific demethylating enzyme by converting methyl-Arg in proteins to citrulline.

To analyze whether histone Arg demethylation occurs, we first focused on PAD4 (the only known nuclear isoform of the PAD protein family), which citrullinates histones at many sites [83] (Figure 5.6). When cellular histone H3 or H4 were treated *in vitro* using purified PAD4 fusion protein, the epitope of H3 Arg17 methylation or H4 Arg3 methylation was lost in western blot experiments, suggesting that PAD4 can react with methyl-Arg in a protein substrate. To further dissect the reaction mechanism, we characterized the released products in the PAD4 reaction. Using volatility assays and cation exchange columns, methylamine was identified as a released product. This led us to propose a demethylimination reaction catalyzed by PAD4 with methyl-Arg containing histones (Figure 5.7).

To test if this reaction happens *in vivo*, we turned our attention to HL-60 granulocytes. In response to DMSO or ATRA (all trans-retinoic acid), human leukemia HL-60 cells differentiate into granulocytes. PAD4 is induced during this differentiation process [84]. Further, histone citrullination was found upon treatment of calcium ionophore to activate PAD4 in HL-60 granulocytes [76, 77]. Consistent with the biochemistry assays *in vitro*, a decrease of histone Arg methylation with concurrent increase of citrullination was observed upon activation of PAD4 by calcium ionophore. Using protein microsequencing and immunological

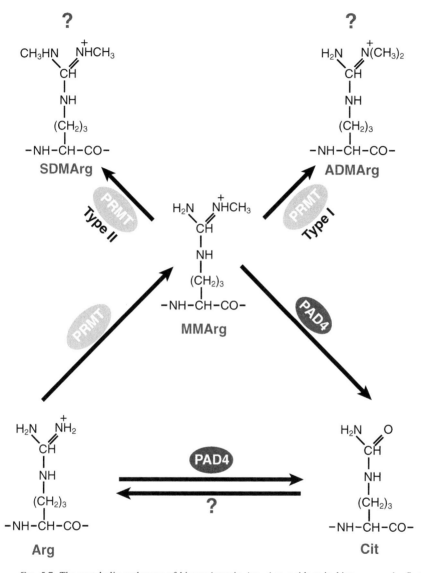

FIG. 5.7. The metabolic pathways of histone/protein Arg. Arg residues in histones can be first converted to monomethyl-Arg and Cit by PRMT and PAD, respectively. Monomethyl-Arg can be further methylated by the type II PRMTs to produce symmetric dimethyl-Arg, or by the type I PRMTs to produce asymmetric dimethyl-Arg. *In vitro*, monomethyl-Arg was found to be converted to Cit by PAD4. However, whether or how dimethyl-Arg will be removed is currently unclear, nor is the fate of citrulline in histones.

methods, the decrease of H3 Arg17 methylation was correlated with the increase of H3 Cit17. As well, the decrease of H4 Arg3 methylation was associated with the production of H4 Cit3. It seems that there is an interconversion of methyl-Arg to Cit in these cells. When PAD4 protein levels were decreased by SiRNA treatment, those granulocytes lost their ability to decrease methyl-Arg and to increase Cit, suggesting that PAD4 is the primary enzyme responsible for the dynamic change of histone Arg methylation.

The function of histone citrullination is currently unknown. However, we favor the view that citrullination of histones may affect global chromatin structure because the positive charge carried by histone Arg residues is lost upon citrullination. As discussed previously, histone Arg methylation mediated by CARM1 and PRMT1 is involved in transcriptional regulation by p53 or nuclear hormone receptors. The ability of PAD4 to decrease histone Arg methylation suggests that the PAD4 activity may be involved in transcriptional regulation by p53 or nuclear hormone receptors. In fact, when tested by transient transfection reporter assays PAD4 can repress an estrogen-inducible reporter gene in human breast cancer MCF-7 cells. On the endogenous pS2 promoter, the conversion of methyl-Arg to Cit on histones is correlated with transcriptional repression [83, 85]. It will be interesting in the future to test how p53 or nuclear-hormone-regulated gene expression is affected in PAD4$^{-/-}$ cells.

Although PRMTs can produce both mono- and di-methyl Arg in histones and other proteins, a mass spectrometry study of cellular histone H4 found only mono-methyl H4 Arg3. Interestingly, in our *in vitro* biochemistry experiments testing the demethylimination activity of PAD4 with histone H4 we found that PAD4 targets primarily mono-methyl Arg in histone H4 (Figure 5.7). It is not clear that the specificity of PAD4 is altered when PAD4 is incorporated with other proteins to form a complex, or when histone H4 is organized in a nucleosome. The antibodies against histone H4 Arg3 methylation were originally generated against dimethyl-Arg H4 peptide. Further, this antibody recognizes both mono- and dimethyl H4 Arg3. The majority of immuno-activity of this antibody with HL-60 granulocytes, suggesting both monomethyl and dimethyl H4 Arg3 epitope have been removed.

In addition to histones, many other nuclear proteins (such as Spt5, STAT1, and p300) are targets of Arg methylation. We speculate that PAD4 may target additional nuclear proteins for demethylimination. In support, a recent report showed that PAD4 can demethyliminate p300 [86]. PAD4 is a nuclear isoform of the PAD enzyme family. Because many cytoplasmic proteins contain methyl-Arg, we propose that a cytoplasmic PAD will be responsible for the demethylimination of cytoplasmic targets. PAD2, a widely expressed cytoplasmic PAD, is an excellent candidate.

Existing evidence suggests that histone Arg residues can exit in different ways of modification, including monomethylation, dimethylation, and citrullination (Figure 5.7). The fate of Cit in histones is currently unknown. Citrullinated histones can be replaced by other histones during the S phase of the cell cycle or by the ATPase-driven chromatin remodeling complexes. Alternatively, histone Cit can be converted to Arg by unknown enzymatic activities. Although such types of enzymes have not been found, enzymes that convert Cit amino acid back to Arg amino acid are well known. Argininosuccinate synthase (AS) and argininosuccinate lyase (AL) are a pair of enzymes involved in the urea cycle or nitric oxide (NO) synthesis pathway, which can convert Cit back to Arg at the amino acid level through an aminotransfer reaction [87]. We postulate that enzymes catalyzing the similar reaction on protein Cit residues may exist.

VIII. Conclusions

Interest in posttranslational modifications of histone proteins is at an all-time high. The potential role of this type of information in the regulating of genomic function is being actively explored by many talented scientists. A "histone code" hypothesis was proposed, which suggest that covalent modifications on histone may affect the chromatin biology to an extent unanticipated previously. Together with other epigenetic information (such as DNA methylation, RNA interference, and micro RNAs), histone modifications may contribute to the "epigenetic code" regulating the genetic code encoded by the four nucleotides A, G, C, and T.

Early embryonic development in a multicellular organism is a recurring theme of cell differentiation. Each differentiated somatic cell plays a unique physiological role due to cell-type-specific gene expression. Because a somatic nucleus can regain totipotency in early mammalian eggs, cellular differentiation is therefore controlled by epigenetic mechanisms that do not involve direct change of the DNA sequence per se.

During tumorigenesis processes, the change of gene expression pattern from a normal cell to a cancer cell is often accompanied by mutations and chromosome instability. However, epigenetic change may precede genetic change to form tumors, especially during early stages of cancer development. It is possible that by catching the tumor cells at an early stage of "epigenetic" change the chance for early tumor diagnosis and for successful treatment will be much higher. Recent findings that SAHA, an HDAC inhibitor, can work as a potent chemotherapy compound to treat cancer patients offered an excellent example that more avenues to fight cancer may become available as our knowledge of epigenetics grows.

REFERENCES

1. Strahl, B.D., and Allis, C.D. (2000). The language of covalent histone modifications. *Nature* 403:41–45.
2. Turner, B.M. (2000). Histone acetylation and an epigenetic code. *Bioessays* 22, 836–45.

3. Turner, B.M. (2002). Cellular memory and the histone code. *Cell* 111:285–291.

4. Jenuwein, T., and Allis, C.D. (2001). Translating the histone code. *Science* 293:1074–1080.

5. Felsenfeld, G., and Groudine, M. (2003). Controlling the double helix. *Nature* 421:448–453.

6. Jaenisch, R., and Bird, A. (2003). Epigenetic regulation of gene expression: how the genome integrates intrinsic and environmental signals. *Nat Genet* 33(Suppl. 2):45–54.

7. Zhang, Y., and Reinberg, D. (2001). Transcription regulation by histone methylation: interplay between different covalent modifications of the core histone tails. *Genes Dev* 15:2343–2360.

8. Luger, K., Mader, A.W., Richmond, R.K., Sargent, D.F., and Richmond T.J. (1997). Crystal structure of the nucleosome core particle at 2.8 A resolution. *Nature* 389:251–260.

9. Whitlock, J.P., Jr. and Simpson, R.T. (1976). Removal of histone H1 exposes a fifty base pair DNA segment between nucleosomes. *Biochemistry* 15:3307–3314.

10. Dorigo, B., Schalch, T., Kulangara, A., Duda, S., Schroeder, R.R., and Richmond, T. (2004). Nucleosome arrays reveal the two-start organization of the chromatin fiber. *Science* 306:1571–1573.

11. Elgin, S.C., and Grewal, S.I. (2003). Heterochromatin: silence is golden. *Curr Biol* 13:R895–R898.

12. Hebbar, P.B., and Archer, T.K. (2002). Nuclear factor 1 is required for both hormone-dependent chromatin remodeling and transcriptional activation of the mouse mammary tumor virus promoter. *Mol Cell Biol* 23:887–898.

13. Thiriet, C., and Hayes, J.J. (2005). Replication-independent core histone dynamics at transcriptionally active loci in vivo. *Genes Dev* 19:677–682.

14. Levchenko, V., and Jackson, V. (2004). Histone release during transcription: NAP1 forms a complex with H2A and H2B and facilitates a topologically dependent release of H3 and H4 from the nucleosome. *Biochemistry* 43:2359–2372.

15. Levchenko, V., Jackson, B. , and Jackson, V. (2005). Histone Release during Transcription: Displacement of the Two H2A-H2B Dimers in the Nucleosome Is Dependent on Different Levels of Transcription-Induced Positive Stress. *Biochemistry* 44:5357–5372.

16. Sanders, S.L., Portoso, M., Mata, J., Bahler, J., Allshire, R.C., and Kouzarides, T. (2004). Methylation of histone H4 lysine 20 controls recruitment of Crb2 to sites of DNA damage. *Cell* 119:603–614.

17. Huyen, Y., Zgheib, O., Ditullio, R.A., Jr., Gorgoulis, V.G., Zacharatos, P., Petty, T.J., Sheston, E.A., Mellert, H.S., Stavridi, E.S., and Halazonetis, T.D. (2004). Methylated lysine 79 of histone H3 targets 53BP1 to DNA double-strand breaks. *Nature* 432:406–411.

18. Paik, W.K., and Kim, S. (1969) Enzymatic methylation of histones. *Arch Biochem Biophys* 134:632–637.

19. Rea, S., Eisenhaber, F., O'Carroll, D., Strahl, B.D., Sun, Z.W., Schmid, M., Opravil, S., Mechtler, K., Ponting, C.P., Allis, C.D., and Jenuwein, T. (2000). Regulation of chromatin structure by site-specific histone H3 methyltransferases. *Nature* 406:593–599.

20. Jacobs, S.A., Taverna, S.D., Zhang, Y., Briggs, S.D., Li. J., Eissenberg, J.C., Allis, C.D., and Khorasanizadeh, S. (2001). Specificity of the HP1 chromo domain for the methylated N-terminus of histone H3. *Embo J* 20:5232–5241.

21. Jacobs, S.A., and Khorasanizadeh, S. (2002). Structure of HP1 chromodomain bound to a lysine 9-methylated histone H3 tail. *Science* 295:2080–2083.

22. Fischle, W., Wang, Y., Jacobs, S., Kim, Y., Allis, C.D., and Khorasanizadeh, S. (2003). Molecular basis for the discrimination of repressive methyl-lysine marks in histone H3 by Polycomb and HP1 chromodomains. *Genes Dev* 17:1870–1881.

23. Rice, J.C., Briggs, S.D., Ueberheide, B., Barber, C.M., Shabanowitz, J., Hunt, D.F., Shinkai, Y., and Allis, C.D. (2003). Histone methyltransferases direct different degrees of methylation to define distinct chromatin domains. *Mol Cell* 12:1591–1598.

24. Peters, A.H., Kubicek, S., Mechtler, K., O'Sullivan, R.J., Derijck, A.A., Perez-Burgos, L., Kohlmaier, A., Opravil, S., Tachibana, M., Shinkai, Y., Martens, J.H., and Jenuwein, T. (2003).

Partitioning and plasticity of repressive histone methylation states in mammalian chromatin. *Mol Cell* 12:1577–1589.

25. Wang, H., An, W., Cao, R., Xia, L., Erdjument-Bromage, H., Chatton, B., Tempst, P., Roeder, R.G., and Zhang, Y. (2003). mAM facilitates conversion by ESET of dimethyl to trimethyl lysine 9 of histone H3 to cause transcriptional repression. *Mol Cell* 12:475–487.

26. Kuzmichev, A., Jenuwein, T., Tempst, P., and Reinberg, D. (2004) Different EZH2-containing complexes target methylation of histone H1 or nucleosomal histone H3. *Mol Cell* 14:183–193.

27. Hess, J.L. (2004). Mechanisms of transformation by MLL. *Crit Rev Eukaryot Gene Expr* 14:235–254.

28. Yokoyama, A., Wang, Z., Wysocka, J., Sanyal, M., Sanyal, M., Aufiero, D.J., Kitabayashi, I., Herr, W, Cleary, M.L. (2004). Leukemia proto-oncoprotein MLL forms a SET1-like histone methyltransferase complex with menin to regulate Hox gene expression. *Mol Cell Biol* 24:5639–5649.

29. Nakamura, T., Mori, T., Tada, S., Krajewski, W., ROzovskaia, T., Wassell, R., Dubois, G., Mazo, A., Corce, C.M., and Canaani, E. (2002). ALL-1 is a histone methyltransferase that assembles a supercomplex of proteins involved in transcriptional regulation. *Mol Cell*, 10:1119–1128.

29b. Dou, Y., Milne, T.A., Tackett, A.J., Smith, E.R., Fukuda, A., Wysocka, J., Allis, C.D., Chait, B.T., Hess, J.L., Roeder, R.G. (2005). Physical association and coordinate function of the H3 K4 methyltransferase MLL1 and dthe H4 K16 acetyltransferase MOF. *Cell* 121:873–885.

30. Nielsen, P.R., Nietlispach, D., Mott, H.R., Callaghan, J., Bannister, A., Kouzarides, T., Murzin, A.G., Murzina, N.V., and Laue, E.D. (2002). Structure of the HP1 chromodomain bound to histone H3 methylated at lysine 9. *Nature* 416:103–107.

31. Nielsen, S.J., Schneider, R., Bauer, U.M., Bannister, A.J., Morrison, A., O'Carroll, D., Firestein, R., Cleary, M., Jenuwein, T., Herrera, R.E., and Kouzarides, T. (2001). Rb targets histone H3 methylation and HP1 to promoters. *Nature* 412:561–565.

32. Chakraborty, S., Sinha, K.K., Senyuk, V., and Nucifora, G. (2003). Suv39H1 interacts with AML1 and abrogates AML1 transactivity. AML1 is methylated in vivo. *Oncogene* 22:5229–5237.

33. Fujita, N., Watanabe, S., Ichimura, T., Tsuruzoe, S., Shinkai, Y., Tachibana, M., Chiba, T., and Nakao, M. (2003). Methyl-CpG binding domain 1 (MBD1) interacts with the Suv39h1-HP1 heterochromatic complex for DNA methylation-based transcriptional repression. *J Biol Chem* 278:24132–24138.

34. Vaute, O., Nicolas, E., Vandel, L., and Trouche, D. (2002). Functional and physical interaction between the histone methyl transferase Suv39H1 and histone deacetylases. *Nucleic Acids Res* 30:475–481.

35. Shao, Z., Raible, F., Mollaaghababa, R., Guyon, J.R., Wu, C.T., Bender, W., and Kingston, R.E. (1999). Stabilization of chromatin structure by PRC1, a Polycomb complex. *Cell* 98:37–46.

36. Cao, R., and Zhang, Y. (2004). SUZ12 is required for both the histone methyltransferase activity and the silencing function of the EED-EZH2 complex. *Mol Cell* 15:57–67.

37. Plath, K., Fang, J., Mlynarczyk-Evans, S.K., Cao, R., Worringer, K.A., Wang, H., de la Cruz, C.C., Otte, A.P., Panning, B., and Zhang, Y. (2003). Role of histone H3 lysine 27 methylation in X inactivation. *Science* 300:131–135.

38. Plath, K., Talbot, D., Hamer, K.M., Otte, A.P., Yang, T.P., Jaenisch, R., and Panning, B. (2004). Developmentally regulated alterations in Polycomb repressive complex 1 proteins on the inactive X chromosome. *J Cell Biol* 167:1025–1035.

38b. Okada, Y., Feng, Q., Lin, Y., Jiang, Q., Li, Y., Coffield, V. M., Su, L., Xu, G., and Zhang, Y. (2005). hDOT1L links in histone methylation to leukemogenesis. *Cell* 121:167–178.

39. Bernstein, B.E., Kamal, M., Lindblad-Toh, K., Bekiranov, S., Bailey, D.K., Huebert, D.J., McMahon, S., Karlsson, E.K., Kulbokas, E.J., 3rd, Gingeras, T.R., Schreiber, S.L., and

Lander, E.S. (2005). Genomic maps and comparative analysis of histone modifications in human and mouse. *Cell* 120:169–181.

40. Ng, H.H., Morshead, K.B., Oettinger, M.A., and Struhl, K. (2003). Lysine-79 of histone H3 is hypomethylated at silenced loci in yeast and mammalian cells: a potential mechanism for position-effect variegation. *Proc Natl Acad Sci USA* 100:1820–1825.

41. Thomas, G., Lange, H.W., and Hempel, K. (1975). Kinetics of histone methylation in vivo and its relation to the cell cycle in Ehrlich ascites tumor cells. *Eur J Biochem* 51:609–615.

42. Borun, T.W., Pearson, D., and Paik, W.K. (1972). Studies of histone methylation during the HeLa S-3 cell cycle. *J Biol Chem* 247:4288–4298.

43. Hempel, K., Thomas, G., Roos, G., Stocker, W., and Lange, H.W. (1979). N epsilon-Methyl groups on the lysine residues in histones turn over independently of the polypeptide backbone. *Hoppe Seylers Z Physiol Chem* 360:869–876.

44. Shi, Y., Lan, F., Matson, C., Mulligan, P., Whetstine, J.R., Cole, P.A., Casero, R.A., and Shi, Y. (2004). Histone demethylation mediated by the nuclear amine oxidase homolog LSD1. *Cell* 119:941–953.

45. Shi, Y, Sawada, J., Sui, G., Affar, el B., Whetstine, J.R., Lan, F., Ogawa, H., Luke, M.P., Nakatani, Y., and Shi, Y. (2003). Coordinated histone modifications mediated by a CtBP co-repressor complex. *Nature*, 422:735–738.

46. Nibu, Y., and Levine, M.S. (2001). CtBP-dependent activities of the short-range Giant repressor in the Drosophila embryo. *Proc Natl Acad Sci USA* 98:6204–6208.

47. Lee, D.Y., Teyssier, C., Strahl, B.D., and Stallcup, M.R. (2005). Role of protein methylation in regulation of transcription. *Endocr Rev* 26:147–170.

48. Frankel, A., Yadav, N., Lee, J., Branscombe, T.L., Clarke, S., and Bedford, M.T. (2002). The novel human protein arginine N-methyltransferase PRMT6 is a nuclear enzyme displaying unique substrate specificity. *J Biol Chem* 277:3537–3543.

49. Boisvert, F.M., Cote, J., Boulanger, M.C., and Richard, S. (2003). A Proteomic Analysis of Arginine-methylated Protein Complexes. *Mol Cell Proteomics* 2:1319–1330.

50. Gary, J.D., and Clarke, S. (1998). RNA and protein interactions modulated by protein arginine methylation. *Prog Nucleic Acid Res Mol Biol* 61:65–131.

51. Mowen, K.A., Tang, J., Zhu, W., Schurter, B.T., Shuai, K., Herschman, H.R., and David, M. (2001) Arginine methylation of STAT1 modulates IFNalpha/beta-induced transcription. *Cell* 104:731–741.

52. Xu, W., Chen, H., Du, K., Asahara, H., Tini, M., Emerson, B.M., Montminy, M., and Evans, R.M. (2001). A transcriptional switch mediated by cofactor methylation. *Science* 294:2507–2511.

53. Boisvert, F.M., Chenard, C.A., and Richard, S. (2005). Protein interfaces in signaling regulated by arginine methylation. *Sci STKE 2005*. 271:re2.

54. Chen, D., Ma, H., Hong, H., Koh, S.S., Huang, S.M., Schurter, B.T., Aswad, D.W., and Stallcup, M.R. (1999). Regulation of transcription by a protein methyltransferase. *Science* 284:2174–2177.

55. Bauer, U.M., Daujat, S., Nielsen, S.J., Nightingale, K., and Kouzarides, T. (2002). Methylation at arginine 17 of histone H3 is linked to gene activation. *EMBO Rep* 3:39–44.

56. Strahl, B.D., Briggs, S.D., Brame, C.J., Caldwell, J.A., Koh, S.S., Ma, H., Cook, R.G., Shabanowitz, J., Hunt, D.F., Stallcup, M.R., and Allis, C.D. (2001). Methylation of histone H4 at arginine 3 occurs in vivo and is mediated by the nuclear receptor coactivator PRMT1. *Curr Biol* 11:996–1000.

57. Wang, H., Huang, Z.Q., Xia, L., Feng, Q., Erdjument-Bromage, H., Strahl, B.D., Briggs, S.D., Allis, C.D., Wong, J., Tempst, P., and Zhang, Y. (2001). Methylation of histone H4 at arginine 3 facilitating transcriptional activation by nuclear hormone receptor. *Science* 293:853–857.

58. Pal, S., Vishwanath, S.N., Erdjument-Bromage, H., Tempst, P., and Sif, S. (2004). Human SWI/SNF-associated PRMT5 methylates histone H3 arginine 8 and negatively regulates expression of ST7 and NM23 tumor suppressor genes. *Mol Cell Biol*, 24:9630–9645.

59. Yadav, N., Lee, J., Kim, J., Shen, J., Hu, M.C., Aldaz, C.M., and Bedford, M.T. (2003). Specific protein methylation defects and gene expression perturbations in coactivator-associated arginine methyltransferase 1-deficient mice. *Proc Natl Acad Sci USA* 100:6464–6468.
60. Xu, W., Cho, H., Kadam, S., Banayo, E.M., Anderson, S., Yate, J.R., 3rd, Emerson, B.M., and Evans, R.M. (2004). A methylation-mediator complex in hormone signaling. *Genes Dev* 18:144–156.
61. Daujat, S., Bauer, U.M., Shah, V., Turner, B., Berger, S., and Kouzarides, T. (2002). Crosstalk between CARM1 methylation and CBP acetylation on histone H3. *Curr Biol* 12:2090–2097.
62. Pawlak, M.R., Scherer, C.A., Chen, J., Roshon, M.J., Ruley, H.E. (2000). Arginine N-methyltransferase 1 is required for early postimplantation mouse development, but cells deficient in the enzyme are viable. *Mol Cell Biol* 20:4859–4869.
63. An, W., Kim, J., and Roeder, R.G. (2004). Ordered cooperative functions of PRMT1, p300, and CARM1 in transcriptional activation by p53. *Cell* 117:735–748.
64. Pollack, B.P., Kotenko, S.V., He, W., Izotova, L.S., Barnoski, B.L., and Pestka, S. (1999). The human homologue of the yeast proteins Skb1 and Hsl7p interacts with Jak kinases and contains protein methyltransferase activity. *J Biol Chem* 274:31531–31542.
65. Branscombe, T.L., Frankel, A., Lee, J.H., Cook, J.R., Yang, Z., Pestka, S., and Clarke, S. (2001). PRMT5 (Janus kinase-binding protein 1) catalyzes the formation of symmetric dimethylarginine residues in proteins. *J Biol Chem* 276:32971–32976.
66. Amente, S., Napolitano, G., Licciardo, P., Monti, M., Pucci, P., Lania, L., and Majello, B. (2005). Identification of proteins interacting with the RNAPII FCP1 phosphatase: FCP1 forms a complex with arginine methyltransferase PRMT5 and it is a substrate for PRMT5-mediated methylation. *FEBS Lett* 579:683–689.
67. Kwak, Y.T., Guo, J., Prajapati, S., Park, K.J., Surabhi, R.M., Miller, B., Gehrig, P., and Gaynor, R.B. (2003). Methylation of SPT5 regulates its interaction with RNA polymerase II and transcriptional elongation properties. *Mol Cell* 11:1055–1066.
68. Brahms, H., Meheus, L., de Brabandere, V., Fischer, U., and Luhrmann, R. (2001). Symmetrical dimethylation of arginine residues in spliceosomal Sm protein B/B′ and the Sm-like protein LSm4, and their interaction with the SMN protein. *RNA* 7:1531–1542.
69. Sprangers, R., Groves, M.R., Sinning, I., Sattler, M. (2003). High-resolution X-ray and NMR structures of the SMN Tudor domain: conformational variation in the binding site for symmetrically dimethylated arginine residues. *J Mol Biol* 327:507–520.
70. Maurer-Stroh, S., Dickens, N.J., Hughes-Davies, L., Kouzarides, T., Eisenhaber, F., and Ponting, C.P. (2003). The Tudor domain 'Royal Family': Tudor, plant Agenet, Chromo, PWWP and MBT domains. *Trends Biochem Sci* 28:69–74.
71. Sarmento, O.F., Digilio, L.C., Wang, Y., Perlin, J., Herr, J.C., Allis, C.D., and Coonrod, S.A. (2004). Dynamic alterations of specific histone modifications during early murine development. *J Cell Sci* 117:4449–4459.
72. Bannister, A.J., Schneider, R., and Kouzarides, T. (2002). Histone methylation: dynamic or static? *Cell* 109:801–806.
73. Vossenaar, E.R., Zendman, A.J., van Venrooij, W.J., and Pruijn, G.J. (2003). PAD, a growing family of citrullinating enzymes: genes, features and involvement in disease. *Bioessays* 25:1106–1118.
74. Akiyama, K., Sakurai, Y., Asou, H., and Senshu, T. (1999). Localization of peptidylarginine deiminase type II in a stage-specific immature oligodendrocyte from rat cerebral hemisphere. *Neurosci Lett* 274:53–55.
75. Nachat, R., Mechin, M.C., Takahara, H., Chavanas, S., Charveron, M., Serre, G., and Simon, M. (2005). Peptidylarginine deiminase isoforms 1–3 are expressed in the epidermis and involved in the deimination of K1 and filaggrin. *J Invest Dermatol* 124:384–393.

76. Nakashima, K., Hagiwara, T., and Yamada, M. (2002). Nuclear localization of peptidylarginine deiminase V and histone deimination in granulocytes. *J Biol Chem* 277:49562–49568.
77. Hagiwara, T., Nakashima, K., Hirano, H., Senshu, T., and Yamada, M. (2002). Deimination of arginine residues in nucleophosmin/B23 and histones in HL-60 granulocytes. *Biochem Biophys Res Commun* 290:979–983.
78. Arita, K., Hashimoto, H., Shimizu, T., Nakashima, K., Yamada, M., and Sato, M. (2004). Structural basis for Ca(2+)-induced activation of human PAD4. *Nat Struct Mol Biol* 11:777–783.
79. Ishida-Yamamoto, A., Senshu, T., Takahashi, H., Akiyama, K., Nomura, K., and Izuka, H. (2000). Decreased deiminated keratin K1 in psoriatic hyperproliferative epidermis. *J Invest Dermatol* 114:701–705.
80. De Keyser, J., Schaaf, M., and Teelken, A. (1999). Peptidylarginine deiminase activity in postmortem white matter of patients with multiple sclerosis. *Neurosci Lett* 260:74–76.
81. Moscarello, M.A., Pritzker, L., Mastronardi, F.G., and Wood, D.D. (2002). Peptidylarginine deiminase: a candidate factor in demyelinating disease. *J Neurochem* 81:335–343.
82. Yamada, R., and Ymamoto, K. (2005). Recent findings on genes associated with inflammatory disease. *Mutat Res* 573:136–151.
83. Wang, Y., Wysocka, J., Sayegh, J., Lee, Y.H., Perlin, J.R., Leonelli, L., Sonbuchner, L.S., McDonald, C.H., Cook, R.G., Clarke, S., Stallcup, M.R., Allis, C.D., and Coonrod, S.A. (2004). Human PAD4 regulates histone arginine methylation levels via demethylimination. *Science* 306:279–283.
84. Nakashima, K., Hagiwara, T., Ishigami, A., Nagata, S., Asaga, H., Kuramoto, M., Senshu, T., and Yamada, M. (1999). Molecular characterization of peptidylarginine deiminase in HL-60 cells induced by retinoic acid and 1alpha,25-dihydroxyvitamin D(3). *J Biol Chem* 274: 27786–27792.
85. Cuthbert, G.L., Daujat, S., Snowden, A.W., Erdjument-Bromage, H., Hagiwara, T., Yamada, M., Schneider, R., Gregory, P.D., Tempst, P., Bannister, A.J., and Kouzarides, T. (2004). Histone deimination antagonizes arginine methylation. *Cell* 118:545–553.
86. Lee, Y.H., Coonrod, S.A., Kraus, W.L., Jelinek, M.A., and Stallcup, M.R. (2005). Regulation of coactivator complex assembly and function by protein arginine methylation and demethylimination. *Proc Natl Acad Sci USA* 102:3611–3616.
87. Hao, G., Xie, L., and Gross, S.S. (2004). Argininosuccinate synthetase is reversibly inactivated by S-nitrosylation in vitro and in vivo. *J Biol Chem* 279:36192–36200.
88. Briggs, S.D., Bryk, M., Strahl, B.D., Cheung, W.L., Davie, J.K., Dent, S.Y., Winston, F., and Allis, C.D. (2001). Histone H3 lysine 4 methylation is mediated by Set1 and required for cell growth and rDNA silencing in Saccharomyces cerevisiae. *Genes Dev* 15:3286–3295.
89. Krogan, N.J., Dover, J., Khorrami, J., Greenblatt, J.F., Schneider, J., Johnston, M., Shilatifard, A. (2002). COMPASS, a histone H3 (Lysine 4) methyltransferase required for telomeric silencing of gene expression. *J Biol Chem* 277:10753–10755.
90. Nagy, P.L., Griesenbeck, J., Kornberg, R.D., and Cleary, M.L. (2002). A trithorax-group complex purified from Saccharomyces cerevisiae is required for methylation of histone H3. *Proc Natl Acad Sci USA* 99:90–94.
91. Roguev, A., Schaft, D., Shevchenko, A., Pijnappel, W.W., Wilm, M., Aasland, R., and Stewart, A.F. (2001). The Saccharomyces cerevisiae Set1 complex includes an Ash2 homologue and methylates histone 3 lysine 4. *Embo J* 20:7137–7148.
92. Nishioka, K., Rice, J.C., Sarma, K., Erdjument-Bromage, H., Werner, J., Wang, Y., Chuikov, S., Valenzuela, P., Tempst, P., Steward, R., Lis, J.T., Allis, C.D., and Reinberg, D. (2002). PR-Set7 is a nucleosome-specific methyltransferase that modifies lysine 20 of histone H4 and is associated with silent chromatin. *Mol Cell* 9:1201–1213.
93. Wang, H., Cao, R., Xia, L., Erdjument-Bromage, H., Borchers, C., Tempst, P., and Zhang, Y. (2001). Purification and functional characterization of a histone H3-lysine 4-specific methyltransferase. *Mol Cell* 8:1207–1217.

94. Milne, T.A., Briggs, S.D., Brock, H.W., Martin, M.E., Gibbs, D., Allis, C.D., and Hess, J.L. (2002). MLL targets SET domain methyltransferase activity to Hox gene promoters. *Mol Cell* 10:1107–1117.

95. Yang, L., Wu, D.Y., Wang, H., Chansky, H.A., Schubach, W.H., Hickstein, D.D., and Zhang, Y. (2002). Molecular cloning of ESET, a novel histone H3-specific methyltransferase that interacts with ERG transcription factor. *Oncogene* 21:148–152.

96. Schotta, G., Ebert, A., Krauss, V., Fischer, A., Hoffmann, J., Rea, S., Jenuwein, T., Dorn, R., and Reuter, G. (2002). Central role of Drosophila SU(VAR)3-9 in histone H3-K9 methylation and heterochromatic gene silencing. *Embo J* 21:1121–1131.

97. Kim, K.C., Geng, L., and Huang, S. (2003). Inactivation of a histone methyltransferase by mutations in human cancers. *Cancer Res* 63:7619–7623.

98. Cao, R., Wang, L., Wang, H., Xia, L., Erdjument-Bromage, H., Tempst, P., Jones, R.S., and Zhang, Y. (2002). Role of histone H3 lysine 27 methylation in Polycomb-group silencing. *Science* 298:1039–1043.

99. Strahl, B.D., Grant, P.A., Briggs, S.D., Sun, Z.W., Bone, J.R., Caldwell, J.A., Mollah, S., Cook, R.G., Shabanowitz, J., Hunt, D.F., and Allis, C.D. (2002). Set2 is a nucleosomal histone H3-selective methyltransferase that mediates transcriptional repression. *Mol Cell Biol* 22:1298–1306.

100. van Leeuwen, F., Gafken, P.R., and Gottschling, D.E. (2002). Dot1p modulates silencing in yeast by methylation of the nucleosome core. *Cell* 109:745–756.

101. Ng, H.H., Xu, R.M., Zhang, Y., and Struhl, K. (2002). Ubiquitination of histone H2B by Rad6 is required for efficient Dot1-mediated methylation of histone H3 lysine 79. *J Biol Chem* 277:34655–34657.

102. Feng, Q., Wang, H., Ng, H.H., Erdjument-Bromage, H., Tempst, P., Struhl, K., and Zhang, Y. (2002). Methylation of H3-lysine 79 is mediated by a new family of HMTases without a SET domain. *Curr Biol* 12:1052–1058.

103. Fang, J., Feng, Q., Ketel, C.S., Wang, H., Cao, R., Xia, L., Erdjument-Bromage, H., Tempst, P., Simon, J.A., and Zhang, Y. (2002). Purification and functional characterization of SET8, a nucleosomal histone H4-lysine 20-specific methyltransferase. *Curr Biol* 12:1086–1099.

104. Wysocka, J., Myers, M.P., Laherty, C.D., Eisenman, R.N., and Herr, W. (2003). Human Sin3 deacetylase and trithorax-related Set1/Ash2 histone H3-K4 methyltransferase are tethered together selectively by the cell-proliferation factor HCF-1. *Genes Dev* 17:896–911.

105. Steele-Perkins, G., Fang, W., Yang, X.H., Van Gele, M., Carling, T., Gu, J., Buyse, I.M., Fletcher, J.A., Liu, J., Bronson, R., Chadwick, R.B., de la Chapelle, A., Zhang, X., Speleman, F., and Huang, S. (2001). Tumor formation and inactivation of RIZ1, an Rb-binding member of a nuclear protein-methyltransferase superfamily. *Genes Dev* 15:2250–2262.

106. Chuikov, S., Kurash, J.K., Wilson, J.R., Xiao, B., Justin, N., Ivanov, G.S., McKinney, K., Tempst, P., Prives, C., Gamblin, S.J., Barlev, N.A., and Reinberg, D. (2004). Regulation of p53 activity through lysine methylation. *Nature* 432:353–360.

107. Tsukada, Y., Fang, J., Erdjument-Bromage, H., Warren, M. E., Borchers, C. H., Jempst, P., Zhang, Y. (2006). Histone demethylation by a family of JmjC domain-containing proteins. *Nature* 439:811–816.

6

Structure of SET Domain Protein Lysine Methyltransferases

BING XIAO • STEVEN J. GAMBLIN • JONATHAN R. WILSON

Division of Molecular Structure
National Institute for Medical Research
The Ridgeway, Mill Hill
London NW71AA UK

I. Abstract

Methylation of lysine residues is now understood to constitute a key component of the complex signaling paradigm referred to as the histone code hypothesis. Rapid progress has been made in the structural and functional biology of the enzymes responsible for this modification. These SET proteins are based on a common fold that appears unique to this family. The structure of the SET domain is such that peptide substrates bind on one surface, whereas the AdoMet cofactor binds on the opposite side of the domain. Remarkably, the target lysine residue gains access to the cofactor by passing through a channel that runs through the SET domain connecting these two binding surfaces.

Different SET enzymes carry out mono-, di-, or tri-methylation of their targets, and these modifications give rise to distinctive biological readouts. Ternary complexes of several SET enzymes reveal how the size and bonding patterns of residues flanking the active site determine the multiplicity of methylation that occurs. Indeed, the methylation multiplicity of some of these enzymes has been engineered by specifically mutating residues close to the active site.

The catalytic activity of SET enzymes depends on adjacent domains at their N- and C-termini. The N-flanking domains seem important for structural stability

155

THE ENZYMES, Vol. XXIV

but the C-flanking domains are necessary for the completion of the active sites of these enzymes. Recent NMR studies have shown that these C-flanking domains are flexible and that their ordering, driven by substrate binding, is an important part of the catalytic cycle of these enzymes.

II. Introduction

Eukaryotic proteins are subject to a plethora of posttranslational modifications (PTM) that play an important role in modulating their physiological functions [1–3]. The types of modifications that occur include phosphorylation [4, 5], acetylation [6, 7], methylation [8, 9], ubiquitination [10, 11], sumoylation [12], and citrullination [13]. These covalent marks vary considerably in terms of their target residue, persistence, and signaling properties. Probably most eukaryotic proteins are subject to some form of PTM, but an important subset of these proteins is modified at multiple sites as a means of coordinating both intracellular and intercellular signaling pathways [13].

One such area of biology that has received considerable attention over the last decade has been the role and mechanism of epigenetic regulation [14, 15]. In particular, significant progress has been made in our understanding of the control of chromatin structure and dynamics and how these events regulate a range of fundamental processes requiring access to DNA [16–19]. The role of many different types of PTM on the histone tails of nucleosomes, together with mechanisms for their interplay and readout, have been brought together as the histone code hypothesis [20–27]. This review is concerned with just one family of enzymes (based on the evolutionarily conserved SET domain) responsible for carrying out most of the known methylation of lysine residues on histones [28–30]. Our task here is to review the rapid progress made in the last few years toward understanding the structural and mechanistic basis of lysine methylation by SET proteins and the biological insights that have emerged from this work.

Lysine residues are subject to hydroxylation at the β CH_2, and acetylation, ubiquitination, neddylation, sumoylation, and methylation at the ε amino group. Moreover, lysine side chains can be mono-, di-, or tri-methylated (Figure 6.1), leading to distinct biological readouts [31–38]. From a chemistry viewpoint it is perhaps instructive to contrast the various states of lysine methylation with those of its acetylation. As indicated in Figure 6.1, lysine ε-amino groups can only accommodate a single acetyl group. This leads to a moiety that is both uncharged and hydrophilic. In contrast, addition of successive methyl groups leads to increased hydrophobicity and basicity of the lysine side chain [15]. However, at the physiological pH ranges encountered in the nucleus and cytosol the methylated lysine species carry essentially the same charge as the unmodified form. Chemically, methylated lysine is a stable species and until recently was thought to be a long-lived mark on proteins. However, recent papers have presented

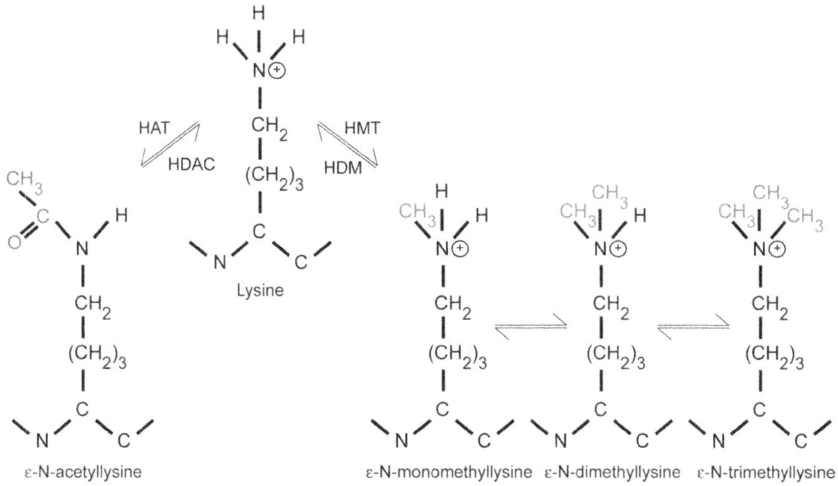

FIG. 6.1. Comparison of acetylation and methylation of lysine side chains. Acetylation is regulated by the activities of histone acetyl transferases (HAT) and histone deacetylases (HDAC). Lysine residues can be modified by a single acetyl group leading to an uncharged but more hydrophilic moiety. Lysine residues are methylated by histone methyl transferases (HMT) and can be demethylated by the action of histone demethylases (HDM). Lysine residues can be mono-, di-, or tri-methylated, leading to increasingly hydrophobic moieties.

evidence for the role of several related proteins in catalyzing biologically pertinent demethylation of lysine side chains [39-41]. Thus, the balance of activities of lysine methyltransferases and lysine demethylases could regulate the overall level of lysine methylation in a manner similar to that achieved by histone acetyltransferases and histone deacetylases for acetyl-lysine [19, 42].

As far as histones are concerned, all known lysine methylation is carried out by SET proteins with the single exception of the modification of K-79 on histone 3 that is carried out by DOT1 [43–47]. The evolutionarily conserved SET domain occurs in numerous chromatin-associated proteins. Its name derives from its identification in the *Drosophila* genes S̲u(var), E̲(z), and T̲rithorax [48] and has proved more persistent than the alternative "tromo" nomenclature [49]. An alignment of a small subset of representative SET domain sequences is shown in Figure 6.2a. For some time the SET domain was suspected of having an important role in chromatin biology before a specific function had been attributed to it. However, in 2000 the key connection was made that SET domains shared limited sequence homology with a group of plant enzymes thought to possess methyltransferase activity. It was then shown that SUV39H1 is a SET-dependent H3-specific histone methyltransferase that modifies K-9 [28]. Subsequently, many more SET domain proteins have been identified as carrying out the methylation of lysine residues on histone tails [29, 50–67].

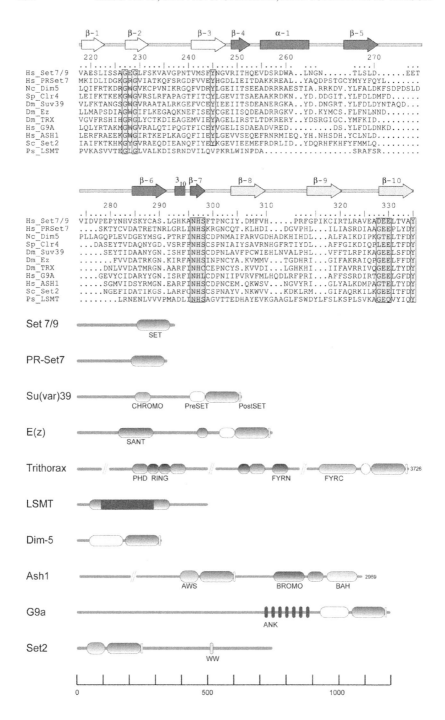

TABLE 6.1

EXAMPLES OF SET DOMAIN METHYLTRANSFERASES INDICATING
THEIR TARGET LYSINE AND PROPOSED BIOLOGICAL READOUT

Name	Organism	Target	Biological Readout	Reference
Set 7/9	*Homo sapiens*	H3 K4*	Activation	64, 68
PR-Set7 (Set8)	*Homo sapiens*	H4 K20	Repression	69, 70
Su(var)39	*Homo sapiens*	H3 K9	Repression	28
E(z)	*D. melanogaster*	H3 K27, H1 K26	Repression	71
Trithorax	*D. melanogaster*	H3 K4	Activation	72
LSMT	*D. melanogaster*	Rubisco	–	72
Dim-5	*N. crassa*	H3 K9	Repression	60
Ash1	*Homo sapiens*	H3 K4	Activation	74, 75
G9a	*Homo sapiens*	H3 K9	Repression	76, 77
Set2	*S. cerevisiae*	H3 K36	Repression	63, 78

*Also shown to modify p53 and Taf10.

The biological consequences of methylation of histone lysine residues, unlike acetylation, vary dramatically according to the location of the target residue and even the extent of methylation that occurs. Some examples of HMTs are given in Table 6.1, together with their cognate sites and the principle effects of the modification. More extensive examples can be found in reviews of SET-containing proteins [16, 29, 30, 56, 59, 66].

SET domains were first identified as histone-specific methyltransferases, but it has subsequently become apparent that they also methylate certain non-histone proteins (e.g. TAF10 and p53 [79, 80]). This has led to the more general suggestion that lysine methylation may well represent a more diverse signaling mark whose scope extends beyond that envisaged in the histone code hypothesis.

In the past few years a host of papers have described the X-ray and NMR structures and dynamics of SET proteins in their apo, binary, and ternary complexes (see Table 6.2) [67, 80–91]. This work has detailed much of the structural and mechanistic basis of SET domain function and has provided considerable insights into this important biological signaling system.

FIG. 6.2. (top) Multiple sequence alignment of some SET domains with certain conserved residues highlighted. The residue numbering is according to full-length Set7/9. The secondary structure elements of Set7/9 are also indicated, but its numbering is with respect to just the core SET domain. The accession codes for the listed sequences are Set7/9, NP_085151; PR-Set7, AAM47033; Dim-5, AAL35215; Clr4, T43745; Su(var)39, P45975; Ez, P42124; TRX, P20659; G9a, S30385; Ash1, NP_060959; Set2, P46995; and LSMT, Q43088 (note that the residues of the insert region, 110 to 220, have been excluded for clarity). (bottom) Some examples of SET domains in the context of larger proteins containing other domains.

TABLE 6.2

PUBLISHED STRUCTURES OF SET DOMAIN METHYLTRANSFERASES

Enzyme	Complex State	K Target	Methylation State	PDB Code	Reference
Set7/9	Apo	H3-K4	Mono	1H3I	86
Set7/9	Binary SAH	H3-K4	Mono	1MT6	81
Set7/9	Apo	H3-K4	Mono	1MUF	81
Set7/9	Binary SAM	H3/K4		1N6A	89
Set7/9	Binary SAM	H3/K4	Mono	1N6C	89
Set7/9	Ternary SAH, H3 (1-9) Me-K4	H3-K4	Mono	1O9S	85
Dim-5	Apo	H3-K9	Tri	1ML9	83
Dim-5	Ternary SAH, H3 (1-16)	H3-K9	Tri	1PEG	82
Clr4	Apo	H3-K9	Tri	1MVH	67
Clr4	Apo	H3-K9	Tri	1MVX	67
PR-Set7/Set8	Ternary SAH, H4 (17-25) Me-K20	H4-K20	Mono	2BQZ	84
PR-Set7/Set8	Ternary SAH, H4 (15-24)	H4-K20	Mono	1ZKK	91
Set7/9	Ternary SAH, p53 (369-377) Me-K372	p53-K372	Mono	1XQH	80
Rubisco LSMT	Ternary* SAH	Rubisco large subunit K14	Tri	1MLV	88
Rubisco LSMT	Ternary SAH, K	Rubisco large subunit K14	Tri	1OZV	87
Rubisco LSMT	Ternary SAH, Me-K	Rubisco large subunit K14	Tri	1P0Y	87
vSET	Apo	H3-K27	—	1N3J	90

*The HEPES molecule represents a pseudo-substrate.

III.   SET Domains in Context

Although protein constructs containing just 20 to 30 additional residues on both the N- and C-terminus of the approximately 130-residue SET domain can be catalytically competent, SET domains generally occur in the context of larger proteins (Figure 6.2b). Moreover, these larger SET domain proteins often occur as subunits of multi-protein complexes, as in (for example) the MLL complex [92–94] or its yeast homolog COMPASS [53, 95, 96] and the PRC2 complex that contains the methyltransferase EZH1 [71, 97]. Unfortunately, structural information is at present only available for the SET module together with its adjacent flanking domains. However, the importance of the biology associated with SET-containing multi-protein complexes are such that structural characterization of these assemblies is likely to be a key long-term objective of many laboratories.

As suggested previously, the evolutionarily conserved core of the SET domain is not catalytically competent of itself but requires at least part of the adjacent domain at its N-terminus and most, if not all, of the adjoining domain at its C-terminus. Some of the first SET domain proteins characterised contained distinctive cysteine-rich domains at the immediate N- and C-terminus of the SET module, and these were not unreasonably described as pre- and post-SET domains (respectively). Now that more SET proteins have been studied, it is apparent that other types of non cysteine-rich domains (also necessary to function) are also located adjacent to the SET domain. We will use here, as elsewhere [98], the generic terms N- and C-flanking domains to describe both cysteine-rich and non cysteine-rich domains. Stewart and his colleagues have shown that classification of SET domains either by the sequences within the SET domain or by characterization of the N- and C-flanking domains leads to the same clade structure [96]. This analysis seems to argue that these variable but catalytically essential modules have coevolved with their partner SET domains over an extended evolutionary period.

IV. Overall Structure

Although there are significant variations among the SET domains whose structures have been determined, there is nonetheless an underlying core structure that consists of 10 β strands arranged into three sheets with extensive connecting loops and about one turn of 3_{10} helix. This core structure is illustrated in Figure 6.3a in Ribbons representation using Set7/9 together with a topology diagram (Figure 6.3b). The organization of this domain is, so far, unique to the SET family.

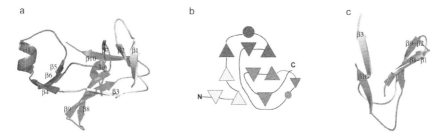

FIG. 6.3. (a) Ribbons representation of the core SET domain illustrated using SET7/9. SET-N is colored yellow, SET-I is in blue, and SET-C in green. The strand bearing the invariant NH motif leading into the threaded loop is colored in red. (b) Topology diagram for SET7/9 colored the same as diagram *a*, with β strands as triangles and helices as circles. The small turn of 3_{10} helix is indicated by the smaller green circle preceding the red strand. (c) Overlap of the SET-N and SET-C regions of SET7/9. (See color plate.)

It is composed of two conserved but not contiguous modules (SET-N and SET-C) separated by a highly variable insert region (SET-I). The SET-N and SET-C modules contain the conserved sequence motifs characteristic of SET domains, and it appears likely that they represent tandem repeats of a precursor motif [58]. This notion is supported by the fact that they each contain a similar subfold (Figure 6.3c). In contrast, the Set-I region is not conserved across the SET family (Figure 6.2a) but appears to be important for mediating contact with the enzyme's protein substrates and thus conferring specificity.

Two particularly striking features of the SET domain structure are the lysine access channel and the pseudo-knot topology at the active site. In all ternary complex structures, and in all but one of the cofactor-bound complexes, there is a pore or channel through the SET domain connecting the substrate and cofactor binding surfaces (particularly apparent in the space-filling representations shown in Figure 6.4). This channel is so constructed that it accommodates the side chain of the target lysine such that its ε-nitrogen is positioned to accept a methyl group from the cofactor. The bottom of this channel, at the cofactor end, constitutes

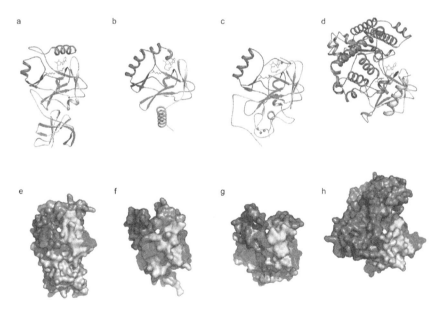

Fig. 6.4. (a–d) Ribbons representation of the ternary complexes of Set7/9, PR-Set7/Set8, Dim-5, and LSMT, respectively. The N-flanking (at the bottom of the figure) and C-flanking domains are colored gray, whereas the SET domain is colored as in Figure 6.2. The target lysine is colored magenta and the cofactor in light blue. For Dim-5, the Zn atoms of the pre- and post-SET domains are colored brown. (e–h) Underneath each of the Ribbons figures are space-filling representations of the same proteins in a somewhat different orientation to illustrate the lysine access channel. The surfaces are colored according to the same key as elsewhere. (See color plate.)

the active site of the enzyme. The size and bonding patterns of a number of the residues that make up this site are responsible for determining whether the enzyme carries out mono-, di-, or tri-methylation of its target lysine residue. Not surprisingly, many of the invariant residues associated with the SET family are located at or adjacent to the active site. The C-terminus of SET C (containing the EELxxxY motif) threads its way through the loop between $\beta 7$ and $\beta 8$ in a topologically unusual way that has been termed a pseudo-knot [99] (Figures 6.3a and b). This nomenclature is used because the topological restraints giving rise to the "knot" in the chain are imposed by hydrogen bonds between two segments of the protein chain.

V. Structure of the N- and C-Flanking Domains

Although at least some proportion of both the N- and C-flanking domains are necessary for functional SET domains, they appear to have quite distinct roles. The structurally characterized N-flanking domains adopt a range of distinct folds, but all seem to provide a stable interface with the SET domain that is important for its stability. In contrast, the equally structurally diverse C-flanking domains make a direct contribution to the active sites.

Of the structures available to date, Set7/9 contains an anti-parallel β repeat N-flanking domain, PR-Set7 and its isoform Set8 contain an α-helix (both are N-terminal truncations of what is predicted to be a larger α-helical domain), Rubisco LSMT also contains an α-helix, and Dim-5 and Clr4 both contain similar pre-SET domains with Cys_9Zn_3 clusters. In spite of the diverse range of structural motifs found in these N-flanking domains, there is some similarity in how they pack against the SET domain (as can be seen in Figures 6.4a through d). Notably, these contact regions are not located particularly close to the enzymes' active sites. Interestingly, the Cys_9Zn_3 clusters that characterize the pre-SET domains are not located at the interface with the SET domain. Instead, they stabilize the core of the pre-SET domain.

Of the structurally characterized C-flanking domains, Set7/9 has a loop and helix, PR-Set7 (and Set8) a loop followed by two helical segments separated by a kink, Dim-5 a post-SET domain with a $Cys_4Zn_1{}^{2+}$ cluster, and LSMT a large (290-residue) α-helical domain (Figures 6.4a through d). It is remarkable how these diverse structures interact with their respective SET domains to fulfill three functions: (1) provision of an aromatic or hydrophobic residue to pack against the adenine ring of the cofactor, (2) creation of a bridging interaction with the absolutely conserved histidine residue of the NH motif, and (3) completion of the lysine access channel. Moreover, it is probable that the C-flanking domains of SET enzymes only become properly ordered following cofactor binding (discussed in a later section).

As shown in Figures 6.5a through d, Trp-352 of Set7/9, Trp-349 of PR-Set7/Set8, Leu-317 of Dim-5, and Phe-302 of LSMT all pack against the adenine ring of the cofactor. Finally, as shown in Figures 6.5i through l, each of the C-flanking domains contributes a residue to the lysine access channel (except for PR-Set7, where an equivalent contribution is made by a residue from the peptide substrate). Thus, the side chains of Tyr-337 of Set7/9, Trp-318 of Dim-5, and Tyr-300 of LSMT occupy an approximately similar position in the access channel as His(-2) of the H4 peptide of PR-Set7/Set8. Figures 6.5e through h show how the key conserved histidine residue interacts with the main chain of the final β strand of the SET domain on one side and with some part of the C-flanking domain on the other. These latter interactions include Glu-356 of Set7/9, the main chain carbonyl of residue 352 of PR-Set7/Set8, the Zn^{2+} cluster of Dim-5, and Glu-304 of LSMT.

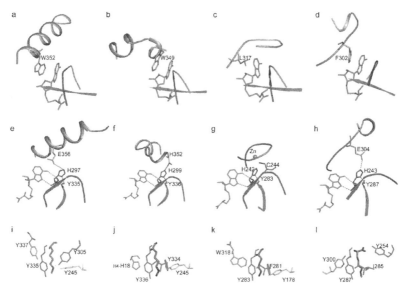

FIG. 6.5. Ribbons representation of three different aspects of C-flanking domains contributing to the active site region of the SET domain for Set7/9 (*a*, *e*, and *i*), PR-Set7/Set8 (*b*, *f*, and *j*), Dim-5 (*c*, *g*, and *k*), and LSMT (*d*, *h*, and *l*). The first row (*a–d*) shows the packing of an aromatic or hydrophobic residue from the C-flanking domain with the adenine ring of the cofactor. The second row (*e–h*) shows interactions of the invariant histidine residue, its main chain groups with cofactor and its side chain with different elements of the C-flanking domains. In the case of Dim-5 there is no direct interaction of the histidine side chain with the post-SET domain, but the cysteine residue following combines with the three cysteine residues in the post-SET domain to complete the coordination to the zinc ion. The third row (*i–l*) shows how the C-flanking domains contribute a residue to the lysine access channel. The exception is for PR-Set7/Set8, where an approximately similar position is occupied by a histidine residue from the peptide (H18 of H4). (See color plate.)

VI. The Cofactor Binding Site

The methyl transfer reaction catalyzed by SET enzymes utilizes S-adenosyl-L-methionine (AdoMet) as the source of what may be regarded as an activated methyl group. The cofactor binding site of SET domains is quite unlike that found in the large family of α/β AdoMet-dependent methyltransferases [100]. Likewise, the U-shaped conformation adopted by the cofactor is quite different from that it adopts when bound to members of generic α/β methyltransferases (Figure 6.6a). AdoMet binds into a pocket on the surface of the SET domain made up of contributions from several structural elements containing conserved sequence motifs. Most notable of these are the GxG motif on $\beta2$ and the NH motif on $\beta7$ (shown in the schematic in Figure 6.6b). It is noteworthy that many of the key interactions with the cofactor are mediated by main chain groups of the SET domain as well as by invariant side chains.

The amino group of the AdoMet hydrogen bonds to the main chain carbonyls of the residues immediately before and between the two conserved glycines of the GxG motif, as well as with the side-chain carbonyl of the invariant asparagine of the NH motif. The conformation adopted by this side chain is itself stabilized by a hydrogen bond with the glutamic acid side chain from the EL/IxxxY motif on the threaded final β strand of the SET domain. The carboxyl of the AdoMet makes a hydrogen bond to the main chain amide of the residue between the conserved glycines (GxG) and to a basic residue, and sometimes to a tyrosine residue contributed from different parts of the SET domain among different family members. Thus, Lys-294 of Set7/9, Arg-228 and Tyr-271 of PR-Set7/Set8, Tyr-204 of Dim-5, and Arg-222 of LSMT all interact with the terminal carboxyl moiety of the cofactor. Another group of conserved interactions with the cofactor are mediated through its adenine ring. We have already mentioned the ring-stacking interaction provided by a residue from the C-flanking domain. In addition, there are hydrogen bonds from the N6 and N7 of the adenine with the main chain carbonyl and amide of the conserved histidine (N<u>H</u>), respectively.

VII. Substrate Binding

Ternary complex structures are available with peptide substrates for Set7/9, Dim-5, and PR-Set7/Set8, as well as with a lysine residue for LSMT. The peptide complexes (Figure 6.7) reveal broadly similar modes of interaction but also some interesting differences that may account for the distinct specificities of these enzymes. The peptide-binding groove is largely formed by part of the α-1 helix and β-5 strand on one side and the loop connecting the last β strand of the SET domain with the C-flanking domain on the other. The substrate peptide adopts a more or less extended β conformation running roughly parallel to

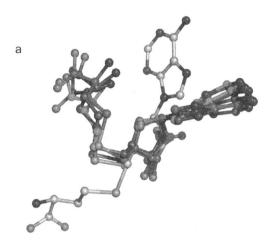

a

b. Set7/9

Asn-296   His-297

Lys-294   Glu-228   Ala-226   Glu-356

c. PR-Set7 (Set8)

Tyr-271   Asn-298   His-299

Lys-226   Arg-228

d. Dim-5

Tyr-204   Asn-241   His-242

Trp-161   Arg-159   Lys-201   Leu-307

e. LSMT

Asn-242   His-243

Glu-80   Leu-82   Arg-222

FIG. 6.6. (a) Conformations of bound cofactor. Similar conformations for AdoHcy from four complexes with SET domains displaying a characteristic U shape are overlapped with a more extended conformation of AdoHcy taken from its complex with the Hhai methyltransferase (pdb accession code 1mht). (b–e) Schematic representations of interactions between cofactor and the active site residues of Set7/9, PR-Set7/Set8, Dim-5, and LSMT, respectively. (See color plate.)

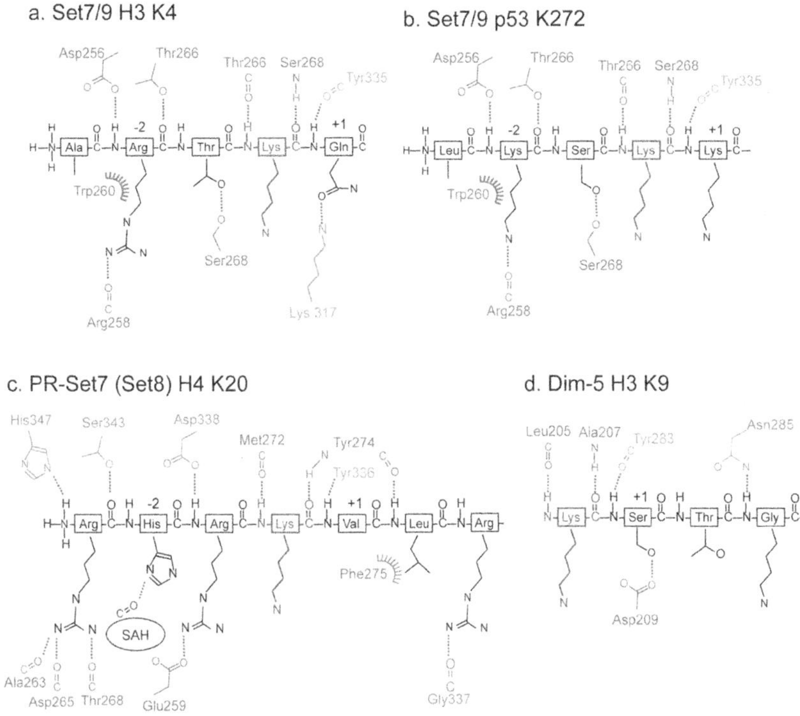

FIG. 6.7. Schematic representation of interactions of peptide substrates with the SET domains of Set7/9 (with H3 and p53 peptides), PR-Set7/Set8, and Dim-5. Interacting residues from the SET domains are colored as elsewhere: SET-I in blue, SET-C in green, and C-flanking in gray. (See color plate.)

β-5 and making some hydrogen bonds with it. This β-5 strand, together with the following β-6 strand and part of the preceding α-1 helix, are components of the highly variable SET-I region.

Apart from lack of sequence conservation, the length of this segment varies by more than 300 residues across family members. Given the variability of this region, and its prominent role in substrate recognition, we can probably regard it as a specificity cassette. This notion can only be taken so far, however, because there are numerous interactions between the SET-I module and nonconserved residues of the core SET domain. This means that it is highly unlikely that SET-I sequences could be swapped between different SET proteins to alter their substrate specificity. Moreover, there are many examples of SET proteins with the same target recognition sequence that have completely different SET-I regions.

Within the common framework of peptide binding described previously there are also substantial variations. These include the extent of the main chain hydrogen bonding with β-5, the extent of the peptide side chain recognition, formation of main chain hydrogen bonds with other β strands of the SET domain, and how extended (or conversely how curved) the substrate peptide is. The sequence of the target peptide sequences of Set7/9, PR-Set7/Set8, and Dim-5 are shown in Figure 6.7, together with schematics of their various interactions.

There are five hydrogen bonds to the main chain groups of the H3 peptide bound to Set7/9, and a series of further hydrogen bonds and van der Waals contacts with (most notably) Gln(-5), Thr(-3), and Arg(-2). Among histones, all three of these residues are unique to the motif containing Lys-4 of H3 and thus provide for specificity for this enzyme. The role of Arg (-2) seems particularly important given its interactions with Asp-256, Trp-260, and the carbonyl group of Arg-258. The H3 peptide makes no interactions with the loop leading into the C-flanking domain or with other β strands of the SET domain. The H3 peptide is also markedly kinked in the binding site.

For Set7/9, there is also a ternary complex available with a p53-derived peptide that has recently been shown to be an alternative substrate for this SET enzyme. Interestingly, the C-terminus of p53 where the methylation occurs is subject to a whole range of PTM, and there is a pattern of repeating basic and hydrophobic residues not entirely dissimilar to the N-terminal tails of histones H3 and H4. The most remarkable feature of the Set7/9/p53 structure is just how similar it is to the complex with H3 peptide despite none of the residues flanking the target lysine being conserved between the two different peptides. The same interactions made by Arg(-2) of H3 are made in a slightly different way by Lys(-2) of p53. In addition, the substitution of Thr(-1) of H3 for Ser(-1) of p53 enables a hydrogen bond to be formed with Ser-268 of the enzyme in both cases.

In the case of PR-Set7 (Set8), there are seven hydrogen bonds between main chain groups on the peptide and residues on β-5 of SET-I and from the C-flanking domain. There are also hydrogen bond and van der Waals interactions with Arg(-3), His(-2), Arg(-1), Leu(+2), and Arg(+3), and the peptide conformation is extended. Of all of these interactions the most remarkable is that made by the (-2) histidine residue, which not only hydrogen bonds with the ribose of the cofactor but packs in such a way as to form one side of the lysine access channel. Mutagenesis studies on the H4 peptide have confirmed that the (-3) arginine, (-2) histidine, and (+2) leucine play a key role in determining the substrate specificity. The consensus binding motif appears to be R-H/F-X-K-X-f, where f is a bulky hydrophobic residue [84, 91].

The ternary complex of Dim-5 with an H3 peptide centered on Lys-9 reveals an extended binding conformation making two hydrogen bonds with the parallel β strand of the Set-I region. In contrast to Set7/9 and PR-Set7 (Set8), there are fewer contacts with the side chains of the substrate peptide. Nonetheless, there is a charged hydrogen bond between Asp-209 of Dim-5 and Ser(+1) of H3. This interaction is particularly interesting given that Ser-10 (at the +1 position) is subject to phosphorylation that then inhibits subsequent methylation of Lys-9 by certain HMTs. The Dim-5 structure is certainly consistent with the notion that phosphorylation of Ser-10 would diminish binding of the H3 peptide to Dim-5 [82].

VIII.   Active Site and Methylation

The active site of SET enzymes contains features in common across the ternary complexes available, but also shows some important differences that appear to be relevant as to whether the enzymes are mono-, di-, or tri-methylases. Whereas the lysine access channel adjacent to the substrate binding cleft is made up of a series of largely nonconserved hydrophobic residues, the part of the lysine access channel that approaches the cofactor pocket contains an invariant tyrosine residue and four main chain carbonyl groups. The arrangement of these groups is shown in Figure 6.8 for Set7/9, PR-Set7, Dim-5, and LSMT. The arrangement

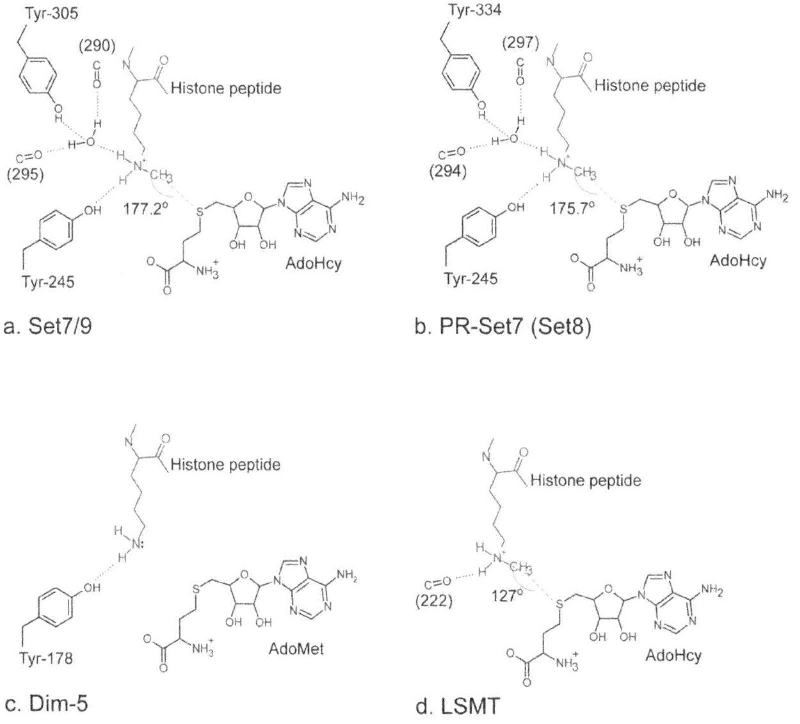

a. Set7/9

b. PR-Set7 (Set8)

c. Dim-5

d. LSMT

FIG. 6.8. Schematic representation of the active sites of Set7/9, PR-Set7/Set8, Dim-5, and LSMT. The substrate lysine is colored in magenta. (See color plate.)

of the lysine side chain and cofactor are such that the sulphur of the cofactor, the carbon of the transferred methyl, and the ε-nitrogen of the lysine are more or less collinear. This geometry positions the lysine ε-amino group for S_N2 nucleophilic attack on the transferable methyl group of the AdoMet [85, 101].

The carbonyl cage environment protects the reaction from solvent and generates, by dipole charges, a negatively charged environment. Particularly interesting regarding the active site of SET enzymes are the carbon-oxygen hydrogen bonds formed between the transferred methyl group and the oxygen atoms of the carbonyl cage and invariant tyrosine residue [87]. This type of hydrogen bonding is made possible by the electropositive nature of the methyl carbon because of its bonding to either the sulfur of AdoMet or the nitrogen of the lysine side chain. It seems that these hydrogen bonds make at least two contributions to catalysis: they stabilize the transition state of the reaction and make the methyl carbon more electrophilic by further polarizing the S-C bond. An additional feature demanded by the mechanism just outlined is that the ε-amino group of the lysine residue be deprotonated. None of the available structures suggests active site residues that have the right chemistry and interactions to act as a general base. Coupled with this is the observation that SET-catalyzed methyl transfer reactions have a high pH optima, suggesting either that the lysine is deprotonated upon entering the active site or that this occurs concomitantly with cofactor binding and the proper ordering of the C-flanking domain.

Two of the available ternary complexes are of enzymes that mono-methylate their substrates–Set7/9 [85] and PR-Set7/Set8 [84, 91]–whereas the other two (Dim-5 [82] and LSMT [87]) catalyze tri-methylation. Importantly, it seems that the observed multiplicity of methylation can be correlated with structural differences between these two classes of enzymes (Figure 6.9). For a second and third methylation reaction to occur following mono-methylation it is necessary for the methyl group to rotate around the ε–carbon-nitrogen bond of the lysine side chain so that the lone pair of electrons made available by further deprotonation of the ε-amine group can be aligned with the methyl-sulfur bond of the AdoMet. In both Set7/9 and PR-Set7/Set8 there is a pair of tyrosine residues that hydrogen bond with the ε-amine group. These seem to prevent this bond rotation from occurring. Tyr-245 of Set7/9 is conserved in all known histone methyltransferases (see Figure 6.2a) and makes a direct hydrogen bond with the lysine.

In contrast, Tyr-305 of Set7/9 and Tyr-334 of PR-Set7/Set8 (presented into the active site from a different piece of secondary structure but ends up with its hydroxyl group in a similar position) both make a water-molecule-mediated interaction with the lysine side chain. The position of these two residues precludes rotation of the methyl group without some reordering of the active site. Moreover, rotation of the methyl group would mean losing at least one of the two hydrogen bonds involved. This would presumably incur a substantial energetic penalty in such a solvent-shielded environment.

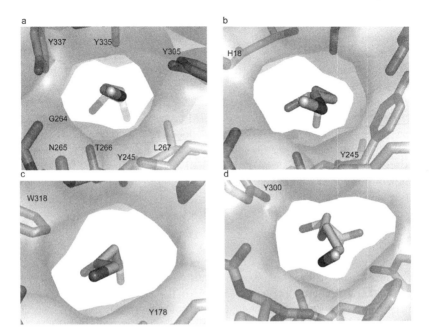

FIG. 6.9. Surface representation of the active site cleft of (a) Set7/9, (b) PR-Set7/Set8, (c) Dim-5, and (d) LSMT. In the upper left of each panel the residue contributed from the C-flanking domain of each of the enzymes to the active site is labeled (compare with Figures *i–l*). The active sites for Set7/9 and PR-Set7/Set8 are noticeably more restricted than those of Dim-5 and LSMT, to the right of the lysine nitrogen, where the attached methyl groups would have to be accommodated for di- or tri-methylation to occur. (See color plate.)

Finally, mutation of Tyr-305 of Set7/9 to phenylalanine leads to an enzyme capable of multiple methylation. In the case of Dim-5, the invariant tyrosine Tyr-178 hydrogen bonds to the lysine amine but the equivalent of Tyr-334 of PR-Set7/Set8 is Phe-281. Thus, there is both a hydroxyl and a water molecule less in the active site. In the case of LSMT, the tyrosine invariant to histone methyltransferases is very approximately replaced by Phe-224 (from a different part of the SET domain) and in place of Tyr-334 of PR-Set7/Set8 is Ile-285. These differences at the active site of LSMT also mean that more space is available and hydrogen bonds are less restrictive, which presumably account for the active site's observed activity.

IX.   Flexibility of the C-Flanking Domain

From the range of published structural studies on SET proteins there is accumulating evidence of a relationship between cofactor binding and ordering of the

C-flanking domain. Crystallographically, ordered C-flanking domains have only been observed in complexes with cofactor. In all but one case, when cofactor is present the C-flanking domain is ordered. The exception is for one of the examples of Set7/9 co-crystallized with cofactor [81], where there is no electron density for the C-flanking domain. Interestingly, superposition of the ternary complex of Set7/9 onto this structure reveals that its crystal-packing arrangement sterically precludes an ordered C-flanking domain. Moreover, a second Set7/9/cofactor binary complex, in a different space group, does contain an ordered C-flanking domain. Taken together, these observations seem consistent with the notion that cofactor binding assists ordering of the C-flanking domain and that further stabilization of this conformation may in some cases be provided by interactions with peptide substrate.

Recently, we have used a series of NMR experiments to probe more directly the dynamics of the C-flanking domain of Set7/9 [84]. First, there are significantly more resonances corresponding to backbone amide protons in 3D trosy-HNCO experiments than can be simply accounted for by the protein construct used, and at least 35 of these peaks disappear upon addition of cofactor. These data strongly suggest structural heterogeneity in the protein in the absence of cofactor. Second, all peaks in the spectra recorded with cofactor overlap with peaks in the spectra of the apo-enzyme. This suggests that certain parts of the enzyme adopt two conformations in the absence of cofactor and that one of these conformations closely resembles that observed when cofactor is bound. Further experiments also indicated that many of the resonances that disappear upon cofactor binding arise from residues experiencing a high degree of internal flexibility and that this largely unstructured population is largely associated with the C-flanking domain.

These data, looking at the effects of cofactor on apo-Set7/9, are also consistent with earlier NMR experiments carried out with Set7/9 using specific $^{13}C/^{15}N$ labeling of the target lysine residue in the context of an otherwise unlabeled peptide monitored using ^{1}H-HSQC experiments. These experiments indicate that the side chain of the target lysine is not ordered in the Set7/9 peptide complex. However, the HSQC signals from the lysine side chain change upon addition of cofactor to those expected for a form with restricted movement.

Taking all of these observations together we arrive at the view that at least for Set7/9 the C-flanking domain exists in an equilibrium between a largely unstructured state and a form that is similar to that found in binary or ternary complexes. This equilibrium is perturbed in the ordered direction upon cofactor binding probably irrespective of the presence or absence of substrate peptide. To what extent these ideas apply to SET proteins in general is difficult to determine at present. However, it seems very likely that the C-flanking domains of SET enzymes are disordered in the absence of substrates and that the range of conserved functions and interactions these modules carry out, albeit by different structural routes, argues for an important role for cofactor in promoting an active conformation.

Set7/9 is different from the other two ternary complexes in that it has no interactions between the substrate peptide and the C-flanking domain. It is therefore possible that peptide substrate binding, as well as cofactor binding, may be important for ordering of the C-flanking domains of some SET proteins.

REFERENCES

1. Mann, M., and Jensen, O.N. (2003). Proteomic analysis of post-translational modifications. *Nat Biotechnol* 21:255–261.
2. Seo, J., and Lee, K.J. (2004). Post-translational modifications and their biological functions: proteomic analysis and systematic approaches. *J Biochem Mol Biol* 37:35–44.
3. Yan, S.C., Grinnell, B.W., and Wold, F. (1989). Post-translational modifications of proteins: some problems left to solve. *Trends Biochem Sci* 14:264–268.
4. Hunter, T. (2000). Signaling–2000 and beyond. *Cell* 100:113–127.
5. Cohen, P. (2002). The origins of protein phosphorylation. *Nat Cell Biol* 4:E127–E130.
6. Roth, S.Y., Denu, J.M., and Allis, C.D. (2001). Histone acetyltransferases. *Annu Rev Biochem* 70:81–120.
7. Polevoda, B., and Sherman, F. (2002). The diversity of acetylated proteins. *Genome Biol* 3 (reviews 0006).
8. Clarke, S. (1993). Protein methylation. *Curr Opin Cell Biol* 5:977–983.
9. Aletta, J.M., Cimato, T.R., and Ettinger, M.J. (1998). Protein methylation: a signal event in post-translational modification. *Trends Biochem Sci* 23:89–91.
10. Hershko, A., and Ciechanover, A. (1998). The ubiquitin system. *Annu Rev Biochem* 67:425–479.
11. Bonifacino, J.S., and Weissman, A.M. (1998). Ubiquitin and the control of protein fate in the secretory and endocytic pathways. *Annu Rev Cell Dev Biol* 14:19–57.
12. Melchior, F. (2000). SUMO–nonclassical ubiquitin. *Annu Rev Cell Dev Biol* 16:591–626.
13. Yang, X.J. (2005). Multisite protein modification and intramolecular signaling. *Oncogene* 24:1653–1662.
14. Berger, S.L. (2002). Histone modifications in transcriptional regulation. *Curr Opin Genet Dev* 12:142–148.
15. Rice, J.C., and Allis, C.D. (2001). Histone methylation versus histone acetylation: new insights into epigenetic regulation. *Curr Opin Cell Biol* 13:263–273.
16. Grewal, S.I., and Moazed, D. (2003). Heterochromatin and epigenetic control of gene expression. *Science* 301:798–802.
17. Workman, J.L., and Kingston, R.E. (1998). Alteration of nucleosome structure as a mechanism of transcriptional regulation. *Annu Rev Biochem* 67:545–579.
18. Fan, H.Y., He, X., Kingston, R.E., and Narlikar, G.J. (2003). Distinct strategies to make nucleosomal DNA accessible. *Mol Cell* 11:1311–1322.
19. Narlikar, G.J., Fan, H.Y., and Kingston, R.E. (2002). Cooperation between complexes that regulate chromatin structure and transcription. *Cell* 108:475–487.
20. Strahl, B.D., and Allis, C.D. (2000). The language of covalent histone modifications. *Nature* 403:41–45.
21. Rice, J.C., and Allis, C.D. (2001). Code of silence. *Nature* 414:258–261.
22. Cosgrove, M.S., and Wolberger, C. (2005). How does the histone code work? *Biochem Cell Biol* 83:468–476.
23. Iizuka, M., and Smith, M.M. (2003). Functional consequences of histone modifications. *Curr Opin Genet Dev* 13:154–160.

24. Loidl, P. (2004). A plant dialect of the histone language. *Trends Plant Sci* 9:84–90.
25. Peterson, C.L., and Laniel, M.A. (2004). Histones and histone modifications. *Curr Biol* 14:R546–R551.
26. Schreiber, S.L., and Bernstein, B.E. (2002). Signaling network model of chromatin. *Cell* 111:771–778.
27. Turner, B.M. (2002). Cellular memory and the histone code. *Cell* 111:285–291.
28. Rea, S., Eisenhaber, F., O'Carroll, D., Strahl, D.B., Sun, Z.W., Schmid, M., Opravil, S., Mechtler, K., Ponting, C.P., Allis, C.D., and Jenuwein, T. (2000). Regulation of chromatin structure by site-specific histone H3 methyltransferases. *Nature* 406:593–599.
29. Lachner, M., and Jenuwein, T. (2002). The many faces of histone lysine methylation. *Curr Opin Cell Biol* 14:286–298.
30. Kouzarides, T. (2002). Histone methylation in transcriptional control. *Curr Opin Genet Dev* 12:198–209.
31. Rice, J.C., Briggs, S.D., Ueberheide, B., Barber, C.M., J. Shabanowitz, J., Hunt, D.F., Shinkai, Y., and Allis C.D. (2003). Histone methyltransferases direct different degrees of methylation to define distinct chromatin domains. *Mol Cell* 12:1591–1598.
32. Freitag, M., Hickey, P.C., Khlafallah, T.K., Read, N.D., and Selker E.U. (2004). HP1 is essential for DNA methylation in neurospora. *Mol Cell* 13:427–434.
33. Min, J., Zhang, Y., and Xu, R.M. (2003). Structural basis for specific binding of Polycomb chromodomain to histone H3 methylated at Lys 27. *Genes Dev* 17:1823–1828.
34. Santos-Rosa, H., Schneider, R., Bannister, A.J., Sherriff, J., Bernstein, B.E., Emre, N.C., Schreiber, S.L., Mellor, J., and Kouzarides, T. (2002). Active genes are tri-methylated at K4 of histone H3. *Nature* 419:407–411.
35. Bannister, A.J., and Kouzarides, T. (2004). Histone methylation: recognizing the methyl mark. *Methods Enzymol* 376:269–288.
36. Biron, V.L., McManus, K.J., Hu, N., Hendzel, M.J., and Underhill, D.A. (2004). Distinct dynamics and distribution of histone methyl-lysine derivatives in mouse development. *Dev Biol* 276:337–351.
37. Stewart, M.D., Li, J., and Wong, J. (2005). Relationship between histone H3 lysine 9 methylation, transcription repression, and heterochromatin protein 1 recruitment. *Mol Cell Biol* 25:2525–2538.
38. Rougeulle, C., Chaumeil, J., Sarma, K., Allis, C.D., Reinberg, D., Avner, P., and Heard, E. (2004). Differential histone H3 Lys-9 and Lys-27 methylation profiles on the X chromosome. *Mol Cell Biol* 24:5475–5484.
39. Metzger, E., Wissmann, M., Yin, N., Muller, J.M., Schneider, R., Peters, A.H., Gunther, T., Buettner, R., and Schule, R. (2005). LSD1 demethylates repressive histone marks to promote androgen-receptor-dependent transcription. *Nature* 437:436–439.
40. Lee, M.G., Wynder, C., Cooch, N., and Shiekhattar, R. (2005). An essential role for CoREST in nucleosomal histone 3 lysine 4 demethylation. *Nature.* 437:432–435.
41. Shi, Y., Lan, F., Matson, C., Mulligan, P., Whetstine, J.R., Cole, P.A., and Casero, R.A. (2004). Histone demethylation mediated by the nuclear amine oxidase homolog LSD1. *Cell* 119:941–953.
42. Yamagoe, S., Kanno, T., Kanno, Y., Sasaki, S., Siegel, R.M., Lenardo, M.J., Humphrey, G., Wang, Y., Nakatani, Y., Howard, B.H., and Ozato, K. (2003). Interaction of histone acetylases and deacetylases in vivo. *Mol Cell Biol* 23:1025–1033.
43. Sawada, K., Yang, Z., Horton, J.R., Collins, R.E., Zhang, X., and Cheng, X. (2004). Structure of the conserved core of the yeast Dot1p, a nucleosomal histone H3 lysine79 methyltransferase. *J Biol Chem* 279:43296–43306.
44. Zhang, W., Hayashizaki, Y., and Kone, B.C. (2004). Structure and regulation of the mDot1 gene, a mouse histone H3 methyltransferase. *Biochem J* 377:641–651.
45. Sawada, K., Yang, Z., Horton, J.R., Collins, R.E., Zhang, X., and Cheng, X. (2004). Structure of the conserved core of the yeast Dot1p, a nucleosomal histone H3 lysine 79 methyltransferase. *J Biol Chem* 279:43296–43306.

46. Min, J., Feng, Q., Li, Z., Zhang, Y., and Xu, R.M. (2003). Structure of the catalytic domain of human DOT1L, a non-SET domain nucleosomal histone methyltransferase. *Cell* 112:711–723.
47. van Leeuwen, F., Gafken, P.R., and Gottschling, D.E. (2002). Dot1p modulates silencing in yeast by methylation of the nucleosome core. *Cell* 109:745–756.
48. Tschiersch, B., Hofmann, A., Krauss, V., Dorn, R., Korge, G., and Reuter, G. (1994). The protein encoded by the Drosophila position-effect variegation suppressor gene Su(var)3-9 combines domains of antagonistic regulators of homeotic gene complexes. *Embo J* 13:3822–3831.
49. Stassen, M.J., Bailey, D., Nelson, S., Chinwalla, V., and Harte, P.J. (1995). The Drosophila trithorax proteins contain a novel variant of the nuclear receptor type DNA binding domain and an ancient conserved motif found in other chromosomal proteins. *Mech Dev* 52:209–223.
50. Stec, I., Wright, T.J., van Ommen, G.J., de Boer, P.A., van Haeringen, A., Moorman, A.F., Altherr, M.R., and den Dunnen, T.J. (1998). WHSC1, a 90 kb SET domain-containing gene, expressed in early development and homologous to a Drosophila dysmorphy gene maps in the Wolf-Hirschhorn syndrome critical region and is fused to IgH in t(4;14) multiple myeloma. *Hum Mol Genet* 7:1071–1082.
51. Huang, S., Shao, G., and Liu, L. (1998). The PR domain of the Rb-binding zinc finger protein RIZ1 is a protein binding interface and is related to the SET domain functioning in chromatin-mediated gene expression. *J Biol Chem* 273:15933–15939.
52. Harte, P.J., Wu, W., Carrasquillo, M.M., and Matera, A.G. (1999). Assignment of a novel bifurcated SET domain gene, SETDB1, to human chromosome band 1q21 by in situ hybridization and radiation hybrids. *Cytogenet Cell Genet* 84:83–86.
53. Miller, T., Krogan, N.J., Dover, J., Erdjument-Bromage, H., Tempst, P., Johnston, M., Greenblatt, J.F., and Shilatifard, A. (2001). COMPASS: a complex of proteins associated with a trithorax-related SET domain protein. *Proc Natl Acad Sci USA* 98:12902–12907.
54. Baumbusch, L.O., Thorstensen, T., Krauss, V., Fischer, A., Naumann, K., Assalkhou, R., Schulz, I., Reuter, G., and Aalen, R.B. (2001). The Arabidopsis thaliana genome contains at least 29 active genes encoding SET domain proteins that can be assigned to four evolutionarily conserved classes. *Nucleic Acids Res* 29:4319–4333.
55. Alvarez-Venegas, R., and Avramova, Z. (2002). SET-domain proteins of the Su(var)3-9, E(z) and trithorax families. *Gene* 285:25–37.
56. Breiling, A., and Orlando, V. (2002). SET domain proteins reSET gene expression. *Nat Struct Biol* 9:894–896.
57. Rayasam, G.V., Wendling, O., Angrand, P.O., Mark, M., Niederreither, K., Song, L., Lerouge, T., Hager, G.L., Chambon, P., and Losson, R. (2003). NSD1 is essential for early post-implantation development and has a catalytically active SET domain. *Embo J* 22:3153–3163.
58. Aravind, L., and Iyer, L.M. (2003). Provenance of SET-domain histone methyltransferases through duplication of a simple structural unit. *Cell Cycle* 2:369–376.
59. Dillon, S.C., Zhang, X., Trievel, R.C., and Cheng, X. (2005). The SET-domain protein superfamily: protein lysine methyltransferases. *Genome Biol* 6:227.
60. Tamaru, H., and Selker, E.U. (2001). A histone H3 methyltransferase controls DNA methylation in Neurospora crassa. *Nature* 414:277–283.
61. Shen, W.H. (2001). NtSET1, a member of a newly identified subgroup of plant SET-domain-containing proteins, is chromatin-associated and its ectopic overexpression inhibits tobacco plant growth. *Plant J* 28:371–383.
62. Briggs, S.D., Bryk, M., Strahl, B.D., Cheung, W.L., Davie, J.K., Dent, S.Y., Winston, F., and Allis, C.D. (2001). Histone H3 lysine 4 methylation is mediated by Set1 and required for cell growth and rDNA silencing in Saccharomyces cerevisiae. *Genes Dev* 15:3286–3295.
63. Strahl, B.D., Grant, P.A., Briggs, S.D., Sun, Z.W., Bone, J.R., Caldwell, J.A., Mollah, S., Cook, R.G., Shabanowitz, J., Hunt, D.F., and Allis, C.D. (2002). Set2 is a nucleosomal histone H3-selective methyltransferase that mediates transcriptional repression. *Mol Cell Biol* 22:1298–1306.

64. Nishioka, K., Chuikov, S.,Sarma, K., Erdjument-Bromage, H., Allis, C.D., Tempst, P., and Reinberg, D. (2002). Set9, a novel histone H3 methyltransferase that facilitates transcription by precluding histone tail modifications required for heterochromatin formation. *Genes Dev* 16:479–489.

65. Nakamura, T., Mori, T., Tada, S., Krajewski, W., Rozovskaia, T., Wassell, R., Dubois, G., Mazo, A., Croce, C.M., and Canaani, E. (2002). ALL-1 is a histone methyltransferase that assembles a supercomplex of proteins involved in transcriptional regulation. *Mol Cell* 10:1119–1128.

66. Springer, N.M., Napoli, C.A., Selinger, D.A., Pandey, R., Cone, K.C., Chandler, V.L., Kaeppler, H.F., and Kaeppler, S.M. (2003). Comparative analysis of SET domain proteins in maize and Arabidopsis reveals multiple duplications preceding the divergence of monocots and dicots. *Plant Physiol* 132:907–925.

67. Min, J., Zhang, X., Cheng, X., Grewal, S.I., and Xu, R.M. (2002). Structure of the SET domain histone lysine methyltransferase Clr4. *Nat Struct Biol* 9:828–832.

68. Wang, H., Cao, R., Xia, L., Erdjument-Bromage, H., Borchers, C., Tempst, P., and Zhang, Y. (2001). Purification and functional characterization of a histone H3-lysine 4-specific methyltransferase. *Mol Cell* 8:1207–1217.

69. Fang, J., Feng, Q., Ketel, C.S., Wang, H., Cao, R., Xia, L., Erdjument-Bromage, H., Tempst, P., Simon, J.A., and Zhang, Y. (2002). Purification and functional characterization of SET8, a nucleosomal histone H4-lysine 20-specific methyltransferase. *Curr Biol* 12:1086–1099.

70. Nishioka, K., Rice, J.C., Sarma, K., Erdjument-Bromage, H., Werner, J., Wang, Y., Chuikov, S., Valenzuela, P., Tempst, P., Steward, R., Lis, T.J., Allis, C.D., and Reinberg, D. (2002). PR-Set7 is a nucleosome-specific methyltransferase that modifies lysine 20 of histone H4 and is associated with silent chromatin. *Mol Cell* 9:1201–1213.

71. Czermin, B., Melfi, R., McCabe, D., Seitz, V., Imhof, A., and Pirrotta, V. (2002). Drosophila enhancer of Zeste/ESC complexes have a histone H3 methyltransferase activity that marks chromosomal Polycomb sites. *Cell* 111:185–196.

72. Ringrose, L., and Paro, R. (2004). Epigenetic regulation of cellular memory by the Polycomb and Trithorax group proteins. *Annu Rev Genet* 38:413–443.

73. Ying, Z., Mulligan, R.M., Janney, N., and Houtz, R.L. (1999). Rubisco small and large subunit N-methyltransferases. Bi- and mono-functional methyltransferases that methylate the small and large subunits of Rubisco. *J Biol Chem* 274:36750–36756.

74. Beisel, C., Imhof, A., Greene, J., Kremmer, E., and Sauer, F. (2002). Histone methylation by the Drosophila epigenetic transcriptional regulator Ash1. *Nature* 419:857–862.

75. Maxon, M.E., and Herskowitz, I. (2001). Ash1p is a site-specific DNA-binding protein that actively represses transcription. *Proc Natl Acad Sci USA* 98:1495–1500.

76. Tachibana, M., Sugimoto, K., Nozaki, M., Ueda, J., Ohta, T., Ohki, M., Fukuda, M., Takeda, N., Niida, H., Kato, H., and Shinkai, Y. (2002). G9a histone methyltransferase plays a dominant role in euchromatic histone H3 lysine 9 methylation and is essential for early embryogenesis. *Genes Dev* 16:1779–1791.

77. Tachibana, M., Ueda, J., Fukuda, M., Takeda, N., Ohta, T., Iwanari, H., Sakihama, T., Kodama, T., Hamakubo, T., and Shinkai, Y. (2005). Histone methyltransferases G9a and GLP form heteromeric complexes and are both crucial for methylation of euchromatin at H3-K9. *Genes Dev* 19:815–826.

78. Schaft, D., Roguev, A., Kotovic, K.M., Shevchenko, A., Sarov, M., Neugebauer, K.M., and Stewart, A.F. (2003). The histone 3 lysine 36 methyltransferase, SET2, is involved in transcriptional elongation. *Nucleic Acids Res* 31:2475–2482.

79. Kouskouti, A., Scheer, E., Staub, A., Tora, L., and Talianidis, I. (2004). Gene-specific modulation of TAF10 function by SET9-mediated methylation. *Mol Cell* 14:175–182.

80. Chuikov, S., Kurash, J.K., Wilson, J.R., Xiao, B., Justin, N., Ivanov, G.S., McKinney, K., Tempst, P., Prives, C., Gamblin, S.J., Barlev, N.A., and Reinberg, D. (2004). Regulation of p53 activity through lysine methylation. *Nature* 432:353–360.

81. Jacobs, S.A., Harp, J.M., Devarakonda, S., Kim, Y., Rastinejad, F., and Khorasanizadeh, S. (2002). The active site of the SET domain is constructed on a knot. *Nat Struct Biol* 9:833–838.
82. Zhang, X., Yang, Z., Khan, S.I., Horton, J.R., Tamaru, H., Selker, E.U., and Cheng, X. (2003). Structural basis for the product specificity of histone lysine methyltransferases. *Mol Cell* 12:177–185.
83. Zhang, X., Tamaru, H., Khan, S.I., Horton, J.R., Keefe, L.J., Selker, E.U., and Cheng, X. (2002). Structure of the Neurospora SET domain protein DIM-5, a histone H3 lysine methyltransferase. *Cell* 111:117–127.
84. Xiao, B., Jing, C., Kelly, G., Walker, P.A., Muskett, F.W., Frenkiel, T.A., Martin, S.R., Sarma, K., Reinberg, D., Gamblin, S.J., and Wilson, J.R. (2005). Specificity and mechanism of the histone methyltransferase Pr-Set7. *Genes Dev* 19:1444–1454.
85. Xiao, B., Jing, C., Wilson, J.R., Walker, P.A., Vasisht, N., Kelly, G., Howell, S., Taylor, I.A., Blackburn, G.M., and Gamblin, S.J. (2003). Structure and catalytic mechanism of the human histone methyltransferase SET7/9. *Nature* 421:652–656.
86. Wilson, J.R., Jing, C., Walker, P.A., Martin, S.R., Howell, S.A., Blackburn, G.M., Gamblin, S.J., and Xiao, B. (2002). Crystal structure and functional analysis of the histone methyltransferase SET7/9. *Cell* 111:105–115.
87. Trievel, R.C., Flynn, E.M., Houtz, R.L., and Hurley, J.H. (2003). Mechanism of multiple lysine methylation by the SET domain enzyme Rubisco LSMT. *Nat Struct Biol* 10:545–552.
88. Trievel, R.C., Beach, B.M., Dirk, L.M., Houtz, R.L., and Hurley, J.H. (2002). Structure and catalytic mechanism of a SET domain protein methyltransferase. *Cell* 111:91–103.
89. Kwon, T., Chang, J.H., Kwak, E., Lee, C.W., Joachimiak, A., Kim, Y.C., Lee, J., and Cho, Y. (2003). Mechanism of histone lysine methyl transfer revealed by the structure of SET7/9-AdoMet. *Embo J* 22:292–303.
90. Manzur, K.L., Farooq, A., Zeng, L., Plotnikova, O., Koch, A.W., Sachchidanand, and Zhou, M.M. (2003). A dimeric viral SET domain methyltransferase specific to Lys27 of histone H3. *Nat Struct Biol* 10:187–196.
91. Couture, J.F., Collazo, E., Brunzelle, J.S., and Trievel, R.C. (2005). Structural and functional analysis of SET8, a histone H4 Lys-20 methyltransferase. *Genes Dev* 19:1455–1465.
92. Milne, T.A., Briggs, S.D., Brock, H.W., Martin, M.E., Gibbs, D., Allis, C.D., and Hess, J.L. (2002). MLL targets SET domain methyltransferase activity to Hox gene promoters. *Mol Cell* 10:1107–1117.
93. Tenney, K., and Shilatifard, A. (2005). A COMPASS in the voyage of defining the role of trithorax/MLL-containing complexes: linking leukemogenesis to covalent modifications of chromatin. *J Cell Biochem* 95:429–436.
94. Yokoyama, A., Wang, A., Wysocka, J., Sanyal, M., Aufiero, D.J., Kitabayashi, I., Herr, W., and Cleary, M.L. (2004). Leukemia proto-oncoprotein MLL forms a SET1-like histone methyltransferase complex with menin to regulate Hox gene expression. *Mol Cell Biol* 24:5639–5649.
95. Krogan, N.J., Dover, J., Khorrami, S., Greenblatt, J.F., Schneider, J., Johnston, M., and Shilatifard, A. (2002). COMPASS, a histone H3 (Lysine 4) methyltransferase required for telomeric silencing of gene expression. *J Biol Chem* 277:10753–10755.
96. Roguev, A., Schaft, D., Shevchenko, A., Pijnappel, WW., Wilm, M., Aasland, R., and Stewart, A.F. (2001). The Saccharomyces cerevisiae Set1 complex includes an Ash2 homologue and methylates histone 3 lysine 4. *Embo J* 20:7137–7148.
97. Su, I.H., Dobenecker, M.W., Dickinson, E., Oser, M., Basavaraj, A., Marqueron, R., Viale, A., Reinberg, D., Wulfing, C., and Tarakhovsky, A. (2005). Polycomb group protein ezh2 controls actin polymerization and cell signaling. *Cell* 121:425–436.
98. Xiao, B., Wilson, J.R., and Gamblin, S.J. (2003). SET domains and histone methylation. *Curr Opin Struct Biol* 13:699–705.

99. Taylor, W.R., Xiao, B., Gamblin, S.J., and Lin, K. (2003). A knot or not a knot? SETting the record 'straight' on proteins. *Comput Biol Chem* 27:11–15.
100. Schubert, H.L., Blumenthal, R.M., and Cheng, X. (2003). Many paths to methyltransfer: a chronicle of convergence. *Trends Biochem Sci* 28:329–335.
101. Blackburn, G.M., Gamblin, S.J., and Wilson, J.R. (2003). Mechanism and control in biological amine methylation. *Helvetica Chimica Acta* 86:4000–4006.

7

Non-Histone Protein Lysine Methyltransferases: Structure and Catalytic Roles

LYNNETTE M. A. DIRK[a] • RAYMOND C. TRIEVEL[b] •
ROBERT L. HOUTZ[a]

[a]Department of Horticulture
University of Kentucky
407 Plant Science Building
Lexington, KY 40546, USA

[b]Department of Biological Chemistry
University of Michigan Medical School
Medical Science Building 1
Ann Arbor, MI 48109, USA

I. Abstract

Non-histone protein lysine methyltransferases (PKMTs) represent an exceptionally diverse and large group of PKMTs. Even accepting the possibility of multiple
protein substrates, if the number of different proteins with methylated lysyl residues
and the number of residues modified is indicative of individual PKMTs there are well
over a hundred uncharacterized PKMTs. Astoundingly, only a handful of PKMTs
have been studied, and of these only a few with identifiable and well-characterized
structure and biochemical properties. Four representative PKMTs responsible for
trimethyllysyl residues in ribosomal protein L11, calmodulin, cytochrome c, and
Rubisco are herein examined for enzymological properties, polypeptide substrate
specificity, functional significance, and structural characteristics.

THE ENZYMES, Vol. XXIV
Copyright © 2006 by Elsevier Inc.

Although representative of non-histone PKMTs, and enzymes for which collectively there is a large amount of information, individually each of the PKMTs discussed in this chapter suffers from a lack of at least some critical information. Other than the obvious commonality in the AdoMet substrate cofactor and methyl group transfer, these enzymes do not have common structural features, polypeptide substrate specificity, or protein sequence. However, there may be a commonality that supports the hypothesis that methylated lysyl residues act as global determinants regulating specific protein-protein interactions.

II. Introduction

It is worth noting that the functional association of protein lysine methyltransferase (PKMT) activity with the SET domain structural motif was a consequence of the molecular and biochemical characterization of the non-histone PKMT ribulose-1,5-bisphosphate carboxylase/oxygenase lysine $^\varepsilon$N-methyltransferase (Rubisco LSMT, EC 2.1.1.127). Moreover, the characterization of Rubisco LSMT was based primarily on earlier studies of two PKMTs that even today remain unidentified in structure: [cytochrome c]-lysine N-methyltransferase (Ctm1p, described earlier as PMIII, EC 2.1.1.59) and calmodulin-lysine N-methyltransferase (CLNMT, EC 2.1.1.60).

In this chapter, we present an overview of the diversity of protein lysine methylation in terms of the polypeptide substrates and, where possible, the biochemical and structural characteristics for the associated enzymes. An immediately obvious observation from this endeavor has been the disparity between the number of proteins with methylated lysyl residues and the identification of the associated PKMTs. Indeed, the far greater number of proteins whose polypeptide sequence is deduced rather than determined suggest that the number of proteins with methylated lysyl residues could be far greater than presented in Table 7.1. For many decades it was accepted that PKMTs were exceptionally specific for the target polypeptide substrate, implying that a different PKMT exists for each of the proteins listed in Table 7.1. However, with the recent demonstration of alternate substrates for at least one histone PKMT (SET7/9, EC 2.1.1.43), the possibility arises that a single PKMT is responsible for several of these modifications.

Surprisingly, compared to the number of proteins with methylated lysyl residues, little information exists overall about the enzymes catalyzing these modifications. Yet, for a few there is a wealth of structural and/or biochemical information that has not been capitalized on in terms of defined cellular function. Although a relatively large number of proteins have been annotated as SET-domain-containing proteins in protein databases, it is now clear that not all of these will have histone lysyl residues as substrates. The unique, as well as fascinating, structural and biochemical aspects of the non-histone PKMTs (coupled with the large number of

Text continued on p. 187

TABLE 7.1

NATURAL OCCURRENCE OF METHYLATED LYSYL RESIDUES

Protein	Residue[a]	Source (Organism/Tissue)	Enzyme	Ref
		ε-N-monomethyllysine		
Actin	U[b]	Acanthamoeba catellanii (amoeba)	U	[1]
NAD-dependent alcohol dehydrogenase, EC 1.1.1.1	11 (~23%), 213 (~50%)	Sulfolobus solfataricus	U	[2]
Aspartate aminotransferase, EC 2.6.1.1	202 (~30%), 384 (~65%)	Sulfolobus solfataricus	U	[3]
Cytochrome c-553	24 (~50%)	Pavlova lutherii (Monochrysis lutheri) Chrysophycean alga	Ctm1p homolog	[4]
Cytochrome c-557	-8[c] (~7%)	Crithidia oncopelti	Ctm1p homolog	[5]
Cytochrome c	55 (~87%)	Hansenula anomala (yeast, Candida pelliculosa)	Ctm1p homolog	[6]
Elongation factor Tu	U	Salmonella typhimurium	U	[7]
	56 (<45%)	Escherichia coli	U	[7–9]
Elongation factor Tu, chloroplast	57 probable	Euglena gracilis	U	[10]
Elongation factor 1A1(eEF1A-1, formerly EF-1α)	U, multiple sites	Mucor racemosus (fungal mycelia not sporangiospores)	U	[11]
Ferredoxin, seven-iron, Zinc-containing, A & B	54 probable	Artemia salina (brine shrimp)	U	[12]
	29	Sulfolobus acidocaldarius, Sulfolobus metallicus, Acidianus infernos	U	[13, 14]
	30	Metallosphaera prunae	U	[14]
	29 (~95%), 101 (~10%)	Acidianus ambivalens (Desulfurolobus ambivalens)	U	[14]
Flagellin	U	Spirillum serpens	U	[15]
	203, 215, 221, 241, 251, 279, 292, 326, 348, 357, 362 and 391	Salmonella typhimurium		[16, 17]
		Salmonella typhimurium	Nml	[18]
		Salmonella typhi	FliU / FliV	[19]
		Salmonella typhimurium	FliB	[20]
		Salmonella choleraesuis	FliU	[21]
		Escherichia coli	LafV	[22]

Continued

TABLE 7.1

Natural Occurrence of Methylated Lysyl Residues—cont'd

Protein	Residue[a]	Source (Organism/Tissue)	Enzyme	Ref
Glutamate dehydrogenase, EC 1.4.1.3	254 (~50%), 260 (~50%), 372 (~50%), 391 (~50%), 392 (~50%), 393 (~50%)	*Sulfolobus solfataricus*	U	[23]
β-Glycosidase	116 (~70%), 135, 273 (~70%), 311 (~70%), 332	*Sulfolobus solfataricus*	U	[24]
Heat shock protein (hsp70A & B)	U(1–2)	*Gallus gallus* (embryonic chicken fibroblast)	U	[25]
Heparin-binding hemagglutinin adhesion (HBHA)	150–199 peptide heterogeneously methylated; 13 Lys in those 50 residues have 20–26 modifications	*Mycobacterium tuberculosis*	U, in cell wall fraction	[26]
Immunity protein for cloacin	12 probable	*Enterobacter cloacae* studied in *Escherichia coli*	U	[27]
Laminin-binding protein (LBP)	110–208 peptide heterogeneously methylated	*Mycobacterium smegmatis*	U, in cell wall fraction	[26]
Myosin	U	*Oryctolagus cuniculus* (rabbit skeletal muscle)	U	[28–30]
Myosin heavy chain, thick filaments of the myofibrils	34 (~60%)	*Oryctolagus cuniculus* (rabbit skeletal muscle)	U	[31]
Rhodopsin [sic]	35	*Gallus gallus* (adult chicken skeletal muscle)	U	[32]
p53 Cellular tumor antigen	U (1–2/10)	*Bos taurus* (bovine retina)	U	[33]
50S ribosomal protein L7/L12 ('A' type)	372	*Homo sapiens* (human)	SET7/9, EC 2.1.1.43	[34]
	83	*Halophilic eubacterium* NRCC 41227	U	[35]
50S ribosomal protein L7/L12 (L8)	76, 87	*Desulfovibrio vulgaris* (strain Miyazaki)	U	[36]
	81	*Escherichia coli*	U	[37]
60S ribosomal protein L29	4 (27% liver; >99% brain, 95% thymus)	*Rattus norvegicus* (rat)	U	[38]
60S ribosomal protein L42 (L44[sic])	40, 54	*Saccharomyces cerevisiae* (baker's yeast)	U	[39]

Protein	Organism		SET7/9, EC 2.1.1.43	Ref.
TAF10 (Transcription initiation factor TFIID subunit 10)	Homo sapiens (human)	189		[40]
Thioredoxin	Chloroflexus aurantiacus (photosynthetic bacterium)	104 (U%)	U	[41]
Tooth matrix protein	Homo sapiens (adult human tooth)	U	U	[42]
ε-N-dimethyllysine				
Actin	Acanthamoeba castellanii (amoeba)	U	U	[1]
Calmodulin	Paramecium tetraurelia	13	CLNMT homolog	[43]
Cytochrome c-557	Crithidia oncopelti	-8c (−14%)	U	[5]
Cytochrome c	Hansenula anomala (yeast, Candida pelliculosa)	55 (−13%)	Ctm1p homolog	[6]
	Thermomyces lanuginosus (Humicola lanuginosa)	72	Ctm1p homolog	[44]
Elongation factor Tu	Mucor racemosus (fungal mycelia not sporangiospores)	U, multiple	U	[11]
	Escherichia coli	56 (−<45%)	U	[7, 8]
Elongation factor 1A1(eEF1A-1, formerly EF-1α)	Saccharomyces cerevisiae (baker's yeast)	316	U	[45]
Ferredoxin-NADP reductase, EC 1.18.1.2	Oryctolagus cuniculus (rabbit reticulocytes)	55, 165	U	[46]
	Chlamydomonas reinhardtii (unicellular green alga)	135	U	[47]
Flagellin	Salmonella typhimurium	U	U	[15]
Heat shock protein (hsp70A & B)	Gallus gallus (embryonic chicken fibroblast)	U (1–2)	U	[25]
Heparin-binding hemagglutinin adhesion (HBHA)	Mycobacterium tuberculosis	150–199 peptide heterogeneously methylated; 13 Lys in those 50 residues have 20–26 modifications	U, in cell wall fraction	[26]
Laminin-binding protein (LBP)	Mycobacterium smegmatis	110–208 peptide heterogeneously methylated	U, in cell wall fraction	[26]
Myofibrillar protein	Oryctolagus cuniculus (rabbit)	U	U	[48]
Myosin II, heavy chain	Acanthamoeba castellanii	188	U	[49]
Rhodopsin [sic]	Bos taurus (bovine retina)	U (4/10)	U	[33]

Continued

TABLE 7.1

Natural Occurrence of Methylated Lysyl Residues—cont'd

Protein	Residue[a]	Source (Organism/Tissue)	Enzyme	Ref
Silaffin 1 precursor, upon processing yields silaffin-1B, silaffin-1A2, and silaffin-1A1	111 probable, 144, 155, 166, 185, 204, 223, 242	Cylindrotheca fusiformis (marine diatom)	U	[50–52]
Thioredoxin	104 (U%)	Chloroflexus aurantiacus (photosynthetic bacteria)	U	[41]
ε-N-trimethyllysine				
Actin	U	Acanthamoeba castellanii (amoeba)	U	[1]
Actobindin	35, 72	Acanthamoeba castellanii (amoeba)	U	[53]
α-Amylase, EC 3.2.1.1	U	Triticum aestivum (wheat)	U	[54]
ATPase subunit c, EC 3.6.3.14, mitochondrial subunit	43	normal Bos taurus (bovine heart), batten–diseased canine (English setters brain and kidneys), human, sheep, mice	U	[55–58]
Calcium vector protein	95, 116	Branchiostoma lanceolatum (common lancelet; amphioxus)	U	[59]
Calmodulin	115	Paramecium tetraurelia; Tetrahymena pyriformis; Rattus norvegius (rat testis); Oryctolagus cuniculus (rabbit skeletal muscle); Bos taurus (bovine brain); Homo sapiens (human brain); Spinacia oleracea (spinach); Triticum aestivum (wheat); Metridium senile (brown or frilled sea anemone)	CLNMT, EC 2.1.1.60	[43, 60–69]
Calmodulin	115, 148	Euglena gracilis	CLNMT homolog	[70]
Citrate synthase, EC 4.1.3.7	368	Sus scrofa (porcine heart)	U	[71, 72]
Cytochrome c	72	Neurospora crassa; Aspergillus niger; Saccharomyces cerevisiae; Enteromorpha sp.; Schizosaccharomyces pombe (fission yeast); Debaryomyces hansenii (D. kloeckeri [sic]) (yeast, Torulaspora hansenii)	U	[73–78]
		Saccharomyces cerevisiae	Ctm1p, EC 2.1.1.59	[79]
	72, 73	Hansenula anomala (yeast, Candida pelliculosa)	Ctm1p homolog	[6]
	86 (~95%)	Thermomyces lanuginosus (Humicola lanuginosa)	Ctm1p homolog	[44]
	72, 86	Abutilon theophrasti (China jute; Indian mallow); Allium porrum (leek); Brassica napus (rape); Brassica oleracea	Ctm1p, EC 2.1.1.59;	[73, 80–91]

Substrate	Species	Methylation site(s)	Methyltransferase (and Ctm1p homolog)	Ref.
(continued)	(cauliflower); *Cannabis sativa* (hemp; marijuana); *Cucurbita maxima* (pumpkin; winter squash); *Fagopyrum esculentum* (common buckwheat); *Ginkgo biloba* (ginkgo); *Gossypium barbadense* (sea-island or Egyptian cotton); *Guizotia abyssinica* (niger; ramtilla); *Helianthus annuus* (common sunflower); *Lycopersicon esculentum* (tomato); *Phaseolus aureus* (mung bean); *Vigna radiata*; *Ricinus communis* (castor bean); *Sesamum indicum* (oriental sesame; gingelly); *Spinacia oleracea* (spinach); *Triticum aestivum* (wheat)			
Cytochrome *c*-557	*Crithidia oncopelti*	-8[c] (~63%), 72	Ctm1p homolog	[5]
Cytochrome *c*-555	*Crithidia fasciculata*	-8[c], 72	Ctm1p homolog	[92]
Cytochrome *c*-558	*Euglena gracilis*	86	Ctm1p homolog	[5]
Elongation factor 1A1(eEF1A-I, formerly EF-1α)	*Mucor racemosus* (fungal mycelia not sporangiospores)	U, multiple sites	U	[11]
	Saccharomyces cerevisiae (baker's yeast)	79	U	[45]
	Artemia salina (brine shrimp)	35, 78, 218, 317	U	[12]
	Oryctolagus cuniculus (rabbit)	36, 79, 318	U	[46]
Elongation factor 1A2 (eEF1A-2)	*Oryctolagus cuniculus* (rabbit skeletal muscle)	55, 165	U	[93]
Ferredoxin-NADP reductase, EC 1.18.1.2	*Chlamydomonas reinhardtii* (unicellular green alga)	83, 89	U	[47]
Src family kinase Fyn	*Homo sapiens* (human)	7 and/or 9	U	[94]
Heat shock protein (hsp70A & B)	*Gallus gallus* (embryonic chicken fibroblast)	U (1–2)	U	[25]
Myosin	*Oryctolagus cuniculus* (rabbit skeletal muscle)	U	U	[30, 48]
Myosin heavy chain, thick filaments of the myofibrils	*Oryctolagus cuniculus* (rabbit skeletal muscle)	129, U	U	[31]
profilin Ia & Ib (acidic) and profilin II (basic)	*Gallus gallus* (chicken gizzard smooth muscle)	127	U	[95]
	Gallus gallus (adult chicken skeletal muscle)	130, 551	U	[32, 96, 97]
	Acanthamoeba castellanii (amoeba)	103	U	[98, 99]
rhodopsin [sic]-associated protein	*Bos taurus* (bovine retina)	U	U	[33]

Continued

TABLE 7.1

NATURAL OCCURRENCE OF METHYLATED LYSYL RESIDUES—cont'd

Protein	Residue[a]	Source (Organism/Tissue)	Enzyme	Ref
ribulose bisphosphate carboxylase large chain chloroplast precursor, EC 4.1.1.39	14	*Nicotiana tabacum* (common tobacco); *Cucumis melo* (muskmelon)		[100]
	14	*Vigna sinensis* (cowpea); *Cucumis sativus* (cucumber); *Glycine max* (soybean); *Solanum tuberosum* (potato); *Petunia X hybrida* (petunia); *Pisum sativum* (garden pea); *Lycopersicon esculentum* (tomato); *Capsicum annuum* (pepper)		[101]
		Pisum sativum (garden pea)	LSMT, EC 2.1.1.127	[102–104]
50S ribosomal protein L11	3, 39	*Escherichia coli* K12	PrmA	[105]
			PrmA	[106–108]
	9, 45	*Thermus thermophilus* (strain HB8)	U	[109]
	U	[PRPL11] *Spinacea oleracea* (spinach chloroplast)	U	[110]
		Bacillus subtillus, B. stearothermophilus	U	[111, 112]
60S ribosomal protein L40 (CEP52)	22	*Rattus norvegicus* (rat liver, brain and thymus)	U	[38]

a. When a percentage in brackets follows the residue number, that percentage is the best estimate of the partial methylation at that site. When a fraction follows the number, that fraction represents the best estimate of how many residues in the protein are modified.

b. Undetermined.

c. By convention, numbering of cytochrome *c* is relative to a conserved Gly that in horse heart mitochondria is the first residue. Thus, negative numbers translate to the sequence in question having an N-terminal extension compared with that mammalian protein.

methylated proteins) makes these enzymes fertile ground for future research. Indeed, remaining studies could potentially provide novel protein structures and/or opportunities for unique molecular genetic discoveries with little more than the application of basic techniques in protein chemistry and cloning.

It is apparent from the following sections that identification of a functional role for site-specific protein trimethyllysyl formation has in many cases been daunting. A general theme, however, we wish to propose is that lysyl methylation is a global mechanism for determining specific protein-protein interactions. This hypothesis, although supported by some of the observations that follow, is more the consequence of the current burgeoning family of protein domains with identifiable binding sites for methylated lysyl residues with important protein-protein interactions. Intense research in the histone modification field has resulted in the identification of several methyllysine binding modules; notably, the Chromo and Tudor domains [113], which jointly belong to the Agenet superfamily [114]. Given the number of protein domains that lack known biological functions, it is highly probable that other methyllysine-binding modules await discovery.

This chapter focuses on four PKMTs for which the largest amount of information is available either for the enzyme or its polypeptide substrate (Figure 7.1), especially within the context of attempting to shed light on mechanistic aspects of polypeptide substrate specificity and on the potential relationship between kinetic reaction mechanisms and processive or distributive methyl group transfer. It is our hope that other researchers will find these enzymes and their function as exciting and rewarding as those aggressively pursuing the functional significance of histone PKMTs.

III. Diversity of Polypeptide Substrates

The voluminous number and diversity of proteins with methylated lysyl residues (Table 7.1), each potentially representing a distinct PKMT substrate, warrants mention. There has been only one previous attempt to consolidate a comprehensive list of proteins with methylated lysyl residues [116], and these authors have been staunch supporters of protein methylation and have provided an ideal starting opportunity for compiling the data presented in Table 7.1. As previously noted, by far, the majority of these proteins do not have an identified or characterized PKMT. Thus, each methylated protein could potentially lead to the discovery of a novel enzyme.

If the PKMT recognition of the target lysyl residue can be predicted (even in part) by primary structure alone, scrutinizing the sequence surrounding the known mono-, di-, and trimethyllysyl residues of these putative substrates may identify alternative substrates for known PKMTs. Although certainly beyond the scope of this chapter to consider each of these PKMT substrates

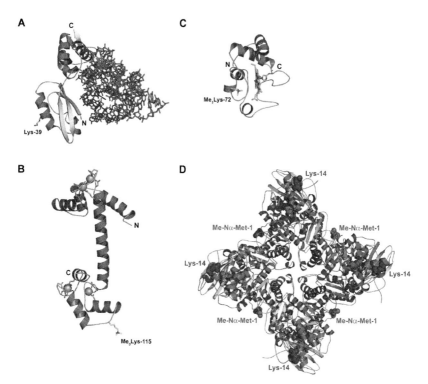

Fig. 7.1. Crystal structures of non-histone proteins that undergo lysine methylation *in vivo*. (A) Structure of the *Thermotoga maritima* ribosomal protein L11 in complex with a 58 nucleotide fragment of 23S ribosomal RNA (blue) with the N- and C-termini of the protein labeled (pdb accession code 1MMS.pdb). The Lys-39 methylation site is depicted in stick representation with yellow carbon atoms, whereas the α-amine and Lys-3 methylation sites are not shown due to the disorder of the protein's N-terminus in the crystal structure. (B) Ca^{2+}-bound form of chicken calmodulin (1UP5.pdb). Trimethyl-Lys-115 is illustrated as in 1a, and the Ca^{2+} cations (orange spheres) and their chelating residues are also displayed. (C) Ribbon diagram of yeast iso-1-cytochrome C in the reduced state (1YCC.pdb). Trimethyl-Lys-72 is displayed as in 1a, whereas the heme prosthetic group is depicted with cyan carbon atoms. (D) Crystal structure of ribulose-1,5 bisphosphate carboxylase/oxygenase (Rubisco) from *Chlamydomonas reinhardtii* (1GK8.pdb). The large subunit Lys-14 and small subunit α-amine methylation sites that are commonly methylated in many plant species are highlighted in blue and magenta, respectively (note that *C. reinhardtii* Rubisco does not undergo Lys-14 trimethylation). The scale of the Rubisco structure was reduced approximately 50% in proportion to the other protein structures illustrated in the figure, and only half of the hexadecameric Rubisco haloenzyme is shown. Ribbon diagram representations were rendered in PyMOL (*http://www.pymol.org*).

individually, other (more interesting or biologically significant) proteins are described briefly as a means of promoting an interest and curiosity in a particular unknown PKMT.

A. VACCINE DEVELOPMENT AGAINST TUBERCULOSIS AND LEPROSY

Both heparin-binding hemagglutinin adhesion (HBHA) and laminin-binding protein (LBP) have likely roles in dissemination of the obligate intracellular pathogens (*Mycobacterium tuberculosis* and *M. lephrae*, respectively) due to epethelial cell binding activity [117]. To date, lysyl methylation in LBP by *M. lephrae* does not appear to be involved in the adhesion properties of the protein [118], but may increase proteolytic resistance [26] against the abundant tracheal trypsin-like proteases [119].

The PKMT responsible for methylation of HBHA is located in the cell wall fraction and by using the much faster-growing *M. smegmatis* the *M. tuberculosis* HBHA was methylated in the foreign organism, presumably with the cell-wall-associated PKMT that catalyzes methylation of the *M. smegmatis* LBP [26]. However, although no direct physical data for the differences in methylation pattern was provided, this *M. smegmatis*-produced HBHA does not elicit the same immunity as purified native HBHA, suggesting that the PKMTs may indeed have unique enzymatic properties [120]. Research is very active in using methylated HBHA as part of an acellular tuberculosis vaccine due to the strong evidence that inoculation with that antigen provides effective protection to mice when subsequently challenged with virulent *M. tuberculosis* [120, 121]. Thus, in this instance a thorough investigation of both PKMTs would potentially yield important and useful information with practical considerations of generating efficacious vaccines against the devastating human diseases tuberculosis and leprosy.

B. METHYLATION AS A MEANS OF STABILIZING PROTEINS

Thermophilic organisms within Archaea are rife with proteins containing methylated lysyl residues, including aspartate aminotransferase, β-glycosidase, NAD-dependent alcohol dehydrogenase, ferredoxin, and glutamate dehydrogenase. Similar to increased resistance to proteolysis, for which nearly every methylated protein has been tested, a generalized functional role in thermostability emerged because of the high frequency of lysyl methylation in this class of organisms. However, for NAD-dependent alcohol dehydrogenase the thermostability was later attributed to nonconserved residues within the sequence [122]. Alternatively, Zappacosta and others [3] postulated a role in protein aging and turnover.

A recent thorough examination of the structural property differences between a methylated and a nonmethylated version of *Sulfolobus solfataricus* β-glycosidase demonstrated that alterations to the surface hydrophobicity and strengthened ionic interactions via salt bridges that occur due to the lysyl methylation are thought to create a more compact structure that is resistant to denaturation and aggregation at the higher temperatures at which these organisms flourish [24]. Thus, perhaps methylation of lysyl residues constitutes a means of not only recruitment but stabilizing of the interactions without causing protein aggregation.

C.   Biomineralization

Posttranslational processing of the silaffin precursor from the marine diatom *Cylindrotheca fusiformis* is a complex progression that includes not only attachment of linear polyamines and cleavage of C-terminal tetrapeptides but a PKMT reaction to add two methyl groups to multiple lysines in the sequence [51]. The result of the complete process is three peptides (silaffin-$1A_1$, -$1A_2$, and -1B), which are capable of precipitating silica. Presumably, the length of the polyamine chain in synergy with the peptide backbone and multivalent anions forms the intricate species-specific pattern of diatoms' skeletons [52].

There is intense biotechnological interest in this biomineralization process for creating specific shapes and sizes for nanotechnological purposes. The functional significance of the dimethylation of these peptides has largely been ignored, and these modifications may be instrumental in creating the fascinating architecture of the diatom's shell. The discovery of the PKMT(s) responsible for methylating silaffin-$1A_1$, -$1A_2$, and -1B may shed light on further means of controlling biomineralization.

D.   Flagellin Protein Lysine Methyltransferase (Nml, FliB, FliU, LafV)

The monomethyl modification of flagellin, and the enzyme responsible, warrant attention. Despite the fact that flagellin from *Salmonella typhimurium* was the first natural occurrence of ε-N-methyl lysine nearly a half century ago [16] and that the evidence for a gene (*nml*) determining the presence of the modification occurred only two years later [18], virtually nothing is known today about the PKMT or the function of the methylation. Twelve of 28 possible lysyl residues that are monomethylated are largely within the central region of the protein [17].

Initially, the flagellin-specific PKMT (FliU and/or FliV) was determined as essential for biosynthesis of *Salmonella* flagella because its mutation led to loss of motility [123], and the introduction of *fliU* together with a downstream gene, *fliV*, restored motility to flagellin-deficient *Salmonella* mutants [19]. However, as will be commonly seen throughout this chapter a lack of this modification in other species (specifically in *E. coli* and other mutant strains of *Salmonella* [124], as well as in *fliU*-mutant *Aeromonas caviae* [125]) undermines assignment of functional significance. Later DNA sequencing of this flagellar gene complex (*fli*) identified *fliU* and *fliV* genes as a single one (*fliB*), and thus what had been predicted as 19 and 20 kDa proteins became a single protein of expected mass of 45.4 kDa [20]. This same gene organization, *fliU* including the previously reported *fliU* and *fliV*, was later determined to exist in *Salmonella cholerasesuis* [21]. Resolution of the true identity of the flagellin-specific PKMT gene will require an enzymatic characterization using the gene products FliU, FliV, and FliB from *S. typhimurium* with a flagellin preparation from a mutant strain that lacks the methylation.

The transfer of the flagellin structural gene (*fliC*) alone to *E. coli* was sufficient to restore motility, but *fliB* was required for the flagellin to be monomethylated [20].

Most recently, though, a *fliB* homolog has been discovered as part of a Flag-2 locus in *E. coli* and is likely part of an ancestral gene cluster responsible for a novel flagellar system [22].

Although the suggestion had been made that immunity against *Salmonella* using the methylated flagellin as an antigen may be altered [20], activation of TLR5 (Toll-like receptor), which detects flagellin in mammalian systems, is unaffected by any posttranslational modifications of flagellin [126]. Thus, the functional significance of lysyl methylation in flagellin remains unknown, and interest wanes because recent reviews of flagellar assembly do not even make mention of FliB in the models [127, 128].

IV.   Non-Histone Protein Methyltransferases

A.   RIBOSOMAL PROTEIN L11 LYSINE METHYLTRANSFERASE (PRMA)

1.   Preliminary Enzymatic Characteristics

Despite modification of a multitude of ribosomal proteins [129], ribosomal protein L11 is the most thoroughly described protein that has lysyl methylation. Indeed, it is the only methylated protein of many from the translational machinery for which the known corresponding PKMT (PrmA) has been crystallized. As noted in Table 7.1, two lysines are trimethylated in both *E. coli* cytoplasm (K3, K39; [105]) and *S. oleracea* chloroplasts (K9, K45; [110]).

Considered conserved in plant plastids [130] and detected in bacteria but not eukaryota [131], the specific lysyl methylation (Figure 7.1c) would appear to endow these organisms with some selective advantage. However, the crystal structure of L11 in complex with a 58-nucleotide fragment of 23S ribosomal RNA [132, 133] reveals that the respective methylation sites do not participate in direct contact with the RNA. Lys-39 is located on the opposite face of the protein from the 23S RNA binding surface (Figure 7.1a), whereas the first few residues of the N-terminus (including the α-amine group and Lys-3) are disordered in the crystal structures, implying that they do not engage in significant interactions with the RNA (Figure 7.1a). Thus, it is unlikely that the posttranslational methylation of L11 plays a direct role in mediating protein nucleotide interactions, although recent structural and modeling studies of the ribosome elude to the function of its N-terminus in elongation factor-G-dependent translocation [134].

Because of the ease of genetic manipulation and of propagating and harvesting material, the majority of information regarding ribosomal proteins and the associated PKMT is derived from bacteria. Nearly three decades ago, PrmA was partially purified by classical protein biochemical techniques. Surprisingly, that study remains the defining enzymological characterization [135]. Thus, partially purified PrmA exhibited a pH optimum of ~8, a K_m for AdoMet of 3.2 μM, a V_{max} of 112.5 pmoles methyl groups transferred/min/mg protein and a molecular mass determined by gel filtration of ~31 kDa (minor methyltransferase activity at 17 and 56 kDa) [135] and by sucrose density centrifugation of ~40 kDa [136]. The protein substrate used for PrmA assays, as with all of the PKMTs described in this chapter, is an unmethylated form of the polypeptide substrate. It is also used for PrmA originated from 50S ribosome subunit preparations from *E. coli* 1500 *rel⁻ met⁻* strains that contain undermethylated ribosomal proteins when grown under methionine-limiting conditions and from *prmA* (*protein methylation*) mutants of *Escherichia coli*. The incorporation of methyl groups into these ribosomal preparations was increased more than twofold after disassociation with lithium chloride and urea and the rRNA removed [135]. The authors claimed that this was evidence that the methylation event occurred prior to 50S assembly. An earlier study of methylation using functional abnormal 50S ribosomes from *E. coli* EA2 (*RC^rel met⁻*) strains grown in the presence of ethionine had shown similar incorporation whether the proteins were part of the particles or disassociated.

Intriguingly, the methyltransferase assay was conducted with significant amounts of both denaturants (final concentration 0.6 M lithium chloride and 0.8 M urea; [136]). Thus, the stage at which L11 becomes methylated is unresolved. Methyl transfer was noted in L5, L3, and L1 proteins to a minor extent (~12%), but clearly the majority was incorporated into L11, as determined by the position of the protein on both 1D and 2D gels [135]. The heterogeneous nature of these ribosomal proteins precluded any kinetic analyses for the peptide substrate. The products of the *in vitro* reaction were identified as monomethyllysyl, trimethyllysyl, and a neutral basic amino acid, which is likely the trimethylated N-alpha amino group. The weakness of all characterization to date remains the L11 methylation status determined only by its position on 2D gels.

2. Functional Significance

E. coli mutants lacking methyl groups in L11 were isolated by three separate research groups. These *prmA* mutations were a result of a specific mutagenic approach for finding bacteria defective in ribosomal protein methylation [106, 137], of an observed altered electrophoretic mobility of L11 (as reported by [107]), and of a revertant of streptomycin-dependent bacteria [138]. Two of the mutants were later shown to have single point mutations in the same codon (*E. coli* Trp 285), resulting in a stop codon with a corresponding loss of nine amino acids from the carboxy terminus and an Arg substitution for *prmA3* and *prmA1*, respectively [107]. However, residual PrmA activity in all mutants left doubt as to the true phenotype of bacteria lacking such activity.

With the creation of the *prmA*-null mutant completely viable and indistinguishable from the parental strains [107], the function of ribosomal L11 lysine methylation at such a high energetic cost to the cell was, and remains, enigmatic. A similar conclusion was reached with studies of the null mutant of *Thermus thermophilus* (PrmA is nonessential) [109]. However, recent studies have emphasized the interactions between the N-terminal domain of prokaryotic L11 and the antibiotic thiostrepton, which inhibits translation of the nascent polypeptide chain by the ribosome [139, 140]. These findings suggest that post-translational methylation of L11, specifically in its N-terminus, may have evolved to block antibiotic interactions that inhibit ribosome translation. This observation is consistent with the prevalence of L11 methylation in prokaryotes and its corresponding absence in eukaryotes.

3. Structural Determination

PrmA, as the only genetic determinant for methyl group transfer to L11 [137], is intriguing from the standpoint that the N-terminal alanyl residue of L11 is known to be trimethylated on the alpha amino group. As suggested by Chang et al. [135], this could be the result of PrmA forming a vital part of a complex that sequentially adds methyl groups to the substrate. Others have proposed that in a single binding event PrmA methylates the two (or three in *T. thermophilus*) target lysines and the alpha amino group. Regardless, opportunities abound for kinetic and mechanistic characterization of a PKMT that varies from the structures known to date. Rather than ribosome preparations, ideally such studies would clone L11 into vectors for induction of protein for use as substrates.

The recent structural determination of PrmA from the archaeon *Thermus thermophilus* [130] has yielded significant insights into the mechanism by which it methylates multiple nitrogens within L11. The overall structure of the enzyme resembles a bola with the globular N- and C-terminal domains connected by a linker composed of an α-helix (Figure 7.2a). The C-terminal domain is the site of catalysis and possesses a mixed α-helical/parallel β-sheet topology that is structurally conserved in the Class I AdoMet-dependent methyltransferase family [141] to which PrmA belongs.

Moreover, the catalytic domain of PrmA shares a high degree of structural homology with that of the yeast histone H3 Lys-79 methyltransferase DOT1 [142] and its human ortholog DOT1L [143] (Figure 7.2b), which are also members of the Class I family. However, the 3D arrangement of the N-terminal domains of DOT1L and PrmA are strikingly different. In DOT1L, the N-terminal α-helical domain directly packs against its C-terminal catalytic domain, burying a significant amount of surface area between the two domains (Figure 7.2b). The linker loop joining the two domains forms a loose hairpin, referred to as the lid [142], which encloses AdoMet within its binding pocket.

Cumulatively, these interactions establish an extensive well-ordered substrate binding site that promotes highly specific interactions with the globular surface surrounding Lys-79 of histone H3 in the nucleosome core particle. In contrast, the flexible linker of PrmA may be a key to its promiscuous specificity for multiple methylation sites in L11. The relatively open AdoMet binding site may permit the enzyme to recognize the α-amine, Lys-3, and Lys-39 sites, whereas the N-terminal domain engages in distal contacts with other regions of L11. Thus, the N-terminal

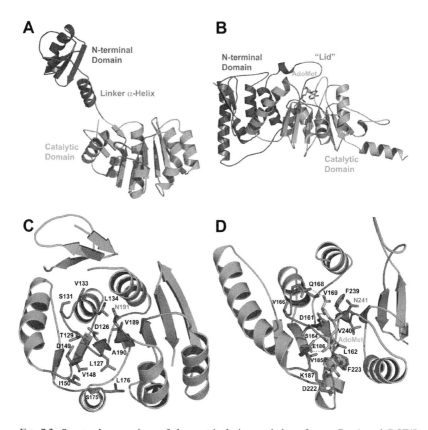

FIG. 7.2. Structural comparison of the protein lysine methyltransferases PrmA and DOT1L (A) Ribbon diagram of *Thermus thermophilus* PrmA (1UFK.pdb) with the N-terminal and C-terminal catalytic domains depicted in blue and green, respectively. The linker α-helix connecting the two domains is illustrated in red, whereas the AdoMet binding pocket is highlighted in magenta. (B) Structure of the non-SET domain human histone H3 Lys-79 methyltransferase DOT1L bound to AdoMet (1NW3.pdb). The N- and C-terminal domains are rendered as in 2a and the lid motif that encloses the AdoMet binding pocket is illustrated in red. (C) AdoMet binding cleft of PrmA. Residues that potentially interact with the cofactor through either van der Waals contacts or hydrogen bonds are depicted, whereas the putative catalytic residue, Asn 191, that aligns the lysine ε-amine for methylation is labeled in red. The protein's color scheme is illustrated as in 2a. (D) Cofactor binding cleft of hDOT1L in complex with AdoMet (denoted by cyan carbons). Residues that contribute to cofactor binding are shown as in 2c and the catalytic Asn 241 is labeled in red. Figures were rendered in PyMOL (*http://www.pymol.org*). (See color plate.)

domain of PrmA may function as a molecular clamp wherein the association of the catalytic domain with a specific methylation site in L11 enables the enzyme's N-terminus to grasp a distal site in the protein substrate in order to promote high-affinity binding. Verification of this hypothetical model awaits further structural and biochemical characterization of the PrmA's substrate specificity.

Motifs identified by an alignment of 34 orthologs [131] can now be related to the tertiary structure of the enzyme. Among the regions exhibiting the highest degree of sequence conservation is the AdoMet binding cleft (Figure 7.2c), which harbors significant structural and sequence similarity to the cofactor binding site of DOT1L [143] (Figure 7.2d). In all Class I methyltransferases, AdoMet binds in an extended confirmation through a series of highly conserved interactions with the catalytic domain. In DOT1L, the cofactor is positioned astride a Gly-rich loop following the first β strand of the parallel β sheet that comprises the core of the catalytic domain. This loop possesses an alternating series of glycines, which is conserved in PrmA. In addition, AdoMet is anchored within its binding cleft in DOT1L through two acidic residues. Asp-161 forms a salt bridge interaction with the cofactor, whereas Glu-186 engages in bifurcated hydrogen bonding with the 2′ and 3′ hydroxyl groups of the ribose moiety. In PrmA, Asp-126 and Asp-149 are in structurally conserved positions to engage in homologous interactions of α-amine group and ribose hydroxyls of AdoMet, respectively.

The structural comparison of the active sites of PrmA and DOT1L also reveals a conserved catalytic asparagine that is positioned proximal to the labile methyl group of AdoMet. In yeast DOT1, Cheng and colleagues have proposed the amide Oδ1 atom of Asn-479 (equivalent to Asn-241 in DOT1L) hydrogen bonds to the ε-amine of the lysine substrate, aligning it for the S_N2 methyltransfer with AdoMet [142]. Furthermore, they note that DNA and protein glutamine methyltransferases exhibit a similar asparagine orientation within their respective active sites, suggesting a conserved mechanism for N-methylation by Class I methyltransferases. In PrmA, Asn-191 is positioned almost identically to Asn-241 in DOT1L, underscoring its conserved role in aligning the α-amine and lysine ε-amine groups of L11 for repetitive rounds of methylation. Taken together, the structural homology observed in the catalytic domains of DOT1L and PrmA reveals a conserved mechanism for lysine ε-amine methylation that may be universally shared by all Class I PKMTs.

4.   Possible Interactors as a Reason for Methylating L11

As a corollary to the emerging consensus through the study of histone PKMTs, protein-protein interactions form the majority of the function for methylation "marks." For ribosomal L11, a direct activation of the enzyme, (p)ppGpp synthetase I (RelA), was dependent on the intact secondary structure of N-terminal proline-rich helix [144]. Such activation, which occurs during amino acid deprivation (part of the so-called stringency response), was removed when L11 Pro22 was mutated to leucine. Regardless of whether this was a direct interaction, it was not influenced by the methylation status of the lysyl residues because the null *prmA* mutant had a normal stringency response [107]. Due to the inherent flexibility of the region in which L11 is located in the 50S subunit of the ribosome [134], it is challenging to definitively assign functions to the N-terminal domain of this ribosomal protein. Nevertheless, the interactions of L11 with elongation factor G and antibiotics such as thiostrepton hint at the possibility that its methylation may have a role in ribosomal translocation during protein synthesis.

Albeit not a new idea [145], the apparent ubiquitous lysyl methylation of various translational factors in multiple organisms (such as L7/L12, L11, EF-Tu, EF1A-1, and EF1A-2) gives rise to the compelling idea that such a posttranslational modification has an important function for the translational apparatus. Yet, given the recent structural determination of the ribosome many of the direct roles that ribosomal proteins had been ascribed are now seen as supportive, aiding in maintaining the structure of the catalytic RNA they surround [146]. Even so, the proximity of methyllysyl-containing proteins during ribosome function (such as the factor binding site, including L11 and L7/L12, making direct contacts with amnioacyl tRNA-EF-Tu complex; [146]) provides tantalizing reasons to conduct research on elucidating the role of methyllysyl modifications.

Although the importance of the translational equipment of a cell is nearly intuitive, the various roles purported for eEF1A-1 and eEF1A-2 deserve notice in the context of this chapter because despite being a relatively abundant cellular protein (~3-10% soluble protein; [45]) the PKMT responsible and the consequence of the methylation are vastly understudied. Indeed, it is the possibility of the involvement of eEF1A-1 with cancer [147] and with growth and differentiation processes of kinetoplastid parasites [148] that may provide a renewal of enthusiasm for posttranslational enzyme investigations.

The role of eEF1A-1 in both of these instances may involve alteration of this protein's usual inhibition of apoptosis. If the function of methylation is truly to recruit specific proteins at a given moment during the cell cycle or development, the list of associations with eEF1A from which to study for the critical interaction is long. These include ribonucleoprotein particles, the cytoskeletal matrix through an actin-binding propensity, and the 26S protease complex [148]. Even with all of these vital functions, the PKMT of eEF1A and the consequence of the methylation event have been all but ignored. The significance of the methylation was recently generalized as increasing the activity of the factor [229].

B.   CALMODULIN-LYSINE N-METHYLTRANSFERASE (CLNMT, EC 2.1.1.60)

Calmodulin-lysine N-methyltransferase (CLNMT, EC 2.1.1.60) is responsible for the formation of trimethyllysyl-115 in calmodulin (Figure 7.1b), and this lysyl residue is methylated in calmodulin from most species [149, 150]. As with nearly all of the PKMTs discussed in this chapter, naturally occurring unmethylated forms of the polypeptide substrate have been used for biochemical studies, but studies with CLNMT were greatly facilitated early on by the development of a synthetic calmodulin gene (VU-1) and bacterial expression [151]. CLNMT has been purified from rat brain cytosol (7,080-fold) [152], the cytosolic fraction from *Paramecium tetraurelia* (6,800-fold) [153], sheep brain (more than 20,000-fold) [154], and rat testes (470-fold) [155].

Homogeneous preparations suggest a monomer with a molecular mass of 33 to 38 kDa [154], although 57 kDa has also been reported [152]. The enzyme has been extensively characterized both in terms of enzyme kinetics and substrate specificity. CLNMT will not methylate the protein substrates of other known PKMTs [152, 156, 157] and exhibits high affinity for calmodulin with bound calcium (K_m 100 nM) [154]. Although there are no specific subcellular localization studies for CLNMT, the enzyme is coordinately regulated with calmodulin methylation *in vivo* and is as widely distributed in tissues and organs, as is calmodulin. The interaction of CLNMT with calmodulin has been extensively investigated in terms of structural alterations in calmodulin as well as the introduction of site-specific mutations. Replacement of Lys-115 with Arg (VU-3 calmodulin) or Ile results in calmodulins that are inactive as substrates [151, 158] but potent inhibitors of CLNMT. Such inhibition suggests that recognition of calmodulin as a substrate by CLNMT involves sites remote from the target lysyl residue. There is extensive information available about the kinetics and substrate specificity for CLNMT, but as of yet no identifiable gene sequence for this enzyme.

1. Kinetic Mechanism Determination

As with many co- and posttranslational processing enzymes, CLNMT is a relatively scarce enzyme with low catalytic turnover. In addition, its purified preparations often require affinity chromatography steps in addition to more conventional techniques. The highest activity reported for CLNMT is ~0.03 s^{-1} (~44 nmoles of methyl groups transferred/min/mg of protein) [154]. Although CLNMT prefers Ca-bound forms of calmodulin as a substrate, the enzyme is capable of methylating Ca-depleted forms of calmodulin with approximately fivefold reduction in enzyme efficiency (k_{cat}/K_m) [154].

Kinetic analyses of CLNMT activity indicate a sequential bi-bi reactant mechanism [154, 155], and product inhibition studies using methylated forms of calmodulin as well as the dead-end inhibitor VU-3 calmodulin (Lys-115 to Arg) suggest that the initial binding of substrates is ordered with AdoMet binding first [155]. However, a thorough examination of double reciprocal velocity plots with a number of inhibitors, targeting either calmodulin-specific or methylation-specific CLNMT activity, revealed complex inhibitor plots suggesting unique structural interactions between CLNMT and its substrates [155]. It is worth noting that kinetic analyses of PKMTs with low k_{cat}s and multiple methylation events is probably additionally complicated by the common absence of product analyses, where radiolabel from [^{3}H-methyl]AdoMet into mono-, di-, and trimethyllysine is individually quantified. This is addressed in the sections following that address cytochrome c and Rubisco LS PKMTs.

2.   *Structural Determinants of Substrate Specificity*

The extensive availability of calmodulin antagonist, the ready availability of high-resolution structures for calmodulin, and studies identifying the interaction of calmodulin with its target proteins as an effector have facilitated the identification of interesting structural requirements in calmodulin for recognition as a substrate by CLNMT. Although these studies identify several commonalities in the structural interaction between CLNMT and calmodulin, that are similar to calmodulin and calmodulin binding proteins, some important differences exist.

Calmodulin contains two N- and C-terminal calcium binding lobes, each with two structurally and sequence-similar EF hand calcium binding sites (I, II, III, and IV) [159-163]. Each EF hand calcium binding site contains a helix-loop-helix structure, and the Lys-115 methylation site is found on a highly conserved solvent-accessible region (LGEKLT) located between helix 6 of EF hand III and helix 7 of EF hand IV (Figure 7.1b). Calmodulin undergoes significant changes in structural conformation in response to binding calcium, most notably exposure of two hydrophobic clefts located in the N- and C-terminal lobes [163]. These regions are directly involved in calmodulin binding to calmodulin-dependent enzymes [164, 165] and contain methionine residues that when oxidized to methionine sulfone residues interfere with calmodulin binding [154].

Similarly, oxidation of these methionine residues nearly abolishes the activity of calmodulin as a substrate for CLNMT, but in relationship to the Lys-115 methylation site the C-terminal hydrophobic cleft is found on the opposite side of calmodulin (providing an early suggestion that substrate recognition involves more than just the target methylation site). The importance of the hydrophobic cleft region in substrate recognition is also supported by *Paramecium* calmodulin mutants that have an Ile-136 to Thr substitution that also perturbs this hydrophobic cleft region and interferes with methylation at Lys-115 [60]. Whereas calmodulin has a structural symmetry between the N- and C-terminal lobe regions (both of which are normally required for the interaction of calmodulin with calmodulin-dependent enzymes), substrate recognition determinants for CLNMT reside solely in the C-terminal lobe, as made evident by CLNMT-catalyzed methylation of a tryptically derived C-terminal lobe fragment (residues 78 through 148) indistinguishable from methylation of intact calmodulin [154].

Additionally, an insightful set of domain swapping and methylation loop replacement studies revealed that introduction of the methylation loop (LGEKLT) between EF hands I and II on the opposite side of calmodulin did not support methylation, and domain exchange between EF hand III or IV by EF hand I or II, respectively, abolished methylation [166]. These studies were followed by more detailed investigations of the structural requirements for calmodulin recognition by CLNMT that reside in the C-terminal lobe region. Site-directed mutagenesis of residues flanking the Lys-115 methylation site identified three residues (Gly-113, Glu-114, and Leu-116) as absolutely essential for methylation of Lys-115 [167]. However, additional structure requirements were identified with α-helices 6 and 7 from EF hands III and IV that were also essential for calmodulin methylation.

These results were structurally corroborated by comparison of the known NMR and X-ray crystal structures for apocalmodulin (without bound calcium) and calcium-bound calmodulin, which demonstrate large changes in the number of surface exposed hydrophobic and charged residues in the C-terminal lobe domain in response to calcium binding (Figure 7.3). Overall, a remarkable amount of information has been reported on the structural determinants in calmodulin for substrate recognition for CLNMT. The current information suggests that there will be more interactions between CLNMT and calmodulin than just those surrounding the target methylation site. However, in the absence of any structural information for CLNMT how these determinants interact with CLNMT to establish substrate specificity can only be speculated. Nevertheless, when a structure for

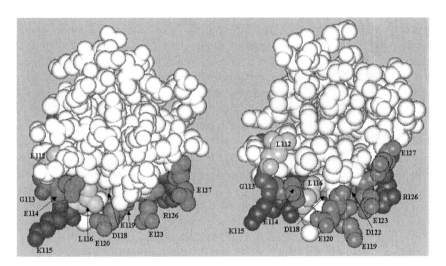

FIG. 7.3. Comparison of the carboxyl-terminal lobe of apoCaM and Ca^{2+}-CaM. The carboxyl-terminal (residues 77 through 148) lobe from the NMR structure of apoCaM (ref) (*left*) and the X-ray crystal structure of Ca^{2+}-CaM (ref) (*right*) is shown. Hydrophobic residues are indicated in yellow. The methylation site (Lys-115) and charged residues in the methylation loop and helix 7 are highlighted in red (negatively charged) or blue (positively charged). Hydrophobic residues in the methylation loop (Leu-116 and Leu-112) are indicated in green. Glycine 113, which facilitates a 90-degree turn in the loop, is shown in pink. The images were generated using a Silicon Graphics Indigo system and Insight II software. *(Reprinted from [167], copyright 2000, with permission from The American Society for Biochemistry and Molecular Biology.)* (See color plate.)

CLNMT does emerge a wealth of information exists that will indubitably guide future studies of CLNMT-calmodulin interactions.

3.   Functional Significance of Lys-115 Methylation

The exquisite substrate specificity of PKMTs, like CLNMT, has always attracted attention to the potential functional ramifications of lysine methylation. An obvious assignment for this modification is always complicated by the apparent functionality of the target polypeptide substrate in a limited number of species where an identical lysyl residue is not methylated. This could be considered a strong indication of a universal variation in functional significance for lysine methylation that is species specific. However, to continue on a theme in this chapter this observation does not exclude the potential species-specific variation in the necessity for lysine methylation as a determinant for protein-protein interactions.

In this regard, the available evidence describing the functional significance of Lys-115 methylation in calmodulin probably provides some of the strongest evidence for this hypothesis. Early comparative studies described similarities and differences among calmodulin isolated from different plant and animal species with regard to their ability to activate myosin light-chain kinase (MLCK) and NAD kinase [168]. Because there was identifiable but limited heterogeneity in the polypeptide sequence (~13 residues) among these calmodulins, variations in enzyme activation were considered a consequence of these differences. However, subsequent studies utilizing VU-1 calmodulin clearly established that methylation at Lys-115 although not affecting the ability of calmodulin to activate cyclic nucleotide phosphodiesterase caused a substantial threefold reduction in the ability to activate NAD kinase [158].

Whereas later studies identified other sequences in calmodulin that affect its ability to activate NAD kinase [169], the effect of Lys-115 methylation exhibited exceptional specificity in that site-directed mutations of the two flanking residues (Glu-114 and Leu-116) did not affect NAD kinase activation (these same residues are determinants for Lys-115 methylation, as described previously) but trimethylation of Lys-115 resulted in drastic reductions [166]. These observations suggest that the methylation status of calmodulin could have significant effects on the *in vivo* regulation of NAD kinase activity. Indeed, studies on the methylation status of plant calmodulin as a function of growth and development demonstrated that the methylation status of calmodulin was regulated in a manner coincident with cell growth [156, 170].

A site-directed mutant of VU-1 calmodulin with a Lys-115 Arg replacement (VU-3 calmodulin) is a potent competitive inhibitor of CLNMT activity [155], cannot be methylated by CLNMT, and moreover retains NAD kinase activator properties identical to unmethylated forms of calmodulin (approximately threefold higher) [158]. As a tool for investigating the potential *in vivo* regulatory properties of methylated versus non-methylated calmodulin on plant NAD kinase,

VU-3 calmodulin was transgenically introduced into tobacco plants (which normally contain trimethylated calmodulin). Transgenic tobacco plants expressing VU-3 calmodulin exhibited several distinct phenotypic characteristics, including an ~60 % decrease in stem internode growth, a tenfold reduction in seed production, and generally reduced seed and pollen viability [171].

Control transgenic plants expressing VU-1 calmodulin exhibited growth characteristics and phenotypes identical to wild-type plants. Later studies by this same laboratory established a firm relationship among transgenic VU-3 tobacco plants, increased NAD kinase activity, increased levels of NADPH, and active oxygen species production in response to environmental stimuli and elicitor challenge, both of which are associated with calcium signal transduction events in plant defense responses [172, 173]. Thus, in plants the differential effects of methylated versus non-methylated calmodulin on NAD kinase activity as a consequence or lack of CLNMT activity may be a tightly regulated system tailored for specific responses to growth and/or environmental conditions. Trimethyllysine-115 has also been reported to block a potential ubiquitination site in calmodulin and to protect the protein from proteasome degradation [174, 175], but other reports have refuted this hypothesis [176, 177].

On a more recent note, methylation of calmodulin (although not affecting NAD kinase binding) apparently turns calmodulin into a "partial agonist," wherein the normal transition from initial protein binding to an active state [178, 179] is prevented, thus trapping bound calmodulin in an intermediate state incapable of target protein activation (Dan Roberts, personal communication). The voluminous studies on trimethyllysine 115 and CLNMT activity seem incongruous with an absence of gene information. It is hoped that the studies mentioned here will encourage others interested in PKMTs to join in the hunt for CLNMT, a significant PKMT that remains unknown in terms of polypeptide and nucleotide sequence.

C. CYTOCHROME C LYSINE N-METHYLTRANSFERASE (CTM1P, EC 2.1.1.59)

Cytochrome c lysine N-methyltransferase (Ctm1p, EC 2.1.1.59), first reported in 1977 [180], catalyzes the formation of trimethyllysine-72 in apocytochrome c in many but not all species (Figure 7.1a). This lysyl modification is prevalent in fungi and plants but conspicuously absent in cytochrome c from vertebrate and invertebrate organisms [74, 116, 181–183]. The formation of trimethyllysyl-72 occurs prior to mitochondrial import, and as expected Ctm1p activity is localized to the cytosol [184, 185]. Ctm1p has been partially purified from wheat germ, *Neurospora crassa* [186], and from *Saccharomyces cerevisiae* [184].

All of these Ctm1p enzyme preparations catalyze formation of trimethyllysine-72 in non-methylated cytochrome c substrates such as horse heart cytochrome c. However, cytochrome c from many plants also contains a trimethyllysyl residue at position 86 [73, 81, 91, 187, 188], but specific enzymatic activity for this modification has not been reported, and the partially purified Ctm1p enzyme from wheat germ only catalyzed Lys-72 methylation [189]. The amino acid sequence surrounding the target methylation site in cytochrome c from residues 68 through 92 is highly evolutionarily conserved [190]. Ctm1p activity measurements, as with all PKMTs discussed in this chapter, rely on species-specific unmethylated forms of cyctochrome c, and again the apparent functional competency of these unmethylated forms is suggestive of a species-specific functional significance for trimethyllysine-72.

1. Enzymatic Properties and Kinetic Mechanism Characterization

There is a preponderance of enzymological, biochemical, and substrate specificity information for Ctm1p, primarily for the *N. crassa* and *S. cerevisiae* enzymes. Both enzymes have distinctly alkaline pH optima and molecular masses in the range of 97 to 120 kDa [180, 184, 186, 189, 191]. The purest preparation of Ctm1p was obtained from *N. crassa*, which was 80% homogeneous after a 3,500-fold enrichment [186]. This enzyme preparation exhibited apparent affinity constants (K_m) of 320 μM (although 1.7 mM has also been reported) for horse heart cytochrome c and 19 μM for AdoMet with a k_{cat} of 0.16 s^{-1} (considering recent identification of the Ctm1p gene in *S. cerevisiae*, see material following, this number may be 0.08).

Quantitative confirmation of trimethyllysine product formation, as well as the distribution of mono and dimethyl forms of lysine during catalysis, is an important consideration for all PKMT enzymes. In this regard, studies of Ctm1p activity have been accompanied by exceptionally reliable product analyses utilizing protein hydrolysis followed by complete amino acid analyses. Moreover, these studies provided some of the first evidence for the relationship between multiple methylation events and the kinetic reaction mechanism for a PKMT enzyme, as well as the first consideration of distributive versus processive addition of methyl groups to cytochrome c.

A profile of product formation including quantitative analyses of mono-, di-, and trimethyllysine during catalysis under short (several minutes) as compared with long (1-hour) incubation times clearly indicated a precursor product relationship between mono- and dimethyllysine with the formation of trimethyllysine, which occurred relatively late in the reaction compared to mono- and dimethyllysine [186]. In these studies, the reaction times relative to turnover rate (0.16 s^{-1}) were favorable for multiple catalytic turnovers, but whether or not methyl group addition from AdoMet was processive or distributive could not be discerned.

As with all of the PKMT enzymes described in this chapter, the k_{cat} values reported are actually the sum of the individual rate constants for formation of each of the individual methylated lysine derivatives (a more detailed discussion is presented in the following section on Rubisco LSMT). Nevertheless, these authors [186] were the first to propose that the addition of methyl groups from

AdoMet to cytochrome c was consistent with a processive mechanism where Ctm1p did not disassociate from cytochrome c between methylation events. This type of mechanism has interesting kinetic implications because it infers the formation of Ctm1p-bound reaction intermediates with partially methylated forms of cytochrome c.

Interestingly, a thorough kinetic reaction mechanism analysis has been reported for *N. crassa* Ctm1p [192]. Double reciprocal velocity plots under conditions favoring multiple catalytic turnovers (5 min) revealed a set of parallel lines similar to those associated with ping-pong reaction mechanisms. Mechanistically, this observation is compatible with processive methyl group addition and Ctm1p-bound methylated cytochrome c intermediates. However, product inhibition studies using AdoHcy (although competitive with AdoMet) were noncompetitive with cytochrome c. In the absence of inhibition studies with methylated forms of cytochrome c, the authors concluded that the reaction mechanism was a hybrid ping-pong with separate binding sites on Ctm1p for cytochrome c and AdoMet. Although a structure for Ctm1p is not yet available, this observation would be consistent with SET domain PKMTs.

2. Substrate Specificity Determinants

An exceptional amount of information is available in regard to the polypeptide substrate specificity requirements for trimethyllysine-72 formation in cytochrome c by Ctm1p. These studies have been based on chemically derived CNBr cleavage fragments from unmethylated cytochrome c, and introduction of site-directed mutants of iso-1-cytochrome c into *S. cerevisiae* strains deficient in cytochrome c. An examination of *N. crassa* Ctm1p activity against five CNBr fragments derived from horse heart cytochrome c led to the first identification of a consensus amino acid sequence required for methylation at Lys-72 [186].

Two CNBr fragments, from 1 to 80 and 66 to 104, containing residues encompassing the Lys-72 methylation site were good substrates for Ctm1p with considerably higher affinities (K_m 7.0 μM and 40 μM, respectively). However, a CNBr fragment containing residues 66 to 80 was completely inactive, providing evidence for the necessity of residues outside the immediate target lysyl methylation site. More interesting was the observation that a CNBr peptide representing residues 1 to 65 without Lys-72 was also a good substrate for Ctm1p. Sequence and size comparisons between these CNBr fragments and cytochrome c identified three major determinants for Ctm1p activity: (1) the consensus sequence Any-Lys-Lys-Any, where the first lysyl residue is the target methylation site; (2) a minimum polypeptide length between 39 and 65 amino acid residues; and (3) tertiary considerations in the folded polypeptide substrate that can obscure potential methylation sites.

These results were partially corroborated by the observation of adjacent lysyl residues at the methylation site in a number of cytochromes c from different plant and fungal species. Following these studies, a number of investigations appeared that capitalized on the genetic flexibility (in terms of aerobic-/anaerobic-regulated cytochrome *c* expression) and transformation capabilities of *S. cerevisiae* to investigate the effects of site-directed mutations in residues surrounding the Lys-72 methylation site and subsequent effects on the formation of trimethylysine-72. There are different amino acid numbering schemes for eukaryotic cytochrome *c* and iso-1-cytochrome *c* from *S. cerevisiae* used in these studies.

Lys-72 is the methylation site in cytochrome *c* that is equivalent to Lys-77 in iso-1-cytochrome *c*, which has five additional N-terminal residues, although the flanking five-amino-acid residue sequence around the target lysyl methylation site is identical [183]. For the sake of simplicity, we will use the numbering system of mammalian cytochrome *c*. Site-directed mutations in cytochrome *c* at residues Pro-71, Lys-73, and Tyr-74 (and a double mutation at 71 and 73) were evaluated as determinants of methylation at Lys-72 after expression in *S. cerevisiae* in the presence of [^{3}H-methyl]-L-methionine, isolation of the mutant proteins, and evaluation of trimethyllysine-72 by amino acid analyses following protein hydrolysis [193].

Although there were a limited number of conservative replacements in this study, all of the substitutions resulted in drastic reductions in the levels of trimethyllysine (two- to tenfold). The only exception was Lys-73 to Met, which was reduced approximately 50%. As might be expected, a Lys-72 to Arg mutation abolished formation of trimethyllysine. Specific identification of polypeptide sequence requirements for methylation at Lys-72 cannot be drawn from these studies, but several other aspects of these mutations on cytochrome *c* stability and heme incorporation were also investigated (as discussed in material following).

The most detailed investigation of the polypeptide sequence requirement for trimethyllysyl-72 formation examined 21 altered iso-1-cytochromes c in *S. cerevisiae* encompassing residues 67 to 77 [183]. From these studies, the flanking Tyr-74 residue was identified as critical for methylation at Lys-72 (with only the conservative replacement of Tyr-74 by Phe supporting methylation). The authors proposed a sequence motif required for methylation of Lys-Any-Tyr, but also observed that placement of this motif in other regions of cytochrome *c* does not create a new methylation site. Thus, in addition to the identified sequence motif the overall conformation of cytochrome *c* is influential in determining methylation at Lys-72. Overall, these substrate specificity studies suggest that a defined sequence motif is recognized by Ctm1p with determinants in the immediate vicinity of the methylation site, but that other structural aspects of cytochrome *c* are also involved in the recognition.

3.   Functional Signficance of Lys-72 Methylation

The functional significance of trimethyllysine-72 in cytochrome *c* has received much attention. Similar to other methylated proteins and enzymes in this chapter, the apparently normal functional activity of unmethylated forms of cytochrome c in many species argues for a species-specific role for this modification. As might be expected, the emphasis in these investigations of the functional significance of Lys-72 methylation has been relative to several well-described biochemical processes ascribed to cytochrome c. Cytochrome c is synthesized in the cytosol as apocytochrome c and subsequently translocated through the outer mitochondrial membrane without an N-terminal transit sequence [194], and the heme group is attached in the mitochondrial intermembrane space [195].

Several *in vitro* mitochondrial binding and uptake studies suggested that the trimethyllysine-72 was required for binding and subsequent import by mitochondria [185, 196]. However, later studies utilizing site-directed mutations of cytochrome c expressed in *S. cerevisiae* (as described previously for substrate specificity studies) revealed little effect of Lys-72 methylation on the amount of cytochrome c found in mitochondria or on heme conjugation [197]. Earlier studies had also provided evidence through pulse-chase studies that trimethyllysine-72 was associated with a significant increase in stability and protection against proteolysis [198]. Again, these studies were refuted by more recent site-directed mutant studies that although generating cytochrome c mutants with altered stability and/or turnover *in vivo* demonstrated no effect of trimethyllysine-72 [183, 197].

A recent report focusing on the newly discovered role of cytochrome c in apoptosis discovered that formation of trimethyllysine-72 in horse heart cytochrome c completely eliminated its pro-apoptotic activity [199]. However, in these same studies non-methylated forms of iso-1-cytochrome c from *S. cerevisiae* (naturally methylated at Lys-72) were also devoid of pro-apoptotic activity. Although providing a potentially intriguing method for the regulation of apoptosis in animals, these studies thus did not identify a specific functional role for trimethyllysine-72. Overall, these studies suggest that the methylation of Lys-72 is dispensable with regard to the biological functions discussed previously. However, the occurrence of trimethyllysine-72 in cytochrome c from many species, and the apparent co-occurrence of Ctm1p activity, clearly do suggest an evolutionarily conserved biological role.

4. Ctm1p Gene Identification and Protein Interactions

There is no structural information available for Ctm1p, but the gene was recently identified in *S. cerevisiae* utilizing a clever genomic strategy for associating biological function with specific genes [79]. The polypeptide sequence does not have significant sequence similarity with other proteins or an identifiable motif associated with AdoMet-dependent protein methyltransferases, but does show a limited tertiary similarity to rRNA methyltransferase from *Streptococcus pneumoniae*. The polypeptide sequence codes for a protein (585 aa) with a molecular mass of 68.3 kDa, which in comparison with the partially purified form of *S. cerevisiae* Ctm1p (97 kDa) suggests that the holoenzyme may be a homodimer.

Disruption of the *Ctm1* results in the expected biochemical phenotype of iso-1-cytochrome *c* without methylation at Lys-72, and this defect is restored by its reintroduction [79]. The most important observations of the *Ctm1*-Δ mutant are the complete absence of any identifiable phenotype and normal levels of cytochrome *c* under a variety of growth and culture conditions. These results corroborate many of the previous biochemical studies, which reached similar conclusions in regard to the dispensable nature of Lys-72 methylation in cytochrome *c*. However, to continue on the theme of lysyl methylation as a global mechanism for protein-protein interactions, cytochrome *c* does interact with several other proteins [200-204] where Lys-72 is important (such as cytochrome c_1, cytochrome *c* oxidase, cytochrome *c* peroxidase, and cytochrome b_2), and the suggestion of the possibility that trimethyllysine-72 could play a role in these other interactions has been made [79]. It will be interesting to see the structure for this enzyme and its relationship to the aforementioned substrate specificity studies as well as its similarity or lack thereof with other PKMTs.

D.   RIBULOSE-1,5 BISPHOSPHATE CARBOXYLASE/OXYGENASE LARGE SUBUNIT
εN-METHYLTRANSFERASE (RUBISCO LSMT, EC 2.1.1.127)

Rubisco LSMT is a SET domain non-histone PKMT responsible for the formation of trimethyllysine-14 in the LS of Rubisco from many plants species. However, similar to the polypeptide substrates for other PKMTs described in this chapter, Rubisco also exists in plant species without trimethyllysine-14 in the LS. Rubisco from higher plants is a large hexadecameric enzyme with eight nuclear-encoded small subunits (SS) and eight chloroplast-encoded large subunits (LS) that is exclusively found in photosynthetic organisms where it catalyzes fixation of CO_2.

Evidence for a chloroplast-localized PKMT specific for Lys-14 in the LS of Rubisco first came from structural characterization of the N-terminal region of the LS of Rubisco. These studies used a combination of chemical and physical analytical techniques to characterize amino acid sequence and posttranslational modifications in the N-terminal region of the LS, which is important to catalysis and highly disordered with low electron density and thus not readily visible by X-ray crystallography.

1.   Enzyme Identification and Kinetic Mechanism

Rubisco LSMT was first partially purified and characterized from tobacco chloroplasts, and the partially purified preparation was used to demonstrate the exclusive formation of trimethyllysine-14 in the LS of Rubisco [102]. Later, Rubisco LSMT was purified to homogeneity (~7,000-fold) from pea chloroplasts using an affinity purification technique that relied on tight and specific binding to PVDP-immobilized forms of unmethylated spinach Rubisco [104]. The native enzyme is a monomer with a molecular mass of ~55 kDa, apparent affinity constants (K_m) for Rubisco and AdoMet of 1.4 and 6.0 μM, and a turnover of ~0.5 s^{-1}.

Rubisco LSMT, like Ctm1p and CLNMT, is a relatively scarce protein and yields from the aforementioned affinity purification are poor (2.8%). Rubisco LSMT does not discriminate between activated and nonactivated forms of Rubisco (which are associated with conformational changes) but is completely inactive on Rubisco complexes with RuBP or the transition state analog, CABP [102]. As expected, Rubisco LSMT does not catalyze methyl group incorporation into Rubisco where the N-terminal region from Ac-Pro-3 to Lys-18 has been proteolytically removed, which confirms Lys-14 as the only lysyl methylation site in Rubisco.

Complexes between Rubisco and RuBP and CABP are associated with relatively large changes in the solvent accessibility of the N-terminal region [205], which probably limit Rubisco LSMT recognition and/or binding. Rubisco LSMT forms a rather tight complex with unmethylated Rubisco with an overall affinity constant of ~0.1 nM, but has little binding affinity for methylated forms of Rubisco (Figure 7.4). The ternary complex between Rubisco LSMT and PVDF-immobilized Rubisco has been characterized and is not dissociable by 0.4 M NaCl or inhibitors such as AdoHcy or AdoEth, but in the presence of AdoMet does undergo a catalytic-dependent disassociation [104].

Recent studies using a bacterially expressed form of pea Rubisco LSMT have capitalized on the ternary complexes formed between Rubisco LSMT and PVDF-immobilized Rubisco, to mechanistically investigate processive versus distributive addition of methyl groups and its relationship to the kinetic reaction mechanism. A preliminary report of this data has been made [206], and a full report will be available soon. The analyses confirm that methyl group addition is processive and accompanied by the formation of Rubisco LSMT-Rubisco bound methylated intermediates whose disassociation is strictly dependent on the formation of trimethyllysine-14 in the LS.

The processive addition of methyl groups is reflected by a complex kinetic reaction mechanism, which under limited turnover conditions favoring only the formation of monomethyllysine-14 is a random or ordered bi-bi mechanism, and under conditions favoring trimethyllysine-14 formation is a ping-pong mechanism. The results are similar to those reported for Ctm1p, and were corroborated by product analyses. It is possible that this mechanism is applicable to all SET domain protein methyltransferases, in that the structurally conserved SET domain has an ideal orientation of substrate binding sites for multiple methyl group transfers without polypeptide disassociation.

2. Gene Identification and Characterization

The cDNA for pea Rubisco LSMT was originally cloned and sequenced using degenerate oligonucleotide primers designed according to amino acid sequences

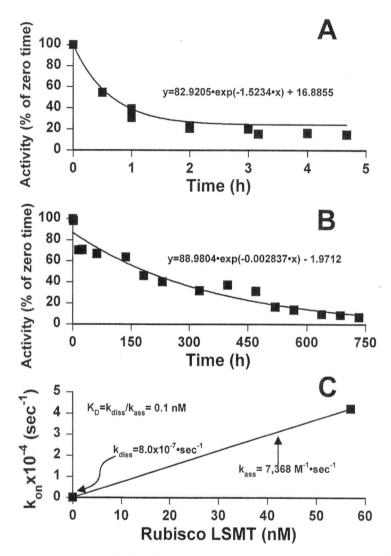

FIG. 7.4. Estimation of the binding affinity of wild-type pea Rubisco LSMT for PVDF-immobilized des(methyl) spinach Rubisco. (A) Loss in Rubisco LSMT activity at 4°C from pea chloroplast lysates over time during incubation with PVDF-immobilized Rubisco. The concentration of Rubisco LSMT was estimated at 58 nM from activity measurements assuming a specific activity of purified pea Rubisco LSMT of 306 nmoles/min/mg protein [104]. (B) Loss in bound Rubisco LSMT activity at 4°C from PVDF-immobilized spinach Rubisco over time as determined by activity measurements of bound Rubisco LSMT. Immunological analyses confirmed that the loss in Rubisco LSMT activity was due to disassociation and not catalytic inactivation. (C) Affinity plot of data from a and b estimating the affinity constant K_D for the binding of pea Rubisco LSMT to spinach Rubisco.

obtained from several polypeptides released after peptic digestion [103]. At the time, the gene and protein sequence did not have significant homology with any other protein or gene, and no identifiable motifs associated with AdoMet-dependent protein methyltransferases. Similar to *CTM*, expression of *Rubisco LSMT* is coordinated with expression of the Rubisco LS in a tissue and light-dependent manner.

The pea Rubisco LSMT cDNA was used to obtain homologous cDNAs as well as identification of the complete gene sequence for *Rubisco LSMT* from tobacco and, surprisingly, spinach, a species without Lys-14 methylation [207]. The gene organization for *Rubisco LSMT* is nearly identical between spinach and tobacco with the exception of an alternative 3′-splice site identified in the spinach gene, which results in the formation of two equally abundant transcripts for *Rubisco LSMT* in spinach [208]. Functional analysis of the promoter for tobacco *LSMT* identified the presence of several light-regulated elements [209]. The occurrence of *Rubisco LSMT* homologs in spinach was an enigma until structural studies discovered the presence of a second methylation site in Rubisco at the N-terminus of the processed form of the SS of Rubisco [210, 211].

Subsequent studies provided evidence that *Rubisco LSMT* homologs, such as those found in spinach, were capable of catalyzing methylation of the N-terminal α-amino group of Met-1 in the SS [208]. However, these studies have not been expanded on given the difficulty in obtaining satisfactory levels of bacterial expression for these constructs. The methylation of Met-1 in the SS is, however, similar to the reports of N-terminal methylation by the PrmA described earlier in this chapter, and appears to be present in the SS of all higher-plant Rubiscos.

3. Functional Significance of Lys-14 Methylation

The functional significance of trimethyllysine-14 in the LS of Rubisco, as well as Met-1 methylation in the SS, remains unknown. The *in vitro* introduction of trimethyllysine-14 in unmethylated forms of Rubisco does not alter any of the numerous kinetic and activation parameters of Rubisco (Table 7.2). Moreover, RNAi-mediated knockouts in tobacco *Rubisco LSMT*, which leads to a loss in methylation of Lys-14 in the LS of tobacco Rubisco, does not result in identifiable changes in phenotype or CO_2 assimilation rates in leaves (unpublished data). Perhaps like Ctm1p and PrmA, the functional significance for Lys-14 methylation in the LS of Rubisco lies with an as yet undetermined protein-protein interaction for Rubisco.

4. Structural Characterization of the Rubisco LSMT

A wealth of structural information is available for Rubisco LSMT as a consequence of the SET domain. The class of protein methyltransferases now referred to as SET domain protein methyltransferases, named for the highly evolutionarily conserved SET structural motif [215]—denoting the genes Su(var)3-9,

TABLE 7.2

EFFECT OF *IN VITRO* STOICHIOMETRIC METHYLATION OF SPINACH RUBISCO BY PEA RUBISCO LSMT ON SPINACH RUBISCO KINETIC PARAMETERS. PURIFIED SPINACH RUBISCO [212] WAS STOCHIOMETRICALLY *IN VITRO* METHYLATED AT LYS-14 IN THE LS USING PURIFIED NATIVE PEA RUBISCO LSMT [104]. AFTER METHYLATION, METHYLATED RUBISCO WAS SEPARATED FROM PEA RUBISCO LSMT BY GEL FILTRATION ON HPLC. RUBISCO CARBOXYLASE ASSAYS WERE ACCORDING TO [213], AND THE ESTIMATION OF THE SPECIFICITY FACTOR ACCORDING TO THOW [214].

Rubisco	$K_m(RuBP)$	$K_m(CO_2)$	$K_{act}(Mg^{++})$	$K_{act}(CO_2)$	k_{cat}[1]	$V_c(N_2)/V_c(O_2)$[2]
			μM		s^{-1}	
Control	18	12	611	24	3.1	1.76
Methylated	17	10	557	22	2.4	1.78

1. Carboxylase activity.
2. Ratio of observed carboxylase activities in the presence of 100% N_2 versus 100% O_2 at limiting CO_2 (5μM), an indirect indication of the CO_2/O_2 specificity factor.

Enhancer of zeste [E(z)], and trithorax (trx)—were discovered by a clever bioinformatics approach using secondary structural similarity searches to potentially identify homologs with defined biochemical function [216].

From this comparison, Rubisco LSMT emerged as the only SET domain protein with defined biochemical characteristics and enzymatic activity, and as a consequence SET domain histone methyltransferases (HMT) were discovered [216]. The previously known influence of SET domain proteins on epigenetic gene expression, coupled with identification as HMTs, fueled an intense research effort to establish the relationship between histone methylation and regulation of gene expression, as well as structure/function relationships for this unique class of protein methyltransferases. This interest resulted in the simultaneous determination of the molecular structure for three different SET domain protein methyltransferases, including Rubisco LSMT (Figure 7.5).

All SET domain protein methyltransferases have a unique and conserved structural motif that contains separate binding sites for the target protein substrate and the methyl donor, AdoMet [217]. The geometric arrangement of substrate binding sites on opposite sides of the SET domain, connected by a narrow channel through which methyl group transfer occurs, has been described as ideal for the multiple transfer of methyl groups without disassociation of the protein substrate [218], and as previously mentioned the authors believe that this hypothesis is applicable to all SET domain protein methyltransferases that catalyze multiple methyl group transfers. The chemical reaction mechanism for methylation by Rubisco LSMT and other SET domain protein methyltransferases is an S_N2 nucleophilic transfer to the unprotonated epsilon amino group of the bound lysyl residue [218].

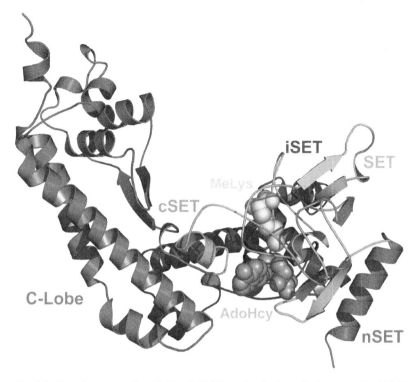

FIG. 7.5. Crystal structure of pea Rubisco LSMT bound to *S*-adenosyl-L-homocysteine (AdoHcy, orange) and monomethyllysine (MeLys, yellow) (1P0Y.pdb). The secondary structures of the nSET, SET, iSET, and cSET regions are delineated in magenta, cyan, blue, and green, respectively, whereas the C-terminal lobe domain is represented in red. The ribbon diagram was rendered using PyMOL (*http://www.pymol.org*). (See color plate.)

Although a conserved active-site tyrosyl residue was originally proposed as a possible residue facilitating proton abstraction from the epsilon amino group, more recent studies clearly exclude this possibility [219]. Both steric constraints in the polypeptide binding site [220] as well as specific hydrogen bonding to the methylated lysyl residue are used by Rubisco LSMT to maintain the ~180-degree geometry necessary for S_N2 reactions, and both act as determinants for the number of methyl groups transferred. This model is supported by the structural and biochemical analyses of ternary complexes between Rubisco LSMT and free as well as monomethyllysine, which are catalytically active as substrates, albeit with low k_{cat}s [219].

The structure of Rubisco LSMT reveals some unique features relative to HMTs, most notably a large amino acid insertion in the SET domain (iSET) and a C-terminal lobe domain that has extensive interaction with Rubisco ending with a C-terminal extension that may have contact with SS pairs [218]. Whereas the bio-chemical significance of these structural differences is unknown, the C-terminal lobe region in Rubisco LSMT is capable of specific binding to Rubisco, an obser-vation supported by the ability of 1,6 Bismaleimidohexane (BMH) (a homobifunc-tional sulfhydryl-specific cross-linking reagent) to cross-link Rubisco LSMT with Cys-459 in the carboxy-terminal region in the LS (unpublished data).

Molecular docking models clearly suggest an extensive region of contacts between Rubisco LSMT and Rubisco (Figure 7.6), as well as the nature of the

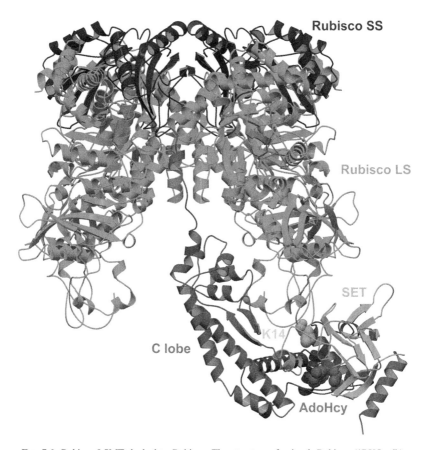

FIG. 7.6. Rubisco LSMT docked to Rubisco. The structure of spinach Rubisco (1RXO.pdb) was docked to LSMT to position Lys-14 (gold) as closely as possible to the HEPES binding site while avoid-ing steric collisions between the two proteins and without altering the conformation of the flexible Rubisco N-terminus. The Rubisco large and small subunits are colored light and dark green, respectively, whereas LSMT is as follows: the nSET, iSET, cSET, and SET regions are colored magenta, blue, green, and cyan, respectively, whereas the C-terminal lobe is red and the domain-swapped extension is gold. AdoHcy is orange. *(Reprinted from [218], copyright 2002, with permission from Elsevier.)* (See color plate.)

polypeptide substrate binding cleft in Rubisco LSMT. The region surrounding the Lys-14 target methylation site is conserved in all higher-plant Rubiscos and suggests that the flanking three amino acids on each side of Lys-14 could in a large part establish substrate specificity (Figure 7.7). With the exception of Phe-13, all of these residues have small side chains, suggesting tolerance for replacement

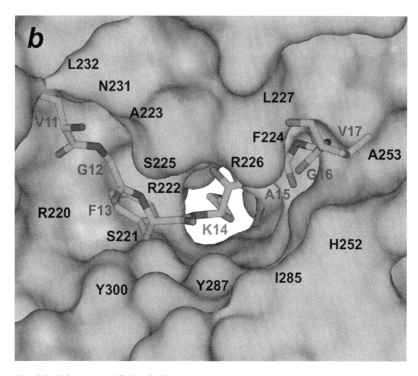

a   **Rubisco Large Subunit N-Terminal Sequences**

Arabidopsis thaliana:	Ac-PQTETKASVGFK$_{14}$AGVKEYKLTYY
Pea (*Pisum sativum*):	Ac-PQTETKAKVGFK$_{14}$AGVKDYKLTYY
Tobacco (*Nicotiana tabacum*):	Ac-PQTETKASVGFK$_{14}$AGVKEYKLTYY
Spinach (*Spinacia oleracea*):	Ac-PQTETKASVGFK$_{14}$AGVKDYKLTYY
Consensus Sequence:	VGFK$_{14}$AGV

b

FIG. 7.7. Substrate specificity of LSMT. (a) Sequence alignment of the N-terminal residues of the large subunits of Rubisco from several plant species. The N-termini begin at Pro3, which are Nα acetylated. The consensus recognition sequence for methylation of Lys14 by LSMT is shown in blue. (b) Modeling of the consensus sequence peptide into the protein substrate-binding cleft in LSMT. The side chain of Arg 226 was removed to provide an unobstructed view of the binding site. Residues that are within van der Waals contact of the peptide model labeled on the molecular surface of LSMT. Molecular surfaces associated with residues that are identical in LSMT sequences are colored cyan. *(Reprinted from [219], copyright 2003, with permission from Nature Publishing Group.)* (See color plate.)

with similar side-chain amino acid residues and thus potential flexibility in substrate specificity. There is considerable interest in the potential for substrate flexibility in SET domain protein methyltransferases because recently, in contrast to the long-standing belief that PKMTs have exquisite polypeptide substrate specificity, two alternative protein substrates, p53 [34] and TAF10 [40], were identified for SET 7/9.

There are currently only a few studies that have investigated the determinants of polypeptide substrate specificity studies for SET domain PKMTs [221–223]. The observation of alternative substrates for a SET domain PKMT may prove important for Rubisco LSMT because there are several other chloroplast-localized proteins that contain methylated lysyl residues such as ribosomal proteins [110, 224–226] and ferredoxin-NADP reductase [47].

5.   Rubisco LSMT Homologs in Other Eukaryotes

Although the strict localization of Rubisco LSMT to chloroplasts and plants possessing active photosynthetic biosynthetic pathways would suggest a phylogenetic limitation to plants, a broad search for Rubisco LSMT homologs suggests otherwise. The sequence alignment depicted in Figure 7.8 indicates that Rubisco LSMT homologs with significant sequence consensus exist in species from humans to fungi. These homologs must have significantly different polypeptide substrate specificities given the restriction of Rubisco to plant species. However, their presence in such a diverse group of organisms implies an evolutionarily conserved role for the Rubisco LSMT class of PKMTs.

Moreover, this observation supports the continuing theme in this chapter that lysyl methylation is a global mechanism as a determinant for protein-protein interactions. Exploring the opportunities presented by the data in Figure 7.8 would be considerably difficult given the uncertainties of protein expression as well as the naturally methylated status of any potential protein substrates. However, as reported by the author of an accompanied chapter in this series newly developed techniques utilizing protein macroarrays can be particularly useful for identification of potential substrates for protein methyltransferases [227, 228].

We recently probed a human protein macroarray utilizing a cloned and expressed form of the mouse Rubisco LSMT homolog identified in Figure 7.8. The results were encouraging and identified three potential substrates for this enzyme. It is our hope that identification and characterization of these polypeptide substrates as well as the Rubisco LSMT homologs will ultimately aid in the identification of the functional significance for Lys-14 methylation in Rubisco.

V.   Conclusions and Future Prospects

The prevalent absence, despite heroic research efforts, to conclusively assign a functional significance to PKMT activity and the associated methylated protein

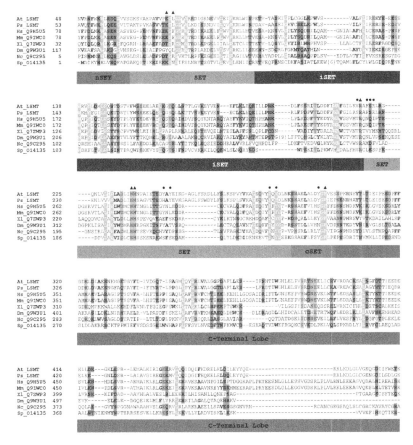

FIG. 7.8. Sequence alignment of plant Rubisco LSMTs with SET domain homologs in the fungal and animal kingdoms. The sequences of Rubisco LSMT homologs from *Arabidopsis thaliana* and the garden pea *Pisum sativum* (Ps) are aligned with SET domain orthologs from *Schizosaccharomyces pombe* (Sp), *Neurospora crassa* (Nc), *Drosophila melanogaster* (Dm), *Xenopus laevis* (Xl), *Mus musculus* (Mm), and *Homo sapiens* (Hs). The ExPASy Swiss-Prot (*http://au.expasy.org*) accession codes and the corresponding sequence numbers for each homolog are listed on the left-hand side of the alignment. The structural domains and motifs of LSMT are illustrated in bars beneath the alignment that are colored according to the scheme in Figure 7.5. Residues that comprise the AdoMet and lysine substrate binding clefts are denoted by black triangles and circles, respectively, shown above the alignment. Sequences were aligned using CLUSTAL, and the annotated alignment figure was generated in CHROMA (*http://www.lg.ndirect.co.uk/chroma/*). (See color plate.)

substrates certainly does not encourage similar endeavors. Today's research environment often suffers from a preoccupation with only those biological processes that can be shown as essential for survivability and growth. However, it is often the more subtle mechanisms in biological science that supercede their apparent insignificance and emerge as much more than observational curiosities.

Indeed, histone methylation and the associated histone PKMTs occupied a similar position not long ago. It is our belief that the significance of non-histone PKMTs, although hidden, is possibly an immense resource for the discovery of many important biological processes that require little more than the attention of dedicated researchers. If this chapter has kindled even the slightest interest in this fascinating field within the general scientific community it has been a success.

Using facile deletion mutation techniques in *Saccharomyces cerevisiae*, the methyltransferase responsible for dimethylation of the ribosomal large subunit protein L23a at two positions (currently K40, 106, 110 are the limited candidates) has been identified and designated as Rkm1 [230]. Similarly, a tour-de-force with multiplicative mutant analysis, relying on published temperature conditional and biochemical phenotypes, elegantly determined an alternative substrate, Dam1, for Set1 from *S. cerevisiae*, though K233 dimethylation identification relies exclusively on immunological evidence [231].

ACKNOWLEDGMENTS

LMAD, RCT, and RLH are grateful for NSF, NIH, and DOE support.

REFERENCES

1. Weihing, R. R., and Korn, E. D. (1970). Epsilon-N-dimethyllysine in amoeba actin. *Nature* 227:1263–1264.
2. Ammendola, S., Raia, C. A., Caruso, C., Camardella, L., Dauria, S., Derosa, M., and Rossi, M. (1992). Thermostable NAD$^+$-dependent alcohol dehydrogenase from *Sulfolobus solfataricus*: Gene and protein sequence determination and relationship to other alcohol dehydrogenases. *Biochemistry* 31:12514–12523.
3. Zappacosta, F., Sannia, G., Savoy, L. A., Marino, G., and Pucci, P. (1994). Post-translational modifications in aspartate aminotransferase from *Sulfolobus solfataricus*: Detection of N-ε-methyllysines by mass spectrometry. *Eur J Biochem* 222:761–767.
4. Laycock, M. V. (1972). Amino acid sequence of cytochrome *c*-553 from chrysophycean alga *Monochrysis lutheri*. *Can J Biochem* 50:1311–1325.
5. Pettigrew, G. W., Leaver, J. L., Meyer, T. E., and Ryle, A. P. (1975). Purification, properties and amino acid sequence of atypical cytochrome *c* from two protozoa, *Euglena gracilis* and *Crithidia oncopelti*. *Biochem J* 147:291–302.
6. Becam, A. M., and Lederer, F. (1981). Amino acid sequence of the cytochrome *c* from the yeast *Hansenula anomala*: Identification of three methylated positions. *Eur J Biochem* 118:295–302.
7. Ames, G.F.L., and Niakido, K. (1979). In vivo methylation of prokaryotic elongation factor-Tu. *J Biol Chem* 254:9947–9950.
8. L'Italien, J. J., and Laursen, R. A. (1979). Location of the site of methylation in elongation factor Tu. *FEBS Lett* 107:359–362.
9. Jones, M. D., Petersen, T. E., Nielsen, K. M., Magnusson, S., Sottrupjensen, L., Gausing, K., and Clark, B.F.C. (1980). The complete amino acid sequence of elongation factor Tu from *Escherichia coli*. *Eur J Biochem* 108:507–526.
10. Toledo, H. and Jerez, C. A. (1990). In vivo and in vitro methylation of the elongation factor EF-Tu from *Euglena gracilis* chloroplast. *FEMS Microbiol Lett* 71:241–246.

11. Hiatt, W. R., Garcia, R., Merrick, W. C., and Sypherd, P. S. (1982). Methylation of elongation factor 1α from the fungus *Mucor*. *Proc Natl Acad Sci USA* 79:3433–3437.

12. Van Hemert, F. J., Amons, R., Pluijms, W. J. M., Vanormondt, H., and Moller, W. (1984). The primary structure of elongation factor EF-1α from the brine shrimp *Artemia*. *EMBO J* 3:1109–1113.

13. Minami, Y., Wakabayashi, S., Wada, K., Matsubara, H., Kerscher, L., and Oesterhelt, D. (1985). Amino acid sequence of a ferredoxin from thermoacidophilic Archaebacterium, *Sulfolobus acidocaldarius*: Presence of an N[6-]monomethyllysine and phyletic consideration of archaebacteria. *J Biochem (Tokyo)* 97:745–753.

14. Gomes, C. M., Faria, A., Carita, J. C., Mendes, J., Regalla, M., Chicau, P., Huber, H., Stetter, K. O., and Teixeira, M. (1998). Di-cluster, seven-iron ferredoxins from hyperthermophilic Sulfolobales. *J Biol Inorg Chem* 3:499–507.

15. Glazer, A. N., DeLange, R. J., and Martinez, R. J. (1969). Identification of epsilon-N-methyllysine in *Spirillum serpens* flagella and of epsilon-N-dimethyllysine in *Salmonella typhimurium* flagella. *Biochim Biophys Acta* 188:164–165.

16. Ambler, R. P., and Rees, M. W. (1959). Epsilon-N-methyl-lysine in bacterial flagellar protein. *Nature* 184:56–57.

17. Kanto, S., Okino, H., Aizawa, S. I., and Yamaguchi, S. (1991). Amino acids responsible for flagellar shape are distributed in terminal regions of flagellin. *J Mol Biol* 219:471–480.

18. Stocker, B. A., Mcdonough, M. W., and Ambler, R. P. (1961). A gene determining presence or absence of ε-N-methyl–lysine in *Salmonella* flagellar protein. *Nature* 189:556–558.

19. Doll, L., and Frankel, G. (1993). *fliU* and *fliV*: two flagellar genes essential for biosynthesis of *Salmonella* and *Escherichia coli* flagella. *J Gen Microbiol* 139:2415–2422.

20. Burnens, A. P., Stanley, J., Sack, R., Hunziker, P., Brodard, I., and Nicolet, J. (1997). The flagellin N-methylase gene *fliB* and an adjacent serovar-specific IS*200* element in *Salmonella typhimurium*. *Microbiology* 143:1539–1547.

21. Ho, K. C., and Chang, G. N. (2000). The *fliU* and *fliV* genes are expressed as a single ORF in *Salmonella choleraesuis*. *Can J Microbiol* 46:1149–1152.

22. Ren, C. P., Beatson, S. A., Parkhill, J., and Pallen, M. J. (2005). The Flag-2 locus, an ancestral gene cluster, is potentially associated with a novel flagellar system from *Escherichia coli*. *J Bacteriol* 187:1430–1440.

23. Maras, B., Consalvi, V., Chiaraluce, R., Politi, L., Derosa, M., Bossa, F., Scandurra, R., and Barra, D. (1992). The protein sequence of glutamate dehydrogenase from *Sulfolobus solfataricus*, a thermoacidophilic archaebacterium: Is the presence of *N*-ε-methyllysine related to thermostability? *Eur J Biochem* 203:81–87.

24. Febbraio, F., Andolfo, A., Tanfani, F., Briante, R., Gentile, F., Formisano, S., Vaccaro, C., Scire, A., Bertoli, E., Pucci, P., and Nucci, R. (2004). Thermal stability and aggregation of *Sulfolobus solfataricus* β-glycosidase are dependent upon the *N*-ε-methylation of specific lysyl residues: Critical role of *in vivo* post-translational modifications. *J Biol Chem* 279:10185–10194.

25. Wang, C., Lazarides, E., Oconnor, C. M., and Clarke, S. (1982). Methylation of chicken fibroblast heat-shock proteins at lysyl and arginyl residues. *J Biol Chem* 257:8356–8362.

26. Pethe, K., Bifani, P., Drobecq, H., Sergheraert, C., Debrie, A. S., Locht, C., and Menozzi, F. D. (2002). Mycobacterial heparin-binding hemagglutinin and laminin-binding protein share antigenic methyllysines that confer resistance to proteolysis. *Proc Natl Acad Sci USA* 99:10759–10764.

27. van den Elzen, P. J. M., Gaastra, W., Spelt, C. E., de Graaf, F. K., Veltkamp, E., and Nijkamp, H.J.J. (1980). Molecular structure of the immunity gene and immunity protein of the bacteriocinogenic plasmid Clo DF13. *Nucleic Acids Res* 8:4349–4363.

28. Hardy, M. F., and Perry, S. V. (1969). *In vitro* methylation of muscle proteins. *Nature* 223:300–302.

29. Huszar, G., and Elzinga, M. (1969). ε-N-Methyl lysine in myosin. *Nature* 223:834–835.

30. Kuehl, W. M., and Adelstein, R. S. (1969). Identification of ε-N-monomethyllysine and ε-N-trimethyllysine. *Biochem Biophys Res Commun* 37:59–65.

31. Tong, S. W., and Elzinga, M. (1983). The sequence of the NH₂ - terminal 204 residue fragment of the heavy chain of rabbit skeletal muscle myosin. *J Biol Chem* 258:3100–3110.

32. Hayashida, M., Maita, T., and Matsuda, G. (1991). The primary structure of skeletal muscle myosin heavy chain: I. Sequence of the amino-terminal 23 kDa fragment. *J Biochem (Tokyo)* 110:54–59.

33. Reporter, M., and Reed, D. W. (1972). Methylation of bovine rhodopsin. *Nat New Biol* 239:201–203.

34. Chuikov, S., Kurash, J. K., Wilson, J. R., Xiao, B., Justin, N., Ivanov, G. S., McKinney, K., Tempst, P., Prives, C., Gamblin, S. J., Barlev, N. A., and Reinberg, D. (2004). Regulation of p53 activity through lysine methylation. *Nature* 432:353–360.

35. Falkenberg, P., Yaguchi, M., Roy, C., Zuker, M., and Matheson, A. T. (1986). The primary structure of the ribosomal-A protein (L12) from the moderate halophile NRCC 41227¹. *Biochem Cell Biol* 64:675–680.

36. Itoh, T., and Otaka, E. (1984). Complete amino acid sequence of an L7/L12-type ribosomal protein from *Desulfovibrio vulgaris*, Miyazaki. *Biochim Biophys Acta* 789:229–233.

37. Arnold, R. J., and Reilly, J. P. (1999). Observation of *Escherichia coli* ribosomal proteins and their posttranslational modifications by mass spectrometry. *Anal Biochem* 269:105–112.

38. Williamson, N. A., Raliegh, J., Morrice, N. A., and Wettenhall, R. E. H. (1997). Post-translational processing of rat ribosomal proteins: Ubiquitous methylation of Lys22 within the zinc-finger motif of RL40 (carboxy-terminal extension protein 52) and tissue-specific methylation of Lys4 in RL29. *Eur J Biochem* 246:786–793.

39. Itoh, T., and Wittmann-Liebold, B. (1978). Primary structure of protein 44 from large subunit of yeast ribosomes. *FEBS Lett* 96:399–402.

40. Kouskouti, A., Scheer, E., Staub, A., Tora, L., and Talianidis, I. (2004). Gene-specific modulation of TAF10 function by SET9-mediated methylation. *Mol Cell* 14:175–182.

41. Biemann, K., and Papayannopoulos, I. A. (1994). Amino acid sequencing of proteins. *Acc Chem Res* 27:370–378.

42. Kalasz, H., Kovacs, G. H., Nagy, J., Tyihak, E., and Barnes, W. T. (1978). Identification of N-methylated basic amino acids from human adult teeth. *J Dent Res* 57:128–132.

43. Schaefer, W. H., Lukas, T. J., Blair, I. A., Schultz, J. E., and Watterson, D. M. (1987). Amino acid sequence of a novel calmodulin from *Paramecium tetraurelia* that contains dimethyllysine in the first domain. *J Biol Chem* 262:1025–1029.

44. Morgan, W. T., Hensley, C. P., and Riehm, J. P. (1972). Proteins of thermophilic fungus *Humicola lanuginosa* 1: Isolation and amino acid sequence of a cytochrome *c*. *J Biol Chem* 247:6555–6565.

45. Cavallius, J., Zoll, W., Chakraburtty, K., and Merrick, W. C. (1993). Characterization of yeast EF-1α: Non-conservation of post-translational modifications. *Biochim Biophys Acta* 1163:75–80.

46. Dever, T. E., Costello, C. E., Owens, C. L., Rosenberry, T. L., and Merrick, W. C. (1989). Location of seven post-translational modifications in rabbit elongation factor 1α including dimethyllysine, trimethyllysine, and glycerylphosphorylethanolamine. *J Biol Chem* 264:20518–20525.

47. Decottignies, P., Lemarechal, P., Jacquot, J. P., Schmitter, J. M., and Gadal, P. (1995). Primary structure and post-translational modification of ferredoxin-NADP reductase from *Chlamydomonas reinhardtii*. *Arch Biochem Biophys* 316:249–259.

48. Hardy, M. F., Harris, C. I., Perry, S. V., and Stone, D. (1970). Occurrence and formation of Nε-methyl-lysines in myosin and myofibrillar proteins. *Biochem J* 120:653–660.

49. Atkinson, M.A.L., Robinson, E. A., Appella, E., and Korn, E. D. (1986). Amino acid sequence of the active site of *Acanthamoeba* myosin II. *J Biol Chem* 261:1844–1848.
50. Kröger, N., Deutzmann, R., and Sumper, M. (1999). Polycationic peptides from diatom biosilica that direct silica nanosphere formation. *Science* 286:1129–1132.
51. Kröger, N., Deutzmann, R., and Sumper, M. (2001). Silica-precipitating peptides from diatoms: The chemical structure of silaffin-1A from *Cylindrotheca fusiformis*. *J Biol Chem* 276:26066–26070.
52. Sumper, M., and Kröger, N. (2004). Silica formation in diatoms: The function of long-chain polyamines and silaffins. *J Mater Chem* 14:2059–2065.
53. Vandekerckhove, J., Van Damme, J., Vancompernolle, K., Bubb, M. R., Lambooy, P. K., and Korn, E. D. (1990). The covalent structure of *Acanthamoeba* actobindin. *J Biol Chem* 265:12801–12805.
54. Motojima, K., and Sakaguchi, K. (1982). Part of the lysyl residues in wheat α-amylase is methylated as N-ε-trimethyl lysine. *Plant Cell Physiol* 23:709–712.
55. Buzy, A., Ryan, E. M., Jennings, K. R., Palmer, D. N., and Griffiths, D. E. (1996). Use of electrospray ionization mass spectrometry and tandem mass spectrometry to study binding of F-0 inhibitors to ceroid lipofuscinosis protein, a model system for subunit c of mitochondrial ATP synthase. *Rapid Commun Mass Spectrom* 10:790–796.
56. Katz, M. L., Christianson, J. S., Norbury, N. E., Gao, C. L., Siakotos, A. N., and Koppang, N. (1994). Lysine methylation of mitochondrial ATP synthase subunit c stored in tissues of dogs with hereditary ceroid lipofuscinosis. *J Biol Chem* 269:9906–9911.
57. Katz, M. L., Gao, C. L., Tompkins, J. A., Bronson, R. T., and Chin, D. T. (1995). Mitochondrial ATP synthase subunit c stored in hereditary ceroid lipofuscinosis contains trimethyl lysine. *Biochem J* 310:887–892.
58. Chen, R. M., Fearnley, I. M., Palmer, D. N., and Walker, J. E. (2004). Lysine 43 is trimethylated in subunit c from bovine mitochondrial ATP synthase and in storage bodies associated with Batten disease. *J Biol Chem* 279:21883–21887.
59. Kobayashi, T., Takagi, T., Konishi, K., and Cox, J. A. (1987). The primary structure of a new *M*r 18,000 calcium vector protein from Amphioxus. *J Biol Chem* 262:2613–2623.
60. Lukas, T. J., Wallen-Friedman, M., Kung, C., and Watterson, D. M. (1989). *In vivo* mutations of calmodulin: A mutant *Paramecium* with altered ion current regulation has an isoleucine-to-threonine change at residue 136 and an altered methylation state at lysine residue 115. *Proc Natl Acad Sci USA* 86:7331–7335.
61. Yazawa, M., Yagi, K., Toda, H., Kondo, K., Narita, K., Yamazaki, R., Sobue, K., Kakiuchi, S., Nagao, S., and Nozawa, Y. (1981). The amino acid sequence of the *Tetrahymena* calmodulin which specifically interacts with guanylate cyclase. *Biochem Biophys Res Commun* 99:1051–1057.
62. Dedman, J. R., Jackson, R. L., Schreiber, W. E., and Means, A. R. (1978). Sequence homology of Ca²⁺-dependent regulator of cyclic nucleotide phosphodiesterase from rat testis with other Ca²⁺-binding proteins. *J Biol Chem* 253:343–346.
63. Grand, R. J. A., Shenolikar, S., and Cohen, P. (1981). The amino acid sequence of the δ–subunit (calmodulin) of rabbit skeletal muscle phosphorylase kinase. *Eur J Biochem* 113:359–367.
64. Watterson, D. M., Iverson, D. B., and Van Eldik, L. J. (1980). Spinach calmodulin: Isolation, characterization, and comparison with vertebrate calmodulins. *Biochemistry* 19:5762–5768.
65. Sasagawa, T., Ericsson, L. H., Walsh, K. A., Schreiber, W. E., Fischer, E. H., and Titani, K. (1982). Complete amino acid sequence of human brain calmodulin. *Biochemistry* 21:2565–2569.
66. Lukas, T. J., Iverson, D. B., Schleicher, M., and Watterson, D. M. (1984). Structural characterization of a higher plant calmodulin. *Plant Physiol* 75:788–795.
67. Toda, H., Yazawa, M., Sakiyama, F., and Yagi, K. (1985). Amino acid sequence of calmodulin from wheat germ. *J Biochem (Tokyo)* 98:833–842.

68. Takagi, T., Nemoto, T., Konishi, K., Yazawa, M., and Yagi, K. (1980). The amino acid sequence of the calmodulin obtained from sea anemone (*Metridium senile*) muscle. *Biochem Biophys Res Commun* 96:377–381.

69. Jamieson, G. A. Jr., Bronson, D. D., Schachat, F. F., and Vanaman, T. C. (1980). Structure and function relationships among calmodulins and troponin C-like proteins from divergent eukaryotic organisms. *Ann NY Acad Sci* 356:1–13.

70. Toda, H., Yazawa, M., and Yagi, K. (1992). Amino acid sequence of calmodulin from *Euglena gracilis*. *Eur J Biochem* 205:653–660.

71. Bloxham, D. P., Parmelee, D. C., Kumar, S., Wade, R. D., Ericsson, L. H., Neurath, H., Walsh, K. A., and Titani, K. (1981). Primary structure of porcine heart citrate synthase. *Proc Natl Acad Sci USA* 78:5381–5385.

72. Bloxham, D. P., Parmelee, D. C., Kumar, S., Walsh, K. A., and Titani, K. (1982). Complete amino acid sequence of porcine heart citrate synthase. *Biochemistry* 21:2028–2036.

73. DeLange, R. J., Glazer, A. N., and Smith, E. L. (1969). Presence and location of an unusual amino acid, ε-*N*-trimethyllysine, in cytochrome *c* of wheat germ and *Neurospora*. *J Biol Chem* 244:1385–1388.

74. DeLange, R. J., Glazer, A. N., and Smith, E. L. (1970). Identification and location of ε-*N*-trimethyllysine in yeast cytochromes *c*. *J Biol Chem* 245:3325–3327.

75. Chin, C. C. Q., Niehaus, W. G., and Wold, F. (1989). Amino acid sequence of cytochrome *c* from *Aspergillus niger*. *J Protein Chem* 8:165–171.

76. Meatyard, B. T. and Boulter, D. (1974). Amino acid sequence of cytochrome *c* from *Enteromorpha intestinalis*. *Phytochemistry* 13:2777–2782.

77. Sugeno, K., Narita, K., and Titani, K. (1971). Amino acid sequence of cytochrome *c* from *Debaryomyces kloeckeri*. *J Biochem (Tokyo)* 70:659–682.

78. Simon-Becam, A. M., Claisse, M., and Lederer, F. (1978). Cytochrome *c* from *Schizosaccharomyces pombe* 2. Amino-acid sequence. *Eur J Biochem* 86:407–416.

79. Polevoda, B., Martzen, M. R., Das, B., Phizicky, E. M., and Sherman, F. (2000). Cytochrome *c* methyltransferase, Ctm1p, of yeast. *J Biol Chem* 275:20508–20513.

80. Thompson, W., Notton, B. A., Richards, M., and Boulter, D. (1971). Amino acid sequence of cytochrome *c* from *Abutilon theophrasti* Medic and *Gossypium barbadense* L. (cotton). *Biochem J* 124:787–791.

81. Brown, R. H. and Boulter, D. (1973). Amino acid sequence of cytochrome *c* from *Allium porrum* (leek). *Biochem J* 131:247–251.

82. Thompson, E. W., Richards, M., and Boulter, D. (1971). Amino acid sequence of sesame (*Sesamum indicum* L.) and castor (*Ricinus communis* L.) cytochrome *c*. *Biochem J* 121:439–446.

83. Wallace, D. G., Brown, R. H., and Boulter, D. (1973). Amino-acid sequence of *Cannabis sativa* cytochrome *c*. *Phytochemistry* 12:2617–2622.

84. Thompson, E. W., Richards, M., and Boulter, D. (1971). Amino acid sequence of cytochrome *c* from *Cucurbita maxima* L. (pumpkin). *Biochem J* 124:779–781.

85. Thompson, E. W., Richards, M., and Boulter, D. (1971). Amino acid sequence of cytochome *c* of *Fagopyrum esculentum* Moench (buckwheat) and *Brassica oleracea* (cauliflower). *Biochem J* 124:783–785.

86. Ramshaw, J.A.M., Richards, M., and Boulter, D. (1971). Amino acid sequence of cytochrome *c* of *Ginkgo biloba*. *Eur J Biochem* 23:475–483.

87. Ramshaw, J.A.M., and Boulter, D. (1975). Amino acid sequence of cytochrome *c* from niger seed, *Guizotia abyssinica*. *Phytochemistry* 14:1945–1949.

88. Ramshaw, J.A.M., Thompson, E. W., and Boulter, D. (1970). Amino acid sequence of *Helianthus annuus* L. (sunflower) cytochrome *c* deduced from chymotryptic peptides. *Biochem J* 119:535–539.

89. Scogin, R., Richards, M., and Boulter, D. (1972). Amino acid sequence of cytochrome *c* from tomato (*Lycopersicon esculentum* Mill). *Arch Biochem Biophys* 150:489–492.

90. Thompson, E. W., Laycock, M. V., Ramshaw, J.A.M., and Boulter, D. (1970). Amino acid sequence of *Phaseolus aureus* L. (mung bean) cytochrome *c*. *Biochem J* 117:183-192.
91. Brown, R. H., Richards, M., Scogin, R., and Boulter, D. (1973). Amino acid sequence of cytochrome *c* from *Spinacea oleracea* (spinach). *Biochem J* 131:253–256.
92. Hill, G. C., and Pettigrew, G. W. (1975). Evidence for amino acid sequence of *Crithidia fasciculata* cytochrome *c*-555. *Eur J Biochem* 57:265–271.
93. Kahns, S., Lund, A., Kristensen, P., Knudsen, C. R., Clark, B. F. C., Cavallius, J., and Merrick, W. C. (1998). The elongation factor 1 A-2 isoform from rabbit: Cloning of the cDNA and characterization of the protein. *Nucleic Acids Res* 26:1884–1890.
94. Liang, X. Q., Lu, Y., Wilkes, M., Neubert, T. A., and Resh, M. D. (2004). The N-terminal SH4 region of the Src family kinase Fyn is modified by methylation and heterogeneous fatty acylation: Role in membrane targeting, cell adhesion, and spreading. *J Biol Chem* 279:8133–8139.
95. Maita, T., Onishi, H., Yajima, E., and Matsuda, G. (1987). Amino acid sequence of the amino-terminal 24 kDa fragment of the heavy chain of chicken gizzard myosin. *J Biochem (Tokyo)* 102:133–145.
96. Komine, Y., Maita, T., and Matsuda, G. (1991). The primary structure of skeletal muscle myosin heavy chain. 2. Sequence of the 50 kDa fragment of subfragment-1. *J Biochem (Tokyo)* 110:60–67.
97. Maita, T., Yajima, E., Nagata, S., Miyanishi, T., Nakayama, S., and Matsuda, G. (1991). The primary structure of skeletal muscle myosin heavy chain. 4. Sequence of the rod, and the complete 1,938-residue sequence of the heavy chain. *J Biochem (Tokyo)* 110:75–87.
98. Ampe, C., Vandekerckhove, J., Brenner, S. L., Tobacman, L., and Korn, E. D. (1985). The amino acid sequence of *Acanthamoeba* profilin. *J Biol Chem* 260:834–840.
99. Ampe, C., Sato, M., Pollard, T. D., and Vandekerckhove, J. (1988). The primary structure of the basic isoform of *Acanthamoeba* profilin. *Eur J Biochem* 170:597–601.
100. Houtz, R. L., Stults, J. T., Mulligan, R. M., and Tolbert, N. E. (1989). Posttranslational modifications in the large subunit of ribulose bisphosphate carboxylase/oxygenase. *Proc Natl Acad Sci USA* 86:1855–1859.
101. Houtz, R. L., Poneleit, L., Jones, S. B., Royer, M., and Stults, J. T. (1992). Posttranslational modifications in the amino-terminal region of the large subunit of ribulose-1,5 bisphosphate carboxylase/oxygenase from several plant species. *Plant Physiol* 98:1170–1174.
102. Houtz, R. L., Royer, M., and Salvucci, M. E. (1991). Partial purification and characterization of ribulose-1,5-bisphosphate carboxylase/oxygenase large subunit $^{\varepsilon}$N-methyltransferase. *Plant Physiol* 97:913–920.
103. Klein, R. R. and Houtz, R. L. (1995). Cloning and developmental expression of pea ribulose-1,5-bisphosphate carboxylase/oxygenase large subunit N-methyltransferase. *Plant Mol Biol* 27:249–261.
104. Wang, P., Royer, M., and Houtz, R. L. (1995). Affinity purification of ribulose-1,5 bisphosphate carboxylase/oxygenase large subunit N^{ε}-methyltransferase. *Protein Expr Purif* 6:528–536.
105. Dognin, M. J. and Wittmann-Liebold, B. (1980). Identification of methylated amino acids during sequence analysis: Application to the *Escherichia coli* ribosomal protein L11. *Hoppe Seylers Z Physiol Chem* 361:1697–1705.
106. Colson, C., and Smith, H. O. (1977). Genetics of ribosomal protein methylation in *Escherichia coli*.1. Mutant deficient in methylation of protein L11. *Mol Gen Genet* 154:167–173.
107. Vanet, A., Plumbridge, J. A., Guerin, M. F., and Alix, J. H. (1994). Ribosomal protein methylation in *Escherichia coli*: The gene *prmA*, encoding the ribosomal protein L11 methyltransferase, is dispensable. *Mol Microbiol* 14:947–958.
108. Alix, J. H. (1989). A rapid procedure for cloning genes from λ libraries by complementation of *E. coli* defective mutants: Application to the *fabE* region of the *E. coli* chromosome. *DNA* 8:779–789.

109. Cameron, D. M., Gregory, S. T., Thompson, J., Suh, M. J., Limbach, P. A., and Dahlberg, A. E. (2004). *Thermus thermophilus* L11 methyltransferase, PrmA, is dispensable for growth and preferentially modifies free ribosomal protein L11 prior to ribosome assembly. *J Bacteriol* 186:5819–5825.

110. Yamaguchi, K., and Subramanian, A. R. (2000). The plastid ribosomal proteins - Identification of all the proteins in the 50 S subunit of an organelle ribosome (chloroplast). *J Biol Chem* 275:28466–28482.

111. Mardones, E., Amaro, A. M., and Jerez, C. A. (1980). Methylation of ribosomal proteins in *Bacillus subtilis. J Bacteriol* 142:355–358.

112. Amaro, A. M., and Jerez, C. A. (1984). Methylation of ribosomal proteins in bacteria - Evidence of conserved modification of the eubacterial 50S subunit. *J Bacteriol* 158:84–93.

113. Daniel, J. A., Pray-Grant, M. G., and Grant, P. A. (2005). Effector proteins for methylated histones: An expanding family. *Cell Cycle* 4:919–926.

114. Maurer-Stroh, S., Dickens, N. J., Hughes-Davies, L., Kouzarides, T., Eisenhaber, F., and Ponting, C. P. (2003). The Tudor domain "Royal Family": Tudor, plant Agenet, Chromo, PWWP and MBT domains. *Trends Biochem Sci* 28:69–74.

115. Wysocka, J., Swigut, T., Milne, T. A., Dou, Y., Zhang, X., Burlingame, A. L., Roeder, R. G., Brivanlou, A. H., and Allis, C. D. (2005). WDR5 associates with histone H3 methylated at K4 and is essential for H3 K4 methylation and vertebrate development. *Cell* 121:859–872.

116. Park, I. K., and Paik, W. K. (1990). The occurrence and analysis of methylated amino acids, in W. K.Paik and S.Kim (eds.), *Protein Methylation*, pp. 1–22, Boca Raton: CRC Press.

117. Pessolani, M. C. V., Marques, M. A. D., Reddy, V. M., Locht, C., and Menozzi, F. D. (2003). Systemic dissemination in tuberculosis and leprosy: do mycobacterial adhesins play a role? *Microbes Infect* 5:677–684.

118. Soares de Lima, C., Zulianello, L., de Melo Marques, M. A., Kim, H., Portugal, M. I., Antunes, S. L., Menozzi, F. D., Henricus, T., Ottenhoff, M., Brennan, P. J., and Pessolani, M. C. V. (2005). Mapping the laminin-binding and adhesive domain of the cell surface-associated Hlp/LBP protein from *Mycobacterium leprae. Microbes Infect* 7:1097–1109.

119. Takahashi, M., Sano, T., Yamaoka, K., Kamimura, T., Umemoto, N., Nishitani, H., and Yasuoka, S. (2001). Localization of human airway trypsin-like protease in the airway: an immunohistochemical study. *Histochem Cell Biol* 115:181–187.

120. Parra, M., Pickett, T., Delogu, G., Dheenadhayalan, V., Debrie, A. S., Locht, C., and Brennan, M. J. (2004). The mycobacterial heparin-binding hemagglutinin is a protective antigen in the mouse aerosol challenge model of tuberculosis. *Infect Immun* 72:6799–6805.

121. Temmerman, S., Pethe, K., Parra, M., Alonso, S., Rouanet, C., Pickett, T., Drowart, A., Debrie, A. S., Delogu, G., Menozzi, F. D., Sergheraert, C., Brennan, M. J., Mascart, F., and Locht, C. (2004). Methylation-dependent T cell immunity to *Mycobacterium tuberculosis* heparin-binding hemagglutinin. *Nat Med* 10:935–941.

122. Cannio, R., Rossi, M., and Bartolucci, S. (1994). A few amino acid substitutions are responsible for the higher thermostability of a novel NAD(+)-dependent bacillar alcohol dehydrogenase. *Eur J Biochem* 222:345–352.

123. Doll, L., and Frankel, G. (1993). Cloning and sequencing of two new *fli* genes, the products of which are essential for *Salmonella* flagellar biosynthesis. *Gene* 126:119–121.

124. Yoshioka, K., Aizawa, S., and Yamaguchi, S. (1995). Flagellar filament structure and cell motility of *Salmonella typhimurium* mutants lacking part of the outer domain of flagellin. *J Bacteriol* 177:1090–1093.

125. Kirov, S. M., Tassell, B. C., Semmler, A. B. T., O'Donovan, L. A., Rabaan, A. A., and Shaw, J. G. (2002). Lateral flagella and swarming motility in *Aeromonas* species. *J Bacteriol* 184:547–555.

126. Ramos, H. C., Rumbo, M., and Sirard, J. C. (2004). Bacterial flagellins: Mediators of pathogenicity and host immune responses in mucosa. *Trends Microbiol* 12:509–517.

127. Macnab, R. M. (2004). Type III flagellar protein export and flagellar assembly. *Biochim Biophys Acta* 1694:207–217.
128. Minamino, T., and Namba, K. (2004). Self-assembly and type III protein export of the bacterial flagellum. *J Mol Microbiol Biotechnol* 7:5–17.
129. Lhoest, J., and Colson, C. (1990). Ribosomal protein methylation, in W. K.Paik and S. Kim (eds.), *Protein Methylation*, pp. 33–58, Boca Raton: CRC Press.
130. Kaminishi, T., Sakai, H., Takemoto-Hori, C., Terada, T., Nakagawa, N., Maoka, N., Kuramitsu, S., Shirouzu, M., and Yokoyama, S. (2003). Crystallization and preliminary X-ray diffraction analysis of ribosomal protein L11 methyltransferase from *Thermus thermophilus* HB8. *Acta Crystallogr D Biol Crystallogr* 59:930–932.
131. Bujnicki, J. M. (2000). Sequence, structural, and evolutionary analysis of prokaryotic ribosomal protein L11 methyltransferases. *Acta Microbiol Pol* 49:19–29.
132. Conn, G. L., Gittis, A. G., Lattman, E. E., Misra, V. K., and Draper, D. E. (2002). A compact RNA tertiary structure contains a buried backbone-K$^+$ complex. *J Mol Biol* 318:963–973.
133. Wimberly, B. T., Guymon, R., McCutcheon, J. P., White, S. W., and Ramakrishnan, V. (1999). A detailed view of a ribosomal active site: the structure of the L11-RNA complex. *Cell* 97:491–502.
134. Agrawal, R. K., Linde, J., Sengupta, J., Nierhaus, K. H., and Frank, J. (2001). Localization of L11 protein on the ribosome and elucidation of its involvement in EF-G-dependent translocation. *J Mol Biol* 311:777–787.
135. Chang, F. N., Cohen, L. B., Navickas, I. J., and Chang, C. N. (1975). Purification and properties of a ribosomal protein methylase from *Escherichia coli* Q13. *Biochemistry* 14:4994–4998.
136. Alix, J. H., and Hayes, D. (1974). Properties of ribosomes and RNA synthesized by *Escherichia coli* grown in presence of ethionine. 3. Methylated proteins in 50 S ribosomes of *Escherichia coli* Ea2. *J Mol Biol* 86:139–159.
137. Colson, C., Lhoest, J., and Urlings, C. (1979). Genetics of ribosomal protein methylation in *Escherichia coli*. 3. Map position of two genes, *Prma* and *Prmb*, governing methylation of proteins L11 and L3. *Mol Gen Genet* 169:245–250.
138. Dabbs, E. R. (1978). Mutational alterations in 50 proteins of *Escherichia coli* ribosome. *Mol Gen Genet* 165:73–78.
139. Bausch, S. L., Poliakova, E., and Draper, D. E. (2005). Interactions of the N-terminal domain of ribosomal protein L11 with thiostrepton and rRNA. *J Biol Chem* 280:29956–29963.
140. Bowen, W. S., Van Dyke, N., Murgola, E. J., Lodmell, J. S., and Hill, W. E. (2005). Interaction of thiostrepton and elongation factor-G with the ribosomal protein L11-binding domain. *J Biol Chem* 280:2934–2943.
141. Schubert, H. L., Blumenthal, R. M., and Cheng, X. (2003). Many paths to methyltransfer: a chronicle of convergence. *Trends Biochem Sci* 28:329–335.
142. Sawada, K., Yang, Z., Horton, J. R., Collins, R. E., Zhang, X., and Cheng, X. (2004). Structure of the conserved core of the yeast Dot1p, a nucleosomal histone H3 lysine 79 methyltransferase. *J Biol Chem* 279:43296–43306.
143. Min, J., Feng, Q., Li, Z., Zhang, Y., and Xu, R. M. (2003). Structure of the catalytic domain of human DOT1L, a non-SET domain nucleosomal histone methyltransferase. *Cell* 112:711–723.
144. Yang, X. M., and Ishiguro, E. E. (2001). Involvement of the N terminus of ribosomal protein L11 in regulation of the RelA protein of *Escherichia coli*. *J Bacteriol* 183:6532–6537.
145. Toledo, H., Amaro, A. M., Sanhueza, S., and Jerez, C. A. (1988). Methylation of proteins from the translational apparatus: an overview. *Arch Biol Med Exp (Santiago)* 21:219–229.
146. Brodersen, D. E., and Nissen, P. (2005). The social life of ribosomal proteins. *FEBS J* 272:2098–2108.
147. Lee, J. M. (2003). The role of protein elongation factor eEF1A2 in ovarian cancer. *Reprod Biol Endocrinol* 1:69.

148. Ouaissi A. (2003). Apoptosis-like death in trypanosomatids: search for putative pathways and genes involved. *Kinetoplastid Biol Dis* 2:5.
149. Siegel, F. L., Vincent, P. L., Neal, T. L., Wright, L. S., Heath, A. A., and Rowe, P. M. (1990). Calmodulin and protein methylation, in W. K.Paik and S. Kim (eds.). *Protein Methylation*, pp. 33–58, Boca Raton: CRC Press.
150. Klee, C. B., and Vanaman, T. C. (1982). Calmodulin. *Adv Protein Chem* 35:213–303.
151. Roberts, D. M., Crea, R., Malecha, M., Alvarado-Urbina, G., Chiarello, R. H., and Watterson, D. M. (1985). Chemical synthesis and expression of a calmodulin gene designed for site-specific mutagenesis. *Biochemistry* 24:5090–5098.
152. Morino, H., Kawamoto, T., Miyake, M., and Kakimoto, Y. (1987). Purification and properties of calmodulin-lysine N-methyltransferase from rat brain cytosol. *J Neurochem* 48:1201–1208.
153. Pech, L. L., and Nelson, D. L. (1994). Purification and characterization of calmodulin (lysine 115) N-methyltransferase from Paramecium tetraurelia. *Biochim Biophys Acta* 1199:183–194.
154. Han, C. H., Richardson, J., Oh, S. H., and Roberts, D. M. (1993). Isolation and kinetic characterization of the calmodulin methyltransferase from sheep brain. *Biochemistry* 32:13974–13980.
155. Wright, L. S., Bertics, P. J., and Siegel, F. L. (1996). Calmodulin N-methyltransferase. Kinetics, mechanism, and inhibitors. *J Biol Chem* 271:12737–12743.
156. Oh, S., and Roberts, D. M. (1990). Analysis of the state of posttranslational calmodulin methylation in developing pea plants. *Plant Physiol* 93:880–887.
157. Rowe, P. M., Wright, L. S., and Siegel, F. L. (1986). Calmodulin N-methyltransferase. *J Biol Chem* 261:7060–7069.
158. Roberts, D. M., Rowe, P. M., Siegel, F. L., Lukas, T. J., and Watterson, D. M. (1986). Trimethyllysine and protein function. Effect of methylation and mutagenesis of lysine 115 of calmodulin on NAD kinase activation. *J Biol Chem* 261:1491–1494.
159. Persechini, A., and Kretsinger, R. H. (1988). The central helix of calmodulin functions as a flexible tether. *J Biol Chem* 263:12175–12178.
160. Persechini, A., Moncrief, N. D., and Kretsinger, R. H. (1989). The EF-hand family of calcium-modulated proteins. *Trends Neurosci* 12:462–467.
161. Chattopadhyaya, R., Meador, W. E., Means, A. R., and Quiocho, F. A. (1992). Calmodulin structure refined at 1.7 A resolution. *J Mol Biol* 228:1177–1192.
162. Barbato, G., Ikura, M., Kay, L. E., Pastor, R. W., and Bax, A. (1992). Backbone dynamics of calmodulin studied by 15N relaxation using inverse detected two-dimensional NMR spectroscopy: The central helix is flexible. *Biochemistry* 31:5269–5278.
163. Babu, Y. S., Bugg, C. E., and Cook, W. J. (1988). Structure of calmodulin refined at 2.2 A resolution. *J Mol Biol* 204:191–204.
164. Ikura, M., Clore, G. M., Gronenborn, A. M., Zhu, G., Klee, C. B., and Bax, A. (1992). Solution structure of a calmodulin-target peptide complex by multidimensional NMR. *Science* 256:632–638.
165. Meador, W. E., Means, A. R., and Quiocho, F. A. (1992). Target enzyme recognition by calmodulin: 2.4 A structure of a calmodulin-peptide complex. *Science* 257:1251–1255.
166. Cobb, J. A., Han, C. H., Wills, D. M., and Roberts, D. M. (1999). Structural elements within the methylation loop (residues 112–117) and EF hands III and IV of calmodulin are required for Lys(115) trimethylation. *Biochem J* 340:417–424.
167. Cobb, J. A., and Roberts, D. M. (2000). Structural requirements for N-trimethylation of lysine 115 of calmodulin. *J Biol Chem* 275:18969–18975.
168. Roberts, D. M., Burgess, W. H., and Watterson, D. M. (1984). Comparison of the NAD kinase and myosin light chain kinase activator properties of vertebrate, higher plant, and algal calmodulins. *Plant Physiol* 75:796–798.
169. Lee, S. H., Seo, H. Y., Kim, J. C., Heo, W. D., Chung, W. S., Lee, K. J., Kim, M. C., Cheong, Y. H., Choi, J. Y., Lim, C. O., and Cho, M. J. (1997). Differential activation of NAD kinase by plant calmodulin isoforms. The critical role of domain I. *J Biol Chem* 272:9252–9259.

170. Oh, S. H., Steiner, H. Y., Dougall, D. K., and Roberts, D. M. (1992). Modulation of calmodulin levels, calmodulin methylation, and calmodulin binding proteins during carrot cell growth and embryogenesis. *Arch Biochem Biophys* 297:28–34.

171. Roberts, D. M., Besl, L., Oh, S. H., Masterson, R. V., Schell, J., and Stacey, G. (1992). Expression of a calmodulin methylation mutant affects the growth and development of transgenic tobacco plants. *Proc Natl Acad Sci USA* 89:8394–8398.

172. Harding, S. A., Oh, S. H., and Roberts, D. M. (1997). Transgenic tobacco expressing a foreign calmodulin gene shows an enhanced production of active oxygen species. *EMBO J* 16:1137–1144.

173. Harding, S. A., and Roberts, D. M. (1998). Incompatible pathogen infection results in enhanced reactive oxygen and cell death responses in transgenic tobacco expressing a hyperactive mutant calmodulin. *Planta* 206:253–258.

174. Gregori, L., Marriott, D., West, C. M., and Chau, V. (1985). Specific recognition of calmodulin from Dictyostelium discoideum by the ATP, ubiquitin-dependent degradative pathway. *J Biol Chem* 260:5232–5235.

175. Gregori, L., Marriott, D., Putkey, J. A., Means, A. R., and Chau, V. (1987). Bacterially synthesized vertebrate calmodulin is a specific substrate for ubiquitination. *J Biol Chem* 262:2562–2567.

176. Ziegenhagen, R., and Jennissen, H. P. (1988). Multiple ubiquitination of vertebrate calmodulin by reticulocyte lysate and inhibition of calmodulin conjugation by phosphorylase kinase. *Biol Chem Hoppe Seyler* 369:1317–1324.

177. Ziegenhagen, R., Goldberg, M., Rakutt, W. D., and Jennissen, H. P. (1990). Multiple ubiquitination of calmodulin results in one polyubiquitin chain linked to calmodulin. *FEBS Lett* 271:71–75.

178. Blumenthal, D. K., and Stull, J. T. (1982). Effects of pH, ionic strength, and temperature on activation by calmodulin and catalytic activity of myosin light chain kinase. *Biochemistry* 21:2386–2391.

179. VanBerkum, M. F., and Means, A. R. (1991). Three amino acid substitutions in domain I of calmodulin prevent the activation of chicken smooth muscle myosin light chain kinase. *J Biol Chem* 266:21488–21495.

180. Nochumson, S., Durban, E., Kim, S., and Paik, W. K. (1977). Cytochrome c-specific protein methylase III from *Neurospora crassa*. *Biochem J* 165:11–18.

181. Paik, W. K., and Kim, S. (1975). Protein methylation: chemical, enzymological and biological significance. *Adv Enzymol Relat Areas Mol Biol* 42:227–286.

182. Paik, W. K., Cho, Y. B., Frost, B., and Kim, S. (1989). Cytochrome c methylation. *Biochem Cell Biol* 67:602–611.

183. Takakura, H., Yamamoto, T., and Sherman, F. (1997). Sequence requirement for trimethylation of yeast cytochrome c. *Biochemistry* 36:2642–2648.

184. DiMaria, P., Polastro, E., DeLange, R. J., Kim, S., and Paik, W. K. (1979). Studies on cytochrome c methylation in yeast. *J Biol Chem* 254:4656–4652.

185. Park, K. S., Frost, B., Tuck, M., Ho, L. L., Kim, S., and Paik, W. K. (1987). Enzymatic methylation of in vitro synthesized apocytochrome c enhances its transport into mitochondria. *J Biol Chem* 262:14702–14708.

186. Durban, E., Nochumson, S., Kim, S., Paik, W. K., and Chan, S. K. (1978). Cytochrome c-specific protein-lysine methyltransferase from *Neurospora crassa*. Purification, characterization, and substrate requirements. *J Biol Chem* 253:1427–1435.

187. Brown, R., and Boulter, D. (1975). A re-examination of the amino acid sequence data of cytochromes c from *Solanum tuberosum* (potato) and *Lycopersicon esculentum* (tomato). *FEBS Lett* 51:66–67.

188. Brown, R. H., and Boulter, D. (1973). Amino acid sequence of cytochrome c from *Nigella damascena* (Love-In-A-Mist). *Biochem J* 133:251–254.

189. DiMaria, P., Kim, S., and Paik, W. K. (1982). Cytochrome *c* specific methylase from wheat germ. *Biochemistry* 21:1036–1044.
190. Janbon, G., Rustchenko, E. P., Klug, S., Scherer, S., and Sherman, F. (1997). Phylogenetic relationships of fungal cytochromes *c*. *Yeast* 13:985–990.
191. Frost, B., and Paik, W. K. (1990). Cytochrome *c* methylation, in W. K.Paik and S. Kim (eds.), *Protein Methylation*, pp. 59–76, Boca Raton: CRC Press.
192. Durban, E., Sangduk, K., Gil-Ja, J., and Paik, W. K. (1983). Cytochrome *c* specific protein-lysine methyltransferase from *Neurospora crassa*: Kinetic mechanism. *Korean J Biochem* 15:19–24.
193. Ceesay, K. J., Rider, L. R., Bergman, L. W., and Tuck, M. T. (1994). The relationship between the trimethylation of lysine 77 and cytochrome *c* metabolism in *Saccharomyces cerevisiae*. *Int J Biochem* 26:721–734.
194. Wang, X., Dumont, M. E., and Sherman, F. (1996). Sequence requirements for mitochondrial import of yeast cytochrome *c*. *J Biol Chem* 271:6594–6604.
195. Thony-Meyer, L. (2002). Cytochrome *c* maturation: a complex pathway for a simple task? *Biochem Soc Trans* 30:633–638.
196. Frost, B., Syed, S. K., Kim, S., and Paik, W. K. (1990). Effect of enzymatic methylation of cytochrome *c* on its function and synthesis. *Int J Biochem* 22:1069–1074.
197. Ceesay, K. J., Bergman, L. W., and Tuck, M. T. (1991). Further investigations regarding the role of tri-methyllysine for cytochrome *c* uptake into mitochondria. *Int J Biochem* 23:761–768.
198. Farooqui, J., DiMaria, P., Kim, S., and Paik, W. K. (1981). Effect of methylation on the stability of cytochrome *c* of *Saccharomyces cerevisiae* in vivo. *J Biol Chem* 256:5041–5045.
199. Kluck, R. M., Ellerby, L. M., Ellerby, H. M., Naiem, S., Yaffe, M. P., Margoliash, E., Bredesen, D., Mauk, A. G., Sherman, F., and Newmeyer, D. D. (2000). Determinants of cytochrome *c* pro-apoptotic activity. The role of lysine 72 trimethylation. *J Biol Chem* 275:16127–16133.
200. Ferguson-Miller, S., Brautigan, D. L., and Margoliash, E. (1978). Definition of cytochrome *c* binding domains by chemical modification. III. Kinetics of reaction of carboxydinitrophenyl cytochromes *c* with cytochrome *c* oxidase. *J Biol Chem* 253:149–159.
201. Rieder, R., and Bosshard, H. R. (1978). The cytochrome *c* oxidase binding site on cytochrome *c*. Differential chemical modification of lysine residues in free and oxidase-bound cytochrome *c*. *J Biol Chem* 253:6045–6053.
202. Rieder, R., and Bosshard, H. R. (1978). Cytochrome *bc*1 and cytochrome oxidase can bind to the same surface domain of the cytochrome *c* molecule. *FEBS Lett* 92:223–226.
203. Rieder, R., and Bosshard, H. R. (1980). Comparison of the binding sites on cytochrome *c* for cytochrome *c* oxidase, cytochrome *bc*1, and cytochrome *c*1. Differential acetylation of lysyl residues in free and complexed cytochrome *c*. *J Biol Chem* 255:4732–4739.
204. Speck, S. H., Ferguson-Miller, S., Osheroff, N., and Margoliash, E. (1979). Definition of cytochrome *c* binding domains by chemical modification: kinetics of reaction with beef mito-chondrial reductase and functional organization of the respiratory chain. *Proc Natl Acad Sci USA* 76:155–159.
205. Duff, A. P., Andrews, T. J., and Curmi, P. M. (2000). The transition between the open and closed states of rubisco is triggered by the inter-phosphate distance of the bound bisphosphate. *J Mol Biol* 298:903–916.
206. Flynn, E. M., Dietzel, K. L., Dirk, L. M. A., Beach, B. M., Hurley, J. H., Trievel, R. C., and Houtz, R. L. (2004). Elucidation of the mechanism for successive methyl group transfers by SET domain containing protein methyltransferases. *29th FEBS Congress* Warsaw, Poland, June 26–July 1, 2004.

207. Ying, Z., Janney, N., and Houtz, R. L. (1996). Organization and characterization of the ribulose-1,5-bisphosphate carboxylase/oxygenase large subunit epsilon N-methyltransferase gene in tobacco. *Plant Mol Biol* 32:663–671.

208. Ying, Z., Mulligan, R. M., Janney, N., and Houtz, R. L. (1999). Rubisco small and large subunit N-methyltransferases. Bi- and mono- functional methyltransferases that methylate the small and large subunits of Rubisco. *J Biol Chem* 274:36750–36756.

209. Mazarei, M., Z.Ying, and R.L.Houtz. (1998). Functional analysis of the Rubisco large subunit N-methyltransferase promoter from tobacco and its regulation by light in soybean hairy roots. *Plant Cell Rep* 17:907–912.

210. Grimm, R., Grimm, M., Eckerskorn, C., Pohlmeyer, K., Rohl, T., and Soll, J. (1997). Postimport methylation of the small subunit of ribulose-1,5-bisphosphate carboxylase in chloroplasts. *FEBS Lett* 408:350–354.

211. Black, M. T., Meyer, D., Widger, W. R., and Cramer, W. A. (1987). Light-regulated methylation of chloroplast proteins. *J Biol Chem* 262:9803–9807.

212. McCurry, S. D., Gee, R., and Tolbert, N. E. (1982). Ribulose-1,5-bisphosphate carboxylase/oxygenase from spinach, tomato or tobacco leaves. *Methods Enzymol* 90:515–521.

213. Pierce, J. W., McCurry, S. D., Mulligan, R. M., and Tolbert, N. E. (1982). Activation and assay of ribulose-1,5-bisphosphate carboxylase/oxygenase. *Methods Enzymol* 89:47–55.

214. Thow, G., Zhu, G., and Spreitzer, R. J. (1994). Complementing substitutions within loop regions 2 and 3 of the α/β-barrel active site influence the CO_2/O_2 specificity of chloroplast ribulose-1,5-bisphosphate carboxylase/oxygenase. *Biochemistry* 33:5109–5114.

215. Jenuwein, T., Laible, G., Dorn, R., and Reuter, G. (1998). SET domain proteins modulate chromatin domains in eu- and heterochromatin. *Cell Mol Life Sci* 54:80–93.

216. Rea, S., Eisenhaber, F., O'Carroll, D., Strahl, B. D., Sun, Z. W., Schmid, M., Opravil, S., Mechtler, K., Ponting, C. P., Allis, C. D., and Jenuwein, T. (2000). Regulation of chromatin structure by site-specific histone H3 methyltransferases. *Nature* 406:593–599.

217. Cheng, X., Collins, R. E., and Zhang, X. (2005). Structural and sequence motifs of protein (histone) methylation enzymes. *Annu Rev Biophys Biomol Struct* 34:267–294.

218. Trievel, R. C., Beach, B. M., Dirk, L. M., Houtz, R. L., and Hurley, J. H. (2002). Structure and catalytic mechanism of a SET domain protein methyltransferase. *Cell* 111:91–103.

219. Trievel, R. C., Flynn, E. M., Houtz, R. L., and Hurley, J. H. (2003). Mechanism of multiple lysine methylation by the SET domain enzyme Rubisco LSMT. *Nat Struct Biol* 10:545–552.

220. Collins, R. E., Tachibana, M., Tamaru, H., Smith, K. M., Jia, D., Zhang, X., Selker, E. U., Shinkai, Y., and Cheng, X. (2005). In vitro and in vivo analyses of a Phe/Tyr switch controlling product specificity of histone lysine methyltransferases. *J Biol Chem* 280:5563–5570.

221. Couture, J. F., Collazo, E., Brunzelle, J. S., and Trievel, R. C. (2005). Structural and functional analysis of SET8, a histone H4 Lys-20 methyltransferase. *Genes Dev* 19:1455–1465.

222. Xiao, B., Jing, C., Wilson, J. R., Walker, P. A., Vasisht, N., Kelly, G., Howell, S., Taylor, I. A., Blackburn, G. M., and Gamblin, S. J. (2003). Structure and catalytic mechanism of the human histone methyltransferase SET7/9. *Nature* 421:652–656.

223. Xiao, B., Jing, C., Kelly, G., Walker, P. A., Muskett, F. W., Frenkiel, T. A., Martin, S. R., Sarma, K., Reinberg, D., Gamblin, S. J., and Wilson, J. R. (2005). Specificity and mechanism of the histone methyltransferase Pr-Set7. *Genes Dev* 19:1444–1454.

224. Yamaguchi, K., von Knoblauch, K., and Subramanian, A. R. (2000). The plastid ribosomal proteins - Identification of all the proteins in the 30 S subunit of an organelle ribosome (chloroplast). *J Biol Chem* 275:28455–28465.

225. Yamaguchi, K., and Subramanian, A. R. (2003). Proteomic identification of all plastid-specific ribosomal proteins in higher plant chloroplast 30S ribosomal subunit. *Eur J Biochem* 270:190–205.

226. Kamp, R. M., Srinivasa, B. R., von Knoblauch, K., and Subramanian, A. R. (1987). Occurrence of a Methylated Protein in Chloroplast Ribosomes. *Biochemistry* 26:5866–5870.
227. Espejo, A., and Bedford, M. T. (2004). Protein-domain microarrays. *Methods Mol Biol* 264:173–181.
228. Lee, J., Cheng, D., and Bedford, M. T. (2004). Techniques in protein methylation. *Methods Mol Biol* 284:195–208.
229. Lamberti, A., Caraglia, M., Longo, O., Marra, M., Abbruzzege, A., and Arcari, P. (2004). The translation elongation factor 1A in tumorigenesis, signal transduction and apoptosis: Review article. *Amino Acids* 26:443–448.
230. Porras-Yakushi, T. R., Whitelegge, J. P., Miranda, T. B., and Clarke S. (2005). A novel SET domain methyltransferase modifies ribosomal protein Rpl23ab in yeast. *J Biol Chem* 280:34590–34598.
231. Zhang, K., Lin, W., Latham, J. A., Riefler, G. M., Schumacher, J. M., Chan, C., Tatchell, K., Hawke, D. H., Kobayashi, R. and Dent, S. Y. R. (2005). The Set1 methyltransferase opposes Ipl1 Aurora kinase functions in chromosome segregation. *Cell* 122:723–734.

8

Demethylation Pathways for Histone Methyllysine Residues

FEDERICO FORNERIS[a] • CLAUDIA BINDA[a] • MARIA ANTONIETTA VANONI[b] • ANDREA MATTEVI[a] • ELENA BATTAGLIOLI[c]

[a]Dipartimento di Genetica e Microbiologia
Università di Pavia
Via Ferrata 1
Pavia 27100, Italy

[b]Dipartimento di Scienze Biomolecolari e Biotecnologie
[c]Dipartimento di Biologia e Genetica per le Scienze Mediche
Università di Milano
Via Celoria 26
Milano 20133, Italy

I. Abstract

Histone lysine methylation is one of the posttranslational modifications involved in transcriptional regulation and chromatin remodeling. The first lysine specific histone demethylase (LSD1) has been recently discovered, which rules out the hypothesis that histone methylation represents a permanent epigenetic mark. LSD1 (previously known as KIAA0601) has been typically found in association with CoREST (a corepressor protein) and histone deacetylases 1 and 2, forming a highly conserved core complex.

After submission of this chapter, Tsukada et al. [1a] have published the discovery of a new histone demethylase (JmjC-domain containing histone demethylase 1) that acts on Ly36 of histone H3. The protein is an oxoglutarate-dependent enzyme that catalyses the demethylation reaction through the hydroxylation of the substrate (Figure 8.6).

THE ENZYMES, Vol. XXIV

These proteins have been shown to be part of several megadalton corepressor complexes, which are proposed to operate in the context of a stable and extended form of repression through silencing of entire chromatin domains. LSD1 is a FAD-dependent protein that specifically catalyzes the demethylation of Lys4 of histone H3 by an oxidative process. The amino acid sequence of the human enzyme (90 kDa) has a modular organization with an N-terminal SWIRM domain, which has been found to mediate protein-protein interactions, and a C-terminal domain similar to FAD-dependent amine oxidases. Three assays based on different events of the demethylation reaction can be used to study LSD1 biochemical properties. The strict substrate specificity of LSD1 suggests the existence of other putative histone lysine demethylases that may use alternative mechanisms for the regulation of this posttranslational modification.

II. Introduction

The nucleosome core particle of chromatin is a compact structure formed by DNA wrapped around the histone octamer [1]. The N-terminal amino acids of histones protrude from the nucleosome like flexible tails and represent targets for various posttranslational modifications such as methylation, acetylation, phosphorylation, and sumoylation [2, 3]. These modifications act in a combinatorial and sequence-dependent manner to yield specific downstream events and are therefore used as a cellular vocabulary for the regulation of different transcription-based processes [4]. Each cell type is characterized by its own pattern of these chromatin marks, which forms the so-called "histone code." In the case of histones H3 and H4, it was shown that their posttranslational modifications are closely related to fundamental cellular events such as transcription activation and repression [5].

Lysine methylation is one of the elements that define the histone code. This modification is correlated with either transcription activation or repression depending on the site of methylation [6]. The apparent stability of this posttranslational modification in bulk histone preparations led to the belief that histone methylation might constitute a permanent epigenetic mark [7, 8]. In contrast, other findings suggested that histone methylation is dynamic and contributes to a number of diverse biological processes, including transcriptional regulation, chromatin condensation, mitosis, and heterochromatin assembly [9, 10]. Recently, it has been shown that in the course of heterochromatin formation loss of histone methylation is a gradual and relatively slow process [11]. Several mechanisms have been proposed to account for changes in histone methylation state, such as demethylating enzymes, histone replacement, and histone clipping [12].

The hunt for a histone lysine demethylating enzyme started more than 40 years ago when Paik and co-workers published in 1964 the purification of ε-Alkyllysinase, an enzyme capable of demethylating free mono- and di-ε-methyllysine [13]. A few years later, the same group reported the detection of

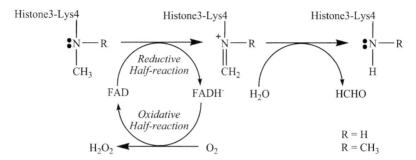

FIG. 8.1. Scheme of the proposed mechanism of the demethylation reaction catalyzed by LSD1.

an enzyme capable of demethylating histones, as well as free methylated lysines [14]. Eventually, the same authors partially purified this enzymatic activity, but were unable to relate it to a specific protein [15]. For more than 30 years after that finding, the molecular identity of a putative histone demethylase has remained elusive, leading to a diffuse scepticism about the existence of this enzyme.

In 2004, the discovery of the first human flavin-dependent lysine-specific histone demethylase LSD1 [16] demonstrated that (similarly to acetylation) histone methylation is a dynamic process, subject to regulation by both methyltransferases (see Chapters 5 and 6) and demethylases. LSD1 is able to catalyze the specific demethylation of Lys4 at histone H3 (H3-K4), which is known to be a modification linked to transcription activation. LSD1 demethylates di- and monomethylated H3-K4 releasing the unmodified lysine residue together with formaldehyde (Figure 8.1). At the same time, Forneris et al. independently demonstrated that this reaction occurs through a FAD-dependent enzymatic oxidation [17].

III. Histone Demethylation by LSD1 Is a Flavin-Dependent Oxidative Process

A. LSD1 IS A FLAVOENZYME

Before the discovery of the histone demethylase activity of LSD1, this protein was already known as KIAA0601 and was isolated as a constituent of large protein complexes involved in transcription regulation [18–21]. Human LSD1/KIAA0601 is an 852-amino-acid protein with a modular domain organization (Figure 8.2).

FIG. 8.2. Domain organization of human LSD1. The scheme refers to the sequence present in the NCBI data bank with accession code NP_055828.

The N-terminal part (residues 150 through 300) has similarity to a conserved domain called the SWIRM domain, which has been found in a number of proteins involved in chromatin regulation [22]. The C-terminal part of LSD1 (residues 301 through 852) has ~20% sequence identity to flavin-dependent amine oxidases such as polyamine oxidase and monoamine oxidase [23]. For this reason, LSD1/KIAA0601 was thought to be a nuclear polyamine oxidase [20]. To unravel the role of LSD1/KIAA0601 in these transcription repression complexes, the protein was expressed in *E. coli* [16]. The purified protein showed a characteristic UV/Vis absorption spectrum with two peaks centered around 384 nm and 458 nm, which is typical of a flavoenzyme in the oxidized state (Figure 8.3). By fluorimetric analysis, the flavin cofactor was determined to be FAD (released upon protein denaturation), demonstrating that the coenzyme is not covalently bound to the protein [17].

The next step in studying LSD1/KIAA0601 was the identification of its enzymatic activity. LSD1/KIAA0601 reactivity was tested by using the typical substrates of polyamine oxidase (such as spermine and spermidine) and monoamine oxidase (aromatic amines including benzylamine, tyramine, and amphetamine) [17]. Because no activity was detected on any of these compounds, the hypothesis

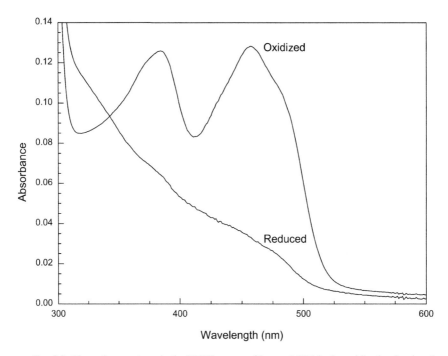

Fig. 8.3. Absorption spectrum in the UV/Vis range of human LSD1 in the oxidized and reduced state.

that LSD1/KIAA0601 could be a nuclear polyamine oxidase was ruled out. Being part of chromatin remodeling complexes, a fascinating hypothesis was that this enzyme could be the long-searched-for histone demethylase. By using peptides corresponding to the N-terminal tails of human histones, LSD1/KIAA0601 was eventually demonstrated to specifically catalyze the removal of methyl groups from mono- and dimethylated Lys4 of histone H3 [16, 17].

B. ENZYMOLOGY OF LSD1

The characteristic spectroscopic properties of LSD1 provide an excellent experimental tool for monitoring the enzyme reactivity. Flavin-dependent amine oxidases catalyze the oxidative cleavage of the α-CH bond of the substrate to form an imine intermediate with the concomitant reduction of the flavin cofactor, which can be detected by monitoring the changes of the visible absorption spectrum of the oxidized FAD [23]. The first step in the LSD1 demethylation reaction is the flavin-mediated two-electron oxidation of the methylated lysine, which leads to a stable two-electron reduced FAD cofactor and formation of the postulated imine compound (Figures 8.1 and 8.3) [17]. It remains to be seen whether imine hydrolysis to yield the final product is catalyzed by the enzyme with release of the demethylated peptide, or hydrolysis is a non-enzymatic process that takes place in solution after dissociation of the enzyme-imine complex.

An intriguing question about LSD1 function concerns the nature of the physiological substrate, which acts as an electron acceptor to complete the catalytic cycle and reoxidize the flavin cofactor. Exposure of the reduced enzyme to air leads to a slow reoxidation of the cofactor, showing that molecular oxygen can act as the electron acceptor of this half-reaction [17]. However, the oxidase activity of LSD1 is not especially pronounced if compared to the reactivity of other flavin-dependent oxidases [24]. Moreover, this process produces hydrogen peroxide, which in the chromatin environment might be harmful by causing oxidative damage to DNA. It has been shown that the reduced flavin cofactor of LSD1 can also be reoxidized by ferricenium, a synthetic electron acceptor [17]. An important feature in this reaction is that the ferricenium-mediated oxidation of LSD1 is essentially instantaneous, clearly differing from the slower oxidation process that uses molecular oxygen as substrate. These data lead to the hypothesis that *in vivo* enzyme reoxidation may not be an oxygen-mediated process but it may involve alternative (and as yet unidentified) electron acceptors.

Three different spectrophotometric methods have been developed to assay LSD1 activity by using peptides corresponding to the N-terminal 21-amino acyl residues of histone H3 mono- or dimethylated on Lys4 [16, 17]. These assays are based on different events related to the demethylation reaction: (1) the formaldehyde dehydrogenase coupled assay [25], (2) the peroxidase-coupled reaction, and (3) ferricenium used as synthetic electron acceptor to assay for the activity of flavin-dependent oxidoreductases [28]. In regard to the formaldehyde dehydrogenase coupled assay, LSD1 catalyzes the release of formaldehyde during the demethylation reaction [16]. This property has been exploited to set up a convenient spectrophotometric assay of the enzyme. In the assay, formaldehyde produced in the reaction is converted by formaldehyde dehydrogenase to formic acid using NAD^+ as electron acceptor, whose reduction to NADH can be spectrophotometrically measured at 340 nm.

In the peroxidase-coupled reaction, LSD1 exhibits an oxidase activity that generates hydrogen peroxide. The latter can be detected by a horseradish peroxidase coupled assay [26]. In the presence of peroxidase and H_2O_2, 4-amino-antipyrine and 3,5-dichloro-2-hydroxybenzene sulphonic acid are oxidized, producing a red chromophore whose concentration can be continuously measured by evaluating its absorbance at 515 nm. By using Amplex Red reagent (10-acetyl-3,7-dihydroxyphenoxazine, from Molecular Probes) instead of 4-amino-antipyrine and 3,5-dichloro-2-hydroxybenzene sulphonic acid, the sensitivity of the assay is increased more than twofold. The red product, resorufin, can be detected either fluorometrically or spectrophotometrically with very low background interference [27].

Ferricenium has been often used as synthetic electron acceptor to assay for the activity of flavin-dependent oxidoreductases [28]. This compound has a characteristic absorption spectrum that changes upon reduction to ferrocene in a one-electron reaction that can be conveniently monitored by UV/Vis absorption spectroscopy.

The results obtained from these assays are consistent with one another, which indicates that there are no artifacts induced by direct interaction between LSD1 and any of the reagents. By using these assays, activity has been detected on the full-length protein [16] as well as on mutants of LSD1 deprived of the first 150 amino acids [17].

C. SUBSTRATE SPECIFICITY

LSD1 does not exhibit any detectable activity with substrates of polyamine oxidase and monoamine oxidase, or with methyllysine, trimethylamine, methylglycine, and methylarginine [17]. This enzyme is also not responsible for the activity that was measured by Paik et al. [14], who first reported on a demethylase acting on histones as well as on free methyllysine. LSD1 is also inactive on trimethylated H3-K4 [16], which is consistent with the chemical nature of the amine oxidation reaction that requires a free lone pair of electrons on the nitrogen atom (Figure 8.1). Moreover, no activity has been detected on other methylated lysines of histones H3 and H4, which are linked to both transcription activation (H3-K36 and H3-K79) and repression (H3-K9, H3-K27, and H4-K20) [16]. These data lead to the conclusion that this enzyme is strictly specific for lysine 4 of histone H3.

IV. LSD1 Is Part of Many Multiprotein Corepressor Complexes

LSD1 has been typically found in association with CoREST (a corepressor protein) and histone deacetylases 1 and 2 forming a highly conserved core complex (Figure 8.4) [18]. These proteins have been shown to be part of several megadalton corepressor complexes [19, 29], which are proposed to operate in the context of a stable and extended form of repression through silencing of entire chromatin domains [20, 30]. The LSD1/CoREST/HDAC1-2 corepressor module is recruited by the neural receptor silencing factor REST/NRSF for the gene expression regulation of neural genes in non-neural cells [31] and mediate plasticity of neuronal gene chromatin throughout neurogenesis [32]. Furthermore, some LSD1-containing corepressor complexes include oncogenes/tumor suppressors, suggesting an involvement of LSD1 in cancer [20, 33]. Recently, by RNAi-based genetic screens deletion mutants in REST/NRSF gene have been found in colorectal cancer cells, which implicates REST as a human tumor suppressor [34].

The fact that LSD1 exists in corepressor complexes raises the hypothesis that the enzyme may play a direct role in transcriptional repression. Indeed, LSD1 has been shown to function as a repressor when directed to a target promoter (Figure 8.5) [16]. The protein lacking the C-terminal amine oxidase domain is not active on methylated H3-K4 and does not show repression activity (Figure 8.5), which suggests that repression mediated by LSD1 requires the C-terminal amine oxidase domain and is related to its demethylase activity [16].

A search in the sequence databases showed that orthologs of LSD1 are found in different eukaryotes, suggesting that this enzyme is evolutionarily conserved. In particular, homologs of both LSD1 and CoREST exist in *C. elegans* [21] and have been found to suppress genes coding for presenilins, which in humans are correlated to Alzheimer's disease. In *A. thaliana*, mutations in a gene coding for a protein 58% identical to LSD1 cause a delay in flowering time, suggesting a role in plant growth [35].

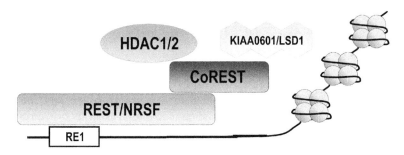

FIG. 8.4. Model of the core corepressor complex composed of CoREST, HDAC1/2, and LSD1, which is recruited by REST/NRSF and involved in different transcriptional repression events.

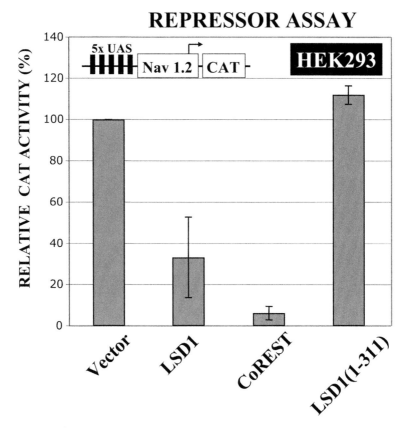

FIG. 8.5. Transcriptional corepression mediated by LSD1. LSD1 exhibits repressor activity when fused to Gal4-DBD and the flavin domain is required for this activity. LSD1 expression vector, a mutant expressing only the LSD1 N-terminal domain (1-311) and the empty vector were transfected into HEK293 cells together with the UAS Nav1.2 chloramphenicol acetyltransferase reporter gene. CoREST repressor activity is shown as control.

V. Other Proposed Mechanisms for Histone Lysine Demethylation

The histone demethylation reaction catalyzed by LSD1 is highly specific. This high specificity suggests that additional histone demethylases are yet to be identified, which would catalyze demethylation reactions at other modified residues on different histones associated with either transcription activation or repression. Based on sequence homology search, at least 10 LSD1-related amine oxidases can be identified in the human genome, and only one is strongly similar in sequence to LSD1. By contrast, there are more than 50 histone methyltransferases that carry out methylation of lysine residues at the N-termini of histones H3 and H4.

Several mechanisms have been proposed for the enzymatic removal of methyl groups from histone lysines. Theoretically, this reaction can be triggered by hydroxylation. Enzymes such as 2-oxoglutarate-dependent dioxygenases can hydroxylate their substrates, leading to demethylation. Enzymes of this class have been shown to chemically remove stable alkylation damage of DNA via AlkB dioxygenases, and might function also for methylated proteins or histone substrates [36, 37]. Histone demethylation mediated by Epe1, a recently identified member of this class of enzymes that affects chromatin integrity, has been proposed [38]. Hydroxylation of the methyl group of mono-, di-, or trimethyllysine would produce an unstable carbinolamine intermediate, which would spontaneously release one of the methyl groups from the modified amino acid residue with the production of formaldehyde (Figure 8.6). Another proposed mechanism is based on radical N-methyl demethylation, which was suggested as a potential demethylation activity of the elongation protein 3 (Elp3) [39]. This protein has sequence similarity to an enzyme superfamily that uses S-adenosyl-L-methionine (SAM) in radical reactions (Figure 8.7). It is interesting to point out that these proposed enzymes would also have the potential to remove trimethylation marks on histone lysines by direct oxidation of the methyl carbon, unlike LSD1 and other putative flavin-dependent demethylases.

An alternative mechanism for the removal of methylated histone residues may simply occur passively through post-replication chromatin assembly and replacement of old methylated histones with "new" unmethylated ones (Figure 8.8). The histone variants (H3.3 and CenH3 in eukaryotes) are the most likely candidates for a replacement-mediated methyl turnover. These histones are almost identical to their standard counterparts except for very few amino acid substitutions (usually at the N-terminal tails, where most of the posttranslational modifications are located).

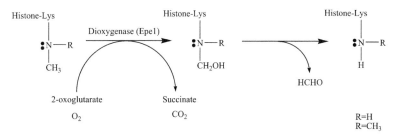

FIG. 8.6. Proposed mechanism for hydroxylation-mediated demethylation [38]. This mechanism would allow for the demethylation of mono-, di-, and trimethylated histone lysines.

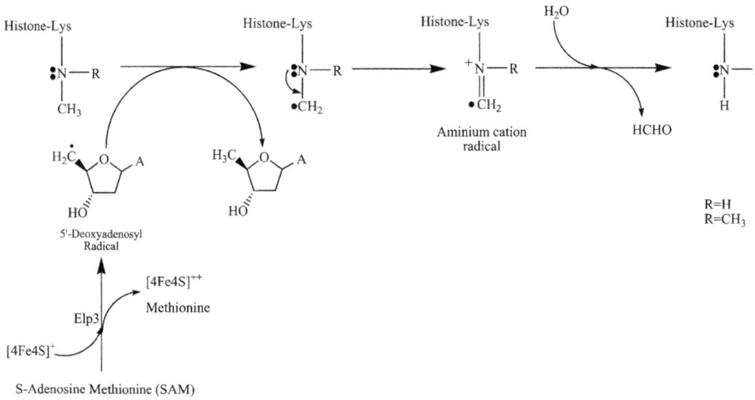

FIG. 8.7. Proposed mechanism for radical N-Methyl demethylation induced by 5′-deoxyadenosyl radical (A stands for adenine) produced by the elongation protein 3 Elp3 [39]. In the mechanism proposed by Chinenov et al., the reaction is triggered by an active site base abstracting a proton and the oxidized FeS cluster is assumed to be regenerated by accepting the electron from the aminium cation radical.

Because the deposition of canonical histones and histone variants occurs during different phases of the cell cycle, the methylation state may be simply regulated by exchange between different variants of the same histone class. For example, there is evidence that histone H3.3 is present only where transcription is ongoing. This suggests that histone H3 replacement may work as a demethylation mechanism during transcription [40].

Another potential mechanism for removing methylation marks from histone tails would require proteolytic processing of histone tails. Histone N-terminal tails are believed to be exposed to solvent and labile to proteolysis [41], and portions of these tails are known to be clipped at precise phases in the cell cycle or at specific stages of development [42]. Alternatively, the clipped histone may not be the final product of the process, but may represent the initial signaling event for the subsequent replacement by a histone variant. In this case, the histone that has to be substituted by a variant would be recognized by the replacement machinery through its clipped N-terminus (Figure 8.8).

VI. Conclusions

The discovery of the first histone lysine demethylase brought a revolution in the concept of lysine methylation, so far considered a static and permanent epigenetic mark. LSD1 specifically demethylates mono- and dimethylated H3-K4,

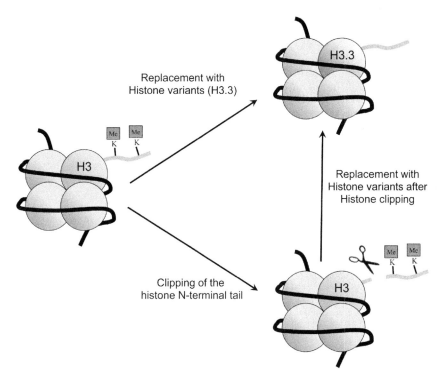

FIG. 8.8. Proposed mechanisms as alternatives to enzymatic histone demethylation: histone replacement, histone clipping, and replacement of targeted histones recognized by their clipped N-terminus (figure adapted from Bannister, 2002 [12]).

which represent only one of the wide range of methylation patterns that constitute the histone code. In this respect, one of the challenges for the future will be the determination of the 3D structure of this enzyme to study the active site details responsible for such specificity.

A large amount of work has to be done to unravel the molecular mechanisms that control histone demethylation. Concerning LSD1, experiments are ongoing to study in more detail electron acceptors, substrate specificity, and formaldehyde neutralization. Moreover, a search for other putative flavin-dependent histone demethylases will increase the understanding of the mechanisms of action and functions of such enzymes. The discovery of LSD1 has also stimulated research on other demethylating enzymes and on alternative mechanisms involved in the regulation of this histone posttranslational modification in the context of chromatin remodeling. It will also be of interest to understand how these different pathways cooperate to accomplish various cellular processes relative to other epigenetic marks.

REFERENCES

1. Luger, K., Mader, A.W., Richmond, R.K., Sargent, D.F., Richmond, T.J. (1997). Crystal structure of the nucleosome core particle at 2.8 A resolution. *Nature* 389:251–260.
1a. Tsukada, Y., Fang, J., Erdjument-Bromage, H., Warren, M.E., Borchers, C.H., Tempst, P. and Zhang, Y. (2006). Histone demethylation by a family of JmjC domain-containing proteins. *Nature* 439:811–816.
2. Allfrey, V.G., Faulkner, R., and Mirsky A.E. (1964). Acetylation and methylation of histones and their possible role in the regulation of rna synthesis. *Proc Natl Acad Sci USA* 61:786–794.
3. Spotswood, H.T., and Turner, B.M. (2002) An increasingly complex code. *J Clin Invest* 110:577–582.
4. Jenuwein T., and Allis, C.D. (2001). Translating the histone code. *Science* 293:1074–1080.
5. Grunstein M. (1997). Histone acetylation in chromatin structure and transcription. *Nature* 389:342–352.
6. Biel, M., Wascholowski, V., and Giannis, A. (2005). Epigenetics–an epicenter of gene regulation: Histones and histone-modifying enzymes. *Angew Chem Int Ed* 44:3186–3216.
7. Byvoet, P. (1972). *In vivo* turnover and distribution of radio-N-methyl in arginine-rich histones from rat tissues. *Arch Biochem Biophys* 152:887–888.
8. Duerre, J.A., and Lee, C.T. (1974). *In vivo* methylation and turnover of rat brain histones. *J Neurochem* 23:541–547.
9. Borun, T.W., Pearson, D., and Paik, W.K. (1972). Studies of histone methylation during the HeLa S-3 cell cycle. *J Biol Chem* 247:4288–4298.
10. Annunziato, A.T., Eason, M.B., and Perry, C.A. (1995). Relationship between methylation and acetylation of arginine-rich histones in cycling and arrested HeLa cells. *Biochemistry* 34:2916–2924.
11. Katan-Khaykovich, Y., and Struhl, K. (2005). Heterochromatin formation involves changes in histone modifications over multiple cell generations. *EMBO J* 24:2138–2149.
12. Bannister J., Schneider, R., and Kouzarides, T. (2002). Histone methylation: dynamic or static? *Cell* 109:801–806.
13. Kim, S., Benoiton, L., and Paik, W.K. (1964). ε-Alkyllysinase. Purification and properties of the enzyme. *J Biol Chem* 239:3790–3796.
14. Paik, W.K., and Kim, S. (1973). Enzymatic demethylation of calf thymus histones. *Biochem Biophys Res Commun* 51:781–788.
15. Paik, W.K., and Kim, S. (1974). ε-Alkyllysinase. New assay method, purification, and biological significance. *Arch Biochem Biophys* 165:369–378.
16. Shi, Y., Lan, F., Matson, C., Mulligan, P., Whetstine, J.R., Cole, P.A., Casero, R.A., and Shi, Y. (2004). Histone demethylation mediated by the nuclear amine oxidase homolog LSD1. *Cell* 119:941–953.
17. Forneris F., Binda, C., Vanoni, M. A., Mattevi, A., and Battaglioli, E., (2005). Histone demethylation catalysed by LSD1 is a flavin-dependent oxidative process. *FEBS Lett* 579:2203–2207.
18. Humphrey, G.W., Wang, Y., Russanova, V.R., Hirai, T., Qin, J., Nakatani, Y., and Howard, B.H. (2001). Stable histone deacetylase complexes distinguished by the presence of SANT domain proteins CoREST/kiaa0071 and Mta-L1. *J Biol Chem* 276:6817–6824.
19. Ballas, N., Battaglioli, E., Atouf, F., Andres, M.E., Chenoweth, J., Anderson, M.E., Burger, C., Moniwa, M., Davie, J.R., Bowers, W.J., Federoff, H.J., Rose, D.W., Rosenfeld, M.G., Brehm, P. and Mandel, G. (2001). Regulation of neuronal traits by a novel transcriptional complex. *Neuron* 31:353–365.
20. Shi, Y., Sawada, J., Sui, G., Affar, B., Whetstine, J.R., Lan, F., Ogawa, H., Luke, M.P., Nakatani, Y. and Shi, Y. (2003). Coordinated histone modifications mediated by a CtBP corepressor complex. *Nature* 422:735–738.

21. Eimer, S., Lakowski, B., Donhauser, R. and Baumeister, R. (2002). Loss of spr-5 bypasses the requirement for the C.elegans presenilin sel-12 by derepressing hop-1. *EMBO J* 21:5787–5796.

22. Aravind, L., and Iyer, L.M. (2002). The SWIRM domain: A conserved module found in chromosomal proteins points to novel chromatin-modifying activities. *Genome Biol* 3:RESEARCH0039.

23. Binda, C., Mattevi, A., and Edmondson, D.E. (2002). Structure-function relationships in flavoenzyme-dependent amine oxidations: A comparison of polyamine oxidase and monoamine oxidase. *J Biol Chem* 277:23973–23976.

24. Massey, V. (1995). Introduction: Flavoprotein structure and mechanism. *FASEB J* 9:473–475.

25. Lizcano, J.M., Unzeta, M., and Tipton, K.F. (2000). A spectrophotometric method for determining the oxidative deamination of methylamine by the amine oxidases. *Anal Biochem* 286:75–79.

26. Smith, T.A., and Barker, J.H. (1988). The di- and polyamine oxidase of plants, in V. Zappia and A. E. Pegg (eds.). *Progress in Polyamine Research: Novel Biochemical, Pharmacological and Clinical Aspects*, pp. 573–589, New York: Plenum Press.

27. Gutheil, W.G., Stefanova, M.E., and Nicholas, R.A. (2000). Fluorescent coupled enzyme assays for D-alanine: application to penicillin-binding protein and vancomycin activity assays. *Anal Biochem* 287:196–202.

28. Lehman, T.C., and Thorpe, C. (1990). Alternate electron acceptors for medium-chain acyl-CoA dehydrogenase: Use of ferricenium salts. *Biochemistry* 29:10594–10602.

29. Battaglioli, E., Andres, M.E., Rose, D.W., Chenoweth, J.G., Rosenfeld, M.G., Anderson, M.E. and Mandel, G. (2002). REST repression of neuronal genes requires components of the hSWI.SNF complex. *J Biol Chem* 277:41038–41045.

30. Lunyak, V.V., Burgess, R., Prefontaine, G.G., Nelson, C., Sze, S.H., Chenoweth, J., Schwartz, P., Pevzner, P.A., Glass, C., Mandel, G., and Rosenfeld, M.G. (2002). Corepressor-dependent silencing of chromosomal regions encoding neuronal genes. *Science* 298:1747–1752.

31. Andres, M.E., Burger, C., Peral-Rubio, M.J., Battaglioli, E., Anderson, M.E., Grimes, J., Dallman, J., Ballas, N., and Mandel, G. (1999). CoREST: A functional corepressor required for regulation of neural-specific gene expression. *Proc Natl Acad Sci USA* 96:9873–9878.

32. Ballas, N., Grunseich, C., Lu, D.D., Speh, J.C., and Mandel, G. (2005). REST and its corepressors mediate plasticity of neuronal gene chromatin throughout neurogenesis. *Cell* 121:645–657.

33. You, A., Tong, J.K., Grozinger, C.M., and Schreiber, S.L. (2001). CoREST is an integral component of the CoREST human histone deacetylase complex. *Proc Natl Acad Sci USA* 98:1454–1458.

34. Westbrook, T.F., Martin, E.S., Schlabach, M.R., Leng, Y., Liang, A.C., Feng, B., Zhao, J.J., Roberts, T.M., Mandel, G., Hannon, G.J., Depinho, R.A., Chin, L., and Elledge, S.J. (2005). A genetic screen for candidate tumor suppressors identifies REST. *Cell* 121:837–848.

35. He, Y.H., Michaels, S.D., and Amasino, R.M. (2003). Regulation of flowering time by histone acetylation in Arabidopsis. *Science* 302:1751–1754.

36. Falnes, P.O., Johansen, R.F., and Seeberg, E. (2002). AlkB-mediated oxidative demethylation reverses DNA damage in *Escherichia coli*. *Nature* 419:178–182.

37. Trewick, S.C., Henshaw, T.F., Hausinger, R.P., Lindahl, T., and Sedgwick, B. (2002). Oxidative demethylation by *Escherichia coli* AlkB directly reverts DNA base damage. *Nature* 419:174–178.

38. Trewick, S.C., McLaughlin, P.J., Allshire, R.C. (2005). Methylation: lost in hydroxylation? *EMBO Rep* 6:315–320.

39. Chinenov, Y. (2002). A second catalytic domain in the Elp3 histone acetyltransferases: A candidate for histone demethylase activity? *Trends Biochem Sci* 27:115–117.

40. Yu, L., and Gorovsky, M.A. (1997). Constitutive expression, not a particular primary sequence, is the important feature of the H3 replacement variant hv2 in *Tetrahymena thermophila*. *Mol Cell Biol* 17:6303–6310.

41. Allis, C.D., Bowen, J.K., Abraham, G.N., Glover, C.V.C., and Gorovsky, M.A. (1980). Proteolytic processing of histone H3 in chromatin: A physiologically regulated event in *Tetrahymena micronuclei*. *Cell* 20:55–64.

42. Lin, R., Cook, R.G., and Allis, C.D. (1991). Proteolytic removal of core histone amino termini and dephosphorylation of histone H1 correlate with the formation of condensed chromatin and transcriptional silencing during *Tetrahymena* macronuclear development. *Genes Dev* 5:1601–1610.

Part III

*Biological Regulation by Protein
Methyl Ester Formation*

9

Structure and Function of Isoprenylcysteine Carboxylmethyltransferase (Icmt): A Key Enzyme in CaaX Processing

JESSICA L. ANDERSON • CHRISTINE A. HRYCYNA

Department of Chemistry
Purdue University
560 Oval Drive
West Lafayette, IN 47907, USA

I. Abstract

Proteins that terminate in a C-terminal CaaX motif undergo three sequential posttranslational modifications: isoprenylation of the cysteine residue, endoproteolysis of the –aaX residues, and methylation of the isoprenylated cysteine by an isoprenylcysteine carboxylmethyltransferase (Icmt). Among the proteins that contain this CaaX sequence are the Ras superfamily of G-proteins and other signal transduction proteins, the nuclear lamins, and the yeast **a**-factor mating pheromone. Icmt is a mechanistically intriguing enzyme that is unique among *S*-adenosyl-L-methionine (SAM)-dependent methyltransferases in that it is an integral membrane enzyme localized to the endoplasmic reticulum. It must also accommodate chemically diverse cofactor and substrate molecules; namely, the hydrophilic methyl-donating cofactor SAM and a lipophilic isoprenylated protein substrate, respectively.

245

THE ENZYMES, Vol XXIV
Copyright © 2006 by Elsevier Inc.
All rights of reproduction in any form reserved.

A complete picture of the cellular consequences of carboxylmethylation of isoprenylated proteins is still emerging. It is known, however, that methylation is critical for the proper localization of the CaaX protein Ras and may be essential for oncogenic transformation. Thus, Icmt may prove to be an excellent chemotherapeutic target. In this review, the structure and function of the family of Icmt enzymes are discussed, as well as recent advances in the development of small-molecule inhibitors of this key enzyme in the CaaX posttranslational processing pathway.

II. Introduction

The posttranslational modifications of eukaryotic proteins have profound influences on their localization and function. Many eukaryotic proteins are modified at their C-terminus by key modifications, including the addition of an isoprenyl lipid moiety, proteolysis, and methylation. The consensus sequence for isoprenylation, proteolysis, and methylation is a C-terminal –CaaX sequence, where C is cysteine, a is an aliphatic residue, and X can be one of several amino acids [1–4]. The C-terminal CaaX sequence undergoes a series of three sequential posttranslational modifications, including isoprenylation of the cysteine residue by either a C_{15}-farnesyl or C_{20}-geranylgeranyl moiety, endoproteolysis of the three terminal amino acids (–aaX), and methylesterification of the α-carboxyl group on the isoprenylated cysteine [1–4] (Figure 9.1). Examples of CaaX proteins include the small GTP-binding proteins such as the Ras and Rho proteins, the γ subunits of heterotrimeric guanine nucleotide binding proteins, the nuclear lamins, several protein phosphatases, and numerous phosphodiesterases [2, 4, 5].

Several physiological roles have been proposed for this trio of modifications. Because they give otherwise soluble proteins a greater hydrophobic character, lipidation and methylation are thought to help guide these polypeptides to their functional sites on intracellular membranes [1–3, 6–13]. Additionally, the methylation reaction has been shown to protect certain precursor polypeptides from further proteolytic digestion once the three terminal amino acids are removed [3, 4, 14, 15]. In some cases, these modifications may also serve as recognition signals for protein-protein interactions, including specific receptor proteins, or may activate cellular proteins or peptides [1, 3, 4, 16]. Many of the modified proteins are involved in signal transduction pathways, and these modifications have been suggested to play a key role in regulating their activities.

The first reaction in the posttranslational CaaX processing pathway involves attachment of a C_{15}-farnesyl or C_{20}-geranylgeranyl moiety via a thioether linkage to the cysteine residue (Figure 9.1). This reaction is carried out by a specific soluble cytosolic isoprenyltransferase, which is either farnesyltransferase or geranylgeranyltransferase depending on the identity of the C-terminal residue (X of the CaaX sequence) [1, 5, 17–20]. The second reaction, endoproteolysis, is performed

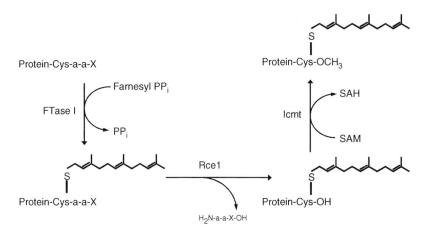

FIG. 9.1. Schematic diagram of the sequential C-terminal posttranslational modifications of eukaryotic CaaX proteins. In this example, farnesylation, proteolysis, and methylation of a model CaaX protein are depicted. FTase I is the farnesyltransferase, Rce1 is the ER membrane-associated endoprotease, and Icmt is the isoprenylcysteine carboxylmethyltransferase. In CaaX proteins that are geranylgeranylated, geranylgeranyltransferase I (GGTase I) is responsible for the initial isoprenylation.

by a membrane-associated isoprenylcysteine-specific endoprotease and takes place on the cytosolic face of the endoplasmic reticulum (ER) [21–23]. The final modification, (α-carboxylmethylesterification of the newly exposed cysteine carboxyl group), is performed by an isoprenylcysteine carboxylmethyltransferase (Icmt), also an integral membrane protein localized to the ER [12, 14, 24–30]. This methyltransferase is the subject of this chapter.

Icmt is a mechanistically intriguing enzyme comprised of multiple membrane spans. It must accommodate chemically diverse methyl donor and acceptor molecules: the hydrophilic cofactor S-adenosyl-L-methionine (SAM) and a lipophilic isoprenylated protein substrate, respectively. Kinetic assays have shown that Icmt follows an ordered Bi Bi mechanism, with SAM binding first and followed by the isoprenylated substrate [31, 32]. Following methylation, the substrate leaves, followed by S-adenosyl-L-homocysteine (SAH) [31, 32]. Strikingly, Icmt shares none of the conserved consensus sequences described for known soluble (protein or nucleic acid) methyltransferases [33], suggesting that Icmt may have an interesting new SAM binding site and novel isoprene binding site that must necessarily be proximal to each other to facilitate the reaction.

In the late 1980s, two pieces of existing evidence pointed to the possibility of a novel posttranslational processing pathway that included a new type of methylation reaction, originally proposed by Clarke et al. [12]. First, the peptide mating pheromones from the jelly fungi *Tremella mesenterica* [34] and *Tremella brasiliensis* [35] were shown to contain an isoprenylated and methylated C-terminal cysteine residue. Furthermore, it was demonstrated that the C-terminal residue of **a**-factor was a farnesylated cysteine that was carboxylmethylated [36, 37]. Second, inspection of the **a**-factor peptide sequence as well as those of numerous mammalian Ras proteins derived from the cDNAs showed that these proteins are initially synthesized with three additional amino acids following the cysteine [38, 39]. It was also noted that the proteins modified by Ram1p, a component of the farnesyltransferase in *S. cerevisiae*, contained a CaaX motif in their predicted amino acid sequences [9].

In their groundbreaking study, Clarke et al. presented evidence that H-Ras is α-carboxylmethylated at the C-terminal cysteine residue [12]. Numerous proteins that contain the CaaX motif were subsequently reported to be similarly modified, including the nuclear lamins [10, 40–43], the yeast and mammalian Ras family of proteins [6, 8, 12, 25, 44–47], related small G proteins [27, 48–52], the α and β subunits of cGMP phosphodiesterases [53–55], and several G-protein γ subunits [56–60]. To prove the existence of methyl esterified proteins in bovine retinal rods, a group of 23- to 29-kDa proteins were proteolyzed and oxidized to yield free cysteic acid methyl ester [61]. This method was also used to show that bovine retinal rod cGMP phosphodiesterase, the G protein G25K, and the γ subunit of heterotrimeric G proteins also had carboxylmethylated C-terminal cysteine residues [55, 58, 60]. In many cases, the methylation of G proteins was found to be enhanced in the presence of GTPγS [4]. This effect is likely due to more efficient methylation of activated G proteins, which further implicates a role for Icmt enzymes in the regulation of G-protein activity. Together, these reports supported the proposal that the CaaX sequence of these proteins directed proteolysis, lipidation, and methylation of these proteins [9, 12].

III.   Characterization of Icmt Proteins

A.   Ste14p from *S. cerevisiae:* The Founding Member of the Icmt Family

To date, the most fully characterized member of the Icmt family is Ste14p from *S. cerevisiae*. Genetic analysis of mating and sterile mutants in *Saccharomyces cerevisiae* allowed us to identify Ste14p as the cellular component implicated in the carboxylmethylation of CaaX proteins [24]. We demonstrated that membrane extracts from a strain overexpressing the wild-type *STE14* gene can catalyze the methylation of the minimal isoprenylated substrate *N*-acetyl-*S*-farnesyl-L-cysteine (AFC) and the synthetic peptide *S*-farnesyl-LARYKC *in vitro* in a concentration-dependent manner, whereas membrane extracts from a *ste14* mutant are unable to do so [24, 25]. This reaction was also shown to require S-adenosyl-L-methionine (SAM) as a cofactor with a K_m of approximately 10 to 20 μM [62]. These data provided the first evidence that the Ste14p protein was the source of C-terminal methyltransferase activity and that this activity was dependent on the concentration of an isoprenylated acceptor substrate. We performed subsequent studies *in vivo* that supported this hypothesis and revealed that Ras1p, Ras2p, and the small G protein G25K were cellular substrates of Ste14p [25, 63].

In addition, we showed that the yeast mating peptide **a**-factor was also an *in vivo* substrate for Ste14p. **a**-factor terminates with a CaaX motif and undergoes posttranslational processing, resulting in a secreted dodecapeptide with a C-terminal cysteine that is farnesylated and α-carboxylmethylated [64]. Defects in the synthesis of **a**-factor cause a detectable MATa cell-specific sterile phenotype and lack of an **a**-factor halo [65]. The lack of a halo is due to the fact that unmethylated **a**-factor is not exported out of the cell by the ABC transporter Ste6p [14]. These data lend credence to the hypothesis that the methyl group serves as a critical determinant for cellular protein-protein interactions. Furthermore, in the *ste14* mutant strain **a**-factor was found to be degraded more rapidly than in the wild-type parental strain, lending evidence to the hypothesis that the methyl group protects against intracellular degradation [14]. In addition, mutant *ste14* strains were found to produce an **a**-factor species identical to base demethylated **a**-factor by thin-layer chromatography [26]. Together, these results clearly indicated that Ste14p plays a critical role in carboxyl methylation and provided the preliminary evidence that Ste14p is the sole component of the C-terminal methyltransferase activity.

The studies described previously suggested that Ste14p is necessary for carboxyl methylation, but do not identify whether it is sufficient for Icmt activity. For example, Ste14p may have been a component of a multimeric enzyme or may simply have been a regulator of methyltransferase activity. To address this issue, we expressed *STE14* as a fusion protein (*TrpE-STE14*) in *E. coli*, an organism that does not have Icmt activity, and showed that membrane extracts prepared from the fusion-expressing *E. coli* strain can mediate methylation of a synthetic farnesylated peptide *in vitro* [63]. Those results provided further compelling evidence that *STE14* encodes the sole component of the Icmt enzyme. Most recently, as ultimate proof that Ste14p functions as a methyltransferase on its own, we have purified Ste14p from yeast and functionally reconstituted its activity in liposomes [66].

To find other substrates for Ste14p, methylation patterns were examined in wild-type and Δ*ste14* deletion strains [67]. Ste14p methyl-accepting substrates were found at 49, 38, 33, 31, and 26 kDa [67]. These methyl-accepting substrates were found solely in the cytosolic fraction, and the methylated forms of all but the 49-kDa species were found solely in the membrane fraction, suggesting that methylation by Ste14p could cause translocation of the methyl-accepting substrates from the cytosol to the cell membranes [67]. Some of the 49-kDa species remained methylated in a Δ*ste14* deletion strain, implicating another methyltransferase activity that methylates a protein at this molecular weight. Very little carboxylmethylation of substrates occurred upon incubation of wild-type membranes, cytosol and [³H-*methyl*]-SAM, suggesting that most of the Ste14p substrates are methylated *in vivo* [67].

B. SEQUENCE ANALYSIS, LOCALIZATION, AND TOPOLOGY OF STE14P

The *STE14* gene, located on chromosome IV, encodes a 239-amino-acid protein with a molecular weight of ~28 kDa [14, 68]. Hydropathy analysis of the amino acid sequence led to a prediction of five or six membrane-spanning segments [14, 68], and biochemical data confirmed localization to the membrane fraction of cells [24, 25, 30, 63]. The localization of Ste14p to the endoplasmic reticulum membrane was subsequently determined by immunofluorescence experiments using a polyclonal anti-Ste14p antibody that recognizes a cytosolic epitope [30].

A more precise topological map of Ste14p was determined by protease protection assays [69]. The 2D hydropathy model reported contains six transmembrane segments with both the N- and C-termini on the cytosolic side of the endoplasmic reticulum and the bulk of the soluble portion of Ste14p cytosolically disposed (Figure 9.2). The model also shows a stretch of 31 amino acids surprisingly

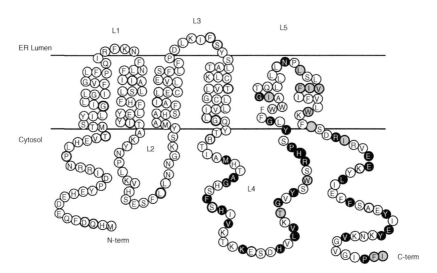

FIG. 9.2. The six-transmembrane segment topology model of Ste14p. In this model [69], the N- and C-termini extend into the cytosol. Transmembrane segments 5 and 6 are proposed to form a helix-turn-helix ("helical hairpin") within the membrane. The conserved residues across the family of Icmts in the C-terminal portion of the protein, defined here as residues 140 through 239, are indicated. Identical residues are black and similar residues are gray. (*Reprinted from* Molecular Biology of the Cell *with the permission of The American Society for Cell Biology [69].*)

predicted to span the membrane twice [69]. In that it usually requires at least 20 amino acids to span the cell membrane once, the possibility that this span (which falls between loop 4 and the C-terminus) contained a helical hairpin was explored. An asparagine-proline pair is located near the middle of the span and these two amino acids have the highest propensity to induce a helical hairpin [70, 71]. Upon mutation of these two residues to leucine, which has a low hairpin-inducing propensity, there was partial translocation of the C-terminus of Ste14p from the cytosol to the lumen of the endoplasmic reticulum. These data support the helical hairpin hypothesis and the 2D topology model [69].

C. CONSENSUS SEQUENCES COMMON TO STE14P AND ITS HOMOLOGS: IMPORTANCE OF THE C-TERMINAL DOMAIN

Ste14p and other members of the Icmt family of methyltransferases are the only major class of membrane-associated methyltransferases described to date. Importantly, none of the four sequence motifs described for known soluble methyltransferases exist in the Ste14p sequence [14, 33, 68], suggesting a novel mode for SAM binding and isoprenylated substrate recognition. Numerous members of the Icmt family have now been sequenced. Alignments of these sequences reveal important regions throughout the length of the protein of possible functional importance (Figure 9.3). The functional importance of conserved amino acids within the Icmt family is highlighted by the finding that mutational analysis identified a number of loss-of-enzymatic-function mutants (G31E, L81F, G132R, S148F, P173L, E213D, E213Q, E214D, and L217S), which lie in residues highly conserved among the Icmt family of enzymes [69].

Interestingly, the mutation of glutamate to aspartate at position 213 or 214 is conservative but results in loss of enzyme activity, suggesting a very important role for these amino acids in Ste14p catalysis. Several of these variants (G31, G132, P173, and E213) completely abrogate not only enzymatic activity but mating and the ability to generate a halo, suggesting that these residues in particular are crucial for the activity or the structural integrity of the Icmt enzymes. The G31E mutation is at the beginning of what is possibly a dimerization domain (GXXXGXXXG), similar to that found in the single-transmembrane protein glycophorin A [72, 73], which may affect Ste14p function by preventing dimerization/oligomerization rather than catalysis directly.

Upon closer inspection of the alignment of the Icmt family shown in Figure 9.3, a novel tripartite consensus motif in the C-terminal half of Ste14p from residues V123 through I239 was identified [69]. This consensus motif consists of region A (LVxxGxYxxxRHPxYxG), followed by a stretch of 30 hydrophobic amino acids that are proposed to make transmembrane segments 5 and 6 and region B (xRxxxEExxLxxxFGxxxxEYxxxVxxxxP) [69]. This motif, which is similar among all members of the Icmt family, emerges as a region of particular

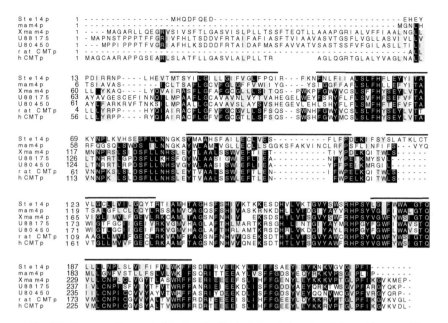

FIG. 9.3. Sequence alignment of the Icmt family depicting the location of the *ste14* mutants identified by random and site-directed mutagenesis. These are alignments of *S. cerevisiae* Ste14p, *S. pombe* mam4p, *X. laevis* Xmam4p, *C. elegans* open reading frames (accession numbers U88175 and U80450), a rat open reading frame (accession number AF0755595.1), and human Icmt (also known as pcCMTp). The bars above the sequence denote the putative transmembrane segments of Ste14p. Black boxes denote amino acid identity and gray boxes denote amino acid similarity. *(Reprinted from Molecular Biology of the Cell with the permission of The American Society for Cell Biology [69].)*

interest for several reasons. First, compared to other regions it shows the highest degree of homology between various members of the family, and from the mutational analysis described previously, several residues in this region are critical for function [69].

Mutation of residues within region A, such as the previously described P173L mutation [69] and a newly constructed R171A mutation (J. L. Anderson and C. A. Hrycyna, unpublished data), have profound effects on the function of Ste14p. Furthermore, site-directed mutagenesis of the highly conserved glutamates in region B of the consensus sequence (mutations E213D,Q and E214D) [69] and R209A (J. L. Anderson and C.A. Hrycyna, unpublished data), revealed that these residues are critical for function as assessed by *in vitro* methyltransferase assays. Second, biochemical evidence demonstrated that insertion of a hemagglutinin-epitope tag at residue 226 [30] or a 10-histidine repeat at the C-terminus [66] disrupts methyltransferase activity *in vitro*, suggesting an essential role of this C-terminal half of Icmt in catalysis or substrate recognition.

D. PURIFICATION AND FUNCTIONAL RECONSTITUTION OF STE14P

Active Ste14p was recently purified, reconstituted, and characterized by our laboratory [66]. A histidine-tagged version of Ste14p (His-Ste14p) was extracted from cell membranes with 1% *n*-dodecyl-β-D-maltoside (DDM). The enzyme retained its enzymatic function upon solubilization in this detergent, the first of its kind to retain full activity during solubilization [66]. His-Ste14p was purified to homogeneity by immobilized metal affinity chromatography (Figures 9.4a and b).

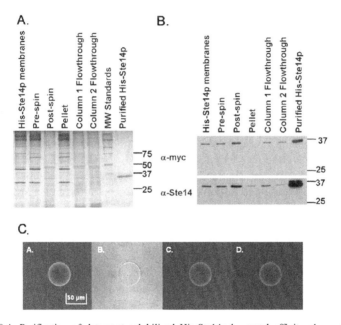

FIG. 9.4. Purification of detergent solubilized His-Ste14p by metal affinity chromatography. Crude membranes containing His-Ste14p were solubilized in 1% n-dodecyl-β-D-maltopyranoside (DDM) (pre-spin) and spun at 300,000 × g for 30 min to pellet unsolubilized protein (pellet). The resultant supernatant (post-spin) was incubated with Talon metal affinity resin (Clontech). Protein was eluted with increasing concentrations of imidazole. (A) Total protein (5 μg) from each fraction and 1 μg total purified His-Ste14p were subjected to 12% SDS-PAGE analysis and silver staining. (B) Total protein (1 μg) from each fraction and 100 ng purified His-Ste14p were subjected to 12 % SDS-PAGE analysis and Ste14p was visualized by immunodetection with either α-*myc* monoclonal antibody (1:10,000) or α-Ste14 polyclonal antibody (1:1000). (C) Confocal microscope images of purified and reconstituted His-Ste14p. Purified His-Ste14p was diluted in buffer containing 100 mM Tris-HCl pH 7.5 and *E. coli* polar lipid extract (Avanti Polar Lipids, Inc.). The sample was stained with an anti-*myc* monoclonal antibody and visualized with a FITC-conjugated goat-anti mouse secondary antibody (panel c) and with Nile Blue, a lipid stain with red fluorescence (panel d). The images were visualized at 60-x magnification. Panel b represents the nonfluorescent confocal transmission image of the vesicle and panel a is the merged image of panels b, c, and d. *(Reproduced with permission from* The Journal of Biological Chemistry *[66].)* (See color plate.)

The activity of the enzyme was greatly enhanced by reconstitution into a variety of different liposomal preparations. As further proof of reconstitution, the enzyme and the lipid were shown to colocalize by confocal microscopy (Figure 9.4c).

These accomplishments allowed for characterization of Ste14p that was not previously possible. There had long been debate as to whether Icmt enzymes preferred farnesylated or geranylgeranylated substrates. These discrepancies were likely due to the differential partitioning of farnesylated versus geranylgeranylated substrates in crude membrane preparations and to the inability of accurately quantifying the amount of substrate available to the enzyme in a reaction. In support of this hypothesis, studies with the Icmt from *T. bruceii* in membrane fractions revealed that altering the amount of membrane protein in the activity assay changed the fold specificity of the enzyme [74]. To address this problem, we performed saturation curves using AFC and AGGC as substrates using purified Ste14p reconstituted in *E. coli* bulk polar lipid liposomes [66].

Because we expected that any excess substrate would alter the kinetic results, we used an approach that removed unincorporated substrate from the liposomes using gel filtration spin column chromatography and then determined an accurate concentration of substrate in each liposomal fraction spectrophotometrically. With the purified and reconstituted enzyme, we first determined K_m values of 3.0 μM and 3.2 μM for AFC and AGGC, respectively [66]. These values are lower than those observed in crude membrane preparations (13.4 μM for AFC and 17.2 μM for AGGC), due to our ability in the purified system to accurately quantify the amount of substrate accessible to the enzyme in each reaction [66]. Upon calculation of the specificity constant ($k_{cat}/K_{m(app)}$), we determined that there is no difference between the specificity constants obtained for AFC (0.99×10^4 M^{-1}s^{-1}) or AGGC (1.0×10^4 M^{-1}s^{-1}), indicating that His-Ste14p does not prefer AGGC over AFC under our assay conditions. Purified and reconstituted His-Ste14p was also shown to methylate His-Ras2p partially purified from a Δ*ste14* deletion strain [66]. These data provided direct evidence that Ste14p can, in fact, methylate *in vivo* substrates *in vitro*.

1. Ste14p and Mammalian Icmts are Putative Metalloenzymes

Purified and reconstituted His-Ste14p was inhibited irreversibly by the hydrophobic metal chelators Zincon, which is specific for zinc or copper, and 1,10-phenanthroline [66]. These observations are in good agreement with previous studies demonstrating that a mammalian Icmt was a metalloenzyme [75]. Similar to mammalian Icmt from rat kidney, Ste14p was relatively unaffected by water-soluble chelators such as EDTA, suggesting that the hydrophobic environment of the protein prevents access to the metal ion by the more hydrophilic chelators [66].

To ascertain whether more hydrophobic chelators would be able to remove the metal from the enzyme, a series of hydrophobic compounds were attached to the metal chelator nitrilotriacetic acid (NTA) via a lysine linker [76]. When cholesterol was attached to NTA in this manner, yielding cholesterol-Lys-NTA, it was a much more potent inhibitor of Ste14p activity ($IC_{50} \approx 75$ μm) than any chelator described previously. Cholesterol alone and a non-chelating version of cholesterol-Lys-NTA did not affect His-Ste14p at concentrations up to 500 μM. These data suggest that the metal ion is located in or near a hydrophobic area of the enzyme, perhaps serving a structural rather than catalytic role. Treatment of mammalian Icmt with 1,10-phenanthroline increased the susceptibility of the enzyme to trypsin, also suggesting that the metal plays a role in stabilizing the structure of Icmt [75].

The 2D topology model of Ste14p proposed by Romano et al. [69] shows that the only three cysteine residues in the protein (at amino acids 99, 121, and 126) are located within predicted transmembrane spanning regions of the protein. To determine if these cysteine residues could possibly be important for Ste14p metal binding and activity, the three residues were all mutated to either alanine or serine to create a cysteine-less version of Ste14p. This cysteine-less Ste14p was expressed at similar levels to the wild-type protein, was enzymatically active both *in vivo* and *in vitro*, and showed similar susceptibility to the cholesterol-Lys-NTA chelator. Together, these data suggest that the cysteines are not essential for activity and that the metal ion is bound elsewhere in the protein, perhaps liganded to histidine residues [76]. The generation of this cysteine-less variant of Ste14p opens the door for numerous experiments aimed at defining important residues in the enzyme as well as determining how Ste14p is folded in the membrane.

2. Other Non-mammalian Icmt Proteins

In the late 1990s, Icmt family enzymes from other non-mammalian organisms were characterized. Icmt family enzymes were discovered in *Schizosacharomyces pombe* and *Xenopus laevis* [77]. Mam4p from *S. pombe* was responsible for the processing of the pheromone M-factor in *h⁻* cells, a role similar to Ste14p in *S. cerevisiae* a-cells. It was also determined that Mam4p was a functional homolog of Ste14p, as it was able to rescue the mating phenotype of Δ*ste14* deletion a-cells *in vivo* [30]. A similar protein from *Xenopus*, named Xmam4p, was able to complement the mating defect in a *S. pombe* Δ*mam4* deletion strain, confirming these enzymes as functional homologs [77].

An Icmt enzyme from *Leishmania donovani* was also characterized [78]. Like Ste14p, *Leishmania* Icmt was able to methylate the simple isoprenylated substrates AFC and AGGC. Icmt substrates in *Leishmania* were found at 95, 68, 46–48, 34, 23, and 14 kDa. These proteins were labeled *in vivo* by both [³H-*methyl*]-SAM and [³H]-mevalonolactone, which labels prenylated proteins. The methylation of these proteins was base labile, indicating formation of a methyl ester. As with Ste14p, GTPγS increased methylation *in vivo*, especially in the 19–26 kDa region (proteins likely to be low-molecular-weight GTPases).

Two proteins similar to Ste14p were also described in *Arabidhopsis thaliana*: *At*STE14A and *At*STE14B [79, 80]. Both proteins were able to rescue the mating defects in a Δ*ste14* deletion strain, have sequence similarity to Ste14p, and are able to methylate AFC and AGGC *in vitro*. *At*STE14B was more catalytically efficient than *At*STE14A and the two Icmt family genes were found to be differentially expressed throughout the plant [80]. A similar biochemical activity was also found in cultured tobacco cells [81].

The Icmt protein from *Trypanosoma brucei*, the protozoan parasite that causes African sleeping sickness, was characterized in 2002 and may prove to be a good therapeutic target for animals infected with the disease [74]. Unlike most other reported Icmt enzymes, *T. brucei* Icmt appears to prefer farnesylated substrates over geranylgeranylated substrates. However, this degree of specificity varied depending on the amount of membrane protein used in the activity assays [74].

3. Mammalian Icmts

Mammalian counterparts of the yeast Ste14p methyltransferase have been described biochemically and genetically [27–29, 31, 75, 82–84]. A homozygous knockout mouse ($Icmt^{-/-}$) was embryonic lethal. However, membrane extracts from fibroblasts derived from the $Icmt^{-/-}$ ES cells could be assayed *in vitro* [85]. These extracts were found to have no detectable Icmt activity, consistent with this gene encoding the sole Icmt activity in mice [83]. Importantly, using these cells Young and co-workers have determined that the methylation of K-Ras proteins is critical for oncogenic transformation [83]. A complete discussion of the genetic approaches used to assess the physiological function of Icmt and carboxylmethylation can be found in Chapter 10 by Young et al. (this volume).

Icmt, the functional human ortholog of the yeast *STE14* gene, was recently identified and cloned from a human myeloid HL60 cell line [27]. Previously designated as *pcCMT*, the *Icmt* cDNA sequence encodes a 33-kDa protein with six predicted hydrophobic segments predicted to cross the membrane seven or eight times [27]. The *Icmt* gene complements the mating defect caused by mutation or deletion of the *STE14* gene when it is expressed in yeast, indicating that the human and yeast proteins are functionally similar [27]. The human *Icmt* gene product has biochemical properties similar to the Ste14p enzyme, including specificity for isoprenylated substrates with K_m values of 7 µM for AFC and 0.6 µM for AGGC [27]. Human Icmt, like Ste14p and the rat liver enzyme [28], is also a membrane-associated protein localized to the ER membrane in cultured cell lines [27].

E. Substrate Specificity of Icmt

Numerous studies have been performed on other mammalian Icmt enzymes in order to elucidate their substrate specificities. Like their yeast counterpart, mammalian Icmts also are SAM-dependent, with K_m values ranging from

1 to 5 µM [29, 86–89]. Stephenson et al. first used the synthetic peptide LARYKC to show that Icmt in rat liver microsomes had the greatest binding affinity for substrates that were S-farnesylated or S-geranylgeranylated, with K_m values of 2.2 µM and 10.9 µM, respectively [29]. Additionally, there was no methylation of non-lipidated substrates or those modified with other unrelated lipids. It was subsequently determined that even simpler small farnesylated molecules such as AFC (N-acetyl-S-farnesyl-L-cysteine) and FTP (S-farnesylthiopropionic acid) were substrates for Icmt, suggesting that the nature of the isoprene is crucial for recognition by the enzyme and not necessarily determinants in the protein structure itself [84, 86, 89–91].

AGGC (N-acetyl-S-geranylgeranyl-L-cysteine) was also determined to be a substrate but AGC (N-acetyl-S-geranyl-L-cysteine) was not, presumably because the isoprene moiety was too short [29, 78, 84, 86, 88, 89]. Membrane fractions from both rat liver and rat brain were also able to methylate AFC and AGGC with equal rates of initial methyl ester formation and K_m values of 25 µM for AFC and 7 µM for AGGC [84]. The Icmt from neutrophil membranes was also shown to accept both AFC and AGGC with K_m values of 11.6 µM and 1.4 µM, respectively.

An AFC analog where the sulfur atom was replaced with a methylene group was also recently reported [92]. It was suggested that the development of an Icmt inhibitor without the labile allylic thioether moiety would lead to Icmt inhibitors with a longer half-life in cell culture. This work extends a previous study that described AFC analogs where the sulfur atom was replaced with oxygen, selenium, sulfoxide, or amino groups [93]. All of these substituted compounds were very poor substrates, as was the carbon-substituted farnesylcysteine molecule [92, 93]. The methylene-substituted farnesylcysteine was also a poor inhibitor of Ste14p [92]. Together, these results emphasize the importance of the sulfur atom in substrate recognition.

F. PURIFICATION ATTEMPTS OF MAMMALIAN ICMTS

An ongoing obstacle in the characterization of mammalian Icmt proteins is the inability to find a suitable detergent for solubilizing these proteins. Unlike the yeast Ste14p (which has 239 amino acids, six putative transmembrane domains, and is solubilized with retention of activity in 1% DDM [66, 69]), most mammalian enzymes have a hydrophobic stretch of amino acids at the N-terminus of the protein that may comprise two additional transmembrane segments [27]. This extra hydrophobicity and the larger size of the mammalian enzyme have precluded purification attempts of the enzymes in active forms. The first attempt to solubilize a mammalian Icmt protein from the membrane was with rat liver Icmt [28].

The enzyme was inactivated by both 0.05% Triton X-100 and Zwittergent 3–14, and was also inactivated by 0.5% of deoxycholate, CHAPS, octylglucopyranoside, and Tween 20. The solubilization of human Icmt from neutrophils also resulted in loss of activity with many detergents [88]. Solubilization with 0.25% CHAPS, followed by detergent removal and reconstitution in PC/PA (10:1), resulted in the retention of 23.2% of activity under their experimental conditions [88]. However, this residual activity may be due to the presence of intact non-solubilized enzyme, as the membranes were not subjected to high-speed centrifugation after treatment with detergent to remove unsolubilized membrane fragments.

The only published attempt to a purify mammalian Icmt to date resulted in a yield of ~2% as measured by enzyme activity and SDS-PAGE analysis followed by staining showed only partial purification [82]. This bovine brain Icmt was first solubilized in 0.5% CHAPS, and then purified by DEAE-cellulose, Superose 6 FPLC, Superdex 75 HPLC, and ND-PAGE. The partially purified sample contained two proteins that were photoaffinity labeled with ^{3}H-SAM. One of these proteins was predicted to be bovine Icmt (30 kDa) and the other was predicted to be protein methylase II methyltransferase [82].

G. CHEMICAL MODIFICATION OF ICMT PROTEINS

Chemical modification reagents have been used to characterize Icmt proteins. Whereas iodoacetamide and iodoacetic acid had little effect on rat liver Icmt activity, NEM was a good inhibitor of the enzyme [28]. The Icmts from rat kidney and bovine adrenal chromaffin cells were also similarly inhibited by NEM [87, 94]. This inhibition is prevented by preincubation with SAM or SAH, suggesting that there may be an important cysteine residue in the active site. Interestingly, cysteine residues do not appear to be important for Ste14p function, as the enzyme is insensitive to NEM treatment and a cysteine-less variant is fully functional [76]. These data represent one of the few significant differences between the yeast and mammalian enzymes.

Phenylglyoxal (PGO), a chemical that modifies arginine residues, inhibited Icmt from rat kidney membranes [87]. SAH could reduce the effects of PGO treatment in a dose-dependent manner, but SAM and AFC could not. There are two arginines located at the C-terminal end of Icmt enzymes that are strictly conserved [69], and based on this inactivation study may play a very important role in catalysis by Icmt. We have found similar results with the yeast Ste14p enzyme (J. L. Anderson and C. A. Hrycyna, unpublished data).

Mammalian and yeast Icmts can also be inhibited by diethylpyrocarbonate (DEPC), which reacts preferentially with histidine but can modify many other amino acids [87] (J. L. Anderson and C. A. Hrycyna, unpublished data). Dicyclohexylcarbodiimide (DCCD), a modifier of free carboxyl groups in proteins, was also found to inhibit the mammalian enzyme [87]. Modification by these two chemicals suggested a possible role for histidine, glutamate, or aspartate in the structure or function of rat kidney Icmt and further implicated the two strictly conserved glutamate residues that occur sequentially near the C-terminus of Icmt proteins [69].

IV.   Cellular Functions of Icmts: Use of Small Molecules to Probe Icmt Function

Numerous studies have addressed the function of Icmt in many cellular processes in organisms. Carboxylmethylation is critical for the proper localization of Ras proteins in yeast and mouse cells, and may also influence the interaction between Ras and other proteins [14, 25, 83, 95]. Icmt has also been found to be essential for early stages of liver development in mice [96], and has been found to modulate endothelial cell apoptosis, possibly by affecting the activity of Ras or other small GTPases [97] and endothelial monolayer permeability via methylation of RhoA [98]. In addition, it has recently been shown that Icmt regulates Rac1 activity by modulating the interaction of Rac1 with RhoGDI via methylation of Rac1 [99].

AFC and AGGC can act as inhibitors of Icmt, both *in vivo* and *in vitro* [4, 27, 84, 100, 101]. Icmt can also be inhibited by both *S*-farnesylthioacetic acid (FTA) and *S*-geranylgeranylthioacetic acid (GGTA), but cannot utilize either as a substrate [84, 89]. These compounds are routinely used to assess the effects of *in vivo* inhibition of Icmt because they are not thought to affect other SAM-dependent methyltransferases. S-adenosyl-L-homocysteine (SAH) and sinefungin are also capable of inhibiting Icmt, but their overlapping specificity with other SAM-dependent methyltransferases make them undesirable for *in vivo* experiments or for use in confirming the identity of a methyltransferase as a member of the Icmt family [29, 31, 88, 97, 100, 102–104].

Most of the studies aimed at understanding the cellular function of Icmt make use of the small molecule inhibitors described previously, such as AFC, AGGC, farnesylthioacetic acid (FTA), or a mixture of adenosine and homocysteine. However, these compounds may not be solely specific for Icmt and the effects of these compounds on other unrelated cellular processes remain unclear [104a]. Therefore, these results are subject to a variety of interpretations and should perhaps be ultimately verified using other approaches. Huzoor-Akbar et al. demonstrated that AFC was able to block the carboxylmethylation of platelet rap1 proteins, causing inhibition of signal transduction caused by several agonists in human platelets [105, 106]. Both AFC and AGGC were shown to inhibit *N*-formyl-methionyl-leucyl-phenylalanine (FMLP)-induced superoxide anion formation, a process that it is suggested to be mediated by Ras-related proteins [101]. Li et al. found that AFC blocked nutrient-induced insulin release from insulin-secreting cells [102, 107]. FTA caused a decrease in the aldosterone-induced Na^+ current in a model of the mammalian renal system [108].

AFC and FTA were both shown to make cells more susceptible to TNFα-induced cell apoptosis [97, 99]. In mouse embryonic fibroblasts, AFC reduced the amount of methylated Rac1 in the cell, as well as a reduction in the amount of Rac1 at the cell membrane, and blocked phosphorylation of p38 MAP kinase upon TNFα stimulation [99]. In pulmonary artery endothelial cells (PAEC), inhibition of Icmt with AGGC or a mixture of adenosine and homocysteine resulted in a decrease in the amounts of activated and membrane-associated Ras proteins, as well as reductions in the amounts of activated downstream effectors of Ras; namely, Akt, ERK-1, and ERK-2 [97]. In the same cell line, these inhibitors were also shown to decrease the carboxylmethylation of RhoA and decrease the amount of RhoA that was bound to GTP [98].

All of these affected pathways are thought to involve G proteins. Much of the data represents direct or indirect evidence that methylation causes localization of G proteins to membranes [52, 97–99, 101, 102, 104, 107, 109]. For example, methylation of the G25K G protein from rabbit brain caused translocation from the cytosol to the membranes [52]. This translocation also occurred in human neutrophils during signal transduction after carboxylmethylation of Ras-related proteins [101]. The methylation state of G proteins also affects their interaction with other proteins (i.e., Rac and RhoGDI [99]) and regulates their half-lives in cells [110]. Finally, G proteins were also implicated as cellular targets of Icmt because their methylation was stimulated by GTPγS [101, 102, 105–108]. These data have led to the hypothesis that either Icmt methylates activated G proteins more efficiently or methylation increases the amount of a given G protein recruited to be in its activated GTP-bound state [97, 98].

V. Development of Novel Small-Molecule Icmt Inhibitors

The farnesyltransferases (FTases) from both yeast and mammalian systems are soluble enzymes that have been purified and characterized and detailed 3D crystal structures have been obtained [17, 111–115]. These studies have led to the development of potent FTase inhibitors by numerous pharmaceutical companies. These compounds have been shown to be effective chemotherapeutically to slow the growth of cancers [116–119], and at least two FTase inhibitors (FTIs) are presently in advanced clinical trials as anticancer agents. The rationale for the use of these compounds was based on the fact that the Ras proteins must be farnesylated in order to transform cells. However, in contrast to the expected results certain *Ras*-positive tumor cells are resistant to FTIs, and many *Ras*-negative tumor cells are sensitive to FTIs [120].

Cells bearing mutant K-Ras genes are generally more resistant to FTI treatment because in the presence of an FTI K-Ras can be geranylgeranylated by GGTase I, and thus still localize and function normally in the cell [121, 122]. The lack of efficacy with K-Ras-driven tumors is a serious issue, as K-Ras is the most commonly mutated form of Ras found in human malignancies. In summary, despite some preliminary clinical success observed with FTIs they have not lived up to their initial promise as selective and specific Ras-targeted anticancer agents. Thus, there has been interest in pursuing enzymes downstream in the CaaX posttranslational pathway, such as Icmt, as molecular targets for the pharmacological inactivation of Ras proteins [123, 124].

Compared to the farnesyltransferases, much less is known about the mechanism of action of the isoprenylcysteine carboxylmethyltransferases and the role of the methyl group in protein localization and function. Importantly, we have shown that a yeast GFP-Ras2p fusion protein normally localized to the plasma membrane (Figure 9.5) is mislocalized in a Δ*ste14* deletion strain (C. A. Hrycyna and S. Michaelis, unpublished data). Young and coworkers have shown that targeted inactivation of the *Icmt* gene in mouse similarly causes striking mislocalization of K-ras [83, 110]. Furthermore, cellular transformation by oncogenic Ras was inhibited in transfected *Icmt*[-/-] ES cells, suggesting a crucial role for the methyl group [83, 110, 123]. Although Ras has recently been shown to be functional in endomembranes [125], it remains to be determined whether this localization can support oncogenesis. It has also been demonstrated that blocking

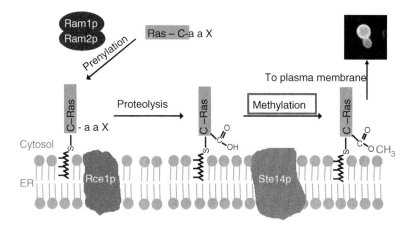

Fɪɢ. 9.5. Normal Ras CaaX protein processing in the yeast *S. cerevisiae*. Initially synthesized as a soluble cytosolic protein, Ras is first farnesylated by the Ram1p/Ram2p complex comprised of two proteins that constitute the farnesyltransferase [9]. Rce1 is the ER membrane-associated endoprotease responsible for cleavage of the three-terminal aaX amino acids [128, 129]. Ste14p is the isoprenylcysteine carboxylmethyltransferase responsible for the third reaction, α-carboxylmethylesterification of the newly exposed farnesylated cysteine residue [14, 24, 25, 68]. When processed normally, yeast Ras proteins are targeted to the plasma membrane. When not methylated, Ras is mislocalized in the cell. Although not depicted, yeast Ras proteins also are palmitoylated [130, 131].

the methylation of K-Ras blocks its association with microtubules, which may be crucial for the localization and biological activity of this Ras variant [126].

A recent study confirms that methylation is required for the proper localization of Ras and extends work demonstrating that the modification is not necessary for localization of the Rho proteins, another class of CaaX proteins [109]. This differential effect was linked to the fact that Ras is farnesylated and the Rho proteins are geranylgeranylated, suggesting that Icmt inhibition will have a much more profound inhibitory effect on the activity of farnesylated proteins, such as K-Ras, than geranylgeranylated proteins. Taken together, these data suggest that (1) inhibition of Icmt should lead to the mislocalization of farnesylated K-Ras but may not have detrimental effects on the function or localization of all other CaaX proteins modified by Icmt, (2) this mislocalization of K-Ras should interfere with its biological activity, and (thus), (3) Icmt inhibitors are intriguing potential anticancer agents.

A. METHOTREXATE

An interesting discovery was made by Casey and coworkers in 2003 that showed that the treatment of cells with methotrexate nearly abolishes Ras methylation *in vivo* [95]. Methotrexate, like other antifolates, is known to increase concentrations of homocysteine and SAH *in vivo*. Because SAH inhibits Icmt in a dose-dependent manner *in vitro*, these data suggest that methotrexate inhibits Ras methylation by SAH inhibition of Icmt. In addition, GFP-tagged forms of H-, N-, and K-Ras were all shown to mislocalize in cells upon treatment with methotrexate [95].

Methotrexate was also able to reduce signaling in Ras pathways, causing a 75% reduction of phosphorylation of p44 MAPK and Akt after EGF stimulation of Icmt$^{+/+}$ cells. Icmt$^{-/-}$ cells were found to be resistant to treatment with methotrexate and there was no reduction in Ras signaling in these cells, suggesting that the ability of methotrexate to kill these cells was dependent on Icmt. Addition of thymidine to the media to compensate for methotrexate inhibition of thymidine synthesis had little effect on the survival of these cells, implying that the lack of thymidine is not responsible for decreased survival of the cells.

Expression of Icmt in Icmt$^{-/-}$ cells restores sensitivity to methotrexate [95]. Icmt$^{+/+}$ cells, Icmt$^{-/-}$ cells, and Icmt$^{-/-}$ cells transfected with Icmt were all equally sensitive to the unrelated chemotherapeutic drug vinblastine, suggesting that the methotrexate results were not generic to all chemotherapeutic drugs. Further proof of Icmt involvement in methotrexate treatment was shown by expressing an H-Ras construct containing a myristoylation signal at the N-terminus and no CaaX box in Icmt$^{+/+}$ cells. This myristoylation was sufficient for localization to the plasma membrane and the cells were resistant to methotrexate [95]. These findings shed new light on this long-used chemotherapeutic drug, showing that whereas many have hypothesized that Icmt inhibiting drugs could be used to treat cancer one has already been used clinically.

B. CYSMETHYNIL

Recently, the development of the indole-based small-molecule Icmt inhibitor cysmethynil (2-[5-(3-methylphenyl)-1-octyl-1H-indol-3-yl]acetamide) was reported [104]. Cysmethynil, which was identified from a library of ~10,000 compounds, inhibits Icmt *in vitro* with an IC_{50} of 0.2–2.4 μM. The compound was also able to inhibit the growth of Icmt$^{+/+}$ cells in a dose-dependent manner, but did not have an effect on Icmt$^{-/-}$ cells except when they had been transfected with Icmt.

Incubation of cells with increasing concentrations of cysmethynil resulted in increasing mislocalization of GFP-K-Ras. An EGF-induced increase in the phosphorylation of MAPK was almost completely blocked and the increase in Akt phosphorylation was partially blocked in cells treated with 1 μM cysmethynil. Overexpression of Icmt (approximately fourfold) was enough to protect cells from cysmethynil blockage of EGF-induced activation of MAPK. Cysmethynil (20 μM) was able to halt the growth of cells expressing endogenous Icmt in soft agar, but could not stop the transformation of cells in which Icmt was overexpressed.

C. AFC ANALOGS

Due to the interest in inactivation of the family of Icmt enzymes, as well as an interest in further elucidating the substrate specificity of Icmt enzymes, a series of prenylcysteine analogs based on the minimal Icmt substrate AFC were developed to potentially inhibit Ste14p as well as human Icmt [127]. The two best inhibitors of the series, illustrated in Figure 9.6, are derivatives of the Icmt substrate AFC. Both have a prenyl substitutent at position 3 of the farnesyl chain and the second inhibitor (4) has the last two isoprene groups of the farnesyl chain replaced with two phenyl rings [127]. The substrate specificity for these two compounds was determined and it was found that Ste14p preferred AFC over compound 3 by 1,000-fold and over compound 4 by nearly 3,500-fold, further emphasizing the isoprenoid as a key recognition element [127].

Ste14p binds compound 3 and AFC comparably but has a much lower V_{max} for 3. This difference is most likely due to the presence of the large substituent at position 3 of the farnesyl group, suggesting that bulk in this position may slow product release from the enzyme. These types of compounds that exhibit a low K_m and a low V_{max} are ideal candidates for lead inhibitors. On the other hand, 3 is a relatively good substrate for the human Icmt enzyme expressed in *S. cerevisiae*. The observation that compound 3 is a better substrate for human Icmt than for Ste14p relative to AFC suggests that human Icmt is more tolerant of changes at

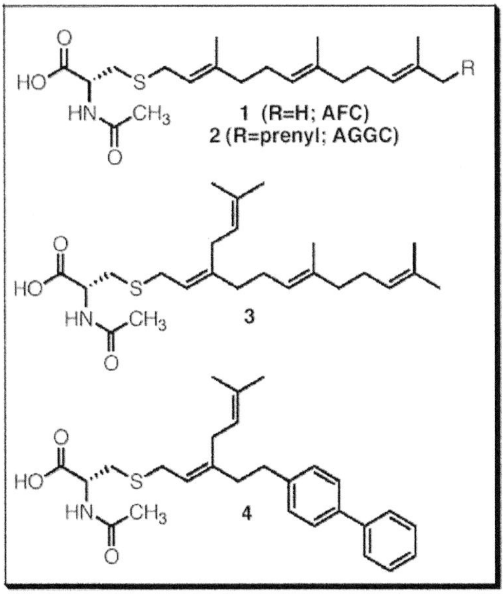

FIG. 9.6. Prenylcysteine analogs synthesized as potential Icmt Inhibitors. The parent compound AFC (1) and AGGC (2) are commonly used as Icmt substrates *in vitro* and are also used to inhibit Icmt *in vivo*. Compound 3 is an isobutenyl derivative of 1. Compound 4 is an isobutenyl biphenyl derivative of 1. *(Reproduced with permission from* The Journal of Biological Chemistry *[127].)*

that position in the farnesyl moiety. However, adding the biphenyl group in addition to the isobutenyl (compound 4) severely affects the ability of the substrate to be recognized by either Icmt enzyme. Although the overall general mechanisms of the yeast and human enzymes are likely very similar, these studies provide the first examples of substrate specificity differences between the yeast and human enzymes. This difference, although subtle, emphasizes the importance of exploring the substrate specificity of the human enzyme for inhibitor development.

These compounds were also examined as inhibitors of AFC methylation. However, in our assay system, if an inhibitor is also a substrate it cannot be tested as an inhibitor because we can not differentiate methylation of AFC versus the inhibitor. As discussed previously, compound 3 was a much better substrate for human Icmt than it was for Ste14p, reaching a V_{max} value ~30% of that for AFC. Therefore, only compound 4 was tested as an inhibitor for human Icmt, and both 3 and 4 were tested against Ste14p. Compound 3 was found to be a competitive inhibitor of Ste14p, with a K_I value of 17.1 ± 1.7 μM. Compound 4 exhibited mixed inhibition for Ste14p, with a competitive K_I of 35.4 ± 3.4 μM and an uncompetitive K_I of 614 ± 148 μM [127].

Compound 4 was also found to be a mixed inhibitor of human Icmt but was less potent, demonstrating a competitive K_I value of 119.3 ± 18.1 μM and an uncompetitive K_I value of 377.2 ± 42.5 µM [127]. Both compounds were also able to inhibit purified and reconstituted Ste14p with a similar trend as seen in crude yeast membranes [127]. The lead compounds 3 and 4 described here are among the first well-characterized inhibitors of Icmt. We are also generating amine-modified AFC analogs as inhibitors of human Icmt and have recently developed several agents that are not substrates and act as low micromolar inhibitors (R. A. Gibbs and C. A. Hrycyna, unpublished data). In the future, these types of inhibitors (if proven to be specific for Icmt in cells) may be valuable lead compounds to serve as tools for probing the isoprenyl binding site of Icmt enzymes and for probing the cellular functions of Icmt, as well as for the development of novel anticancer agents.

REFERENCES

1. Clarke, S. (1992). Protein isoprenylation and methylation at carboxyl-terminal cysteine residues. *Annu Rev Biochem* 61:355–386.
2. Zhang, F. L., and Casey, P. J. (1996). Protein prenylation: Molecular mechanisms and functional consequences. *Annu Rev Biochem* 65:241–269.
3. Young, S. G., Ambroziak, P., Kim, E., and Clarke, S. (2000). "Post- isoprenylation Protein Processing: CXXX (CaaX) Endoproteases and Isoprenylcysteine Carboxyl Methyltransferase," in F. Tamanoi and D. S. Sigman (eds.). *Protein Lipidation: A Volume of The Enzymes* (3d ed.). pp. 21, 155–213, San Diego: Academic Press.
4. Hrycyna, C. A., and Clarke, S. (1993). Modification of eukaryotic signaling proteins by C-terminal methylation reactions. *Pharmacol Ther* 59:281–300.
5. Casey, P. J., and Seabra, M. C. (1996). Protein prenyltransferases. *J Biol Chem* 271:5289–5292.
6. Deschenes, R. J., Stimmel, J. B., Clarke, S., Stock, J., and Broach, J. R. (1989). RAS2 protein of Saccharomyces cerevisiae is methyl-esterified at its carboxyl terminus. *J Biol Chem* 264:11865–11873.
7. Glomset, J. A., Gelb, M. H., and Farnsworth, C. C. (1990). Prenyl proteins in eukaryotic cells: a new type of membrane anchor. *Trends Biochem Sci* 15:139–142.
8. Gutierrez, L., Magee, A. I., Marshall, C. J., and Hancock, J. F. (1989). Posttranslational processing of p21ras is two-step and involves carboxyl-methylation and carboxy-terminal proteolysis. *EMBO J* 8:1093–1098.
9. Powers, S., Michaelis, S., Broek, D., Santa Anna, S., Field, J., Herskowitz, I., and Wigler, M. (1986). RAM, a gene of yeast required for a functional modification of RAS proteins and for production of mating pheromone a-factor. *Cell* 47:413–422.
10. Wolda, S. L., and Glomset, J. A. (1988). Evidence for modification of lamin B by a product of mevalonic acid. *J Biol Chem* 263:5997–6000.
11. Hancock, J. F., Cadwallader, K., and Marshall, C. J. (1991). Methylation and proteolysis are essential for efficient membrane binding of prenylated p21K-ras(B). *EMBO J* 10:641–646.
12. Clarke, S., Vogel, J. P., Deschenes, R. J., and Stock, J. (1988). Posttranslational modification of the Ha-ras oncogene protein: evidence for a third class of protein carboxyl methyltransferases. *Proc Natl Acad Sci USA* 85:4643–4647.
13. He, B., Chen, P., Chen, S. Y., Vancura, K. L., Michaelis, S., and Powers, S. (1991). RAM2, an essential gene of yeast, and RAM1 encode the two polypeptide components of the farnesyltransferase that prenylates a-factor and Ras proteins. *Proc Natl Acad Sci USA* 88:11373–11377.

14. Sapperstein, S., Berkower, C., and Michaelis, S. (1994). Nucleotide sequence of the yeast STE14 gene, which encodes farnesylcysteine carboxyl methyltransferase, and demonstration of its essential role in a-factor export. *Mol Cell Biol* 14:1438–1449.

15. Backlund, P. S. Jr. (1997). Posttranslational processing of RhoA. Carboxyl methylation of the carboxyl-terminal prenylcysteine increases the half-life of Rhoa. *J Biol Chem* 272:33175–33180.

16. Marcus, S., Caldwell, G. A., Miller, D., Xue, C. B., Naider, F., and Becker, J. M. (1991). Significance of C-terminal cysteine modifications to the biological activity of the Saccharomyces cerevisiae a-factor mating pheromone. *Mol Cell Biol* 11:3603–3612.

17. Casey, P. J. (1992). Biochemistry of protein prenylation. *J Lipid Res* 33:1731–1740.

18. Gelb, M. H., Scholten, J. D., and Sebolt-Leopold, J. S. (1998). Protein prenylation: From discovery to prospects for cancer treatment. *Curr Opin Chem Biol* 2:40–48.

19. Liang, P. H., Ko, T. P., and Wang, A. H. (2002). Structure, mechanism and function of prenyl-transferases. *Eur J Biochem* 269:3339–3354.

20. Park, H. W., and Beese, L. S. (1997). Protein farnesyltransferase. *Curr Opin Struct Biol* 7:873–880.

21. Boyartchuk, V. L., Ashby, M. N., and Rine, J. (1997). Modulation of Ras and a-factor function by carboxyl-terminal proteolysis. *Science* 275:1796–800.

22. Schmidt, W. K., Tam, A., Fujimura-Kamada, K., and Michaelis, S. (1998). Endoplasmic reticulum membrane localization of Rce1p and Ste24p, yeast proteases involved in carboxyl-terminal CAAX protein processing and amino-terminal a-factor cleavage. *Proc Natl Acad Sci USA* 95:11175–11180.

23. Tam, A., Schmidt, W. K., and Michaelis, S. (2001). The multispanning membrane protein Ste24p catalyzes CaaX proteolysis and NH2-terminal processing of the yeast a-factor precursor. *J Biol Chem* 276:46798–46806.

24. Hrycyna, C. A., and Clarke, S. (1990). Farnesyl cysteine C-terminal methyltransferase activity is dependent upon the STE14 gene product in Saccharomyces cerevisiae. *Mol Cell Biol* 10:5071–5076.

25. Hrycyna, C. A., Sapperstein, S. K., Clarke, S., and Michaelis, S. (1991). The Saccharomyces cerevisiae STE14 gene encodes a methyltransferase that mediates C-terminal methylation of a-factor and RAS proteins. *Embo J* 10:1699–1709.

26. Marr, R. S., Blair, L. C., and Thorner, J. (1990). Saccharomyces cerevisiae STE14 gene is required for COOH-terminal methylation of a-factor mating pheromone. *J Biol Chem* 265:20057–20060.

27. Dai, Q., Choy, E., Chiu, V., Romano, J., Slivka, S. R., Steitz, S. A., Michaelis, S., and Philips, M. R. (1998). Mammalian prenylcysteine carboxyl methyltransferase is in the endoplasmic reticulum. *J Biol Chem* 273:15030–15034.

28. Stephenson, R. C., and Clarke, S. (1992). Characterization of a rat liver protein carboxyl methyltransferase involved in the maturation of proteins with the -CXXX C-terminal sequence motif. *J Biol Chem* 267:13314–13319.

29. Stephenson, R. C., and Clarke, S. (1990). Identification of a C-terminal protein carboxyl methyltransferase in rat liver membranes utilizing a synthetic farnesyl cysteine-containing peptide substrate. *J Biol Chem* 265:16248–16254.

30. Romano, J. D., Schmidt, W. K., and Michaelis, S. (1998). The Saccharomyces cerevisiae prenylcysteine carboxyl methyltransferase Ste14p is in the endoplasmic reticulum membrane. *Mol Biol Cell* 9:2231–2247.

31. Shi, Y. Q., and Rando, R. R. (1992). Kinetic mechanism of isoprenylated protein methyltransferase. *J Biol Chem* 267:9547–9551.

32. Baron, R. A., and Casey, P. J. (2004). Analysis of the kinetic mechanism of recombinant human isoprenylcysteine carboxylmethyltransferase (Icmt). *BMC Biochem* 5:19.

33. Kagan, R. M., and Clarke, S. (1994). Widespread occurrence of three sequence motifs in diverse S-adenosylmethionine-dependent methyltransferases suggests a common structure for these enzymes. *Arch Biochem Biophys* 310:417–427.

34. Sakagami, Y., Yoshida, M., Isogai, A., and Suzuki, A. (1981). Peptidal sex hormones inducing conjugation tube formation in compatible mating-type cells of *Tremella mesenterica*. *Science* 212:1525–1526.

35. Ishibashi, Y., Sakagami, Y., Isogai, A., and Suzuki, A. (1984). Structures of tremerogens *A*-9291-I and *A*-9291-VIII: Peptidyl sex hormones of *Tremella brasiliensis*. *Biochemistry* 23:1399–1404.

36. Betz, R., Crabb, J. W., Meyer, H. E., Wittig, R., and Duntze, W. (1987). Amino acid sequences of a-factor mating peptides from Saccharomyces cerevisiae. *J Biol Chem* 262:546–548.

37. Anderegg, R. J., Betz, R., Carr, S. A., Crabb, J. W., and Duntze, W. (1988). Structure of Saccharomyces cerevisiae mating hormone a-factor. Identification of S-farnesyl cysteine as a structural component. *J Biol Chem* 263:18236–18240.

38. Barbacid, M. (1987). Ras Genes. *Annu Rev Biochem* 56:779–827.

39. Brake, A. J., Brenner, C., Najarian, R., Laybourn, P., and Merryweather, J. (1985). "Structure of Genes Encoding Precursors of the Yeast Peptide Mating Pheromone **a**-Factor" in *Protein Transport and Secretion*, M. J. Gething, ed., pp. 103–108.

40. Chelsky, D., Olson, J. F., and Koshland, D. J. (1987). Cell cycle-dependent methyl esterification of lamin B. *J Biol Chem* 262:4303–4309.

41. Chelsky, D., Sobotka, C., and O'Neill, C. L. (1989). Lamin B methylation and assembly into the nuclear envelope. *J Biol Chem* 264:7637–7643.

42. Farnsworth, C. C., Wolda, S. L., Gelb, M. H., and Glomset, J. A. (1989). Human lamin B contains a farnesylated cysteine residue. *J Biol Chem* 264:20422–20429.

43. Holtz, D., Tanaka, R. A., Hartwig, J., and McKeon, F. (1989). The CaaX motif of lamin A functions in conjunction with the nuclear localization signal to target assembly to the nuclear envelope. *Cell* 59:969–977.

44. Casey, P. J., Solski, P. A., Der, C. J., and Buss, J. E. (1989). p21ras is modified by a farnesyl isoprenoid. *Proc Natl Acad Sci USA* 86:8323–8327.

45. Fujiyama, A., and Tamanoi, F. (1990). RAS2 protein of Saccharomyces cerevisiae undergoes removal of methionine at N-terminus and removal of three amino acids at C-terminus. *J Biol Chem* 265:3362–3368.

46. Hancock, J. F., Magee, A. I., Childs, J. E., and Marshall, C. J. (1989). All ras proteins are polyisoprenylated but only some are palmitoylated. *Cell* 57:1167–1177.

47. Schafer, W. R., Kim, R., Sterne, R., Thorner, J., Kim, S. H., and Rine, J. (1989). Genetic and pharmacological suppression of oncogenic mutations in ras genes of yeast and humans. *Science* 245:379–385.

48. Buss, J. E., Quilliam, L. A., Kato, K., Casey, P. J., Solski, P. A., Wong, G., Clark, R., McCormick, F., Bokoch, G. M., and Der, C. J. (1991). The COOH-terminal domain of the Rap1A (Krev-1) protein is isoprenylated and supports transformation by an H-Ras:Rap1A chimeric protein. *Mol Cell Biol* 11:1523–1530.

49. Farnsworth, C. C., Kawata, M., Yoshida, Y., Takai, Y., Gelb, M. H., and Glomset, J. A. (1991). C-terminus of the small GTP-binding protein smg p25A contains two geranylgeranylated cysteine residues and a methyl ester. *Proc Natl Acad Sci USA* 88:6196–6200.

50. Kawata, M., Farnsworth, C. C., Yoshida, Y., Gelb, M. H., Glomset, J. A., and Takai, Y. (1990). Posttranslationally processed structure of the human platelet protein smg p21B: evidence for geranylgeranylation and carboxyl methylation of the C-terminal cysteine. *Proc Natl Acad Sci USA*. 87:8960–8964.

51. Yamane, H. K., Farnsworth, C. C., Xie, H. Y., Evans, T., Howald, W. N., Gelb, M. H., Glomset, J. A., Clarke, S., and Fung, B. K.-K. (1991). Membrane-binding domain of the small G protein G25K contains an S-(all-trans-geranylgeranyl)cysteine methyl ester at its carboxyl terminus. *Proc Natl Acad Sci USA* 88:286–290.

52. Backlund, P. S. J. (1992). GTP-stimulated carboxyl methylation of a soluble form of the GTP-binding protein G25K in brain. *J Biol Chem* 267:18432–18439.

53. Anant, J. S., Ong, O. C., Xie, H. Y., Clarke, S., O'Brien, P. J., and Fung, B. K.-K. (1992). In vivo differential prenylation of retinal cyclic GMP phosphodiesterase catalytic subunits. *J Biol Chem* 267:687–690.

54. Catty, P., and Deterre, P. (1991). Activation and solubilization of the retinal cGMP-specific phosphodiesterase by limited proteolysis. Role of the C-terminal domain of the beta-subunit. *Eur J Biochem* 199:263–269.

55. Ong, O. C., Ota, I. M., Clarke, S., and Fung, B. K.-K. (1989). The membrane binding domain of rod cGMP phosphodiesterase is posttranslationally modified by methyl esterification at a C-terminal cysteine. *Proc Natl Acad Sci USA* 86:9238–9242.

56. Backlund, P. S. J., Simonds, W. F., and Spiegel, A. M. (1990). Carboxyl methylation and COOH-terminal processing of the brain G-protein gamma-subunit. *J Biol Chem* 265:15572–15576.

57. Fukada, Y., Takao, T., Ohguro, H., Yoshizawa, T., Akino, T., and Shimonishi, Y. (1990). Farnesylated gamma-subunit of photoreceptor G protein indispensable for GTP-binding. *Nature* 346:658–660.

58. Fung, B. K.-K., Yamane, H. K., Ota, I. M., and Clarke, S. (1990). The gamma subunit of brain G-proteins is methyl esterified at a C-terminal cysteine. *Febs Lett* 260:313–317.

59. Lai, R. K., Perez, S. D., Canada, F. J., and Rando, R. R. (1990). The gamma subunit of transducin is farnesylated. *Proc Natl Acad Sci USA* 87:7673–7677.

60. Yamane, H. K., Farnsworth, C. C., Xie, H. Y., Howald, W., Fung, B. K.-K., Clarke, S., Gelb, M. H., and Glomset, J. A. (1990). Brain G protein gamma subunits contain an all-trans-geranylgeranylcysteine methyl ester at their carboxyl termini. *Proc Natl Acad Sci USA* 87:5868–5872.

61. Ota, I. M., and Clarke, S. (1989). Enzymatic methylation of 23-29-kDa bovine retinal rod outer segment membrane proteins. Evidence for methyl ester formation at carboxyl-terminal cysteinyl residues. *J Biol Chem* 264:12879–12884.

62. Romano, J. D. (2000). Localization and Topological Analysis of Ste14, the *Saccharomyces cerevisiae* Prenylcysteine Carboxylmethyltransferase, *Ph.D. Dissertation,* The Johns Hopkins University School of Medicine, 314 pages.

63. Hrycyna, C. A., Wait, S. J., Backlund, P. S., Jr., and Michaelis, S. (1995). Yeast STE14 methyltransferase, expressed as TrpE-STE14 fusion protein in Escherichia coli, for in vitro carboxylmethylation of prenylated polypeptides. *Methods Enzymol* 250:251–266.

64. Chen, P., Sapperstein, S. K., Choi, J. D., and Michaelis, S. (1997). Biogenesis of the Saccharomyces cerevisiae mating pheromone a-factor. *J Cell Biol* 136:251–269.

65. Michaelis, S., Chen, P., Berkower, C., Sapperstein, S., and Kistler, A. (1992). Biogenesis of yeast a-factor involves prenylation, methylation and a novel export mechanism. *Antonie Van Leeuwenhoek* 61:115–117.

66. Anderson, J. L., Frase, H., Michaelis, S., and Hrycyna, C. A. (2005). Purification, functional reconstitution, and characterization of the Saccharomyces cerevisiae isoprenylcysteine carboxylmethyltransferase Ste14p. *J Biol Chem* 280:7336–7345.

67. Hrycyna, C. A., Yang, M. C., and Clarke, S. (1994). Protein carboxyl methylation in Saccharomyces cerevisiae: evidence for STE14-dependent and STE14-independent pathways. *Biochemistry* 33:9806–9812.

68. Ashby, M. N., Errada, P. R., Boyartchuk, V. L., and Rine, J. (1993). Isolation and DNA sequence of the STE14 gene encoding farnesyl cysteine: carboxyl methyltransferase. *Yeast* 9:907–913.

69. Romano, J. D., and Michaelis, S. (2001). Topological and Mutational Analysis of Saccharomyces cerevisiae Ste14p, Founding Member of the Isoprenylcysteine Carboxyl Methyltransferase Family. *Mol Biol Cell* 12:1957–1971.

70. Monne, M., Hermansson, M., and von Heijne, G. (1999). A turn propensity scale for transmembrane helices. *J Mol Biol* 288:141–145.

71. Monne, M., Nilsson, I., Elofsson, A., and von Heijne, G. (1999). Turns in transmembrane helices: Determination of the minimal length of a "helical hairpin" and derivation of a fine-grained turn propensity scale. *J Mol Biol* 293:807–814.
72. Russ, W. P., and Engelman, D. M. (2000). The GxxxG motif: A framework for transmembrane helix-helix association. *J Mol Biol* 296:911–919.
73. Schneider, D., and Engelman, D. M. (2004). Motifs of two small residues can assist but are not sufficient to mediate transmembrane helix interactions. *J Mol Biol* 343:799–804.
74. Buckner, F. S., Kateete, D. P., Lubega, G. W., Van Voorhis, W. C., and Yokoyama, K. (2002). Trypanosoma brucei prenylated-protein carboxyl methyltransferase prefers farnesylated substrates. *Biochem J* 367:809–816.
75. Desrosiers, R. R., Nguyen, Q. T., and Beliveau, R. (1999). The carboxyl methyltransferase modifying G proteins is a metalloenzyme. *Biochem Biophys Res Commun* 261:790–797.
76. Hodges, H. B., Zhou, M., Haldar, S., Anderson, J. L., Thompson, D. H., and Hrycyna, C. A. (2005). Inhibition of membrane-associated methyltransferases by a cholesterol-based metal chelator. *Bioconjug Chem* 16:490–493.
77. Imai, Y., Davey, J., Kawagishi-Kobayashi, M., and Yamamoto, M. (1997). Genes encoding farnesyl cysteine carboxyl methyltransferase in Schizosaccharomyces pombe and Xenopus laevis. *Mol Cell Biol* 17:1543–1551.
78. Hasne, M. P., and Lawrence, F. (1999). Characterization of prenylated protein methyltransferase in Leishmania. *Biochem J* 342(3):513–518.
79. Crowell, D. N., and Kennedy, M. (2001). Identification and functional expression in yeast of a prenylcysteine alpha-carboxyl methyltransferase gene from Arabidopsis thaliana. *Plant Mol Biol* 45:469–476.
80. Narasimha Chary, S., Bultema, R. L., Packard, C. E., and Crowell, D. N. (2002). Prenylcysteine alpha-carboxyl methyltransferase expression and function in Arabidopsis thaliana. *Plant J* 32:735–747.
81. Crowell, D. N., Sen, S. E., and Randall, S. K. (1998). Prenylcysteine alpha-carboxyl methyl-transferase in suspension-cultured tobacco cells. *Plant Physiol* 118:115–123.
82. Yoo, B. C., Kang, M. S., Kim, S., Lee, Y. S., Choi, S. Y., Ryu, C. K., Park, G. H., and Han, J. S. (1998). Partial purification of protein farnesyl cysteine carboxyl methyltransferase from bovine brain. *Exp Mol Med* 30:227–234.
83. Bergo, M. O., Leung, G. K., Ambroziak, P., Otto, J. C., Casey, P. J., and Young, S. G. (2000). Targeted inactivation of the isoprenylcysteine carboxyl methyltransferase gene causes mislocalization of K-Ras in mammalian cells. *J Biol Chem* 275:17605–17610.
84. Volker, C., Lane, P., Kwee, C., Johnson, M., and Stock, J. (1991). A single activity carboxyl methylates both farnesyl and geranylgeranyl cysteine residues. *Febs Lett* 295:189–194.
85. Bergo, M. O., Leung, G. K., Ambroziak, P., Otto, J. C., Casey, P. J., Gomes, A. Q., Seabra, M. C., and Young, S. G. (2001). Isoprenylcysteine carboxyl methyltransferase deficiency in mice. *J Biol Chem* 276:5841–5845.
86. Perez-Sala, D., Gilbert, B. A., Tan, E. W., and Rando, R. R. (1992). Prenylated protein methyl-transferases do not distinguish between farnesylated and geranylgeranylated substrates. *Biochem J* 284:835–840.
87. Boivin, D., Lin, W., and Beliveau, R. (1997). Essential arginine residues in isoprenylcysteine protein carboxyl methyltransferase. *Biochem Cell Biol* 75:63–69.
88. Pillinger, M. H., Volker, C., Stock, J. B., Weissmann, G., and Philips, M. R. (1994). Characterization of a plasma membrane-associated prenylcysteine- directed alpha carboxyl methyltransferase in human neutrophils. *J Biol Chem* 269:1486–1492.
89. Tan, E. W., Perez, S. D., Canada, F. J., and Rando, R. R. (1991). Identifying the recognition unit for G protein methylation. *J Biol Chem* 266:10719–10722.

90. Tan, E. W., and Rando, R. R. (1992). Identification of an isoprenylated cysteine methyl ester hydrolase activity in bovine rod outer segment membranes. *Biochemistry* 31:5572–5578.

91. Volker, C., Miller, R. A., McCleary, W. R., Rao, A., Poenie, M., Backer, J. M., and Stock, J. B. (1991). Effects of farnesylcysteine analogs on protein carboxyl methylation and signal transduction. *J Biol Chem* 266:21515–21522.

92. Henriksen, B. S., Anderson, J. L., Hrycyna, C. A., and Gibbs, R. A. (2005). Synthesis of Desthio Prenylcysteine Analogs: Sulfur is Important for Biological Activity. *Bioorganic & Medicinal Chemistry Letters* 15:5080–5083.

93. Tan, E. W., Perezsala, D., and Rando, R. R. (1991). Heteroatom Requirements For Substrate Recognition By GTP-Binding Protein Methyltransferase. *J Amer Chem Soc* 113:6299–6300.

94. De Busser, H. M., Van Dessel, G. A., and Lagrou, A. R. (2000). Identification of prenylcysteine carboxymethyltransferase in bovine adrenal chromaffin cells. *Int J Biochem Cell Biol* 32:1007–1016.

95. Winter-Vann, A. M., Kamen, B. A., Bergo, M. O., Young, S. G., Melnyk, S., James, S. J., and Casey, P. J. (2003). Targeting Ras signaling through inhibition of carboxyl methylation: an unexpected property of methotrexate. *Proc Natl Acad Sci USA* 100:6529–6534.

96. Lin, X., Jung, J., Kang, D., Xu, B., Zaret, K. S., and Zoghbi, H. (2002). Prenylcysteine carboxylmethyltransferase is essential for the earliest stages of liver development in mice. *Gastroenterology* 123:345–351.

97. Kramer, K., Harrington, E. O., Lu, Q., Bellas, R., Newton, J., Sheahan, K. L., and Rounds, S. (2003). Isoprenylcysteine carboxyl methyltransferase activity modulates endothelial cell apoptosis. *Mol Biol Cell* 14:848–857.

98. Lu, Q., Harrington, E. O., Hai, C. M., Newton, J., Garber, M., Hirase, T., and Rounds, S. (2004). Isoprenylcysteine carboxyl methyltransferase modulates endothelial monolayer permeability: involvement of RhoA carboxyl methylation. *Circ Res* 94:306–315.

99. Papaharalambus, C., Sajjad, W., Syed, A., Zhang, C., Bergo, M. O., Alexander, R. W., and Ahmad, M. (2005). Tumor necrosis factor alpha stimulation of Rac1 activity. Role of isoprenylcysteine carboxylmethyltransferase. *J Biol Chem* 280:18790–18796.

100. Perez-Sala, D., Tan, E. W., Canada, F. J., and Rando, R. R. (1991). Methylation and demethylation reactions of guanine nucleotide-binding proteins of retinal rod outer segments. *Proc Natl Acad Sci USA* 88:3043–3046.

101. Philips, M. R., Pillinger, M. H., Staud, R., Volker, C., Rosenfeld, M. G., Weissmann, G., and Stock, J. B. (1993). Carboxyl methylation of Ras-related proteins during signal transduction in neutrophils. *Science* 259:977–980.

102. Li, Y., Kowluru, A., and Metz, S. A. (1996). Characterization of prenylcysteine methyltransferase in insulin-secreting cells. *Biochem J* 316(1):345–351.

103. Ratter, F., Gassner, C., Shatrov, V., and Lehmann, V. (1999). Modulation of tumor necrosis factor-alpha-mediated cytotoxicity by changes of the cellular methylation state: mechanism and in vivo relevance. *Int Immunol* 11:519–527.

104. Winter-Vann, A. M., Baron, R. A., Wong, W., dela Cruz, J., York, J. D., Gooden, D. M., Bergo, M. O., Young, S. G., Toone, E. J., and Casey, P. J. (2005). A small-molecule inhibitor of isoprenylcysteine carboxyl methyltransferase with antitumor activity in cancer cells. *Proc Natl Acad Sci USA* 102:4336–4341.

104a. Ma, Y-T., Gilbert, B. A. and Rando, R. R. (1995) Farnesylcysteine analogs to probe role of prenylated protein methyltransferase. *Methods Enzymol* 250:226–234.

105. Huzoor-Akbar, Wang, W., Kornhauser, R., Volker, C., and Stock, J. B. (1993). Protein prenylcysteine analog inhibits agonist-receptor-mediated signal transduction in human platelets. *Proc Natl Acad Sci USA* 90:868–872.

106. Huzoor-Akbar, Winegar, D. A., and Lapetina, E. G. (1991). Carboxyl methylation of platelet rap1 proteins is stimulated by guanosine 5'-(3-O-thio)triphosphate. *J Biol Chem* 266:4387–4391.

107. Kowluru, A., Seavey, S. E., Li, G., Sorenson, R. L., Weinhaus, A. J., Nesher, R., Rabaglia, M. E., Vadakekalam, J., and Metz, S. A. (1996). Glucose- and GTP-dependent stimulation of the carboxyl methylation of CDC42 in rodent and human pancreatic islets and pure beta cells. Evidence for an essential role of GTP-binding proteins in nutrient-induced insulin secretion. *J Clin Invest* 98:540–555.
108. Stockand, J. D., Edinger, R. S., Al-Baldawi, N., Sariban-Sohraby, S., Al-Khalili, O., Eaton, D. C., and Johnson, J. P. (1999). Isoprenylcysteine-O-carboxyl methyltransferase regulates aldosterone-sensitive Na(+) reabsorption. *J Biol Chem* 274:26912–26916.
109. Michaelson, D., Ali, W., Chiu, V. K., Bergo, M., Silletti, J., Wright, L., Young, S. G., and Philips, M. (2005). Postprenylation CaaX Processing Is Required for Proper Localization of Ras but Not Rho GTPases. *Mol Biol Cell* 16:1606–1616.
110. Bergo, M. O., Gavino, B. J., Hong, C., Beigneux, A. P., McMahon, M., Casey, P. J., and Young, S. G. (2004). Inactivation of Icmt inhibits transformation by oncogenic K-Ras and B-Raf. *J Clin Invest* 113:539–550.
111. Reiss, Y., Seabra, M. C., Brown, M. S., and Goldstein, J. L. (1992). p21ras farnesyltransferase: purification and properties of the enzyme. *Biochem Soc Trans* 20:487–488.
112. Schaber, M. D., O'Hara, M. B., Garsky, V. M., Mosser, S. C., Bergstrom, J. D., Moores, S. L., Marshall, M. S., Friedman, P. A., Dixon, R. A., and Gibbs, J. B. (1990). Polyisoprenylation of Ras in vitro by a farnesyl-protein transferase. *J Biol Chem* 265:14701–14704.
113. Seabra, M. C., Reiss, Y., Casey, P. J., Brown, M. S., and Goldstein, J. L. (1991). Protein farne-syltransferase and geranylgeranyltransferase share a common alpha subunit. *Cell* 65:429–434.
114. Long, S. B., Casey, P. J., and Beese, L. S. (2000). The basis for K-Ras4B binding specificity to protein farnesyltransferase revealed by 2 A resolution ternary complex structures. *Structure Fold Des* 8:209–222.
115. Park, H. W., Boduluri, S. R., Moomaw, J. F., Casey, P. J., and Beese, L. S. (1997). Crystal structure of protein farnesyltransferase at 2.25 angstrom resolution. *Science* 275:1800–1804.
116. Kohl, N. E., Wilson, F. R., Thomas, T. J., Bock, R. L., Mosser, S. D., Oliff, A., and Gibbs, J. B. (1995). Inhibition of Ras function in vitro and in vivo using inhibitors of farnesyl-protein transferase. *Methods Enzymol* 255:378–386.
117. James, G. L., Goldstein, J. L., Brown, M. S., Rawson, T. E., Somers, T. C., McDowell, R. S., Crowley, C. W., Lucas, B. K., Levinson, A. D., and Marsters, J. C., Jr. (1993). Benzodiazepine peptidomimetics: potent inhibitors of Ras farnesylation in animal cells. *Science* 260:1937–1942.
118. Gibbs, J. B., and Oliff, A. (1997). The potential of farnesyltransferase inhibitors as cancer chemotherapeutics. *Annu Rev Pharmacol Toxicol* 37:143–166.
119. Gibbs, J. B., Oliff, A., and Kohl, N. E. (1994). Farnesyltransferase inhibitors: Ras research yields a potential cancer therapeutic. *Cell* 77:175–178.
120. Sepp-Lorenzino, L., Ma, Z., Rands, E., Kohl, N. E., Gibbs, J. B., Oliff, A., and Rosen, N. (1995). A peptidomimetic inhibitor of farnesyl:protein transferase blocks the anchorage-dependent and independent growth of human tumor cell lines. *Cancer Res* 15:5302–5309.
121. Rowell, C. A., Kowalczyk, J. J., Lewis, M. D., and Garcia, A. M. (1997). Direct demonstra-tion of geranylgeranylation and farnesylation of Ki-Ras in vivo. *J Biol Chem* 272:14093–14097.
122. Whyte, D. B., Kirschmeier, P., Hockenberry, T. N., Nunez-Oliva, I., James, L., Catino, J. J., Bishop, W. R., and Pai, J. K. (1997). K- and N-Ras are geranylgeranylated in cells treated with farnesyl protein transferase inhibitors. *J Biol Chem* 272:14459–14464.
123. Clarke, S., and Tamanoi, F. (2004). Fighting cancer by disrupting C-terminal methylation of signaling proteins. *J Clin Invest* 113:513–515.
124. Winter-Vann, A. M., and Casey, P. J. (2005). Post-prenylation-processing enzymes as new targets in oncogenesis. *Nat Rev Cancer* 5:405–412.

125. Chiu, V. K., Bivona, T., Hach, A., Sajous, J. B., Silletti, J., Wiener, H., Johnson, R. L., 2nd, Cox, A. D., and Philips, M. R. (2002). Ras signalling on the endoplasmic reticulum and the Golgi. *Nat Cell Biol* 4:343–350.

126. Chen, Z., Otto, J. C., Bergo, M. O., Young, S. G., and Casey, P. J. (2000). The C-Terminal polylysine region and methylation of K-ras are critical for the interaction between K-ras and microtubules. *J Biol Chem* 275:41251–41257.

127. Anderson, J. L., Henriksen, B. S., Gibbs, R. A., and Hrycyna, C. A. (2005). The isoprenoid substrate specificity of isoprenylcysteine carboxylmethyltransferase: Development of novel inhibitors. *J Biol Chem* 280:29454–29461.

128. Ashby, M. N., King, D. S., and Rine, J. (1992). Endoproteolytic processing of a farnesylated peptide in vitro. *Proc Natl Acad Sci USA* 89:4613–4617.

129. Hrycyna, C. A., and Clarke, S. (1992). Maturation of isoprenylated proteins in Saccharomyces cerevisiae. Multiple activities catalyze the cleavage of the three carboxyl- terminal amino acids from farnesylated substrates in vitro. *J Biol Chem* 267:10457–10464.

130. Deschenes, R. J., and Broach, J. R. (1987). Fatty acylation is important but not essential for Saccharomyces cerevisiae RAS function. *Mol Cell Biol* 7:2344–2351.

131. Dong, X., Mitchell, D. A., Lobo, S., Zhao, L., Bartels, D. J., and Deschenes, R. J. (2003). Palmitoylation and plasma membrane localization of Ras2p by a nonclassical trafficking pathway in Saccharomyces cerevisiae. *Mol Cell Biol* 23:6574–6584.

10

Genetic Approaches to Understanding the Physiologic Importance of the Carboxyl Methylation of Isoprenylated Proteins

STEPHEN G. YOUNG[a] • STEVEN G. CLARKE[b] • MARTIN O. BERGO[c] • MARK PHILLIPS[d] • LOREN G. FONG[a]

[a]Division of Cardiology
Department of Internal Medicine
[b]Department of Chemistry and Biochemistry and the Molecular Biology Institute
University of California, Los Angeles
405 Hilgard Avenue
Los Angeles, CA 90095, USA

[c]Wallenberg Laboratory, Department of Internal Medicine
Sahlgrenska University Hospital
SE-431 80 Mölndal
Grothenburg S-41345, Sweden

[d]Department of Medicine
Cell Biology and Pharmacology
New York University School of Medicine
530 First Avenue
New York, NY 10016, USA

I. Abstract

This chapter examines recent studies on the physiologic importance of the carboxyl methylation of isoprenylated proteins, focusing largely on what has

THE ENZYMES, Vol. XXIV
Copyright © 2006 by Elsevier Inc.

been learned from cells lacking the *Icmt* methyltransferase. Proteins terminating with a *CaaX* motif (e.g., the nuclear lamins, the Ras family of proteins) undergo posttranslational modification of a carboxyl-terminal cysteine with an isoprenyl lipid (a process generally called protein isoprenylation or protein prenylation). Following this lipidation step, *CaaX* proteins generally undergo two additional processing steps: endoproteolytic release of the last three residues of the protein (i.e., the −*aaX* of the *CaaX* motif) and methylesterification of the newly exposed isoprenylcysteine α-carboxyl group.

The *CaaX* proteins are not, however, the only prenylated proteins that undergo carboxyl methylation. A subset of the Rab family of proteins, those terminating with a *CXC* motif, undergo methylesterification of a carboxyl-terminal geranylgeranyl-cysteine. The methylation of *CaaX* proteins and the *CXC* Rab proteins is carried out by a single membrane methyltransferase of the endoplasmic reticulum, Icmt (for isoprenylcysteine carboxyl methyltransferase). Many studies have shown that protein prenylation is essential for the proper intracellular targeting and function of numerous intracellular proteins, but the physiologic importance of the carboxyl methylation step has remained less certain. Here, we review recent studies that have shed light on the importance of carboxyl methylation of prenylated proteins.

II. Introduction

Proteins terminating with a *CaaX* sequence undergo a series of posttranslational processing steps, beginning with protein prenylation, which is the attachment of a 15-carbon farnesyl or 20-carbon geranylgeranyl lipid to the thiol group of a carboxyl-terminal cysteine (the *C* of the *CaaX* motif) [1]. Next, the last three amino acids of the protein (i.e., the −*aaX* of the *CaaX* motif) are removed by Rce1, a prenylprotein-specific endoprotease within the endoplasmic reticulum (ER) membrane [2–4]. Third, the carboxyl group of the newly exposed isoprenyl-cysteine is methylated by Icmt [4–6]. These processing reactions have been studied most intensely with **a**-factor, a farnesylated mating pheromone in *Saccharomyces cerevisiae* [2, 7–9], and with yeast and mammalian Ras proteins, small GTP-binding proteins involved in signal transduction [10, 11]. However, many cellular proteins with diverse functions contain a *CaaX* motif and undergo prenylation and the "postisoprenylation" modifications [4].

The enzymes responsible for the prenylation of the *CaaX* proteins, protein farnesyltransferase and protein geranylgeranyltransferase I, have been characterized extensively in both yeast and higher organisms [1]. During the past few years, significant progress has been made in understanding the *CaaX* endoproteases and the *CaaX* methyltransferase. Two genes from *Saccharomyces cerevisiae*, *RCE1* and *STE24*, were identified as playing a role in the endoproteolytic removal of the −*aaX* from Ras2p and the precursor of the yeast mating pheromone **a**-factor [2]. Rce1p (for Ras and **a**-factor converting enzyme 1) is essential for cleaving the −*aaX* from yeast Ras proteins, and is also capable of releasing the −*aaX* from yeast **a**-factor.

Ste24p is a zinc metalloproteinase that plays dual roles in the endoproteolytic processing of **a**-factor. Along with Rce1p, Ste24p is capable of carrying out the carboxyl-terminal cleavage reaction in **a**-factor (i.e., the release of the −*aaX*) [2, 9]. In addition, Ste24p cleaves seven amino acids from the amino terminus of the protein [9]. Ste24p plays no role in the endoproteolytic processing of the yeast Ras proteins [2]. In fact, **a**-factor is the only well-established substrate for Ste24p in yeast [2]. *RCE1* and *STE24* have orthologs in mammals. In mice, Rce1 is critical for the endoproteolytic processing of many *CaaX* proteins [12, 13]. Zmpste24 is critical for the endoproteolytic processing of prelamin A [14, 15].

After endoproteolytic release of the −*aaX* from *CaaX* proteins, the carboxyl group of the isoprenylcysteine is methylated by Icmt, an integral membrane protein of the endoplasmic reticulum. The enzyme responsible for the carboxyl methylation of prenylcysteines was first identified in yeast—the product of the *STE14* gene [16]. In yeast, Ste14p methylates many *CaaX* proteins, including **a**-factor and the Ras proteins. Ste14p contains six transmembrane helices [4, 17]. The mammalian protein, Icmt, also contains multiple membrane-spanning helices and is located in the ER [6, 18]. Icmt is essential for methylating all *CaaX* proteins. In addition, Icmt methylates a subset of the Rab GTPases—those terminating with a *CXC* motif [19].

The fact that Icmt methylates many key cellular proteins, including several regulatory GTPases (both heterotrimeric and monomeric), has suggested the possibility that carboxyl methylation is functionally important. In this chapter, we will review some relatively recent information on the functional importance of methylation of prenylcysteines in cells.

III. STE14 from *S. cerevisiae*

Methyltransferase activity against farnesylated peptides was initially detected in a crude membrane fraction of wild-type *S. cerevisiae* [16]. Little or no activity was detected in cytosolic fractions, and nonfarnesylated peptides were not methyl acceptors. This methyltransferase activity was absent in *STE14*-deficient yeast [16]. Transformation of wild-type and mutant cells with a plasmid containing the *STE14* gene resulted in the production of an active methyltransferase [16]. These experiments and subsequent studies confirmed that the *S. cerevisiae STE14* gene was the structural gene for isoprenylcysteine carboxyl methyltransferase [protein-*S* isoprenylcysteine *O*-methyltransferase (E. C. number 2.1.1.100.)] [5]. Ste14p contains 239 amino acid residues and its catalytic activity appears to be dependent on a bound metal ion [20]. As judged by immunofluorescence microscopy and cellular fractionation experiments, Ste14p is located in the ER membrane and has the active site facing the cytosol [17].

Neither the intracellular nor the extracellular forms of *a*-mating factor were methylated in *STE14*-deficient (*ste14Δ*) yeast [5]. Similarly, when *RAS1* and *RAS2* were overexpressed in *ste14Δ* yeast no methyl esters could be detected on either Ras protein [5]. Metabolic labeling experiments revealed that Ste14p was responsible for methylating a large number of polypeptides [21]. Thus, it was quickly established that Ste14p methylates a variety of yeast proteins aside from the Ras proteins and **a**-factor. Pulse-chase experiments revealed virtually no turnover of methyl esters in some proteins and only slow turnover in other proteins [21]. These results suggest, at least in yeast, that the methylation of prenylcysteines is fairly complete and is not generally readily reversible.

The availability of *S. cerevisiae* mutants lacking Ste14p provided an opportunity to address the functional importance of prenylcysteine methylation. Methylation of **a**-factor appeared to be important for several reasons. First, wild-type **a**-factor was ~200-fold more active in binding to the **a**-factor receptor, compared with the nonmethylated **a**-factor produced by *ste14Δ* yeast [5]. That result was not particularly surprising, given that *ste14Δ* mutants were originally identified because of a sterile phenotype [22]. In addition, the lack of **a**-factor methylation results in enhanced proteolytic degradation inside cells and an essentially complete block in the export of the protein from cells [23].

Aside from the loss of mating ability, *ste14Δ* yeast are viable and grow well [5, 23]—a somewhat surprising result considering the vital importance of Ras1p and Ras2p for growth and survival. Biochemical studies revealed that the post-translational processing of the Ras proteins in *ste14Δ* yeast was perturbed, and there were reduced amounts of Ras proteins within the membrane fraction. However, genetic studies suggested that *STE14* deficiency did not have a major impact on phenotypes elicited by an activated Ras protein, suggesting that defective posttranslational processing was not accompanied by major effects on Ras function [5].

The methylation of *CaaX* proteins means that these proteins terminate with a carboxyl methyl ester rather than an α-carboxylate anion. This change would be expected to render the carboxyl terminus of the protein more hydrophobic. Indeed, methylated prenylcysteine compounds partition more efficiently into phospholipid vesicles [24] and into *n*-octanol [25, 26] than do nonmethylated forms. Also, the nonmethylated Ras proteins in *ste14Δ* yeast bound bind to membranes with lower avidity. It seems possible, however, that carboxyl methylation is important beyond an effect on hydrophobicity, at least for some proteins. Support for this idea comes from studies with **a**-factor. As noted earlier, both the secretion of **a**-factor *via* the Ste6p transporter [23] and its interaction with its G protein-coupled receptor [27] are dependent on carboxyl methylation, suggesting that methylation is important for protein-protein interactions. Also, the increased proteolytic susceptibility of **a**-factor in cells lacking Ste14p [23] suggests that the methylation event may confer protection against intracellular proteases.

IV. Biochemical Studies of Icmt Function in Mammals

After Ste14p was characterized in *S. cerevisiae*, a 426-bp mouse cDNA with homology to yeast *STE14* appeared in the EST databases. This sequence was used to clone the human *ICMT* cDNA [6]. This 3595-bp cDNA encoded a 284-amino acid protein that is 26% identical to yeast Ste14p. The single *Icmt* gene in the mouse genome is on chromosome 4; the single human *ICMT* gene is on chromosome 1p36.21 and is expressed as two mRNA variants by alternative splicing. The major variant 1 encodes the full-length polypeptide (Genbank locus NM_01245; gi:24797154); the minor variant 2 has an additional exon inserted into the codon for Tyr-96 and appears to encode an N-terminal truncated protein that begins with Met-97 (Genbank locus NM_170705; gi:24797155). At this point, the potential truncated protein has not been characterized and it is not known whether it would be catalytically active.

Transfection of the human *ICMT* clone into *ste14Δ* yeast largely restored the mating defect, indicating that the human enzyme could carry out the carboxyl methylation of **a**-factor. Like Ste14p, the human enzyme is predicted to contain multiple transmembrane helices. When tagged with GFP, Icmt is located exclusively in the ER without any expression on the plasma membrane (Figure 10.1) [6]. Transient transfection of the human *ICMT* cDNA into Cos-1 cells strikingly increased the methylation of *N*-acetyl-*S*-farnesylcysteine (AFC), *N*-acetyl-*S*-geranylgeranylcysteine (AGGC), and several mammalian *CaaX* proteins [6].

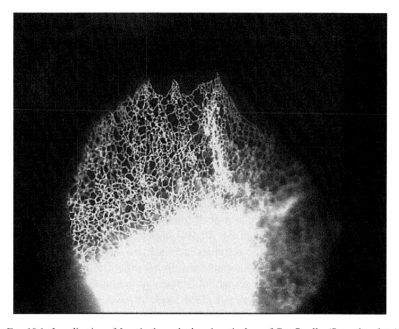

FIG. 10.1. Localization of *Icmt* in the endoplasmic reticulum of Cos-7 cells. (See color plate.)

Several studies have compared the activity of proteins with and without the carboxyl-terminal methyl ester and quantified the effect of this modification on protein function. In general, these studies have shown that methylation either does not affect, or affects minimally, the activities of isoprenylated proteins. One approach has been to use isolated proteins and membrane fractions from cells preincubated with general methyltransferase inhibitors. With that approach, it has been possible to show that methylation of the γ subunit of the βγ complex enhances the ability of pertussis toxin to ADP-ribosylate the α subunit by about twofold while not enhancing the membrane attachment of the βγ subunit itself [28, 29].

Another approach has been to use pig liver esterase to catalyze the hydrolysis of the carboxyl-terminal isoprenylcysteine methyl esters in the intact protein [30]. With this approach, it was possible to show that the methylated transducin βγ was about twofold more efficient than the nonmethylated form [30]. Also, the methylated transducin βγ was more than 10-fold more potent in activating phosphatidylinositol-specific phospholipase C and phosphoinositide 3-kinase. Similar experiments showed that methylation of the γ subunit of transducin βγ can enhance its affinity for phosducin by about twofold [31].

Functional studies of the mammalian smg-25A/rab3A protein (an isoprenylated *CXC* protein) have also been reported [32]. When the properties of the nonmethylated protein were compared with those of the fully methylated cellular-derived species, no differences were observed in membrane-binding properties or in its interactions with its GDP/GTP exchange protein [32]. Interestingly, RhoA, a small GTP-binding protein, was found to have a decreased half-life when methylation was inhibited [33], suggesting as with yeast **a**-factor that methylation may stabilize at least some isoprenylated species against proteolytic degradation [34].

Although protein methyl esters are in theory good substrates for specific esterases that might reverse the modification, there is little evidence to date that methyl esters on isoprenylated proteins can be physiologically hydrolyzed to a free carboxyl group. Initial pulse-chase experiments have indicated slow if any turnover [21, 35] and it appears that polypeptides such as the γ subunit of transducin are fully modified when isolated [36]. Methyl esters are readily cleaved by pig liver esterase *in vitro* [25], but it has not been established that the enzyme catalyzes this type of reaction *in vivo*.

The hydrolysis of *N*-acetyl-farnesylcysteine methyl ester and methylated transducin can be catalyzed by membrane fractions of bovine retinal rods [37, 38], and at least two activities have been found in rabbit brain that catalyze the hydrolysis of methyl esters from the small G protein G25K [39], but the *in vivo* significance of these enzymatic activities are uncertain. On the other hand, there is evidence that Rho proteins may not be fully modified in the cell, indicating either that methylation occurs slowly after translation and the initial carboxyl-terminal processing or that there is indeed some turnover of the methyl group [40, 41]. If reversible methylation were to occur, one could imagine that protein function and biological responses could be modulated by the relative activity of methyltransferases and methylesterases. This is a ripe area for further study.

V. Inactivating *Icmt* in Mouse Embryonic Stem Cells

The functional importance of prenylation/endoproteolysis/methylation has been studied by expressing mutant forms of Ras that cannot be prenylated (i.e., Ras proteins with cysteine to serine substitutions in the *CaaX* motif) [42, 43]. However, such studies are obviously not useful for dissecting the biological role of methylation.

To determine unambiguously the biological role of carboxyl methylation, our group used gene-targeting in mouse embryonic stem (ES) cells to inactivate *Icmt*. A sequence-replacement vector was used to delete exon 1 sequences (containing the translational start site) and the first 65 amino acids of the protein (containing the first two membrane-spanning helices). Southern blots identified multiple ES cell clones that were heterozygous for the knockout mutation (*Icmt$^{+/-}$*)]. Subsequently, *Icmt$^{-/-}$* cells were obtained by high-G418 selection. The *Icmt$^{-/-}$* ES cells grew more slowly than *Icmt$^{+/-}$* or *Icmt$^{+/+}$* cells.

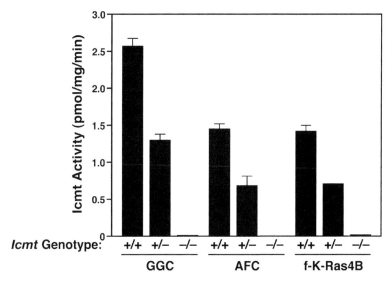

FIG. 10.2. Isoprenylcysteine carboxyl methyltransferase activity is absent in *Icmt$^{-/-}$* cells. *Icmt$^{+/+}$*, *Icmt$^{+/-}$*, and *Icmt$^{-/-}$* extracts were mixed with S-adenosyl-L-[*methyl*-14C]methionine (10 μM) and three different methyl-accepting substrates: N-acetyl-S-geranylgeranyl-L-cysteine (AGGC), N-acetyl-S-farnesyl-L-cysteine (AFC), and farnesyl-K-Ras4B. Methylation (pmol/mg total cell protein/min) was measured with a base-hydrolysis assay [12, 44]. Similar results were observed for three independent *Icmt$^{+/-}$* clones and two independent *Icmt$^{-/-}$* clones. *(Reproduced with permission from* The Journal of Biological Chemistry *[47].)*

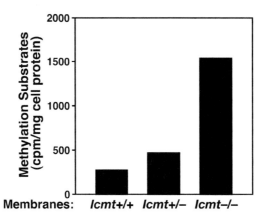

FIG. 10.3. Icmt substrates accumulate in *Icmt*[-/-] cells. Accumulation of Icmt substrates documented with recombinant Ste14p. *Icmt*[+/+], *Icmt*[+/-], and *Icmt*[-/-] extracts (60 μg) were mixed with S-adenosyl-L-[*methyl*-14C]methionine (10 μM) and yeast Ste14p. The bar graph shows little base-labile methylation in *Icmt*[+/+] extracts, but substantial methylation of *Icmt*[-/-] extracts (indicating an accumulation of Icmt substrates). *(Reproduced with permission from* The Journal of Biological Chemistry *[47].)*

Whole-cell extracts from *Icmt*[+/+] cells were capable of using S-adenosyl-L-[*methyl*-14C]methionine to methylate both AGGC and AFC, but extracts from the *Icmt*[-/-] cells were not (Figure 10.2). The results were identical when farnesyl-K-Ras was used as the substrate (Figure 10.2). These studies strongly suggested that Icmt was the sole protein with isoprenylcysteine methyltransferase activity, at least in ES cells.

To determine if nonmethylated Icmt substrates accumulate in the absence of Icmt, we incubated whole-cell extracts of *Icmt*[+/+], *Icmt*[+/-], and *Icmt*[-/-] cells with S-adenosyl-L-[*methyl*-14C]methionine and *Sf9* membranes expressing yeast *STE14*. We then measured base-labile methylation as an index of the amount of Icmt substrates in cells. Very low levels of methylation were observed with the *Icmt*[+/+] extracts, and some of that could easily have been due to other carboxyl methyltransferase activities (e.g., Pcmt1 [44, 45]). In contrast, we observed a substantial increase in "methylatable protein substrates" in the *Icmt*[-/-] cells (Figure 10.3). These studies established that Icmt protein substrates accumulate in the absence of Icmt.

To determine whether the absence of Icmt altered the intracellular distribution of K-Ras, we compared the relative levels of K-Ras in the membrane (P100) and cytosolic (S100) fractions of *Icmt*[+/+] and *Icmt*[-/-] ES cells (Figure 10.4). In *Icmt*[+/+] ES cells, the vast majority of K-Ras was in the P100 fraction, whereas a significant proportion of the K-Ras in *Icmt*[-/-] cells was present in the S100 fraction. Also, a significant fraction of a GFP–K-Ras-tail fusion protein (GFP fused to the last 20 amino acids of K-Ras, including the *CaaX* motif) was trapped within the cytoplasm of *Icmt*[-/-] cells, although some clearly reached the plasma

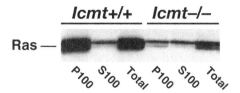

FIG. 10.4. P100/S100 distribution of K-Ras in *Icmt*$^{+/+}$ and *Icmt*$^{-/-}$ ES cells. Membrane (P100) and cytosolic (S100) fractions from *Icmt*$^{+/+}$ and *Icmt*$^{-/-}$ ES cells were isolated by ultracentrifugation. Ras proteins were immunoprecipitated from the P100 and S100 fractions (as well as from the total cell lysates) with antibody Y13-259. The immune complexes were then analyzed by western blotting with the K-Ras-specific antibody Ab-1. *(Reproduced with permission from* The Journal of Biological Chemistry *[47].)*

membrane (Figure 10.5). In contrast, nearly all of the fusion protein in *Icmt*$^{+/+}$ cells was localized along the plasma membrane.

VI. *Icmt* Knockout Mice

Previously, we had demonstrated that homozygous *Rce1* knockout mice died very late in embryonic development [between embryonic (E) day 15 and 20], and a few *Rce1* knockout mice were even born alive [12]. Because several studies had suggested the existence of more than one isoprenylcysteine methyltransferase activity [18, 46], our *a priori* prediction was that *Icmt*-deficient mice might be affected less severely than the *Rce1*-deficient mice. To test this prediction, we used the *Icmt*$^{+/-}$ ES cells to generate *Icmt*$^{+/-}$ mice, which were then intercrossed to generate *Icmt*$^{-/-}$ mice [47].

Contrary to our expectation, the *Icmt*$^{-/-}$ mice died *earlier*, not later, than the *Rce1* knockout mice. Genotyping of embryos at various stages of development

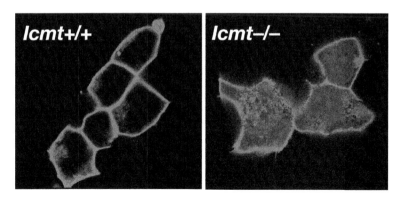

FIG. 10.5. Mislocalization of a green fluorescent protein–K-Ras fusion protein in *Icmt*$^{-/-}$ cells. Confocal images of *Icmt*$^{+/+}$ (right) and *Icmt*$^{-/-}$ (left) ES cells that had been transfected with a GFP–K-Ras-tail fusion plasmid. *(Reproduced with permission from* The Journal of Biological Chemistry *[47].)* (See color plate.)

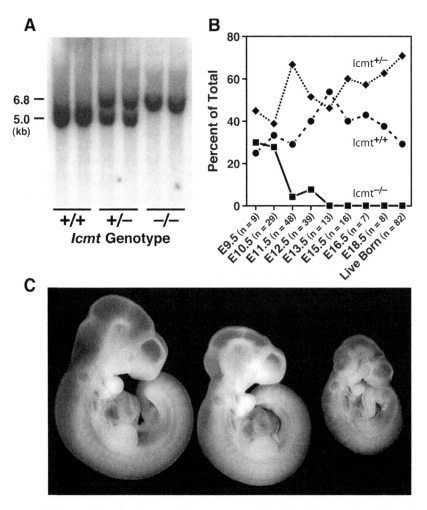

FIG. 10.6. Death of *Icmt*⁻/⁻ embryos at midgestation. (A) Southern blot of the genomic DNA from the yolk sacs of two *Icmt*⁺/⁺ embryos, two *Icmt*⁺/⁻ embryos, and two *Icmt*⁻/⁻ embryos. Genomic DNA was digested with *Bam*HI; the Southern blot was hybridized with a 5-flanking probe [47]. The *Bam*HI fragment in the wild-type allele is 5.0 kb, whereas it is 6.8 kb in the targeted allele. (B) Percentages of *Icmt*⁺/⁺, *Icmt*⁺/⁻, and *Icmt*⁻/⁻ embryos surviving at different time points. (C) *Icmt*⁺/⁺, *Icmt*⁺/⁻, and *Icmt*⁻/⁻ embryos at E11.5. *(Reproduced with permission from* The Journal of Biological Chemistry *[78].)* (See color plate.)

revealed that *Icmt*⁻/⁻ embryos began to die at embryonic day 10.5 (Figure 10.6). By E11.5, there were only a few viable *Icmt*⁻/⁻ embryos; surviving *Icmt*⁻/⁻ embryos had beating hearts and red blood cells, but were much smaller than heterozygous and wild-type embryos (Figure 10.6). Virtually all of the *Icmt*⁻/⁻ mice died by E12.5, several days before the first of the homozygous *Rce1* knockout embryos started

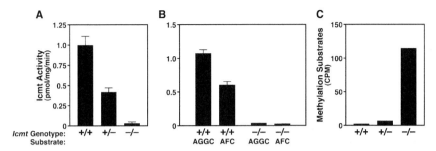

FIG. 10.7. Methyltransferase activity in embryo extracts. (A) Ability of extracts from $Icmt^{+/+}$, $Icmt^{+/-}$, and $Icmt^{-/-}$ embryos to methylate farnesyl-K-Ras4B. (B) Methylation of N-acetyl-S-geranylgeranyl-L-cysteine (AGGC) and N-acetyl-S-farnesyl-L-cysteine (AFC) by embryo extracts. The bars show the mean values; error bars show the standard deviation. In some cases, the standard deviations were low so that the error bar is not visible above the bar. (C) Accumulation of methylation substrates in embryo extracts. Extracts (100 µg) from $Icmt^{+/+}$, $Icmt^{+/-}$, and $Icmt^{-/-}$ embryos ($n = 2$) were mixed with S-adenosyl-L-[$methyl$-^{14}C]methionine (10 µM) and yeast Ste14p; methylation was assessed with a base-hydrolysis assay. The bar graph shows the average of values for the two embryos of each genotype. *(Reproduced with permission from* The Journal of Biological Chemistry *[78].)*

to die [12]. Another research group generated $Icmt^{-/-}$ mice and concluded that the embryos died as a consequence of defective development of the liver [48].

To gain insights into the importance of $Icmt$ in different organs, we generated chimeric mice ($n = 10$) from $Icmt^{-/-}$ ES cells. Interestingly, the $Icmt^{-/-}$ cells contributed significantly to the development of skeletal muscle, but made a negligible contribution to the formation of the brain, liver, and testis.

To determine if $Icmt^{-/-}$ embryos retained any biochemical activity capable of methylating isoprenylated proteins, we tested the capacity of extracts of $Icmt^{-/-}$ embryos to methylate farnesyl-K-Ras and two small-molecule substrates, AFC and AGGC (Figure 10.7). No enzymatic activity above background levels was identified in $Icmt^{-/-}$ embryos. We also measured methyltransferase activity against small-molecule substrates in extracts from $Icmt^{+/-}$ embryos and tissues (liver, brain, and heart). Activities were invariably reduced by 50% (data not shown), not <50%, as might be the case if there had been redundant methyltransferase activities. Also, there was a substantial accumulation of methyltransferase substrates in extracts from $Icmt^{-/-}$ embryos (i.e., an accumulation of cellular proteins that could be methylated by yeast Ste14p) (Figure 10.7), again strongly suggesting that Icmt is the only enzymatic activity capable of methylating isoprenylcysteine residues.

VII. Altered Electrophoretic Mobility of Ras Proteins in $Icmt^{-/-}$ Fibroblasts

The electrophoretic mobility of the Ras proteins in $Rce1^{-/-}$ mouse embryonic fibroblasts (MEF) is abnormally slow, reflecting an absence of endoproteolytic

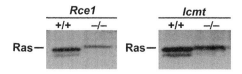

FIG. 10.8. Abnormal migration of the Ras proteins in *Icmt*$^{-/-}$ cells. *(Reproduced with permission from* Methods in Enzymology *[79].)*

processing [12]. This result was perhaps not surprising, given the fact that the Ras proteins in *Rce1*$^{-/-}$ MEFs are three amino acid longer than they are in *Rce1*$^{+/+}$ cells. Remarkably, however, the Ras proteins in *Icmt*$^{-/-}$ MEFs also exhibited retarded electrophoretic mobility (Figure 10.8). One could imagine several possible explanations for the retarded migration of the Ras proteins in *Icmt*$^{-/-}$ MEFs. One explanation is simply that the replacement of the α-carboxylate anion with a methyl ester increases electrophoretic mobility.

Another explanation is that the absence of methylation in *Icmt*$^{-/-}$ MEFs leads to reduced palmitoylation of the Ras proteins. The latter explanation has never been excluded experimentally, but we are somewhat skeptical about this explanation because Ras palmitoylation was not reduced in *Rce1*-deficient cells [49]. Finally, the absence of the isoprenyl group can reduce the electrophoretic mobility of at least the yeast RAS polypeptides (50). If the absence of the methyl ester leads to the proteolytic clipping of the "unprotected" C-terminus, the removal of the farnesylcysteine residue could also contribute to the retarded mobility on SDS gels [34, 51].

VIII.   Mislocalization of Ras Proteins in *Icmt*$^{-/-}$ Fibroblasts

As noted previously, Ras proteins tagged with green fluorescent proteins are mislocalized in mouse ES cells, as judged by confocal immunofluorescence microscopy. We have made similar observations in fibroblasts prepared from *Icmt*$^{-/-}$ embryos. We transfected *Icmt*$^{-/-}$ and *Icmt*$^{+/+}$ MEFs with a GFP–K-Ras-tail fusion plasmid (a GFP plasmid linked in frame to sequences encoding the last 20 amino acids of K-Ras) and then examined the cells by confocal microscopy. In *Icmt*$^{+/+}$ cells, the fluorescence was, as expected, localized to the plasma membrane (Figure 10.9). In the *Icmt*$^{-/-}$ cells, most of the fluorescence was cytosolic or was associated with internal membrane compartments, but small amounts of fluorescence were located at the plasma membrane in the *Icmt*$^{-/-}$ cells (Figure 10.9).

In a follow-up study, the localization of GFP fusion proteins for all three Ras proteins (K-Ras, N-Ras, and H-Ras) was examined [41]. In these cases, the GFP fusion proteins encoded GFP linked in-frame to the entire coding sequences of K-Ras, N-Ras, and H-Ras. In the case of the N-Ras and H-Ras fusions, none

Icmt⁺/⁺

Icmt⁻/⁻

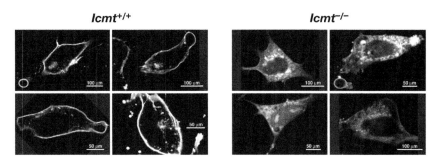

FIG. 10.9. Confocal micrographs of spontaneously immortalized $Icmt^{+/+}$ and $Icmt^{-/-}$ fibroblasts that had been transfected with a GFP–K-Ras-tail fusion construct. Several of these images have been published in *The Journal of Clinical Investigation* [55] and are reproduced with permission.

reached the plasma membrane; all of the fluorescence was in the cytosol and in internal membrane compartments (e.g., Golgi apparatus) (Figure 10.10). In the case of the K-Ras fusion, most of the fluorescence was located in the cytosol and in internal membrane compartments, but some was clearly detectable at the plasma membrane.

Interestingly, the impact of carboxyl methylation on the localization of the GFP-Ras fusions depends on the fact that these proteins are farnesylated. When mutations were introduced into the *CaaX* motif so as to create GFP-Ras fusions that were geranylgeranylated, the impact of *Icmt* deficiency on intracellular localization was no longer detectable [41]. Thus, geranylgeranylated GFP–N-Ras and geranylgeranylated GFP–K-Ras are largely localized to the plasma membrane in $Icmt^{-/-}$ MEFs and are *not* mislocalized to the cytosol (i.e., a pattern indistinguishable from that in $Icmt^{+/+}$ cells) (Figure 10.11).

H-Ras   N-Ras   K-Ras   H-Ras61L H-Ras12V

Icmt ⁺/⁺

Icmt ⁻/⁻

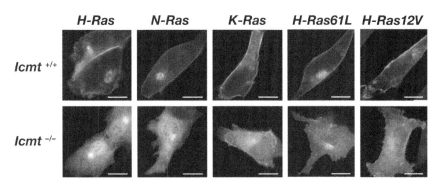

FIG. 10.10. Postprenylation *CaaX* processing is required for proper localization of all three Ras isoforms. MEFs expressing (+/+) or deficient in (−/−) *Icmt* were transfected with GFP-tagged forms of H-, N- and K-Ras as indicated and imaged 24 h after transfection. *(Reproduced with permission from* Molecular Biology of the Cell *[41].)*

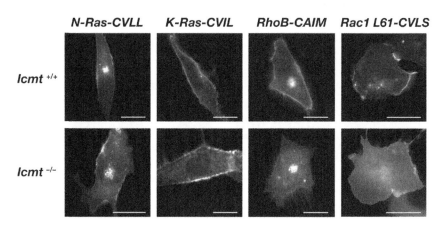

FIG. 10.11. A prenyl swap reverses the sensitivity of Ras proteins to mislocalization in *Icmt*-deficient cells. MEFs expressing (+/+) or deficient (−/−) in *Icmt* or *Rce1* were transfected with GFP-tagged Ras proteins with mutations in their *CaaX* sequences that direct geranylgeranylation of Ras proteins. *(Reproduced with permission from* Molecular Biology of the Cell *[41].)*

Consistent with these observations, the intracellular localization of RhoB and Rac1 are not affected in *Icmt*$^{−/−}$ MEFs, likely because those proteins are geranylgeranylated. However, we suspected that RhoB and Rac1 localization would be sensitive to the absence of methylation if those proteins were farnesylated. This suspicion was confirmed. When the *CaaX* motifs of RhoB and Rac1 were mutated so that these proteins were farnesylated, the localization of these proteins was sensitive to methylation. Farnesylated GFP–Rac1 and farnesylated GFP–RhoB were located mainly along the plasma membrane in wild-type cells. In *Icmt*$^{−/−}$ MEFs, these proteins were mislocalized away from the plasma membrane [41] (Figure 10.11).

IX.   Impact of Methylation on the Binding of K-Ras to Microtubules

Thissen et al. [52] identified a specific and prenylation-dependent interaction between tubulin/microtubules and K-Ras, but not H-Ras or several other small GTPases [52]. Later, Chen and co-workers [53] demonstrated that the interaction between K-Ras and microtubules is dependent on both the carboxyl-terminal polylysine domain of K-Ras as well as K-Ras endoproteolysis and methylation [53]. Partially processed K-Ras that was farnesylated but not clipped (i.e., retained the −aaX residues) bound to microtubules. However, endoproteolytic removal of the −aaX from K-Ras abolished all binding to microtubules. Surprisingly, the binding of K-Ras to microtubules was restored by methylation of the carboxyl-terminal prenylcysteine.

Consistent with these results, localization of the GFP–K-Ras fusion was paclitaxel-sensitive in cells lacking *Rce1*, while no paclitaxel effect was observed in cells lacking the methyltransferase Icmt [53]. These studies suggested the possibility of a functional consequence of prenylcysteine methylation [53]. However, it should be noted that the laboratory of John Hancock was not able to identify any obvious colocalization of K-Ras and microtubules in living cells [54], although a functional microtubule network was required for K-Ras to transit the cell and reach the plasma membrane.

X. Icmt Is Also Responsible for Methylating the *CXC* Rab Proteins

The Rab proteins undergo geranylgeranylation by protein geranylgeranyl-transferase type II at a pair of carboxyl-terminal cysteines. A subset of these proteins, those terminating in *CXC* (i.e., Rab3B, Rab3D, and Rab6) undergo methylation of the carboxyl-terminal geranylgeranylcysteine [19]. It had been suggested that the *CaaX* proteins and *CXC* Rab proteins are methylated by distinct methyltransferase activities [46]. If this were true, then the *CXC* Rab proteins might undergo carboxyl methylation in *Icmt*-deficient cells. This was not the case. Recombinant, *in vitro* prenylated Rab proteins failed to be methylated by extracts from *Icmt*$^{-/-}$ embryos, but were readily methylated by extracts from *Icmt*$^{+/+}$ embryos (Figure 10.12). Thus, Icmt is responsible for methylating both the *CaaX* proteins and the *CXC* Rab proteins. It is conceivable that the more severe embryonic lethal phenotype of *Icmt* deficiency in mice, compared with *Rce1* deficiency, relates to the fact that Icmt is required for Rab processing whereas Rce1 is not.

XI. Assessing the Impact of *Icmt* Deficiency on Ras Transformation

To gain insights into the importance of *Icmt* for cell growth and Ras-transformation, we used gene targeting to create a conditional allele of *Icmt* (Figure 10.13a). The targeted ES cells were used to create chimeric mice, which were bred to produce heterozygous mice (*Icmt*$^{+/flx}$). When *Icmt*$^{+/flx}$ mice were intercrossed, homozygous mice (*Icmt*$^{flx/flx}$) were born at the predicted mendelian ratios and were entirely healthy. *Cre*-mediated excision of the floxed segment of *Icmt* in *Icmt*$^{flx/flx}$ cells yielded cells with two knockout alleles (*Icmt*$^{\Delta}$) (Figure 10.13b). As expected, *Icmt*$^{\Delta/\Delta}$ cells lacked Icmt enzymatic activity (Figure 10.13c) and had

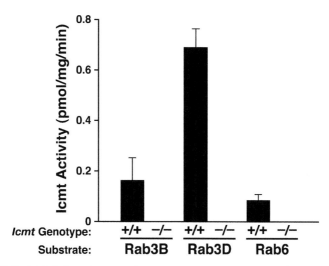

FIG. 10.12. Methylation of *CXC* Rab proteins by extracts from $Icmt^{+/+}$ and $Icmt^{-/-}$ cells. Recombinant Rab3B, Rab3D, and Rab6 were geranylgeranylated with recombinant geranylgeranyl-transferase II and tested as methylation substrates with extracts from $Icmt^{+/+}$ and $Icmt^{-/-}$ fibroblasts. The bars show the mean values; error bars show the standard deviation. *(Reproduced with permission from* The Journal of Biological Chemistry *[78].)*

an accumulation of Icmt substrates (Figure 10.13d). Similar to cells homozgygous for the conventional knockout allele (i.e., $Icmt^{-/-}$ cells) [47], the Ras proteins in $Icmt^{\Delta/\Delta}$ cells were mislocalized [55].

Interestingly, inactivation of *Icmt* in K-Ras-transfected $Icmt^{flx/flx}$ fibroblasts resulted in an accumulation of total Ras proteins and GTP-bound Ras proteins within cells (Figure 10.14). The higher levels of the Ras proteins in $Icmt^{\Delta/\Delta}$ cells, compared with the parental $Icmt^{flx/flx}$ cells, was apparently caused by a decreased turnover of Ras proteins [55]. In contrast to the results with the Ras proteins, we found that *Icmt* deficiency resulted in reduced amounts of RhoA in cells. The steady-state levels of total RhoA and GTP-bound RhoA in K-Ras-transfected $Icmt^{\Delta/\Delta}$ fibroblasts were only about 5 to 10% of those in the parental K-Ras-$Icmt^{flx/flx}$ cells (Figure 10.15). This finding was clearly due to absence of Icmt activity, in that K-Ras-$Icmt^{\Delta/\Delta}$ fibroblasts expressing the human *ICMT* cDNA exhibited normal levels of RhoA [55] (Figure 10.15).

The low levels of RhoA protein in K-Ras-$Icmt^{\Delta/\Delta}$ fibroblasts were not accompanied by a comparable reduction in the levels of RhoA mRNA [55] (Figure 10.15). To determine if accelerated Rho turnover was responsible for the low levels of RhoA in the K-Ras-transfected $Icmt^{\Delta/\Delta}$ fibroblasts, we used metabolic labeling/pulse-chase experiments to assess the half-life of Rho proteins [55]. These studies revealed that the Rho proteins disappeared more quickly in the K-Ras-$Icmt^{\Delta/\Delta}$ cells

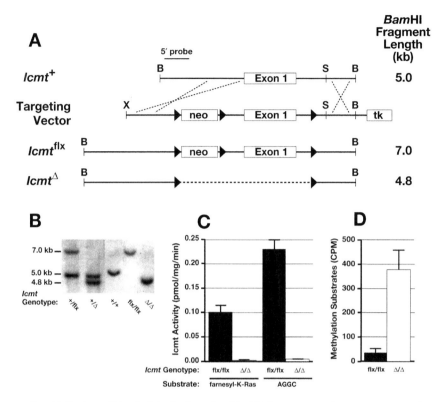

FIG. 10.13. A conditional allele for *Icmt* (*Icmt*flx). (A) A sequence-replacement gene-targeting vector designed to flank exon 1 and upstream sequences with *loxP* sites. (B) Southern blot identification of the *Icmt*+, *Icmt*flx, and *Icmt*Δ alleles with *Bam*HI-cleaved genomic DNA and the 5'-flanking probe. (C) Icmt activity in extracts of *Icmt*flx/flx and *Icmt*Δ/Δ fibroblasts, as judged by a base-hydrolysis vapor-diffusion assay. Assays used *S*-adenosyl-L-[*methyl*-14C]methionine as the methyl donor and either farnesyl-K-Ras or *N*-acetyl-*S*-geranylgeranyl-L-cysteine (AGGC) as substrates. (D) Accumulation of Icmt substrates in *Icmt*Δ/Δ cells. Recombinant yeast Ste14p was added to extracts of *Icmt*flx/flx and *Icmt*Δ/Δ cells along with *S*-adenosyl-L-[*methyl*-14C]methionine; methylation of protein substrates was measured with the base-hydrolysis vapor-diffusion assay. *(Reproduced with permission from* The Journal of Clinical Investigation *[55].)*

than in parental K-Ras-*Icmt*flx/flx cells (Rho half-life, 22.0 ± 9.4 h in K-Ras-*Icmt*flx/flx cells *vs.* 2.8 ± 0.4 h in the K-Ras-*Icmt*Δ/Δ cells) [55]. Thus, the absence of carboxyl methylation in K-Ras-*Icmt*Δ/Δ cells is associated with accelerated turnover of the Rho proteins. This result was entirely consistent with the earlier biochemical studies of Backlund [56].

As one would predict, *Icmt*flx/flx fibroblasts that had been transfected with an activated form of K-Ras (K-Ras-*Icmt*flx/flx) exhibited a transformed phenotype

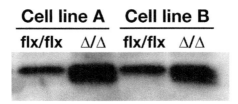

Fig. 10.14. Increased steady-state levels of K-Ras in *Icmt*-deficient fibroblasts. GTP-bound Ras proteins in K-Ras-*Icmt*$^{flx/flx}$ fibroblasts and derivative K-Ras-*Icmt*$^{\Delta/\Delta}$ fibroblasts. GTP-bound Ras proteins were precipitated from K-Ras-*Icmt*$^{flx/flx}$ and K-Ras-*Icmt*$^{\Delta/\Delta}$ cells with the Ras-binding domain of Raf, and a western blot was performed with a pan-Ras antibody. *(Reproduced with permission from The Journal of Clinical Investigation [55].)*

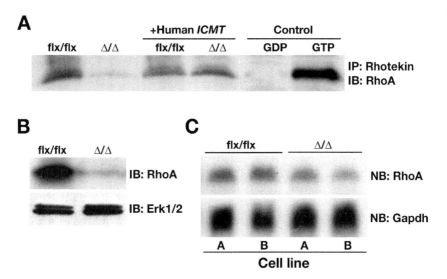

Fig. 10.15. Low steady-state levels of RhoA in *Icmt*-deficient fibroblasts. (A) GTP-bound Rho proteins were immunoprecipitated from K-Ras-*Icmt*$^{flx/flx}$ fibroblasts and derivative K-Ras-*Icmt*$^{\Delta/\Delta}$ fibroblasts with equal amounts of Rothekin-GST (Rho Activation Kit, Upstate). The GTP-bound Rho proteins were resolved by SDS-PAGE and detected with a RhoA-specific antibody (26C4 monoclonal, Santa Cruz Biotechnology). Similar results were obtained when using a pan-Rho antibody. (B) Cell extracts from K-Ras-*Icmt*$^{flx/flx}$ fibroblasts and the derivative K-Ras-*Icmt*$^{\Delta/\Delta}$ fibroblasts were analyzed by immunoblotting with a RhoA-specific antibody. The blot was stripped and incubated with an anti-Erk1/2 antibody as a loading control. (C) Northern blot of total cellular RNA from K-Ras-*Icmt*$^{flx/flx}$ fibroblasts and the derivative K-Ras-*Icmt*$^{\Delta/\Delta}$ fibroblasts was hybridized with a mouse *RhoA* cDNA probe. IB, immunoblot; IP, immunoprecipitation; NB, northern blot. *(Reproduced with permission from The Journal of Clinical Investigation [55].)*

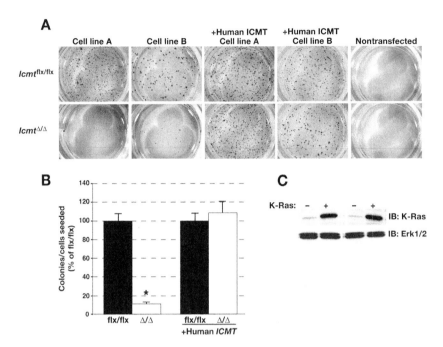

FIG. 10.16. Reduced capacity of K-Ras-transfected *Icmt*-deficient fibroblasts to form colonies in soft agar. (A) K-Ras-*Icmt*$^{flx/flx}$ and derivative K-Ras-*Icmt*$^{\Delta/\Delta}$ fibroblasts (2,000 cells each) were mixed with 0.35% agarose and poured onto plates containing a 0.70% agarose base. Colonies were stained and photographed 21 days later. Nontransfected cells did not form colonies in soft agar. (B) Number of colonies formed in soft agar in four independent experiments, Data in each experiment were normalized to the number of colonies that formed with the parental K-Ras-*Icmt*$^{flx/flx}$ fibroblasts. Inactivation of *Icmt* significantly reduced the number of colonies that formed in soft agar (*$P < 0.0001$). In the cells expressing human *ICMT*, inactivation of mouse *Icmt* did not affect colony formation ($P = 0.63$). (C) Western blot showing higher K-Ras expression levels in K-Ras-transfected cells (+) compared with nontransfected cells (−). The blot was stripped and incubated with an anti-Erk1/2 antibody as a loading control. IB, immunoblot. *(Reproduced with permission from* The Journal of Clinical Investigation *[55].)*

(formed numerous colonies in soft agar) (Figure 10.16a) [55]. To assess the consequences of *Icmt* deficiency on K-Ras transformation, we used a Cre adenovirus to inactivate *Icmt* in K-Ras-*Icmt*$^{flx/flx}$ cells, generating K-Ras-*Icmt*$^{\Delta/\Delta}$ cell lines. In multiple experiments, the K-Ras-*Icmt*$^{\Delta/\Delta}$ cells yielded 90 to 95% fewer colonies in soft agar than the parental K-Ras-*Icmt*$^{flx/flx}$ cells (Figures 10.16a and b). When these experiments were repeated with K-Ras-*Icmt*$^{flx/flx}$ cells that had been transfected with a human *ICMT* cDNA, there was no effect of the *Icmt* inactivation on the ability of the cells to grow in soft agar (Figures 10.16a and b) [55]. Thus, inactivation of *Icmt* inhibited the ability of K-Ras transfected cells to grow in soft agar, except when the cells had been transfected with human *ICMT*.

A

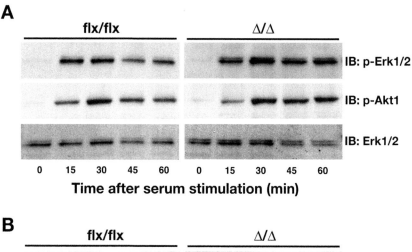

Time after serum stimulation (min)

B

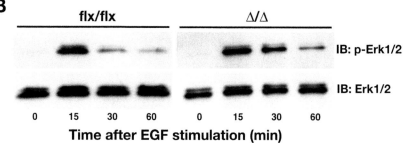

Time after EGF stimulation (min)

FIG. 10.17. IGrowth factor–stimulated Erk1/2 and Akt1 phosphorylation in *Icmt*-deficient fibroblasts. (A) Nontransfected *Icmt*^flx/flx and the derivative *Icmt*^Δ/Δ fibroblasts were seeded at equal density and serum-starved overnight. Serum-containing medium was then added to the cells. Cells were harvested at the indicated time-points and analyzed by immunoblotting with antibodies against phosphorylated Erk1/2 (p-Erk1/2), phosphorylated Akt1 (p-Akt1), and total Erk1/2. (B) *Icmt*^flx/flx and the derivative *Icmt*^Δ/Δ fibroblasts were seeded at equal density and serum-starved overnight. Medium (0.5% serum) supplemented with EGF (50 ng/ml) was added to the cells. Cells were harvested at the indicated time-points and analyzed by immunoblotting with antibodies against p-Erk1/2 and total Erk1/2. IB, immunoblot. *(Reproduced with permission from* The Journal of Clinical Investigation *[55].)*

We suspected that the reduction in the transformed phenotype in K-Ras-*Icmt*^Δ/Δ cell lines, compared with the parental K-Ras-*Icmt*^flx/flx cells, may have been accompanied by a reduction in serum- or epidermal growth factor (EGF)-stimulated phosphorylation of the downstream Ras effectors, Erk1/2 and Akt1. However, this was not the case (Figures 10.17a and b) [55]. Thus, the absence of carboxyl methylation had prominent effects on Ras targeting to membranes and on the expression of the transformed phenotype, but apparently had little or no effect on the growth factor stimulated activation of Erk1/2 and Akt. Subsequent studies showed that the K-Ras-*Icmt*^Δ/Δ cell lines expressed higher levels of the regulatory protein p21^{Cip1}, a protein that binds to cyclins and inhibits cell-cycle progression [57, 58]. The higher levels of p21^{Cip1}, which may have been a consequence of the low levels of RhoA, likely played a role in reducing the ability of the K-Ras-*Icmt*$^{\Delta/\Delta}$ cell lines to grow in soft agar. When we compared the ability of p21^{Cip1}-deficient K-Ras-*Icmt*$^{flx/flx}$ cells and p21^{Cip1}-deficient K-Ras-*Icmt*$^{\Delta/\Delta}$ fibroblasts to grow in soft agar, we observed no difference [55].

XII. An Effect of *Icmt* Deficiency on Prelamin A Processing

In recent years, considerable interest has been focused on the posttranslational processing of prelamin A. Prelamin A, a farnesylated *CaaX* protein, is converted to lamin A, a major component of the nuclear lamina, by a series of posttranslational processing reactions. First, the cysteine of the carboxyl-terminal *CaaX*-box of prelamin A is farnesylated by FTase. Second, the –*aaX* is clipped off. Third, the newly exposed carboxyl-terminal farnesylcysteine is methylated by Icmt. Finally, the carboxyl-terminal 15 amino acids, including the farnesylcysteine methyl ester, are clipped off by Zmpste24 and degraded, leaving mature lamin A.

Several years ago, to assess the importance of *Zmpste24* in mammals, we produced *Zmpste24* knockout mice (*Zmpste24*$^{-/-}$) and demonstrated that the cells and tissues from those mice lacked the ability to convert prelamin A to mature lamin A (Figure 10.18a) [15, 59]. Similar findings were reported by Pendás et al. [14]. Western blots of *Zmpste24*$^{-/-}$ fibroblast extracts with an antibody against the carboxyl terminus of prelamin A (i.e., an antibody against the 15 amino acids that are normally clipped off in the final cleavage reaction) revealed a striking accumulation of prelamin A (Figure 10.18a). Western blots with an antibody against the amino terminus of lamin A revealed fully processed lamin A (72 kDa) in *Zmpste24*$^{+/+}$ fibroblasts and the larger prelamin A (74 kDa) in *Zmpste24*$^{-/-}$ fibroblasts (Figure 10.18a) [15, 60].

In keeping with the results of earlier studies on prelamin A [61], we found that a farnesyltransferase inhibitor completely blocked the processing of prelamin A to mature lamin A. In our initial description of the phenotype of the *Zmpste24*-deficient mice, we reported that *Icmt* deficiency completely blocked prelamin A processing to mature lamin A. That result was consistent with the biochemical studies by Sinensky's laboratory, which showed that the carboxyl-terminal methyl ester on prelamin A was required for the final endoproteolytic processing step [62]. However, more recent experiments on immortalized *Icmt*$^{-/-}$ cell lines have shown that the prelamin A processing defect is probably only partial (Figures 10.18b and c). With different immortalized *Icmt*$^{-/-}$ cell lines, we invariably observe a striking increase in the amount of prelamin A, as judged by western blots with an antibody against the carboxyl-terminal portion of prelamin A.

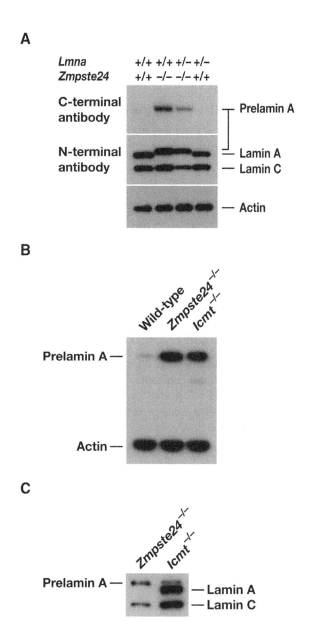

Fig. 10.18. Increased amounts of prelamin A in fibroblasts lacking *Icmt*. (A) Western blots of extracts from wild-type, *Zmpste24⁻/⁻*, and *Zmpste24⁻/⁻Lmna⁺/⁻* MEFs with a carboxyl-terminal prelamin A antibody and an amino-terminal lamin A/C antibody. (*Lmna* is the gene encoding prelamin A and the alternatively spliced product lamin C, which is not a *CaaX* protein.) In the presence of heterozygosity for *Lmna* deficiency (*Lmna⁺/⁻*), the amount of prelamin A and lamin C in *Zmpste24*-deficient mice is reduced by approximately 50%. (B) Western blot of wild-type, *Zmpste24⁻/⁻*, and *Icmt⁻/⁻* fibroblasts with a prelamin A-specific antibody and an antibody against β-actin. (C) Western blot of *Zmpste24⁻/⁻* and *Icmt⁻/⁻* fibroblasts with a lamin A/C-specific antibody. *(Panel a reproduced with permission from* The Proceedings of the National Academy of Science USA *[60].)*

These studies clearly establish that the processing of prelamin A in *Icmt*$^{-/-}$ cells is abnormal. However, in western blots with an antibody against the amino-terminal portion of lamin A/C, we invariably observe that some mature lamin A is produced in *Icmt*$^{-/-}$ cells. Thus, *Icmt* deficiency produces a partial, but not a complete, blockade in the processing of prelamin A to mature lamin A.

The accumulation of prelamin A in cells has a number of adverse consequences, leading to a weakened nuclear envelope and grossly misshapen nuclei at the cellular level [60], and a host of precocious aging-like phenotypes at the whole-animal level, both in mouse models and in human patients [15, 60, 63, 64]. It is conceivable that the minor degree of prelamin A accumulation in *Icmt*-deficient cells contributes to slow growth, but the slight accumulation of prelamin A probably does not explain the embryonic lethality of *Icmt*-deficient knockout mice, given that *Zmpste24*-deficient mice survive development quite normally [15]. However, it is conceivable that the failure to methylate lamin B1 and lamin B2 could impair the function of those proteins. Interestingly, Maske and co-workers [65] found increased numbers of misshapen nuclei in *Icmt*-deficient fibroblasts.

XIII. Pharmacologic Inhibition of Icmt

The fact that *Cre*-mediated inactivation of *Icmt* reduced the transformed phenotype of Ras-transformed cells [55] suggested that *Icmt* might be a target for treatment of cancer. The possibility that the isoprenylcysteine methyltransferase might be a good target for antitumor agents has been reviewed in several recent papers [34, 66, 67].

Interestingly, one already well-established anticancer drug might act in part by reducing the activity of Icmt [68]. Work from Winter-Vann and co-workers [69] has suggested that methotrexate, one of the oldest anticancer chemotherapeutic agents, may work in part by blocking the activity of Icmt. The efficacy of methotrexate is generally attributed to a decrease in the synthesis of thymidylate by reducing the concentration of its N^5,N^{10}-methylene tetrahydrofolate precursor when the folate pool is trapped as dihydrofolate by direct inhibition of dihydrofolate reductase [68]. However, the same N^5,N^{10}-methylene tetrahydrofolate precursor is also required for the conversion of homocysteine to methionine, and an intracellular build-up of homocysteine would be expected to also increase the concentration of S-adenosyl-L-homocysteine, a potent inhibitor of most methyltransferases, including Icmt [69, 70].

Treatment of mouse embryonic fibroblasts or DKOB8 human colon cancer cells with 1 µM methotrexate resulted in a "marked" increase in intracellular S-adenosyl-L-homocysteine [69]. Because carboxyl methylation of Ras is important for proper plasma membrane localization and function, Winter-Vann et al. [69] examined the possible role of Icmt in the antiproliferative effect of methotrexate.

After methotrexate treatment of DKOB8 cells, methotrexate reduced methylation of Ras by nearly 90%. This reduced methylation was accompanied by a mislocalization of Ras to the cytosol and decreased EGF-dependent phosphorylation of p42/44 MAPK and Akt. Immortalized *Icmt*$^{-/-}$ fibroblasts lacking *Icmt* were resistant to methotrexate, but transfection of a human *ICMT* cDNA into those cells reversed that resistance (i.e., rendered them sensitive to growth inhibition with methotrexate).

Of note, Winter-Vann et al. [69] found no defect in the activation of EGF-dependent phosphorylation of p42/44 MAPK and Akt in immortalized *Icmt*$^{-/-}$ cells. They speculated that during the process of immortalization *Icmt*$^{-/-}$ cells adapted by developing an alternate mechanism to respond to EGF, independent of the Ras proteins. Of course, any drug-induced elevation of intracellular *S*-adenosyl-L-homocysteine would be expected to also inhibit the ~300 other types of mammalian methyltransferases [71], further complicating the interpretation of these results.

During the past few years, a number of pharmaceutical companies have worked to create specific *Icmt* inhibitor drugs, but as far as we know no Icmt inhibitor drug has been tested in clinical trials. Very recently, the laboratory of Patrick Casey identified an Icmt inhibitor, 2-[5-(3-methylphenyl)-1-octyl-1H-indol-3-yl]acetamide (cysmethynil) [72]. Cysmethynil mislocalized Ras proteins in cells in a dose-dependent fashion, impaired EGF signaling, retarded cell growth in mouse fibroblasts, and blocked the anchorage-independent growth of a human colon cancer cell line [72]. The effect of cysmethynil treatment could be reduced by overexpressing *ICMT* in the cells. The drug had little effect on the growth of *Icmt*$^{-/-}$ cells [72]. Interestingly, there was an apparent time-dependent activation of the inhibitory effect of this compound. It is possible that nonspecific amidases in cellular extracts may convert the amide form of the drug to a more potent carboxylate form, accounting for this effect.

The results with cysmethynil were potentially exciting because they tended to support the results of genetic experiments with *Icmt* inactivation [55], particularly with regard to the effect of reduced Icmt activity on the transformed cell phenotype. However, there was an important discrepancy. With either the *Icmt*$^{-/-}$ cells from the conventional *Icmt* knockout mice or the *Icmt*$^{\Delta/\Delta}$ cells derived from the *Icmt*$^{flx/flx}$ mice, we have never observed a defect in serum- or growth factor stimulated Erk1/2 activation [55]. In contrast, treatment of cells with cysmethynil (or the small-molecule competitive inhibitors AFC and AGGC) inhibited this pathway [72, 73]. We do not understand this discrepancy. One possibility is that the Icmt inhibitors are not truly specific and that the signaling differences reflect other biological effects. Alternatively, it is possible that the knockout cells are not good model systems for studying Ras signaling, perhaps because *Icmt* deficiency elicits adaptations that mask effects on ERK1/2 signaling. Currently, these different results are difficult to reconcile and warrant further study.

A significant amount of work has been done utilizing derivatives of farnesyl-cysteine [51, 74] as competitive inhibitors of Icmt. Although these compounds are generally cell permeable, they appear to have additional targets in addition to the methyltransferase, possibly interfering with protein-protein interactions based on isoprenyl group recognition [74, 75]. Another route to inhibiting the isoprenylcysteine methyltransferase is by chelating the essential metal ion required for catalysis [76, 77]. Recently, the laboratory of Christine Hrycyna has developed a membrane soluble analog of cholesterol that contains a high-affinity metal chelating group based on lysine nitrilotriacetic acid [20]. This compound inhibits the yeast isoprenylcysteine methyltransferase with an IC-50 value of about 50 μM and represents an additional approach to developing pharmaceutical agents active on Icmt.

ACKNOWLEDGMENTS

Supported by National Institutes of Health (NIH) Grants CA099506, AR050200, AI054384, and GM26020, and grants from the Progeria Research Foundation and the Swedish Cancer Society. We thank Mr. Brian Young for assistance with artwork.

REFERENCES

1. Zhang, F.L., and Casey, P.J. (1996). Protein prenylation: Molecular mechanisms and functional consequences. *Annu Rev Biochem* 65:241–269.
2. Boyartchuk, V.L., Ashby, M.N., and Rine, J. (1997). Modulation of Ras and a-factor function by carboxyl-terminal proteolysis. *Science* 275:1796–1800.
3. Ashby, M.N. (1998). CaaX converting enzymes. *Curr Opin Lipidol* 9:99–102.
4. Young, S.G., Ambroziak, P., Kim, E., and Clarke, S. (2000). Postisoprenylation protein processing: CXXX (CaaX) endoproteases and isoprenylcysteine carboxyl methyltransferase, in F. Tamanoi and D.S. Sigman (eds.). *The Enzymes*, pp. 155–213, San Diego: Academic Press.
5. Hrycyna, C.A., Sapperstein, S.K., Clarke, S., and Michaelis, S. (1991). The *Saccharomyces cerevisiae STE14* gene encodes a methyltransferase that mediates C-terminal methylation of a-factor and Ras proteins. *EMBO J* 10:1699–1709.
6. Dai, Q., Choy, E., Chiu, V., Romano, J., Slivka, S.R., Steitz, S.A., Michaelis, S., and Philips, M.R. (1998). Mammalian prenylcysteine carboxyl methyltransferase is in the endoplasmic reticulum. *J Biol Chem* 273:15030–15034.
7. Boyartchuk, V.L., and Rine, J. (1998). Roles of prenyl protein proteases in maturation of *Saccharomyces cerevisiae* a-factor. *Genetics* 150:95–101.
8. Fujimura-Kamada, K., Nouvet, F.J., and Michaelis, S. (1997). A novel membrane-associated metalloprotease, Ste24p, is required for the first step of NH$_2$-terminal processing of the yeast a-factor precursor. *J Cell Biol* 136:271–285.
9. Tam, A., Nouvet, F.J., Fujimura-Kamada, K., Slunt, H., Sisodia, S.S., and Michaelis, S. (1998). Dual roles for Ste24p in yeast a-factor maturation: NH$_2$-terminal proteolysis and COOH-terminal CaaX processing. *J Cell Biol* 142:635–649.
10. McCormick, F. (1993). How receptors turn Ras on. *Nature* 363:15–16.
11. McCormick, F. (1998). Signal transduction: Why Ras needs Rho. *Nature* 394:220–221.

12. Kim, E., Ambroziak, P., Otto, J.C., Taylor, B., Ashby, M., Shannon, K., Casey, P.J., and Young, S.G. (1999). Disruption of the mouse *Rce1* gene results in defective Ras processing and mislocalization of Ras within cells. *J Biol Chem* 274:8383–8390.

13. Otto, J.C., Kim, E., Young, S.G., and Casey, P.J. (1999). Cloning and characterization of a mammalian prenyl protein-specific protease. *J Biol Chem* 274:8379–8382.

14. Pendás, A.M., Zhou, Z., Cadiñanos, J., Freije, J.M.P., Wang, J., Hultenby, K., Astudillo, A., Wernerson, A., Rodríguez, F., Tryggvason, K., and Lopéz-Otín, C. (2002). Defective prelamin A processing and muscular and adipocyte alterations in Zmpste24 metalloproteinase–deficient mice. *Nat Genet* 31:94–99.

15. Bergo, M.O., Gavino, B., Ross, J., Schmidt, W.K., Hong, C., Kendall, L.V., Mohr, A., Meta, M., Genant, H., Jiang, Y., Wisner, E.R., van Bruggen, N., Carano, R.A.D., Michaelis, S., Griffey, S.M., and Young, S.G. (2002). *Zmpste24* deficiency in mice causes spontaneous bone fractures, muscle weakness, and a prelamin A processing defect. *Proc Natl Acad Sci USA* 99:13049–13054.

16. Hrycyna, C.A., and Clarke, S. (1990). Farnesyl cysteine C-terminal methyltransferase activity is dependent upon the *STE14* gene product in *Saccharomyces cerevisiae*. *Mol Cell Biol* 10:5071–5076.

17. Romano, J.D., Schmidt, W.K., and Michaelis, S. (1998). The *Saccharomyces cerevisiae* prenylcysteine carboxyl methyltransferase Ste14p is in the endoplasmic reticulum membrane. *Mol Biol Cell* 9:2231–2247.

18. Pillinger, M.H., Volker, C., Stock, J.B., Weissmann, G., and Philips, M.R. (1994). Characterization of a plasma membrane-associated prenylcysteine-directed α carboxyl methyltransferase in human neutrophils. *J Biol Chem* 269:1486–1492.

19. Smeland, T.E., Seabra, M.C., Goldstein, J.L., and Brown, M.S. (1994). Geranylgeranylated Rab proteins terminating in Cys-Ala-Cys, but not Cys-Cys, are carboxyl-methylated by bovine brain membranes *in vitro*. *Proc Natl Acad Sci USA* 91:10712–10716.

20. Hodges, H.B., Zhou, M., Haldar, S., Anderson, J.L., Thompson, D.H., and Hrycyna, C.A. (2005). Inhibition of membrane-associated methyltransferases by a cholesterol-based metal chelator. *Bioconjug Chem* 16:490–493.

21. Hrycyna, C.A., Yang, M.C., and Clarke, S. (1994). Protein carboxyl methylation in *Saccharomyces cerevisiae*: Evidence for STE14-dependent and STE14-independent pathways. *Biochemistry* 33:9806–9812.

22. Herskowitz, I. (1987). Functional inactivation of genes by dominant negative mutations. *Nature* 329:219–222.

23. Sapperstein, S., Berkower, C., and Michaelis, S. (1994). Nucleotide sequence of the yeast *STE14* gene, which encodes farnesylcysteine carboxyl methyltransferase, and demonstration of its essential role in a-factor export. *Mol Cell Biol* 14:1438–1449.

24. Silvius, J.R., and l'Heureux, F. (1994). Fluorimetric evaluation of the affinities of isoprenylated peptides for lipid bilayers. *Biochemistry* 33:3014–3022.

25. Parish, C.A., and Rando, R.R. (1996). Isoprenylation/methylation of proteins enhances membrane association by a hydrophobic mechanism. *Biochemistry* 35:8473–8477.

26. Parish, C.A., Smrcka, A.V., and Rando, R.R. (1996). The role of G protein methylation in the function of a geranylgeranylated βγ isoform. *Biochemistry* 35:7499–7505.

27. Marcus, S., Caldwell, G.A., Miller, D., Xue, C.-B., Naider, F., and Becker, J.M. (1991). Significance of C-terminal cysteine modifications to the biological activity of the *Saccharomyces cerevisiae* a-factor mating pheromone. *Mol Cell Biol* 11:3603–3612.

28. Lederer, E.D., Jacobs, A.A., Hoffman, J.L., Harding, G.B., Robishaw, J.D., and McLeish, K.R. (1994). Role of carboxylmethylation in chemoattractant receptor-stimulated G protein activation and functional responses. *Biochem Biophys Res Commun* 200:1604–1614.

29. Rosenberg, S.J., Rane, M.J., Dean, W.L., Corpier, C.L., Hoffman, J.L., and McLeish, K.R. (1998). Effect of γ subunit carboxyl methylation on the interaction of G protein α subunits with βγ subunits of defined composition. *Cell Signal* 10:131–136.

30. Parish, C.A., and Rando, R.R. (1994). Functional significance of G protein carboxymethylation. *Biochemistry* 33:9986–9991.

31. Chen, F., and Lee, R.H. (1997). Phosducin and βγ-transducin interaction I: Effects of post-translational modifications. *Biochem Biophys Res Commun* 233:370–374.

32. Musha, T., Kawata, M., and Takai, Y. (1992). The geranylgeranyl moiety but not the methyl moiety of the *smg*-25A/*rab*3A protein is essential for the interactions with membrane and its inhibitory GDP/GTP exchange protein. *J Biol Chem* 267:9821–9825.

33. Backlund, P.S. Jr. (1992). GTP-stimulated carboxyl methylation of a soluble form of the GTP-binding protein G25K in brain. *J Biol Chem* 267:18432–18439.

34. Clarke, S., and Tamanoi, F. (2004). Fighting cancer by disrupting C-terminal methylation of signaling proteins. *J Clin Invest* 113:513–515.

35. Gutierrez, L., Magee, A.I., Marshall, C.J., and Hancock, J.F. (1989). Posttranslational processing of p21*ras* is two-step and involves carboxyl-methylation and carboxy-terminal proteolysis. *EMBO J* 8:1093–1098.

36. Fukada, Y., Matsuda, T., Kokame, K., Takao, T., Shimonishi, Y., Akino, T., and Yoshizawa, T. (1994). Effects of carboxyl methylation of photoreceptor G protein γ-subunit in visual transduction. *J Biol Chem* 269:5163–5170.

37. Pérez-Sala, D., Tan, E.W., Cañada, F.J., and Rando, R.R. (1991). Methylation and demethylation reactions of guanine nucleotide-binding proteins of retinal rod outer segments. *Proc Natl Acad Sci USA* 88:3043–3046.

38. Tan, E.W., and Rando, R.R. (1992). Identification of an isoprenylated cysteine methyl ester hydrolase activity in bovine rod outer segment membranes. *Biochemistry* 31:5572–5578.

39. Dunten, R.L., Wait, S.J., and Backlund, P.S. Jr. (1995). Fractionation and characterization of protein C-terminal prenylcysteine methylesterase activities from rabbit brain. *Biochem Biophys Res Commun* 208:174–182.

40. Philips, M.R., Pillinger, M.H., Staud, R., Volker, C., Rosenfeld, M.G., Weissmann, G., and Stock, J.B. (1993). Carboxyl methylation of ras-related proteins during signal transduction in neutrophils. *Science* 259:977–980.

41. Michaelson, D., Ali, W., Chiu, V.K., Bergo, M., Silletti, J., Wright, L., Young, S.G., and Philips, M. (2005). Postprenylation CaaX processing is required for proper localization of Ras but not Rho GTPases. *Mol Biol Cell* 16:1606–1616.

42. Hancock, J.F., Paterson, H., and Marshall, C.J. (1990). A polybasic domain or palmitoylation is required in addition to the CaaX motif to localize p21*ras* to the plasma membrane. *Cell* 63:133–139.

43. Kato, K., Cox, A.D., Hisaka, M.M., Graham, S.M., Buss, J.E., and Der, C.J. (1992). Isoprenoid addition to Ras protein is the critical modification for its membrane association and transforming activity. *Proc Natl Acad Sci USA* 89:6403–6407.

44. Kim, E., Lowenson, J.D., MacLaren, D.C., Clarke, S., and Young, S.G. (1997). Deficiency of a protein-repair enzyme results in the accumulation of altered proteins, retardation of growth, and fatal seizures in mice. *Proc Natl Acad Sci USA* 94:6132–6137.

45. Lowenson, J.D., Kim, E., Young, S.G., and Clarke, S. (2001). Limited accumulation of damaged proteins in L-isoaspartyl (D-aspartyl) *O*-methyltransferase-deficient mice. *J Biol Chem* 276:20695–20702.

46. Giner, J.-L., and Rando, R.R. (1994). Novel methyltransferase activity modifying the carboxy terminal bis(geranylgeranyl)-Cys-Ala-Cys structure of small GTP-binding proteins. *Biochemistry* 33:15116–15123.

47. Bergo, M.O., Leung, G.K., Ambroziak, P., Otto, J.C., Casey, P.J., and Young, S.G. (2000). Targeted inactivation of the isoprenylcysteine carboxyl methyltransferase gene causes mislocalization of K-Ras in mammalian cells. *J Biol Chem* 275:17605–17610.

48. Lin, X., Jung, J., Kang, D., Xu, B., Zaret, K.S., and Zoghbi, H. (2002). Prenylcysteine carboxylmethyltransferase is essential for the earliest stages of liver development in mice. *Gastroenterology* 123:345–351.

49. Bergo, M.O., Ambroziak, P., Gregory, C., George, A., Otto, J.C., Kim, E., Nagase, H., Casey, P.J., Balmain, A., and Young, S.G. (2002). Absence of the CaaX endoprotease Rce1: Effects on cell growth and transformation. *Mol Cell Biol* 22:171–181.

50. Fujiyama, A., Matsumoto, K., and Tamanoi, F. (1987). A novel yeast mutant defective in the processing of ras proteins: Assessment of the effect of the mutation on processing steps. *EMBO J* 6:223–228.

51. Hrycyna, C.A., and Clarke, S. (1993). Modification of eukaryotic signaling proteins by C-terminal methylation reactions. *Pharmacol Ther* 59:281–300.

52. Thissen, J.A., Gross, J.M., Subramanian, K., Meyer, T., and Casey, P.J. (1997). Prenylation-dependent association of Ki-Ras with microtubules. Evidence for a role in subcellular trafficking. *J Biol Chem* 272:30362–30370.

53. Chen, Z., Otto, J.C., Bergo, M.O., Young, S.G., and Casey, P.J. (2000). The C-terminal polylysine region and methylation of K-Ras are critical for the interaction between K-Ras and microtubules. *J Biol Chem* 275:41251–41257.

54. Apolloni, A., Prior, I.A., Lindsay, M., Parton, R.G., and Hancock, J.F. (2000). H-Ras but not K-Ras traffics to the plasma membrane through the exocytic pathway. *Mol Cell Biol* 20:2475–2487.

55. Bergo, M.O., Gavino, B.J., Hong, C., Beigneux, A.P., McMahon, M., Casey, P.J., and Young, S.G. (2004). Inactivation of Icmt inhibits transformation by oncogenic K-Ras and B-Raf. *J Clin Invest* 113:539–550.

56. Backlund, P.S. Jr. (1997). Posttranslational processing of RhoA. Carboxyl methylation of the carboxyl-terminal prenylcysteine increases the half-life of RhoA. *J Biol Chem* 272:33175–33180.

57. Olson, M.F., Paterson, H.F., and Marshall, C.J. (1998). Signals from Ras and Rho GTPases interact to regulate expression of p21[Waf1/Cip1]. *Nature* 394:295–299.

58. Sahai, E., Olson, M.F., and Marshall, C.J. (2001). Cross-talk between Ras and Rho signalling pathways in transformation favours proliferation and increased motility. *EMBO J* 20:755–766.

59. Leung, G.K., Schmidt, W.K., Bergo, M.O., Gavino, B., Wong, D.H., Tam, A., Ashby, M.N., Michaelis, S., and Young, S.G. (2001). Biochemical studies of *Zmpste24*-deficient mice. *J Biol Chem* 276:29051–29058.

60. Fong, L.G., Ng, J.K., Meta, M., Cote, N., Yang, S.H., Stewart, C.L., Sullivan, T., Burghardt, A., Majumdar, S., Reue, K., Bergo, M.O., and Young, S.G. (2004). Heterozygosity for Lmna deficiency eliminates the progeria-like phenotypes in Zmpste24-deficient mice. *Proc Natl Acad Sci USA* 101:18111–18116.

61. Beck, L.A., Hosick, T.J., and Sinensky, M. (1990). Isoprenylation is required for the processing of the lamin A precursor. *J Cell Biol* 110:1489–1499.

62. Kilic, F., Dalton, M.B., Burrell, S.K., Mayer, J.P., Patterson, S.D., and Sinensky, M. (1997). *In vitro* assay and characterization of the farnesylation-dependent prelamin A endoprotease. *J Biol Chem* 272:5298–5304.

63. Eriksson, M., Brown, W.T., Gordon, L.B., Glynn, M.W., Singer, J., Scott, L., Erdos, M.R., Robbins, C.M., Moses, T.Y., Berglund, P., Dutra, A., Pak, E., Durkin, S., Csoka, A.B., Boehnke, M., Glover, T.W., and Collins, F.S. (2003). Recurrent *de novo* point mutations in lamin A cause Hutchinson–Gilford progeria syndrome. *Nature* 423:293–298.

64. Moulson, C.L., Go, G., Gardner, J.M., Wal, A.C.v.d., Smitt, J.H.S., Hagen, J.M.V., and Miner, J.H. (2005). Homozygous and compound heterozygous mutations in *ZMPSTE24* cause the laminopathy restrictive dermopathy. *J Invest Derm* (in press).

65. Maske, C.P., Hollinshead, M.S., Higbee, N.C., Bergo, M.O., Young, S.G., and Vaux, D.J. (2003). A carboxyl-terminal interaction of lamin B1 is dependent on the CaaX endoprotease Rce1 and carboxymethylation. *J Cell Biol* 162:1223–1232.
66. Winter-Vann, A.M., and Casey, P.J. (2005). Post-prenylation-processing enzymes as new targets in oncogenesis. *Nat Rev Cancer* 5:405–412.
67. Walker, K., and Olson, M.F. (2005). Targeting Ras and Rho GTPases as opportunities for cancer therapeutics. *Curr Opin Gen Develop* 15:62–68.
68. Philips, M.R. (2004). Methotrexate and Ras methylation: a new trick for an old drug? *Sci STKE* 2004:pe13.
69. Winter-Vann, A.M., Kamen, B.A., Bergo, M.O., Young, S.G., Melnyk, S., James, S.J., and Casey, P.J. (2003). Targeting Ras signaling through inhibition of carboxyl methylation: An unexpected property of methotrexate. *Proc Natl Acad Sci USA* 100:6529–6534.
70. Stephenson, R.C., and Clarke, S. (1990). Identification of a C-terminal protein carboxyl methyltransferase in rat liver membranes utilizing a synthetic farnesyl cysteine-containing peptide substrate. *J Biol Chem* 265:16248–16254.
71. Katz, J.E., Dlakic, M., and Clarke, S. (2003). Automated identification of putative methyltransferases from genomic open reading frames. *Mol Cell Proteomics* 2:525–540.
72. Winter-Vann, A.M., Baron, R.A., Wong, W., dela Cruz, J., York, J.D., Gooden, D.M., Bergo, M.O., Young, S.G., Toone, E.J., and Casey, P.J. (2005). A small-molecule inhibitor of isoprenylcysteine carboxyl methyltransferase with antitumor activity in cancer cells. *Proc Natl Acad Sci USA* 102:4336–4341.
73. Chiu, V.K., Silletti, J., Dinsell, V., Wiener, H., Loukeris, K., Ou, G., Philips, M.R., and Pillinger, M.H. (2004). Carboxyl methylation of Ras regulates membrane targeting and effector engagement. *J Biol Chem* 279:7346–7352.
74. Volker, C., and Stock, J.B. (1995). Carboxyl methylation of Ras-related proteins. *Methods Enzymol* 255:65–82.
75. Scheer, A., and Gierschik, P. (1993). Farnesylcysteine analogues inhibit chemotactic peptide receptor-mediated G-protein activation in human HL-60 granulocyte membranes. *FEBS Lett* 319:110–114.
76. Anderson, J.L., Frase, H., Michaelis, S., and Hrycyna, C.A. (2005). Purification, functional reconstitution, and characterization of the Saccharomyces cerevisiae isoprenylcysteine carboxylmethyltransferase Ste14p. *J Biol Chem* 280:7336–7345.
77. Desrosiers, R.R., Nguyen, Q.T., and Beliveau, R. (1999). The carboxyl methyltransferase modifying G proteins is a metalloenzyme. *Biochem Biophys Res Commun* 261:790–797.
78. Bergo, M.O., Leung, G.K., Ambroziak, P., Otto, J.C., Casey, P.J., Gomes, A.Q., Seabra, M.C., and Young, S.G. (2001). Isoprenylcysteine carboxyl methyltransferase deficiency in mice. *J Biol Chem* 276:5841–5845.
79. Svensson, A., Casey, P.J., Young, S.G., and Bergo, M.O. (2005). Genetic and Pharmacologic Analyses of the Role of Icmt in Ras Membrane Association and Function. *Methods in Enzymology* (in press).

11

Reversible Methylation of Protein Phosphatase 2A

SARI LONGIN • JOZEF GORIS

Afdeling Biochemie, Faculteit Geneeskunde
Campus Gasthuisberg O&N
Katholieke Universiteit Leuven
Herestraat 49 bus 901
Leuven B-3000, Belgium

I. Abstract

PP2A has been shown to be methylated at the C-terminal leucine residue of the catalytic subunit by a specific 38 kDa methyltransferase (LCMT1) and demethylated by a specific 44-kDa methylesterase (PME-1). This reversible methylation does not seem to drastically change the PP2A activity but is shown to be a modulating factor in the binding of the third regulatory subunit. The structure of LCMT1 is solved and a model for the catalysis of the methylation reaction is presented.

By purifying the PP2A-methylesterase, inactive dimeric (PP2Ai$_D$) and trimeric (PP2Ai$_{T55}$) holoenzymes were found to be associated with PME-1. Activation of this inactive complex is possible by the action of a ubiquitous and highly conserved activatory protein, PTPA. The function of PME-1 in this system seems to be independent of its demethylating activity. A large proportion of cellular PP2A is found methylated and the subject of regulation. Aberrant (de)methylation

THE ENZYMES, Vol. XXIV

seems to be involved in the causes of diseases such as Alzheimer's disease and diabetes.

II. PP2A Can Be Reversibly Methylated on Its C-terminal Leucine Residue: The Discovery

Some 12 years ago, it was discovered that incubation of brain extracts with *S*-adenosyl [methyl ^{3}H]methionine (AdoMet)[1] results in methylation of a 36-kDa protein on its C-terminal leucine [1]. This reaction was catalyzed by a methyltransferase of about 38 kDa that could be distinguished from three other known classes of methyltransferases. Moreover, the 36-kDa protein substrate was a component of a higher-molecular-weight complex that was estimated at 178 kDa by gel filtration. During the same year another group identified the 36-kDa substrate as the catalytic subunit of PP2A [2] and partially purified the methyltransferase.

These results also explained previous observations that a 36-kDa SV40 small tumor antigen-associated cellular protein (later identified as PP2A$_C$ [3]) was a major acceptor of the methylgroup when AdoMet was added to SV40 infected cell extracts [4]. By studying the differences in immunoreactivity of PP2A$_C$ after alkali or ethanol treatment, Favre et al. estimated that in MCF7 cells 70 to 90% of PP2A$_C$ exists in the methylated state [5]. Soon it became clear that this methylation was reversible by the action of a methylesterase [6]. After these basic facts were known, the search for the functional change that could be the consequence of this methylation was started. Even today, the answers to this quest are only partial and further research is still necessary to fully understand the consequences of what was quoted as "a novel strategy of the cell to modulate the function of PP2A" [6].

PP2A is one of the major serine/threonine phosphatases present in all eukaryotic cells. By dephosphorylation of key regulatory proteins, PP2A is implicated in the regulation of many cellular processes including apoptosis, transcription, translation, DNA replication, Wnt signaling, tumourigenesis, and regulation of the cell cycle [7, 8]. The complex composition and regulation of PP2A provide the basis for the appropriate regulation of these numerous cellular processes. The core structure of PP2A consists of a 36-kDa catalytic subunit (PP2A$_C$/C$_{36}$) and an anchoring 65-kDa regulatory subunit (PR65/A). This core structure (PP2A$_D$) can exist as such or it can be associated with a third variable subunit. Three major classes of variable subunits are currently described: PR55/B, PR61/B' and PR72/B'.

Each of these subunits exists in at least two isoforms leading to a multitude of assembly combinations of holoenzymes partially explaining the multiple and

---

1. The abbreviations used are PP2A, protein phosphatase 2A; PP2A$_C$/C$_{36}$, PP2A catalytic subunit; PR65/A, PP2A anchoring subunit; PP2A$_D$, dimeric core of PP2A; PP2Ai, inactive form of PP2A; PP2A$_{Tx}$, trimeric form of PP2A where x indicates the type of third B subunit; LCMT1/2, Leucine Carboxyl Methyltransferase 1/2; PME-1, PP2A methylesterase 1; AdoMet, *S*-adenosyl methionine; and ASBD, A subunit binding domain.

diverse cellular functions of PP2A. No sequence homology, except for the presence of two A subunit binding domains (ASBD), is found between the different B subunit subfamilies [9]. Moreover, the diverse number of B subunits target PP2A to different substrates and specific cellular compartments. Furthermore, PP2A can be associated with and regulated by a still growing number of other cellular and viral proteins.

Some proteins directly interact with the catalytic subunit, even in the absence of the PR65/A subunit. For example, Tap42/α4 (an important player in the TOR pathway from yeast to mammals) forms a complex with PP2A$_C$ and displaces PR65/A and PR55/B from the core dimer and trimer [10, 11]. Alpha4 is also a linker between PP2A$_C$ and MID1, which is mutated in Opitz syndrome and encodes a ubiquitine ligase that targets PP2A$_C$ for degradation [12]. In addition, the catalytic subunit of PP2A can undergo posttranslational modifications such as phosphorylation and methylation, the topic of this chapter.

III. Methylation of PP2A by a Leucine Carboxyl Methyltransferase (LCMT1)

A. PURIFICATION AND BIOCHEMICAL CHARACTERIZATION OF LCMT1

AdoMet-dependent methyltransferases are a large and diverse class of enzymes that catalyze the transfer of methyl groups from AdoMet to a wide range of substrates, including nucleic acids, proteins, and small molecules. Methylation of proteins is a rare event, irreversible when occurring at amino groups but reversible when carboxyl groups are involved, suggesting a regulatory role for this latter type of methylation [13, 14]. To date, there exist four types of carboxyl AdoMet-dependent methyltransferases, classified according to the type of amino acid that is modified. The first type of AdoMet-dependent methyltransferases methylates glutamate residues and plays a role in chemotaxis and regulation of sensory transduction in bacteria [15]. Second, methylation of L-isoaspartate or D-aspartate is thought to be involved in the repair of damaged proteins generated during the aging process in bacteria and eukaryotes [16]. The third type involves methylation of isoprenyl-cysteine residues of nuclear lamins, low-molecular-weight GTP-binding proteins, and G proteins, and is believed to play an important role in signal transduction [17].

Finally, a unique type of reversible methylation occurs at the carboxyl group of the C-terminal leucine amino acid (L309) of PP2A$_C$ [1, 2, 5, 6] (see II). This methylation reaction is catalyzed by a specific AdoMet-dependent cytosolic Leucine Carboxyl MethylTransferase 1 (LCMT1) [18]. Characterization of this enzymatic activity [1, 2, 5, 6] demonstrated that the properties of this enzyme are clearly distinct from any other known protein carboxyl methyltransferase, leading to a fourth type of protein carboxyl methyltransferase. The methylation of PP2A$_C$ is alkaline and ethanol labile *in vitro* [5] and is inhibited by okadaic acid (OA) and mycrocystin-LR, probably by binding of OA and mycrocystin-LR to

TABLE 11.1

CARBOXYL TERMINAL TAILS OF DIFFERENT MAMMALIAN AND YEAST PP2A-LIKE ENZYMES

	Protein	Organism	Sequence of Last Six C-terminal Residues
PP2A	$PP2A_{C\alpha}$	H. sapiens	TPDYFL
	Pph21	S. cerevisiae	TPDYFL
	Pph22	S. cerevisiae	TPDYFL
PP2A-like	Pph3	S. cerevisiae	QMDYFL
	PP4	H. sapiens	VADYFL
	PP6	H. sapiens	TTPYFL
	Sit4	S. cerevisiae	RAGYFL
	Ppg	S. cerevisiae	HVDYFL

the carboxyl terminus of $PP2A_C$, thereby preventing access of LCMT1 to the carboxyl terminal Leu309 of $PP2A_C$ [19,20].

In *Xenopus* oocyte extracts, cAMP moderately stimulates PP2A methylation, whereas Ca^{2+} and calmodulin have no effect and sinefungin (a nonhydrolyzable analog of S-adenosyl-L-homocysteine) inhibits methylation [18, 20]. In contrast with these observations, it was reported that in rat pancreatic cells Ca^{2+} stimulated methylation and cAMP had no influence [21]. Furthermore, it was shown that a synthetic C-terminal octapeptide based on the sequence of $PP2A_C$ is neither a substrate nor an inhibitor of LCMT1, suggesting that this enzyme recognizes aspects of the tertiary and/or quaternary structure of the native phosphatase [6] (see III.E.2).

The six C-terminal residues (TPDYFL) are absolutely conserved between all known $PP2A_C$ subunits (see Table 11.1) and contain also a tyrosine (Y307) and a threonine (T304) residue, both of which might be a target for inhibitory phosphorylations [22-24]. In addition, the YFL tripeptide is extremely well conserved in all of the known PP2A-like enzymes (Pph3, PP4, PP6/Sit4, Ppg) in most diverse phyla (see Table 11.1) and it was shown that PP4 can be methylated by a yet unknown methyltransferase [25]. Furthermore, in yeast it was shown that only Pph21 and Pph22 (and not Pph3, a distantly related phosphatase) are methylated by PPM1, the yeast LCMT1 [26], suggesting that LCMT1 only recognizes substrates with the conserved TPDYFL sequence at their C-termini (see Table 11.1).

B. CLONING OF LCMT1

In 1999, De Baere et al. isolated the cDNA encoding the human LCMT1 [18]. LCMT1 was purified from porcine brain and the human homolog was cloned based on tryptic peptides from the porcine form [18]. The cDNA encodes for a protein of 334 amino acids with a calculated M_r of 38 305 and a predicted pI of 5. 72. Expression of this protein in bacteria shows that its biophysical, catalytic, and immunological properties are indistinguishable from the native enzyme.

C. STRUCTURE OF LCMT1

Recently, the structure of the yeast homolog of LCMT1, PPM1, has been determined [26]. The structure reveals a conserved core methyltransferase fold, common to the class I AdoMet-dependent methyltransferases. This core methyltransferase fold comprises a mixed seven-stranded β sheet with several α helices packing on both sides of the β sheet [28]. The structural and sequential similarity of PPM1 to other methyltransferases is limited to this AdoMet-dependent methyltransferase fold. To achieve substrate specificity, several insertions and variations are added to this core fold. In the case of PPM1, four different regions (I through IV) are added to the core domain, which show no sequence similarity to any methyltransferase reported to date and create a PP2A-specific binding site [27].

The AdoMet binding site is located within a deep pocket at the center of the protein, with contributions from region I and the core methyltransferase fold containing the conserved GXG AdoMet binding motif characteristic of class I AdoMet-dependent methyltransferases (often referred to as motif I) [18, 27, 29]. Several arguments suggest that a funnel-shaped cavity perpendicular to the AdoMet binding tunnel probably represents the $PP2A_C$ binding site. First, the Cε methyl donor group of AdoMet is situated at the bottom of this cavity. Second, because of the restricted geometry of the cavity and limited access to the methyl donor, the acceptor substrate needs to penetrate deeply inside the cavity, precluding methylation of residues belonging to a structured protein region. Based on the known crystal structure of protein phosphatase 1 (PP1) [30], the six C-terminal residues of $PP2A_C$ are probably unstructured [31]. This permits the C-terminal leucine residue to enter deep inside the cavity in proximity of the methyl group of AdoMet. Moreover, the minimal length to span the cavity is six residues, and together with the absolute conservation of the six C-terminal residues of $PP2A_C$, and the conservation of residues lining the cavity of PPM1, this cavity is therefore considered the binding site for $PP2A_C$.

D. IDENTIFICATION OF LCMT2

Database screening revealed the existence of a sequence with a high similarity to LCMT1, but with a C-terminal extension of 350 amino acids (LCMT2) [18]. Both the AdoMet binding motifs and other conserved regions of LCMT1 are present in the N-terminal part of LCMT2 (see Figure 11.1). Therefore, LCMT2 is considered to represent a homolog of LCMT1. The C-terminal extension has an internal repeated sequence, with as hallmark the occurrence of three to four apolar residues followed by two glycines, showing significant sequence similarity with members of the kelch domain family.

The kelch domain has been shown to have actin binding capacities [32]. Therefore, LCMT2 might be directed to this cellular location to perform its function, although to date no methyltransferase activity against PP2A has been found

LCMT1

LCMT2

FIG. 11.1. Domain structure of LCMT. LCMT1 is an AdoMet-dependent methyltransferase that contains three conserved motifs responsible for AdoMet binding (black boxes). LCMT1 shows no obvious sequence similarity to other known protein methyltransferases except for the AdoMet binding motifs. Through database screening, a sequence was found with similarity to LCMT1, containing an additional C-terminal extension of 350 amino acids (gray shading) and three AdoMet binding motifs (black boxes). This protein was called LCMT2. The C-terminal extension contains kelch-like glycine-rich repeats consisting of three to four apolar residues followed by two glycines (white boxes). The kelch domain has been shown to have actin binding capacities. To date, no methyltransferase activity against PP2A$_C$ has been found for LCMT2.

for LCMT2 [26, 33, 34]. This could be caused by the fact that in PPM2 three amino acids lining the cavity of the putative PP2A$_C$ binding site are different, suggesting that these mutations could lead to a loss of binding specificity for PP2A, to a decreased efficiency of the methyl transfer catalysis, or to direct PPM2 activity to an unidentified target [27].

E.   EFFECT OF METHYLATION ON PP2A

1.   *Methylation Has No Drastic Effect on PP2A Activity*

Conflicting data exist on the effect of PP2A$_C$ methylation on its catalytic activity. Two groups observed a moderate increase in phosphatase activity [5, 21], another found no direct effect on phosphatase activity [18], and a third observed a decrease in activity [35]. Therefore, it is believed that methylation of PP2A$_C$ may affect other characteristics of PP2A.

2.   *Methylation Is a Modulating Factor for the Formation of Heterotrimeric PP2A Holoenzymes*

a.   **Methylation of PP2A$_C$ Is Important for Binding of the Third Regulatory B Subunits**

Many observations have shown that methylation of PP2A$_C$ is a modulating factor in third B-subunit binding in mammals [36–38] as well as in yeast [31, 33, 34, 39, 40]. In this respect, it was shown that methylation of the carboxyl terminus of $PP2A_C$ is important for association of the PR55/B subunit, but in mammals not for the formation of the core dimer ($PP2A_D$) [36, 37, 38]. In contrast, binding of SG2NA, striatin, or polyoma virus middle tumor antigen is unaffected or even increased by demethylation of $PP2A_C$ [38]. Mutational analysis demonstrated that mutation of the C-terminal leucine residue into an alanine (L309A) or deletion of the last nine residues of $PP2A_C$ (T301Δ) or the C-terminal leucine 309 (L309Δ) abolishes PP2A methylation and PR55 binding but not A/PR65 binding, showing that these amino acids are not essential for proper folding to allow formation of $PP2A_D$ [36, 38, 41] (see Table 11.2).

When a mutated form of mammalian $PP2A_C$ lacking the conserved Leu309 (L309Δ) is expressed in a yeast mutant lacking endogenous $PP2A_C$, no difference in cell growth is observed under normal conditions from the wild-type form of $PP2A_C$, showing that Leu309 is apparently dispensable for cell growth [31, 39]. Moreover, yeast cells deleted for PPM1, PPM2, or both, grow normally at room temperature, suggesting that $PP2A_C$ methylation is not essential for cell growth at this temperature [34, 40]. It was also shown that deletion of PPM1 or mutation of the C-terminal leucine into an alanine residue inhibits cdc55 (yeast PR55/B subunit)

TABLE 11.2

ROLE OF THE DIFFERENT CONSERVED AMINO ACIDS IN THE C-TERMINAL TAIL OF THE
C SUBUNIT IN THE METHYLATION OF PP2A

C Subunit[a]	Methylation Competence[b]	PR55/B Subunit Binding[c]	PR65/A Subunit Binding[f]
wt	+++	+++	+++
L309Δ	−	−	+++
T301Δ	−	−	+++
L309A	−	−	+++
L309Q	n.d.	−	+++
Y307E	−	−	+++
Y307F	++	++++[d]/−[e]	+++
Y307Q	−	−	+
Y307K	−	−	+++
T304D	+++	−	+++
T304A	+++	++++	+++
T304N	+++	−	+++
T304K	+++	+	+++
T301D	+++	+	+++

[a]For a description of the C subunit mutants and their meaning see text.

[b]The data for the methylation competence were taken from Yu et al. [38], except for L309A, which was taken from Bryant et al. [36].

[c]The data for the PR55/B subunit binding were taken from Ogris et al. [41], except for L309Δ, L309A, and L309Q, which were taken from Yu et al. [38], Bryant et al. [36], and Chung et al. [42], respectively.

[d,e]The data for the Y307F are from Ogris et al. [41] and Chung et al. [42], respectively.

[f]The data for the PR65/A subunit binding were taken from Ogris et al. [41], except for L309Δ, L309A, and L309Q, which were taken from Yu et al. [38], Bryant et al. [36], and Chung et al. [42], respectively.

binding [33, 34, 40], diminishes Tpd3 (yeast PR65/A subunit) [33, 34] and Rts1 (yeast PR61/B' subunit) binding [34], and causes a similar phenotype as deletion of cdc55 and Rts1 (including a spindle checkpoint defect [33, 34, 40]). Surprisingly, mutation of the C-terminal leucine to an isoleucine reverses this spindle checkpoint defect [40], suggesting that hydrophobic interactions might mediate the interaction between PP2A$_C$ and the third subunits.

b. The Conserved C-terminal Amino Acids of PP2A$_C$ Play a Role in Methylation of PP2A

In addition to the methylation of the C-terminal leucine residue, several amino acids in the well-conserved PP2A$_C$ terminus have been shown to be important for the binding of the third subunits to the core dimer and to play a role in the methylation of PP2A$_C$ [38, 41]. PP2A$_C$ mutants in which threonine 301 is substituted with aspartate (T301D) [and threonine 304 with aspartate (T304D), alanine (T304A), asparagine (T304N), or lysine (T304K)] bind with the PR65 subunit and are methylated at a high level, although these mutants are defective in PR55 binding except for the T304A and T301D mutants (which still bind PR55 [38, 41]) (see Table 11.2).

This indicates that threonine 301 and 304 are not essential for methylation of PP2A$_C$. However, because changing this T304 to certain amino acids such as aspartate, asparagine, and lysine abolishes B subunit binding this residue is clearly in a position to interfere directly or indirectly with B subunit binding. It also indicates that PR55 binding is not important for maintaining a high methylation level of PP2A$_C$ [38]. On the other hand, replacing Y307 with the negatively charged amino acid glutamic acid (Y307E) or with the positively charged amino acids glutamine (Y307Q) or lysine (Y307K) abolishes methylation of PP2A$_C$ and PR55 binding, but not PR65 binding [38, 41].

Substitution of Y307 with phenylalanine results in an intermediate level of methylation, and a higher PR55 binding, than wild-type PP2A$_C$ (see Table 11.2). This indicates that tyrosine 307 may be involved in PP2A methylation and B subunit binding, and suggests an important but not essential role for the hydroxyl group of Y307 in PP2A methylation. The finding that three non-aromatic amino acid substitutions no longer bind detectable B subunit and no longer undergo methylation suggests that the aromatic ring may be important [38, 41]. In this regard, it was suggested that the PP2A binding cavity of PPM1 might not be able to harbor a phosphorylated tyrosine and therefore be a possible mechanism for modulating the methylation of PP2A and consequently the binding of the B subunits to PP2A [27].

The fact that the Y307F mutant simultaneously reduces recognition by LCMT1 while strengthening PR55 subunit binding (see Table 11.2) supports a model in which hydrophobic interactions between the C-terminus of PP2A$_C$ and the PR55 subunit play a critical role in stabilizing the heterotrimeric PP2A holoenzymes [38]. In contrast with these results, Chung et al. showed that $PP2A_C$ subunits containing the single mutation Y307F or L309Q formed mostly AC dimers, whereas C subunits containing the double mutation Y307F/L309Q were bound to the α4 protein instead of the PR65 subunit [42]. Experiments in yeast showed that the importance of the $PP2A_C$ C-terminal residues for interaction with the B subunits is highly conserved between mammals and yeast, although there are some minor differences [34].

c. A Role for Active Site Residues of $PP2A_C$ in PP2A Methylation

It was demonstrated that the active site residues of $PP2A_C$ play a role in the methylation of $PP2A_C$ [38]. Analysis of active site point mutations (H59Q, H118Q, D85N, and R88A) revealed that these active site residues are essential for obtaining a high steady-state level of $PP2A_C$ methylation and binding of PR55.

Although two of the active site mutants (H59Q and H118Q) form a stable complex with PME-1 [43] (see IV), the other two active site mutants do not, excluding the possibility that loss of PR55 subunit binding is due to a stable PME-1 association. Moreover, these data suggest that proper recognition by LCMT1 and/or PME-1 requires both the C-terminal residues of $PP2A_C$ and an additional structure, probably formed by residues or coordinated metals in or near the active site [38].

d. Methylation Status of Tissue-Isolated PP2A Forms Is Type-Specific

In addition, $PP2A_D$ as isolated from tissues was found to be fully demethylated [18, 36, 37], whereas $PP2A_{T61}$ [37] and $PP2A_{T72}$ [18] were fully methylated, and $PP2A_{T55}$ were sometimes methylated and sometimes not [18, 36, 37].

Taken together, the reversible carboxyl methylation of $PP2A_C$ affects the holoenzyme composition of PP2A, providing an important regulatory mechanism in that B subunits target PP2A to different substrates and specific cellular compartments. *In vivo* evidence for this hypothesis is provided by a transient and reversible interconversion of holoenzyme forms (from $PP2A_{T61}$ to $PP2A_{T55}$) during the initial stages of retinoic acid-induced granulocyte differentiation, which coincides with increased methylation of $PP2A_C$ [35]. However, whether the increased methylation of $PP2A_C$ is the consequence rather than the cause of this interconversion remains to be determined.

Moreover, it can be suggested that methylation is important for heterodimer and heterotrimer formation in yeast, whereas in mammalian cells methylation is required for stable trimer but not dimer formation. In both yeasts and mammals, the carboxyl terminal (as well as active site residues) is important for the binding of third subunits. Replacement of threonine 304 and tyrosine 307 with a charged amino acid abolishes binding of the PR55 subunit, whereas introduction of a conservative amino acid change at either of these positions has little (if any) effect.

Because threonine 304 and tyrosine 307 are completely conserved in all organisms, these residues could have important functions and phosphorylation of these residues might also regulate heterotrimer or heterodimer formation. If methylation is required for binding of the PR61 and PR72 subfamilies remains to be determined. Moreover, methylation seems to provide a selection mechanism for recruiting proteins with ASBD motifs because binding of SG2NA, striatin, and polyoma virus middle-tumor antigen requires no $PP2A_C$ methylation [9, 44].

IV.   PP2A-methylesterase (PME-1)

Because reversibility of PP2A methylation could drastically increase the importance of this posttranslational modification, immediately after methylation of $PP2A_C$ was proven efforts were made to prove that an enzymatic activity could be responsible for such esterase activity [6]. Soon thereafter, a monomeric 46-kDa soluble protein was purified from bovine brain that could specifically demethylate [^{3}H] methylated $PP2A_D$ and that ran as two peaks on a Mono Q column [45]. This methylesterase could be distinguished from any other methylesterase based on its substrate selectivity. It could not demethylate protein substrates that were methylated by three distinct types of methyltransferases: iso-aspartyl methylesters in ovalbumin, the methylated carboxy-terminal prenylated cysteine residues on Ras, and the methylated glutamyl site chains in bacterial chemoreceptor proteins.

A final proof of reversible methylation of PP2A was provided by a reconstituted system composed of purified transferase, esterase, and PP2A. By adding [^{3}H]AdoMet, an amount of [^{3}H]methanol was formed that was calculated to correspond to 40 rounds of methylation/demethylation [45]. The final cloning of PME-1 came via a completely different approach. By sequencing several proteins that stably associated with two inactive mutants of $PP2A_C$ made by substituting two putative active site histidines by glutamines (H59Q and H118Q), a 44-kDa protein was identified and cloned. After expression of human PME-1 in bacteria, the lysates were proven to contain PP2A-methylesterase activity [43].

The methylesterase activity associated with PP2A, which represented the majority of PP2A methylesterase (>90% in brain), was also purified, partially sequenced [46], and proven to be identical with the sequence of PME-1 associated with the PP2A-inactive mutant forms [43]. Moreover, during the purification of the PP2A-linked methylesterase PME-1 was separated from the co-purifying PP2A activity in the last purification step but still contained the PR65/A and C_{36} PP2A subunits, showing almost no activity. This inactive form of PP2A (PP2Ai) could be activated by ATP/Mg^{2+} and PTPA, a protein previously characterized as phosphotyrosyl phosphatase activator [47–49] but that can also activate the ser/thr phosphatase activity of this special form of PP2A [7, 46]. It was therefore renamed *phosphatase two A phosphatase activator*. Because PME-1 could not inactivate active PP2A$_D$, it was concluded that PME-1 stabilizes a pool of PP2A in an inactive conformation (PP2Ai) that might be induced by an unknown factor/mechanism.

V.   The PP2A-PME-PTPA Interplay

Reactivation of the inactive PP2A pool, associated with PME-1, revealed a novel function for both PME-1 and PTPA. PME-1 seems to perform this "stabilizing" function independently of any demethylating activity, and PTPA was shown to activate the ser/thr activity of PP2Ai in addition to the tyrosyl phosphatase (PTPase) activity [46]. Previously, it was shown that PTPA performed this function on "active" PP2A without influencing the ser/thr activity shown to be stable in contrast to the PTPase activity more labile [47, 50] and activatable by PTPA in the presence of ATP/Mg^{2+} [47]. Therefore, the active site of PP2A$_C$ seems to exist in several conformations (attainable for tyrosyl-phosphorylated substrates or not, with or without ser/thr phosphatase activity), and PTPA seems to govern these conformational changes. Moreover, it is now proven that PTPA has a peptidyl-prolyl cis/trans isomerase activity with some model peptide substrates, highly suggestive for a working mechanism of PTPA [91].

PTPA is an ubiquitous, highly conserved protein [48, 49, 51]. Human PTPA is encoded by a single gene, mapped to chromosome 9q34 [52], giving rise to four different functional proteins by alternative splicing [53]. Promoter analysis revealed that expression of the gene requires binding of the ubiquitous transcription factor Yin Yang 1 (YY1) to two functional *cis* elements in the minimal promoter that contains neither a TATA-box nor a CAAT-box, indicative for housekeeping (or essential) genes [54]. Interestingly, PTPA expression can significantly be downregulated by the tumour suppressor protein p53 in normal conditions and after UVB irradiation leading to p53 upregulation. This p53-mediated suppression occurs through a negative control of YY1 [55].

Most of the functional information on PTPA came from yeast. In *Saccharomyces cerevisiae*, PTPA is encoded by two genes, *YPA1/RRD1* and *YPA2/RRD2*. Deletion of both genes is lethal [56, 57], but can be rescued by overexpression of Pph22, one of the yeast homologs of PP2A$_C$ [57]. In general, single disruption of *YPA1* results in a more severe phenotype than single disruption of *YPA2* [56, 57], suggesting that the two proteins are not completely redundant. The phenotype of single *YPA1* mutants is pleiotropic and resembles the phenotype of PP2A-deficient strains in specific aspects such as aberrant bud morphology, abnormal actin distribution, and similar growth defects in various growth conditions.

YPA1 mutants progress more rapidly to S phase after G1 arrest [56] and are defective in at least two mitogen-activated protein kinase (MAPK) pathways (the osmosensing or HOG1 pathway and the PKC/MPK1 pathway [57]). Moreover, *YPA1* mutants (and *YPA2* somewhat less) are rapamycin resistant [56, 57], suggesting that PTPA may be implicated in the TOR (target of rapamycin) pathway. Tap42 (PP2A associating protein) is known as a component of this TOR pathway [58] and interacts not only with the catalytic subunit of PP2A but also with Sit4 [59] and all PP2A-like phosphatases in yeast [60].

In addition, Ypa1 and Ypa2 themselves directly interact with PP2A specifically: Ypa1 with Pph3, Sit4, and Ppg1, and Ypa2 with Pph21 and Pph22. They do not compete with Tap42 but instead Ypa1, Sit4, and Tap42 co-immunoprecipitate [60]. This observation was confirmed [61] and extended: Ypa2/Rrd2 interacts with the Tap42-PP2A$_C$ complex. Moreover, rapamycin treatment resulted in the dissociation of the trimeric PTPA-Tap42-phosphatase complexes and release of a functional dimeric PTPA-phosphatase, activated toward specific substrates [61].

We sought a correlation between the capacity of Ypa to bind to phosphatases and to activate the inactive phosphatases complexed with yeast PME-1 and did not find such correlation [62]. Therefore, PTPA might have two cellular functions: to activate PP2Ai and function as a type of structural element that might direct the PTPA-phosphatase complexes to some specific substrates. It must be stressed that the affinity for PTPA was highly increased in several PP2A inactive mutants (D575, H595, H59Q, and R89Q) [60] and that activation of the PP2Ai–PME-1 complex by PTPA resulted in a dissociation of the complex concomitant with the activation [46]. Moreover, no active PME-1–PP2A complex could be demonstrated, even though co-purified with the PME-1–PP2A$_i$ complex in most of the several purification steps. Therefore, the interplay of PP2A–PME-1 and PTPA has still not fully released its secrets.

VI.   Regulation of PP2A Methylation

A substantial amount of PP2A$_C$ is in the methylated form, ranging from 70 to 93%, depending on the method used to evaluate the level of methylation [5, 38]. Methylation is important for binding of the B subunits targeting PP2A to different substrates and specific cellular compartments. Thus, methylation of PP2A can serve as an important regulatory mechanism, provided the methylation of PP2A is tightly regulated.

There seems to be two possible levels of regulation: either the (de)methylating enzymes, LCMT1 or PME-1, can be regulated or the conserved C-terminal tail of PP2A$_C$ can be a site for regulation. However, to date it is not known how methylation of PP2A is regulated. Nevertheless, it has been demonstrated that the methylation of PP2A is regulated or even deregulated in certain diseases.

A. Reversible PP2A Methylation Fluctuates During the Cell Cycle

The role of PP2A in the control of cell cycle progression, and in particular its role during mitosis, has been extensively studied [7]. However, little is known about the biochemical basis of the regulation of PP2A activity during the cell cycle. Intriguingly, both the expression and the activity of PP2A were found to be constant throughout the cell cycle [63]. In contrast, the microtubule-associated [64] and the SV40 large T-specific PP2A activity are regulated during the cell cycle [65].

Both studies suggested an increase in PP2A activity during S phase. Furthermore, methylation of $PP2A_C$ varies during cell cycle progression [66]. In particular, $PP2A_C$ is methylated throughout cycling, but temporary decreases in methylation are observed at the G0/G1 boundary in the cytoplasm, and at the G1/S boundary in the nucleus [66]. In agreement with these observations, Zhu et al. demonstrated that the C-terminus of $PP2A_C$ is transiently methylated during S phase in HL-60 cells [35]. The mechanism of this cell-cycle-dependent regulation and the physiological consequences of this oscillating methylation are unknown but suggest cell-cycle-specific holoenzyme rearrangements (see III.E.2).

B. Different Subcellular Localization of LCMT1 and PME-1

To prevent a continuous futile methylation and demethylation of $PP2A_C$, the methylating and demethylating enzymes of PP2A must be strictly controlled. One way of regulation can be established by a different subcellular localization of LCMT1 and PME-1. By making GFP-fusing proteins of LCMT1 and PME-1, we found that LCMT1 is most abundant in the cytoplasm, whereas PME-1 is predominantly localized in the nucleus (our own undocumented results, published as an abstract [67, 68]). *In vivo* immunofluorescence with antibodies against endogenous LCMT1 and PME-1 confirmed these specific localizations. Furthermore, PME-1 has a functional nuclear localization signal, suggesting a specific role for PME-1 in the nucleus, and probably during the cell cycle (see VI.A).

C. Demethylated PP2A Is Involved in the Progression of Alzheimer's Disease

The high expression of both $PP2A_C$ isoforms in brain and the brain-specific expression of some members of the PR55/B and PR61/B' subunit families suggest that PP2A has unique functions in neuronal cells [7]. In this respect, it was shown that PP2A is involved in the regulation of the phosphorylation state of neuronal-specific microtubule-associated proteins (MAPs), including tau and MAP2, which bind to microtubules and regulate microtubule stability. Moreover, accumulating evidence indicates that PP2A has a role in the progression of Alzheimer's disease (AD) [69].

AD is characterized by the presence of two histopathological hallmarks called amyloid-β containing senile plaques and neurofibrillary tangles. The latter consist of abnormally hyperphosphorylated tau, assembled in paired helical filaments. A pool of PP2A, mainly composed of $PP2A_{T55\alpha}$, is associated with microtubules [64], and a $PP2A_{T55\alpha/\beta}$ can bind and dephosphorylate the microtubule- associated tau protein [70]. Moreover, the mRNA levels of $PP2A_{C\alpha}$, PR55γ and PR61ε (but not of PR55α or PR61δ) are reduced in AD brains [71]. In addition, a significant loss of the Bα subunit expression in AD-affected regions is observed, suggesting that the changes in PR55α expression levels occur at the posttranslational level [72].

Recently, it has been shown that LCMT1 expression and $PP2A_C$ methylation levels are decreased in AD [73]. This suggests an important role for methylated PP2A in AD. Indeed, if LCMT1 expression levels are reduced in AD it might compromise $PP2A_C$ methylation leading to a net loss of $PP2A_{T55\alpha}$ and an accumulation of hyperphosphorylated tau. Moreover, LCMT1 is an AdoMet-dependent methyltransferase and the levels of AdoMet are reduced in AD [74]. In addition, epidemiological evidence indicates that elevated plasma homocysteine is a risk factor for AD, because accumulation of homocysteine leads to an increase of S-adenosyl-L-homocysteine, a potent competitive inhibitor of virtually all methyltransferases [75]. Furthermore, it has been proposed that cerebral ischemia may be a risk factor for AD [76], and following ischemic damage $PP2A_C$ becomes demethylated in specific rat brain regions [77].

D.   METHYLATION OF PP2A HAS A NEGATIVE ROLE IN INSULIN SECRETION

Several studies showed that inhibition of PP2A is associated with a stimulation of insulin secretion, probably because key signaling proteins of the insuline exocytotic cascade are retained in their phosphorylated state, leading to a stimulated insulin secretion [78]. In this respect, it was shown that inhibition of PP2A demethylation using ebelactone (an inhibitor of methylesterases) inhibited insulin secretion from isolated islets and INS-1 cell lines [21]. Moreover, phosphorylated derivatives of nucleotides or hexoses inhibited both the methylation of $PP2A_C$ and PP2A activity, whereas divalent metal ions appear to stimulate $PP2A_C$ methylation without significantly affecting the activity of PP2A [79]. In addition, in insulin-secreting pancreatic β cells it was demonstrated that glucose and calcium reduce the methylation of $PP2A_C$ [78].

Taken together, these data suggest that methylation of PP2A has a key regulatory role in the sequence of events leading to insulin secretion, because a reduction in methylation of $PP2A_C$ probably results in a decrease of holoenzyme assembly.

Interestingly, protein phosphatases may also positively modulate functions of several key proteins involved in the intermediary metabolism of the islet β cell. For instance, aetyl-CoA carboxylase (AAC), a lipogenic enzyme, is inactivated by phosphorylation and reactivated upon dephosphorylation by a magnesium- and glutamate-activated protein phosphatase that appears to be PP2A [81, 82]. Furthermore, magnesium (but not glutamate) stimulates the methylation of PP2A, suggesting that methylation may be necessary for the magnesium- and glutamate-mediated activation of ACC [83].

E. PP2A METHYLATION IS POSSIBLY INVOLVED IN THE ADENOSINE A1
    RECEPTOR ACTIVATION IN THE HEART

Adenosine receptors represent a family of G-protein coupled receptors that are ubiquitously expressed in a wide variety of tissues. Four receptor subtypes (termed A1, A2A, A2B, and A3) have been identified based on their pharmacological profile and cloning [84]. β-adrenergic stimulation of the heart increases the production of adenosine, which in turn reverses these adrenergic effects, including a reduction in heart rate and atrial contractility, and an attenuation of the stimulatory actions of catecholamines on the heart [85].

These antiadrenergic effects are mediated by the adenosine A1 receptor. It was hypothesized that activation of the adenosine A1 receptor leads to an increased methylation of PP2A and causes a translocation of PP2A to a particulate fraction in ventricular myocytes. Increases in localized PP2A activity could then lead to a dephosphorylation of key cardiac protein, such as phospholamban and troponin I, responsible for the antiadrenergic effects of the adenosine A1 receptor [86].

VII. Conclusions and Perspectives

Methylation of the catalytic subunit of PP2A seems to be a necessary event in the formation of trimeric holoenzymes [31, 33, 34, 36, 37, 38], and impairment of methylation leads to decreased third-subunit binding [37]. Because methylated trimers with PR55/B, PR61/B', and PR72 were found [46], methylation does not seem to be a selection criterion for the type of third subunit, although five-times-higher affinities of PR61/B' than PR55/B subunits have been reported for methylated PP2A$_D$ [37].

Only the binding of SG2NA, striatin, or polyomavirus middle-tumor antigen was not decreased [38]. The first two were classified as PP2A subunits [87], but are missing the two ASBD motifs present in all other third regulatory subunits [9]. SG2NA and striatin would therefore be better classified in the long list of PP2A-interacting proteins [7]. The association of these proteins is probably governed by other rules and elements in their primary sequence. A similar case can be made for PME-1: in purified PP2Ai, PME-1 can be present in stoichiometric amounts, and in this complex PP2A is fully demethylated.

Methylation of PP2A has been shown to be regulated during the cell cycle [66]. Retinoic acid induced differentiation of HL-60 cells [35], and it was demonstrated that glucose and Ca^{2+} reduced the level of methylation of $PP2A_C$ in insulin-secreting pancreatic β cells [80]. At the moment, it is still not known whether the final methylation level is regulated at the level of LCMT1, PME-1, or $PP2A_C$. However, the cellular location indicates that in basal conditions LCMT1 and PME-1 cannot be active at the same time because they have a different location of the cytosol and the nucleus, respectively. Translocation of PP2A to the nucleus or PME-1 to the cytosol are therefore potential regulatory mechanisms.

The C-terminal-conserved five amino acids (TPDYFL) are also targets for phosphorylation that might strongly influence methylation, in that the potential binding pocket in LCMT1 for binding the C-terminus of PP2A would be drastically influenced by introducing the negative charge of phosphate [27]. There is still some controversy whether the third B subunits would preclude demethylation. Demethylation by PME-1 of methylated core dimer is inhibited by addition of PR61ε/B′ε or PR55α/B subunits, and methylated trimeric holoenzymes are much more resistant to demethylation than methylated dimer [37]. This led these authors to the conclusion that "holoenzyme formation essentially locks the methyl group for the lifetime of the complex." However, by adding recombinant LCMT1, AdoMet, and PME-1 to methylated trimers several rounds of methylation/demethylation were observed, suggesting that PME-1 can be active on these trimers [46].

The unexpected novel function of PME-1 in stabilizing an inactive conformation of PP2A that can be activated by PTPA still needs to be fully explored but opens the possibility of an extra regulatory circuit for PP2A. It is still not known what might make this inactive PP2A. However, any speculation on phosphorylation being the sole underlying mechanism of inactivation is doubtful because inactivation by tyrosyl phosphorylaton at Tyr307 (as well as by phosphorylation at an undefined threonine residue) are known to be reversed by autodephosphorylation [22, 23, 24] and no (spontaneous) activation of PP2Ai is ever observed.

PP2A is known to contain Fe^{2+} and Zn^{2+} at its catalytic center [88], and by losing its metal ions it becomes Mn^{2+} stimulated [89]. Most of the PP2Ai preparations also contain a Mn^{2+}-stimulated form of PP2A, but this could nicely be distinguished from PP2Ai [46]. Although we cannot yet fully exclude the possibility that PP2A is synthesized in its inactive conformation and that PTPA would operate as a type of chaperone [62], we rather favor the hypothesis that PP2A exists in a dynamic equilibrium of active and inactive forms. Indeed, the first hypothesis would imply that all PP2A molecules would be affected if no PTPA is present and this seems not to be the case [90].

ACKNOWLEDGMENTS

We highly appreciate the expert technical assistance of Marie-Rose Verbiest and Maria Veeckmans. This work was supported by the Geconcerteerde Onderzoeksacties van de Vlaamse Gemeenschap, the F.W.O.-Vlaanderen, and the Interuniversity Attraction Poles.

REFERENCES

1. Xie, H., and Clarke, S. (1993). Methyl esterification of C-terminal leucine residues in cytosolic 36-kDa polypeptides of bovine brain. A novel eukaryotic protein carboxyl methylation reaction. *J Biol Chem* 268:13364–13371.
2. Lee, J., and Stock, J. (1993). Protein phosphatase 2A catalytic subunit is methyl-esterified at its carboxyl terminus by a novel methyltransferase. *J Biol Chem* 268:19192–19195.
3. Pallas, D.C., Shahrik, L.K., Martin, B.L., Jaspers, S., Miller, T.B., Brautigan, D.L., and Roberts, T.M. (1990). Polyoma small and middle T antigens and SV40 small t antigen form stable complexes with protein phosphatase 2A. *Cell* 60:167–176.
4. Rundell, K. (1987). Complete interaction of cellular 56,000- and 32,000-Mr proteins with simian virus 40 small-t antigen in productively infected cells. *J Virol* 61:1240–1243.
5. Favre, B., Zolnierowicz, S., Turowski, P., and Hemmings, B. A. (1994). The catalytic subunit of protein phosphatase 2A is carboxyl-methylated in vivo. *J Biol Chem* 269:16311–16317.
6. Xie, H., and Clarke, S. (1994). Protein phosphatase 2A is reversibly modified by methyl esterification at its C-terminal leucine residue in bovine brain. *J Biol Chem* 269:1981–1984.
7. Janssens, V., and Goris, J. (2001). Protein phosphatase 2A: A highly regulated family of serine/threonine phosphatases implicated in cell growth and signalling. *Biochem J* 353:417–439.
8. Janssens, V., Goris, J., and Van Hoof, C. (2005). PP2A: The expected tumor suppressor. *Curr Opin Genet Dev* 15:34–41.
9. Li, X., and Virshup, D. M. (2002). Two conserved domains in regulatory B subunits mediate binding to the A subunit of protein phosphatase 2A. *Eur J Biochem* 269:546–552.
10. Murata, K., Wu, J., and Brautigan DL. (1997). B cell receptor–associated protein alpha4 displays rapamycin-sensitive binding directly to the catalytic subunit of protein phosphatase 2A. *Proc Natl Acad Sci USA* 30:10624–10629.
11. Jacinto, E., and Hall, M.N. (2003). Tor signalling in bugs, brain and brawn. *Nat Rev Mol Cell Biol* 4:117–126.
12. Schweiger, S., and Schneider, R. (2003). The MID1/PP2A complex: A key to the pathogenesis of Opitz BBB/G syndrome. *Bioessays* 25:356–366.
13. Clarke, S. (1985). Protein carboxyl methyltransferases: Two distinct classes of enzymes. *Annu Rev Biochem* 54:479–506.
14 Aletta, J.M., Cimato, T.R., and Ettinger, M.J. (1998). Protein methylation: A signal event in posttranslational modification. *Trends Biochem Sci* 23:89–91.
15. Koshland, D.E. Jr. (1988). Chemotaxis as a model second-messenger system. *Biochemistry* 27:5829–5834.
16. McFadden, P.N., and Clarke, S. (1987). Conversion of isoaspartyl peptides to normal peptides: implications for the cellular repair of damaged proteins. *Proc Natl Acad Sci USA* 84:2595–2599.
17. Philips, M.R., Pillinger, M.H., Staud, R., Volker, C., Rosenfeld, M.G., Weissmann, G., and Stock, J.B. (1993). Carboxyl methylation of Ras-related proteins during signal transduction in neutrophils. *Science* 259:977–980.
18. De Baere, I., Derua, R., Janssens, V., Van Hoof, C., Waelkens, E., Merlevede, W., and Goris, J. (1999). Purification of porcine brain protein phosphatase 2A leucine carboxyl methyltransferase and cloning of the human homologue. *Biochemistry* 38:16539–16547.

19. Li, M., and Damuni, Z. (1994). Okadaic acid and microcystin-LR directly inhibit the methylation of protein phosphatase 2A by its specific methyltransferase. *Biochem Biophys Res Commun* 202:1023–1030.

20. Floer, M., and Stock, J. (1994). Carboxyl methylation of protein phosphatase 2A from Xenopus eggs is stimulated by cAMP and inhibited by okadaic acid. *Biochem Biophys Res Commun* 198:372–379.

21. Kowluru, A., Seavey, S.E., Rabaglia, M.E., Nesher, R., and Metz, S.A. (1996). Carboxylmethylation of the catalytic subunit of protein phosphatase 2A in insulin-secreting cells: evidence for functional consequences on enzyme activity and insulin secretion. *Endocrinology* 137:2315–2323.

22. Guo, H., and Damuni, Z. (1993). Autophosphorylation-activated protein kinase phosphorylates and inactivates protein phosphatase 2A. *Proc Natl Acad Sci USA* 90:2500–2504.

23. Chen, J., Martin, B.L. and Brautigan, D.L. (1992). Regulation of protein serine-threonine phosphatase type-2A by tyrosine phosphorylation. *Science* 257, 1261–4.

24. Chen, J., Parsons, S., and Brautigan, D.L. (1994). Tyrosine phosphorylation of protein phosphatase 2A in response to growth stimulation and v-src transformation of fibroblasts. *J Biol Chem* 269:7957–7962.

25. Kloeker, S., Bryant, J.C., Strack, S., Colbran, R.J., and Wadzinski, BE. (1997). Carboxymethylation of nuclear protein serine/threonine phosphatase X. *Biochem J* 327:481–486.

26. Kalhor, H.R., Luk, K., Ramos, A., Zobel-Thropp, P., and Clarke, S. (2001). Protein phosphatase methyltransferase 1 (Ppm1p) is the sole activity responsible for modification of the major forms of protein phosphatase 2A in yeast. *Arch Biochem Biophys* 395:239–245.

27. Leulliot, N., Quevillon-Cheruel, S., Sorel, I., de La Sierra-Gallay, I.L., Collinet, B., Graille, M., Blondeau, K., Bettache, N., Poupon, A., Janin, J., and van Tilburgh, H. (2003). Structure of protein phosphatase methyltransferase 1 (PPM1), a leucine carboxyl methyltransferase involved in the regulation of protein phosphatase 2A activity. *J Biol Chem* 279:8351–8358.

28. Schubert, H.L., Blumenthal, R.M., and Cheng, X. (2003). Many paths to methyltransfer: a chronicle of convergence. *Trends Biochem Sci* 28:329–335.

29. Martin, J.L., and McMillan, F.M. (2002). SAM (dependent) I AM: The S-adenosylmethionine-dependent methyltransferase fold. *Curr Opin Struct Biol* 12:783–793.

30. Egloff, M.P., Cohen, P.T., Reinemer, P., and Barford, D. (1995). Crystal structure of the catalytic subunit of human protein phosphatase 1 and its complex with tungstate. *J Mol Biol* 254:942–959.

31. Evans, D.R., Myles, T., Hofsteenge, J., and Hemmings, B.A. (1999). Functional expression of human PP2Ac in yeast permits the identification of novel C-terminal and dominant-negative mutant forms. *J Biol Chem* 274:24038–24046.

32. Xue, F., and Cooley, L. (1993). *kelch* encodes a component of intercellular bridges in *Drosophila* egg chambers. *Cell* 72:681–693.

33. Wu, J., Tolstykh, T., Lee, J., Boyd, K., Stock, J.B., and Broach, J.R. (2000). Carboxyl methylation of the phosphoprotein phosphatase 2A catalytic subunit promotes its functional association with regulatory subunits in vivo. *EMBO J* 19:5672–5681.

34. Wei, H., Ashby, D.G., Moreno, C.S., Ogris, E., Yeong, F.M., Corbett, A.H., and Pallas, D.C. (2001). Carboxymethylation of the PP2A catalytic subunit in Saccharomyces cerevisiae is required for efficient interaction with the B-type subunits Cdc55p and Rts1p. *J Biol Chem* 276:1570.

35. Zhu, T., Matsuzawa, S., Mizuno, Y., Kamibayashi, C., Mumby, M.C., Andjelkovic, N., Hemmings, B.A., Onoe, K., and Kikuchi, K. (1997). The interconversion of protein phosphatase 2A between PP2A1 and PP2A0 during retinoic acid-induced granulocytic differentiation and a modification on the catalytic subunit in S phase of HL-60 cells. *Arch Biochem Biophys* 339:210–217.

36. Bryant, J.C., Westphal, R.S., and Wadzinski, B.E. (1999). Methylated C-terminal leucine residue of PP2A catalytic subunit is important for binding of regulatory Balpha subunit. *Biochem J* 339:241–246.

37. Tolstykh, T., Lee, J., Vafai, S. and Stock, J.B. (2000). Carboxyl methylation regulates phospho-protein phosphatase 2A by controlling the association of regulatory B subunits. *EMBO J* 19:5682–91.

38. Yu, X.X., Du, X., Moreno, C.S., Green, R.E., Ogris, E., Feng, Q., Chou, L., McQuoid, M.J., and Pallas, D.C. (2001). Methylation of the protein phosphatase 2A catalytic subunit is essential for association of Balpha regulatory subunit but not SG2NA, striatin, or polyomavirus middle tumor antigen. *Mol Biol Cell* 12:185–199.

39. Evans, D.R., and Hemmings, B.A. (2000). Important role for phylogenetically invariant PP2Ac alpha active site and C-terminal residues revealed by mutational analysis in Saccharomyces cerevisiae. *Genetics* 156:21–29.

40. Evans, D.R., and Hemmings, B.A. (2000). Mutation of the C-terminal leucine residue of PP2Ac inhibits PR55/B subunit binding and confers supersensitivity to microtubule destabilization in Saccharomyces cerevisiae. *Mol Gen Genet* 264:425–432.

41. Ogris, E., Gibson, D.M., and Pallas, D.C. (1997). Protein phosphatase 2A subunit assembly: The catalytic subunit carboxy terminus is important for binding cellular B subunit but not polyomavirus middle tumor antigen. *Oncogene* 15:911–917.

42. Chung, H., Nairn, A.C., Murata, K., and Brautigan, D.L. (1999). Mutation of Tyr307 and Leu309 in the protein phosphatase 2A catalytic subunit favors association with the alpha 4 subunit which promotes dephosphorylation of elongation factor-2. *Biochemistry* 38:10371–10376.

43. Ogris, E., Du, X., Nelson, K.C., Mak, E.K., Yu, X.X., Lane, W.S., and Pallas, D.C. (1999). A protein phosphatase methylesterase (PME-1) is one of several novel proteins stably associating with two inactive mutants of protein phosphatase 2A. *J Biol Chem* 274:14382–14391.

44. Strack, S., Ruediger, R., Walter, G., Dagda, R.K., Barwacz, C.A., and Cribbs, J.T. (2002). Protein phosphatase 2A holoenzyme assembly: Identification of contacts between B-family regulatory and scaffolding A subunits. *J Biol Chem* 277:20750–20755.

45. Lee, J., Chen, Y., Tolstykh, T., and Stock, J. (1996). A specific protein carboxyl methylesterase that demethylates phosphoprotein phosphatase 2A in bovine brain. *Proc Natl Acad Sci USA* 93:6043–6047.

46. Longin, S., Jordens, J., Martens, E., Stevens, I., Janssens, V., Rondelez, E., De Baere, I., Derua, R., Waelkens, E., Goris, J., and Van Hoof, C. (2004). An inactive protein phosphatase 2A population is associated with methylesterase and can be re-activated by the phosphotyrosyl phosphatase activator. *Biochem J* 380:111–119.

47. Cayla, X., Goris, J., Hermann, J., Hendrix, P., Ozon, R., and Merlevede, W. (1990). Isolation and characterization of a tyrosyl phosphatase activator from rabbit skeletal muscle and Xenopus laevis oocytes. *Biochemistry* 29:658–667.

48. Cayla, X., Van Hoof, C., Bosch, M., Waelkens, E., Vandekerckhove, J., Peeters, B., Merlevede, W., and Goris; J. (1994). Molecular cloning, expression, and characterization of PTPA, a protein that activates the tyrosyl phosphatase activity of protein phosphatase 2A. *J Biol Chem* 269:15668–15675.

49. Van Hoof, C., Janssens, V., Dinishiotu, A., Merlevede, W., and Goris, J. (1998). Functional analysis of conserved domains in the phosphotyrosyl phosphatase activator. Molecular cloning of the homologues from Drosophila melanogaster and Saccharomyces cerevisiae. *Biochemistry* 37:12899–12908.

50. Chernoff, J., Li, H.C., Cheng, Y.S., and Chen, L.B. (1983). Characterization of a phosphotyrosyl protein phosphatase activity associated with a phosphoseryl protein phosphatase of Mr = 95,000 from bovine heart. *J Biol Chem* 258:7852–7857.

51. Van Hoof, C., Cayla, X., Bosch, M., Merlevede, W., and Goris, J. (1994). The phosphotyrosyl phosphatase activator of protein phosphatase 2A. A novel purification method, immunological and enzymic characterization. *Eur J Biochem* 226:899–907.

52. Van Hoof, C., Aly, M.S., Garcia, A., Cayla, X., Cassiman, J.J., Merlevede, W., and Goris, J. (1995). Structure and chromosomal localization of the human gene of the phosphotyrosyl phosphatase activator (PTPA) of protein phosphatase 2A. *Genomics* 28:261–272.

53. Janssens, V., Van Hoof, C., Martens, E., De Baere, I., Merlevede, W., and Goris, J. (2000). Identification and characterization of alternative splice products encoded by the human phosphotyrosyl phosphatase activator gene. *Eur J Biochem* 267:4406–4413.

54. Janssens, V., Van Hoof, C., De Baere, I., Merlevede, W., and Goris, J. (1999). Functional analysis of the promoter region of the human phosphotyrosine phosphatase activator gene: Yin Yang 1 is essential for core promoter activity. *Biochem J* 344:755–763.

55. Janssens, V., Van Hoof, C., De Baere, I., Merlevede, W., and Goris, J. (2000). The phosphotyrosyl phosphatase activator gene is a novel p53 target gene. *J Biol Chem* 275:20488–20495.

56. Van Hoof, C., Janssens, V., De Baere, I., de Winde, J.H., Winderickx, J., Dumortier, F., Thevelein, J.M., Merlevede, W., and Goris, J. (2000). The Saccharomyces cerevisiae homologue YPA1 of the mammalian phosphotyrosyl phosphatase activator of protein phosphatase 2A controls progression through the G1 phase of the yeast cell cycle. *J Mol Biol* 302:103–120.

57. Rempola, B., Kaniak, A., Migdalski, A., Rytka, J., Slonimski, P.P., and di Rago, J.P. (2000). Functional analysis of RRD1 (YIL153w) and RRD2 (YPL152w), which encode two putative activators of the phosphotyrosyl phosphatase activity of PP2A in Saccharomyces cerevisiae. *Mol Gen Genet* 262:1081–1092.

58. Di Como, C.J., and Arndt, K.T. (1996). Nutrients, via the Tor proteins, stimulate the association of Tap42 with type 2A phosphatases. *Genes Dev* 10:1904–1916.

59. Mitchell, D.A., and Sprague, G.F. Jr. (2001). The phosphotyrosyl phosphatase activator, Ncs1p (Rrd1p), functions with Cla4p to regulate the G(2)/M transition in Saccharomyces cerevisiae. *Mol Cell Biol* 21:488–500.

60. Van Hoof, C., Martens, E., Longin, S., Jordens, J., Stevens, I., Janssens, V., and Goris, J. (2005). Specific interactions of PP2A and PP2A-like phosphatases with the yeast PTPA homologues, Ypa1 and Ypa2. *Biochem J* 386:93–102.

61. Zheng, Y., and Jiang, Y. (2005). The yeast phosphotyrosyl phosphatase activator is part of the Tap42-phosphatase complexes. *Mol Biol Cell* 16:2119–2127.

62. Fellner, T., Lackner, D.H., Hombauer, H., Piribauer, P., Mudrak, I., Zaragoza, K., Juno, C., and Ogris, E. (2003). A novel and essential mechanism determining specificity and activity of protein phosphatase 2A (PP2A) in vivo. *Genes Dev* 17:2138–2150.

63. Ruediger, R., Van Wart Hood, J.E., Mumby, M., and Walter, G. (1991). Constant expression and activity of protein phosphatase 2A in synchronized cells. *Mol Cell Biol* 11:4282–4285.

64. Sontag, E., Nunbhakdi-Craig, V., Bloom, G.S., and Mumby, M.C. (1995). A novel pool of protein phosphatase 2A is associated with microtubules and is regulated during the cell cycle. *J Cell Biol* 128:1131–1144.

65. Ludlow, J.W. (1992). Selective ability of S-phase cell extracts to dephosphorylate SV40 large T antigen in vitro. *Oncogene* 7:1011–1014.

66. Turowski, P., Fernandez, A., Favre, B., Lamb, N.J., and Hemmings, B.A. (1995). Differential methylation and altered conformation of cytoplasmic and nuclear forms of protein phosphatase 2A during cell cycle progression. *J Cell Biol* 129:397–410.

67. Longin, S., Stevens, I., Janssens, V., Martens, E., Zwaenepoel, K., Jordens, J., Goris, J., and Van Hoof, C. (2004). The methylating and demethylating enzymes of protein phosphatase 2A have a different subcellular localisation. The Belgian Society of Biochemistry and Molecular Biology (BMB), Brussels, Belgium, 12 November 2004.

68. Longin, S., Stevens, I., Martens, E., Zwaenepoel, K., Jordens, J., Goris, J., and Janssens, V. (2005). Different subcellular localisation of the methylating and demethylating enzymes of protein phosphatase 2A prevents a futile cycle. Identification of the functional NLS of PME-1. EMBO Workshop, Meiotic Divisions and Checkpoints, Cargèse, France, 16-20 March 2005.

69. Tian, Q., and Wang, J. (2002). Role of serine/threonine protein phosphatase in Alzheimer's disease. *Neurosignals* 11:262–269.

70. Sontag, E., Nunbhakdi-Craig, V., Lee, G., Bloom, G. S., and Mumby, M. C. (1996). Regulation of the phosphorylation state and microtubule-binding activity of Tau by protein phosphatase 2A. *Neuron* 17:1201–1207.

71. Vogelsberg-Ragaglia, V., Schuck, T., Trojanowski, J.Q., and Lee, V.M. (2001). PP2A mRNA expression is quantitatively decreased in Alzheimer's disease hippocampus. *Exp Neurol* 168:402–412.

72. Sontag, E., Luangpirom, A., Hladik, C., Mudrak, I., Ogris, E., Speciale, S., and White, C.L. III (2004). Altered expression levels of the protein phosphatase 2A ABalphaC enzyme are associated with Alzheimer disease pathology. *J Neuropathol Exp Neurol* 63:287–301.

73. Sontag, E., Hladik, C., Montgomery, L., Luangpirom, A., Mudrak, I., Ogris, E., and White, C.L. III (1994). Downregulation of protein phosphatase 2A carboxyl methylation and methyltransferase may contribute to Alzheimer disease pathogenesis. *J Neuropathol Exp Neurol* 63:1080–1091.

74. Morrison, L.D., Smith, D.D., and Kish, S.J. (1996). Brain S-adenosylmethionine levels are severely decreased in Alzheimer's disease. *J Neurochem* 67:1328–1331.

75. Vafai, S.B., and Stock, J.B. (2002). Protein phosphatase 2A methylation: a link between elevated plasma homocysteine and Alzheimer's Disease. *FEBS Lett* 518:1–4.

76. Kalaria, R.N. (2000). The role of cerebral ischemia in Alzheimer's disease. *Neurobiol Aging* 21:321–330.

77. Martin de la Vega, C., Burda, J., Toledo Lobo, M.V., and Salinas, M. (2002). Cerebral postischemic reperfusion-induced demethylation of the protein phosphatase 2A catalytic subunit. *J Neurosci Res* 69:540–549.

78. Sim, A.T., Baldwin, M.L., Rostas, J.A., Holst, J., and Ludowyke, R.I. (2003). The role of serine/threonine protein phosphatases in exocytosis. *Biochem J* 373:641–659.

79. Kowluru, A., and Metz, S.A. (1998). Purine nucleotide- and sugar phosphate-induced inhibition of the carboxyl methylation and catalysis of protein phosphatase-2A in insulin-secreting cells: protection by divalent cations. *Biosci Rep* 18:171–186.

80. Palanivel, R., Veluthakal, R., and Kowluru, A. (2004). Regulation by glucose and calcium of the carboxylmethylation of the catalytic subunit of protein phosphatase 2A in insulin-secreting INS-1 cells. *Am J Physiol Endocrinol Metab* 286:E1032–E1041.

81. Gaussin, V., Hue, L., Stalmans, W., and Bollen, M. (1996). Activation of hepatic acetyl-CoA carboxylase by glutamate and Mg2+ is mediated by protein phosphatase-2A. *Biochem J* 316:217–224.

82. Kowluru, A., Chen, H.Q., Modrick, L.M., and Stefanelli, C. (2001). Activation of acetyl-CoA carboxylase by a glutamate- and magnesium-sensitive protein phosphatase in the islet beta-cell. *Diabetes* 50:1580–1587.

83. Palanivel, R., Veluthakal, R., McDonald, P., and Kowluru, A. (2005). Further Evidence for the Regulation of Acetyl-CoA Carboxylase Activity by a Glutamate- and Magnesium-Activated Protein Phosphatase in the Pancreatic beta Cell: Defective Regulation in the Diabetic GK Rat Islet. *Endocrine* 26:71–78.

84. Yaar, R., Jones, M.R., Chen, J.F., and Ravid, K. (2005). Animal models for the study of adenosine receptor function. *J Cell Physiol* 202:9–20.

85. Hutchinson, S.A., and Scammells, P.J. (2004). A(1) adenosine receptor agonists: medicinal chemistry and therapeutic potential. *Curr Pharm Des* 10:2021–2039.

86. Liu, Q., and Hofmann, P.A. (2003). Modulation of protein phosphatase 2a by adenosine A1 receptors in cardiomyocytes: Role for p38 MAPK. *Am J Physiol Heart Circ Physiol* 285:H97–H103.

87. Moreno, C.S., Park, S., Nelson, K., Ashby, D., Hubalek, F., Lane, W.S., and Pallas, D.C. (2000). WD40 repeat proteins striatin and S/G(2) nuclear autoantigen are members of a novel family of calmodulin-binding proteins that associate with protein phosphatase 2A. *J Biol Chem* 275:5257–5263.

88. Nishito, Y., Usui, H., Shinzawa-Itoh, K., Inoue, R., Tanabe, O., Nagase, T., Murakami, T., and Takeda, M. (1999). Direct metal analyses of Mn2+-dependent and -independent protein phosphatase 2A from human erythrocytes detect zinc and iron only in the Mn2+-independent one. *FEBS Lett* 447:29–33.

89. Nishito, Y., Usui, H., Tanabe, O., Shimizu, M., and Takeda, M. (1999). Interconversion of Mn(2+)-dependent and -independent protein phosphatase 2A from human erythrocytes: role of Zn(2+) and Fe(2+) in protein phosphatase 2A. *J Biochem (Tokyo)* 126:632–638.

90. Van Hoof, C., Janssens, V., De Baere, I., Stark, M.J., de Winde, J.H., Winderickx, J., Thevelein, J.M., Merlevede, W., and Goris J. (2001). The Saccharomyces cerevisiae phospho-tyrosyl phosphatase activator proteins are required for a subset of the functions disrupted by protein phosphatase 2A mutations. *Exp Cell Res* 264:372–387.

91. Jordens, J., Janssens, V., Longin, S., Stevens, I., Martens, E., Bultynck, G., Engelborghs, Y., Lescrinier, E., Waelkens, E., Goris, J. and Van Hoof, C. (2006) The protein phosphatase 2A phosphatase activator is a novel peptidyl-prolyl *cis/trans* isomerase. *J. Biol. Chem.* 281:6349–57.

12

Reversible Methylation of Glutamate Residues in the Receptor Proteins of Bacterial Sensory Systems

FRANCES M. ANTOMMATTEI • ROBERT M. WEIS

Department of Chemistry
University of Massachusetts
710 North Pleasant Street
Amherst, MA 01003, USA

I. Abstract

The methyltransferase CheR catalyzes methyl group transfer from *S*-adenosyl-L-methionine to specific glutamic acid side chains of bacterial chemoreceptors, referred to as the methyl-accepting chemotaxis proteins (MCPs). A second enzyme, the methylesterase CheB, catalyzes ester hydrolysis. Together, CheR and CheB facilitate a reversible receptor methylation process that is essential for sensory adaptation. This property of adaptation has been most extensively studied in free-swimming *Escherichia coli* and *Salmonella*, where it serves as a rudimentary short-term memory during chemotaxis in gradients of attractants and repellents.

The methylation-demethylation process allows the bacterium to compare and respond to changes in the current concentration relative to those of the past three to four seconds. The feedback loop in which CheR participates facilitates perfect or near-perfect adaptation over a large range of chemoeffector concentrations, generating the means by which the cell remains responsive to small changes in

THE ENZYMES, Vol. XXIV
Copyright © 2006 by Elsevier Inc.

chemoeffector concentration. The structures of *Salmonella* CheR and the methyl-accepting domain of the serine receptor (Tsr) from *E. coli*, with biochemical data, paint a relatively detailed picture of receptor methylation, which proceeds by transmethylation in *E. coli* and *Salmonella*. Beyond *E. coli* and *Salmonella*, the diversity of receptor organization and the roles of methylation are still emerging. This diversity is likely to be large, given the wide range of ecological niches that prokaryotes occupy.

II. Introduction

The title of this chapter is intended to convey the broadening scope of the CheR function. This *S*-adenosyl-L-methionine (AdoMet) dependent transferase was first studied and is still mainly studied in the *Escherichia coli* chemotaxis system (and in the system of its close cousin *Salmonella enterica* serovar Typhimurium). Consequently, the perspective of this chapter is decidedly "*E. coli*-centric." *E. coli* chemotaxis is a remarkable phenomenon that provides the impetus for continued investigation: to generate a complete molecule-based description of the signaling pathway and motility apparatus of the cell. This objective has, through the years, benefited from a multifaceted approach to elucidate the chemical, structural, dynamical, and energetic properties of the pathway, and the macromolecular protein assemblies that constitute this behavior.

This chapter paints a current picture of CheR function, and suggests what the future may hold. It is as much the story of the substrate for CheR (the methyl-accepting chemotaxis proteins, MCPs) as it is of CheR itself. It begins with a review of receptor methylation and the salient features of the *E. coli* chemotactic behavior, emphasizing the role played by methylation. The structural data of representative examples for CheR and MCP are then presented and comparisons are drawn to proteins identified through genome sequencing. This is followed by a presentation of the mechanism of receptor methylation in this structural context, again using the best-studied examples (i.e., the MCPs from *E. coli* and *Salmonella*). Throughout the chapter, emphasis is placed on the role of receptor methylation in chemotactic function and its molecular basis. The MCPs are clustered in the membrane, often at the cell pole. As a substrate of CheR, the complexity of these receptor assemblies represents a continuing challenge to prepare and study samples in a manner that is sufficiently well defined to capture the salient features of the system in an unambiguous and quantitative manner. Nonetheless, significant recent progress has been made. The chapter concludes with prospects for the future.

Numerous review articles and book chapters have addressed receptor methylation either as the main focus [57, 188, 194] or as part of a more general review on chemotactic signaling in bacteria [62, 137, 150, 184, 191]. The bacterial motor receives the output of the signaling pathway. The contribution by Berg in volume 23 of this series [21] provides an in-depth treatment of the current state of affairs in understanding the structure and function of the motor, and also covers bacterial swimming behavior and signaling. The present contribution is neither an exhaustive retrospective of receptor methylation nor simply an update. Rather, the attempt has been made to find the middle ground by summarizing the important past findings in the field, and emphasizing recent advances.

III.   The Signaling Pathway

The synthesis of genetic, biochemical, behavioral, structural, and theoretical modes of investigation have generated a relatively detailed picture of chemotactic signal transduction in *E. coli* and *Salmonella*. Due to the relative simplicity of chemotactic signaling in these two organisms, most components of the *E. coli* and *Salmonella* pathways are universal to the chemotaxis and chemotaxis-like pathways of *Bacteria* and *Archaea*. Differences from the well-studied *E. coli* system lie mainly in the finding of additional protein modules (domains) and the signaling logic, which is manifested in a diversity of domain organization and biochemical circuitry. Szurmant and Ordal have recently reviewed the current state of this diversity for chemotaxis pathways in *Bacteria* and *Archaea* [198]. Here, the primary focus is the *E. coli* pathway.

A.   A BRIEF HISTORY OF RECEPTOR METHYLATION

Of the two forms of protein modification involved in the chemotaxis signaling pathway, receptor methylation was detected [101] and studied extensively well before protein phosphorylation [85, 221]. Early studies established *S*-adenosyl-L-methionine as the source of the methyl group [11], *cheR* as the gene encoding for the methyltransferase [189], and the transmembrane receptors as targets of methylation [101]. This latter feature gave rise to the term *methyl-accepting chemotaxis protein* (MCP). Critical to the role of methylation is the fact that MCPs are multiply-methylated [33, 45, 61]. It is a property that can be conveniently visualized by methylation-dependent shifts in protein mobility on SDS-polyacrylamide gels (Figure 12.1) [29].

The physical basis for this mobility change has not been systematically investigated, but it is certainly related to the charge on the glutamate side chain. Modifications of this carboxyl group that eliminate the negative charge, either genetically (e.g. through the substitution of a glutaminyl residue in place of the glutamyl residue), or through the reaction catalyzed by CheR to form the glutamyl γ-methyl ester, produce similar increases in the electrophoretic mobility. (Figure 12.1). The identification of more-extensively modified MCPs with more-rapidly migrating protein bands has proven to be a simple but effective method to establish the

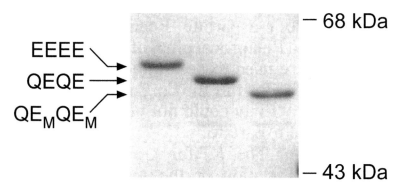

FIG. 12.1. Modification-dependent migration of MCPs on an SDS polyacrylamide gel. This photograph of a Coomassie-stained SDS-polyacrylamide gel of the *E. coli* aspartate receptor (Tar) was taken from Figure 1a of Borkovich et al. [29]. The three samples of Tar, in different levels of modification, were isolated from three different *E. coli* strains, either methyltransferase-minus (*cheR*), methylesterase-minus (*cheB*), or transferase-minus/esterase-minus (*cheRB*). The four major methylation sites in the Tar protein (Q295, E302, Q309, and E491) isolated from the *cheRB* background (*middle*) had the gene-encoded glutamines and glutamates at sites 1 and 3, and 2 and 4, respectively (QEQE). The *cheRB* background had neither transferase nor esterase activity. Tar isolated from *cheR* cells (*left*) had unmodified glutamate residues at the four major sites (EEEE), giving evidence that CheB has deamidase as well as esterase activity. The gene-encoded glutaminyl residues at sites 1 and 3 were converted to methyl-accepting glutamates. Tar protein isolated from transferase-plus/esterase-minus (*cheB*) cells (*right*) had glutamines at sites 1 and 3, and γ-glutamate methyl esters at sites two and four (QE_MQE_M). Note that the glutaminyl and the γ-glutamyl methyl ester modifications produce similar increases in protein mobility.

influences of attractants and repellants on methylation level, decipher the logic of the signaling pathways, and determine the requirements for efficient receptor methylation, demethylation, and deamidation.

The location of the glutamyl γ-methyl esters in some of the *E. coli* and *Salmonella* MCPs were determined with a combination of trypsin digestion, peptide mapping, and sequencing of methylated peptides [93, 94, 96, 200, 201]. Since then, sequence homology among the MCPs has been an effective predictor of the location of methylation sites [109]. Figure 12.2 depicts the locations of the major sites in the MCPs from *E. coli* and *Salmonella enterica* (serovar Typhimurium). The locations of these sites are representative of a pattern in which the major sites are found on two methylations helices (MHs). In *E. coli* and *Salmonella* three of the four major sites are clustered on MH1, whereas the fourth site and a (minor) fifth site reside in a cluster on MH2. From data such as these, the consensus sequence for methylation (Glu-*Glu*-Xxx-Xxx-Ala-Thr/Ser)[1]

---

1. Xxx represents unconserved amino acids and the methyl-accepting glutamate is in italics.

```
                297    304    311                    493       503
                 |      |      |                      |         |
Ec Tsr   290  DLSSRTEQQAASLEETAASMEQLTATVKQNA    QQNAALVEESAAAAAALEEQASRLTE   510
St Tsr   290  DLSSRTEQQAASLEETAASMEQLTATVKQNA    QQNASLVEESAAAAAALEEQASRLTQ   510
Ec Tar   288  DLSSRTEQQASALEETAASMEQLTATVKQNA    QQNASLVQESAAAAAALEEQASRLTQ   508
St Tar   288  DLSSRTEQQASALEETAASMEQLTATVKQNA    QQNASLVQESAAAAAALEEQASRLTQ   508
Ec Trg   298  DLSSRTEEQAAAIEQTAASMEQLTATVKQNA    QQNASLVEEASAAAVSLEEQAARLTE   518
St Trg   302  DLSSRTEQQAAAIEQTAASMEQLTATVKQNA    QQNASLVEEASAAARSLEEQAARLTQ   522
Ec Tap   287  DLSSRTEQQAASLAQTAASMEQLTATVGQNA    QQNASLVEEAAVATEQLANQADHLSS   506
St Tcp   289  DLSSRTEQQASALEETAASMEQLTATVRQNT    QQNASLVEESAAAAAALEDQANELRQ   519
Ec Aer   280  ELNEHTQQTVDNVQQTVATMNQMAASVKQNS    QKNAELVEESAQVSAMVKHRASRLED   500
St Aer   280  DLNEHTRQTVENVQETVTTMNQMAESVKLNS    QKNAALVEESAQVSAMVKHRASRLED   500
```

FIG. 12.2. Sequences of the methylation regions in the *E. coli* (*Ec*) and *S. typhimurium* (*St*) MCPs: Tsr, serine receptor; Tar, aspartate receptor; Trg, ribose/galactose receptor; Tap, dipeptide receptor; Tcp, citrate receptor; and Aer, aerotaxis receptor. The numbering across the top indicates the major modification sites in *E. coli* Tsr. Methyl-accepting residues are in bold.

was generated, and it was postulated that the methylation sites reside on one face of an α helix [202]. The crystal structure for the cytoplasmic domain of the serine receptor [97] (described in the section IV.B.1) has confirmed these findings, led to the creation of a molecular model for the MCP dimer, and paved the way for detailed investigations of receptor function, including transferase-receptor interactions.

The pre-genome-era sequencing efforts facilitated the identification and characterization of the MCP family in *E. coli* and *Salmonella*. The aspartate receptor (Tar), the serine receptor (Tsr), the ribose/galactose receptor (Trg), the dipeptide receptor (Tap), and the citrate receptor (Tcp) are all named for their primary attractant ligands [28, 103, 164, 223], although the receptors often exhibit specific sensitivities to more than one ligand. For example, *E. coli* Tar also senses cobalt and nickel as repellents and the maltose binding protein-maltose (MalE-maltose) complex as an attractant [159]. The basic domain organization of all of these MCPs is the same: a ligand binding domain located in the periplasm, two transmembrane segments, and a methyl-accepting cytoplasmic domain involved in signal regulation, where the methylation sites are located. This organization is discussed in the section IV.B.1.

Completion of the *E. coli* and *Salmonella* genomes presaged things to come [27, 132]. Aer, a receptor known to be involved in oxygen taxis (or more generally, energy taxis), was found by an analysis of these genomes [23, 160]. Notably, the Aer protein represents one of the first examples of an MCP-like

---

2. The Htr protein of *Halobacter salinarium* is another early example of an unorthodox MCP organization [224]. In the case of Htr, the sensory input is provided by a separate protein, sensory rhodopsin, SR and it is a physical association between HTR and SR that generates a photoreceptor that can sense light and regulate CheA [86].

protein with an unorthodox domain organization,[2] where both the sensing (input) domain and the methyl-accepting (output) domain are located in the cytoplasm. However, recent evidence indicates that Aer is not methylated, unlike its "orthodox" relatives [24]. Thus, the presence of a "methyl-accepting" (MA) domain does not automatically indicate that the protein is a substrate of CheR. The analysis of aligned MA domain sequences provides a straightforward means to determine if a receptor like Aer is a CheR substrate. The poor agreement between the consensus sequence and sites one through three of Aer (Figure 12.2) indicates that the requirements for effective interactions between the active site of CheR and the methylation sites are not met. This conclusion is supported by the experimental observation that Aer is not methylated in the cell under conditions where Tsr can be methylated [24].

The trickle of genome sequence projects has since become a torrent. The significant number of completed genome sequences has made it feasible to compare the prevalence of signal transduction gene products in different bacterial species [71]. *E. coli* K12 has just four MCPs and one MCP-like protein, a relatively small number compared to the number of MCP-like genes identified in bacterial genome sequences, in which upward of 50 MA domain-containing ORFs have been identified,[3] and genomes with 20 or more are common [2, 71]. Why the range? A plausible explanation is based on the diversity of ecological niches. *E. coli* K12 occupies a simpler, less-diverse niche and therefore has a smaller requirement for generating chemotactic signals in response to environmental cues. Bacteria that live in complex (soil and marine) environments are faced with more diverse, constantly changing environmental conditions. Multiple chemotactic pathways and a more diverse array of MCPs represent one way to effectively meet the challenges presented by these environmental conditions.

B.   THE *E. COLI* CHEMOSENSORY SYSTEM

E. coli chemotaxis and motility has been the subject of a number of reviews [21, 32, 198, 204]. Recent reviews of MCP function emphasize the manner and extent to which the signaling proteins in the *E. coli* system assemble into functional clusters [37, 150, 184, 207]. Figure 12.3a is a cartoon synopsis of the bacterial chemotaxis signaling pathway, highlighting the role played by receptor methylation. The activity of the signaling pathway is depicted as a balance between two states

---

3. The current record, of more than 70 *mcp* genes, is found in the draft genome of *Magnetospirillum magnetotacticum* MS-1 (DOE Joint Genome Institute, *http://www.jgi.doe.gov/*).

of the MCP/CheW/CheA signaling complex: kinase activating and kinase inhibiting, which provides a simple but effective qualitative framework for understanding pathway behavior.

In the active state, CheA (a member of the histidine protein kinase superfamily [77, 193, 216]) rapidly phosphorylates two proteins, one that consists of a single receiver domain (the response regulator CheY) and one in which the receiver domain regulates the activity of a catalytic domain (the methylesterase CheB). The phosphorylated forms of CheY and CheB (CheY~P and CheB~P) are active and labile. CheY~P binds to FliM in the motor switch complex, which serves to elicit cell tumbling by enhancing the CW bias of the motor. As a substrate of the phosphatase, CheZ, CheY~P is rapidly hydrolyzed to CheY and inorganic phosphate. CheB~P has much greater esterase activity relative to unphosphorylated form [6], a key feature of the negative feedback loop that facilitates adaptation. Sensitivity to environmental signals arises by regulating CheA activity through the MCPs in the context of a ternary complex with CheA and the SH3-like adaptor protein, CheW [73, 168]. MCPs form clusters in the membrane [127], which is a consequence of ternary complex formation and is a key feature of kinase regulation. As Figure 12.3a depicts, the MCP cluster can be heterogeneous with respect to both ligand binding specificity and the level of covalent modification.

Although the detailed nature of (i.e., the differences between) the signaling complexes in the activating and inhibiting states is not known, and represents an active area of investigation, it is certain that ligand binding and methylation levels are two principal factors influencing the balance between the activating and inhibiting states. Attractant binding shifts the balance toward the inhibited state; removal of attractant shifts the balance toward activation. The binding of ligands that elicit a repellent response (e.g., cobalt and nickel ions sensed through Tar) will shift the balance toward kinase activation. The second factor, covalent modification, also strongly influences the balance. Increases in the level of modification shift the balance toward CheA activation. As described in material following, the covalent modification level (controlled in part by a negative feedback loop) is used to achieve adaptation by countering the effects of ligand. CheR-catalyzed receptor methylation is central to this adaptation function.

What are the stimulus-response characteristics of the signaling pathway? This picture emerges when changes in the balance of the activated and inhibited states are considered in the context of a stimulus-response experiment (Figure 12.3b). A stimulus in the form of a step-like increase in the attractant concentration is shown, and the responses of a bacterium are depicted qualitatively as the time-dependent changes in the clockwise (CW) bias of the motor (~ tumble frequency), the CheY-phosphate concentration, and the level of receptor methylation. Their prestimulus values are set by the CheA activity of the adapted state, a state that

a

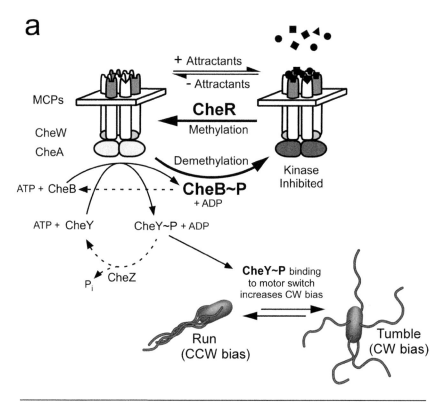

MCPs

CheW
CheA

+ Attractants
− Attractants

CheR
Methylation

Demethylation

Kinase
Inhibited

ATP + CheB

CheB~P
+ ADP

ATP + CheY   CheY~P + ADP

P$_i$ CheZ

CheY~P binding
to motor switch
increases CW bias

Run
(CCW bias)

Tumble
(CW bias)

b

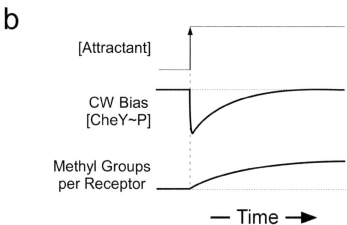

[Attractant]

CW Bias
[CheY~P]

Methyl Groups
per Receptor

— Time ➤

is characterized by equivalent (but opposing) overall rates of receptor methylation and demethylation (catalyzed by CheR and CheB~P, respectively), which maintains a steady-state level of receptor methylation and CheA activity.

An abrupt increase in attractant concentration rapidly decreases CheA activity by shifting the balance toward the inhibiting state. The CheY~P and CheB~P concentrations, and the CW bias (tumble frequency), all decrease as a result. Favorable stimuli like the one in this example serve to prolong the smooth-swimming run of the bacterium along the same trajectory. Over a longer period of time, the overall level of receptor methylation increases to a higher steady-state level, which resets the signaling system to the prestimulus value of CW (tumble) bias, and by inference the prestimulus CheA activity and CheY~P concentration.

The path toward the higher methylation level is influenced by the transient decrease in the methylesterase activity, and a sustained increase in the CheR-catalyzed methyl group transfer produced by attractant binding. A decrease in the attractant concentration (not shown) produces qualitatively opposite effects on [CheY~P], CW bias, and receptor methylation levels, although quantitative differences in the response threshold and the time course of adaptation exist. The net effect is to bias the swimming motility toward favorable conditions (i.e., larger concentrations of attractants and smaller concentrations of repellents).

Receptor methylation facilitates perfect or near-perfect adaptation over a wide range of attractant concentrations (~ nM to mM), enabling the signaling pathway

---

FIG. 12.3. The *E. coli* chemotaxis signaling pathway (a) and the response-adaptation timeline (b). (a) The biochemistry of the chemotaxis signaling pathway depicted with a two-state signaling model. Heterogeneous clusters of MCPs form complexes with the adaptor protein CheW and the protein kinase CheA to form a signaling complex. CheA autophosphorylation and phosphate transfer to the substrates CheY and CheB are rapid in the active-state signaling complex (*left*) and much slower in the inhibited state (*right*). The phosphorylated forms of response regulator CheY and the methylesterase CheB (CheY~P and CheB~P, respectively) are active in eliciting tumbles (CW motor bias) and receptor demethylation (and deamidation), respectively. Processes that influence the balance between activating and inhibiting states are shown associated with arrows (attractant binding/unbinding, CheR-catalyzed receptor methylation, CheB~P-catalyzed receptor demethylation). Attractant binding serves to inhibit CheA and also serves to increase the receptor methylation rate. (b) A timeline for response and adaptation to a positive stimulus. An abrupt increase in the attractant concentration (↑) generates a rapid decrease in the CW bias, which is a consequence of attractant-mediated CheA inhibition, *vis-á-vis* Figure 12.3a, and the subsequent rapid decrease in [Chey~P]. Adaptation is manifested in the restoration of the CW bias (and [CheY~P]) to the prestimulus level over a longer period of time, which results from an increase in the steady-state level of receptor methylation brought about by attractant binding. The time course of adaptation depends on the stimulus size: a few seconds in the limit of a small stimulus; several minutes for a large, saturating stimulus.

to maintain sensitivity to small changes in concentration (~0.1%) in this range. Critical to effective chemotaxis is the time-scale of adaptation. This depends on several factors, including the sign and magnitude of the stimulus, and the relative amounts of receptor and signaling components. For example, elevating the level of receptor expression lengthens the period of time required to achieve adaptation, and elevating the level of CheR shortens it [3, 164]. The time scale of temporal comparisons to physiologically relevant (small, nonsaturating) stimuli is about four seconds. *E. coli* "remembers" the chemoeffector concentration of the past four seconds [170].

Such a time scale is compatible with the physical constraints placed on free-swimming *E. coli*. A temporal comparison on a smaller time scale (shorter memory) provides insufficient averaging to distinguish chemical gradients from local concentration fluctuations [22]. Temporal comparisons over an excessively long time period are unproductive for a different reason: rotational diffusion of the bacterial cell results in spatial "memory loss" [20].

In this limit, the bacterium compares and responds to differences in the concentration along uncorrelated swimming trajectories. In essence, without knowing where it is coming from the bacterium cannot know where it should be going. Consequently, chemotactic efficiency depends in an important way on the factors that influence the rates of receptor methylation and demethylation. *E. coli* strains without the methylation/demethylation system (*cheRB*) can respond to attractants and repellents, but are unable to make the appropriate temporal comparisons and cannot undergo efficient chemotaxis [170, 211, 213].

C.   COVALENT MODIFICATION AND REGULATION OF CHEA ACTIVITY

Some of the experiments cited in support of a balance between CheA-activating and CheA-inhibiting signaling states has been obtained through measurements of kinase activity in reconstituted systems [29–31, 116, 117]. Figure 12.4 shows representative data for these types of experiments. The kinase activities of Tsr-CheW-CheA complexes are plotted as a function of the ligand concentration for complexes formed with the Tsr protein in fixed levels of modification at the major sites: low (EEEE), intermediate (QEQE), and high (QQQQ) [117]. As in studies conducted with Tar [30, 31], the Tsr molecules shown in Figure 12.4 were isolated in fixed levels of amidation, in that the effect generated by the glutaminyl residues is qualitatively similar to the effect of glutamyl γ-methyl ester formation [29, 60] and homogeneous samples (at least with respect to the modification level) were more readily prepared in this way.

Similar qualitative trends were observed in all these studies and include (1) CheA activity that increases with the level of modification in the absence of attractant ligand, (2) the need for larger concentrations of ligand to produce half-maximal inhibition (IC_{50}) as covalent modification increases, and (3) inhibition

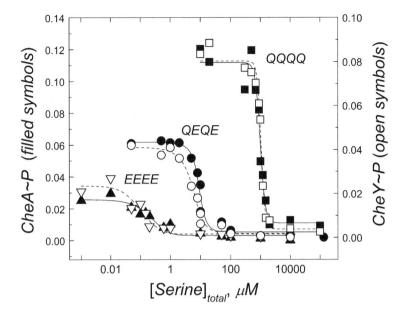

FIG. 12.4. Kinase inhibition as a function of MCP covalent modification. CheA activity was measured in reconstituted Tsr:CheW:CheA-signaling complexes as a function of the serine concentration and the level of covalent modification on Tsr, which was altered by glutamine incorporation at the major methylation sites: four glutamates at the major sites (EEEE), glutamines at sites one and three (QEQE), and glutamines at all four sites (QQQQ). Adapted from Figure 3 of Li and Weis [117].

is cooperative. Fits of the data in Figure 12.4 to the Hill equation generated estimates for the cooperativity coefficients approaching 10, with signaling complexes containing highly-modified (QQQQ) Tsr [117].

Data such as these provided the first biochemical evidence of significant cooperative interactions among receptors in the regulation of CheA, and efforts to understand the molecular origins of this cooperativity are a current focus of research. The correspondence of data collected in experiments with reconstituted signaling complexes (as in Figure 12.4) and subsequent experiments conducted *in vivo* are largely consistent. Sourjik and Berg developed an elegant approach based on *in vivo* FRET to measure dose-response (sensitivity) curves of CheA activity inside living *E. coli* as a function receptor modification levels and receptor composition (e.g. Tar alone, Tsr alone, or Tar-Tsr mixtures [186, 187]). The same qualitative trends were observed: (1) the CheA activity increased with receptor modification, (2) the IC_{50} increased with receptor modification, and (3) the inhibition was cooperative. Moreover, the *in vivo* FRET studies were among the first to provide significant new evidence for interactions between receptors of different ligand specificity [74, 107] and to explore some of the factors that lead to changes in the degree of cooperativity [187].

Quantitative differences are found in the parameters estimated in different studies (IC_{50}; the cooperativity coefficient, n; and the shift in IC_{50} with changes in modification), for which there is not always an explanation. In reconstituted systems, some of these differences can result from variations in the MCP preparations that influence kinase activity and complex assembly, and thus understanding the basis for these differences may lead to important insights. Invariably, the kinase activity of receptor-CheW-CheA complexes in the absence of ligand has been observed to increase with the level of covalent modification (as in Figure 12.4).

On the other hand, the kinase activity was the same for all levels of covalent modification (EEEE, QEQE, and QQQQ) when the complexes were generated by template-directed self-assembly of the Tar cytoplasmic domain (sans ligand-binding, transmembrane, and HAMP domains), CheW and CheA [179]. The difference between the two systems not only reflects the possible dependence of signaling properties on the mode of assembly but provides insight into the properties of the MA domain in the absence of the ligand-binding and HAMP domains.

The theoretical models that invoke cooperativity to explain the gain and sensitivity characteristics of receptors often assume an underlying structure for the receptor array [1, 31, 34, 35, 133, 134, 174]. Thus, establishing the properties of the array is important (e.g., the detailed arrangement of proteins and the energetics of interactions among the components). Receptor methylation is certain to play a prominent role in modulating these interactions.

IV.   The Methyltransferase and Its Substrate

A.   THE METHYLTRANSFERASE

1.   The Structure of Salmonella CheR

Two crystal structures of *Salmonella* CheR (EC 2.1.1.80) have been determined: in one structure, CheR is bound to *S*-adenosyl-L-homocysteine (AdoHcy) [55] and in the other structure, CheR is bound to both AdoHcy and an acetylated pentapeptide from the C-terminal of Tar (ac-NWETF) [56]. The structures reveal the two-domain organization of CheR, giving insight to the mechanism of substrate recognition and catalysis (Figure 12.5). The N-terminal domain, which consists of four α helices, plays a key role in methylation helix recognition. The C-terminal domain contains the binding site for *S*-adenosyl-L-methionine (AdoMet), although the N-terminal domain and the N-domain/C-domain linker also make contact with AdoHcy [55].

AdoMet-dependent methyltransferases are grouped into five classes, and the α-β-α architecture of the CheR C-terminal domain defines it as a member of class I [167]. The interaction between AdoMet and CheR has been reviewed [57],

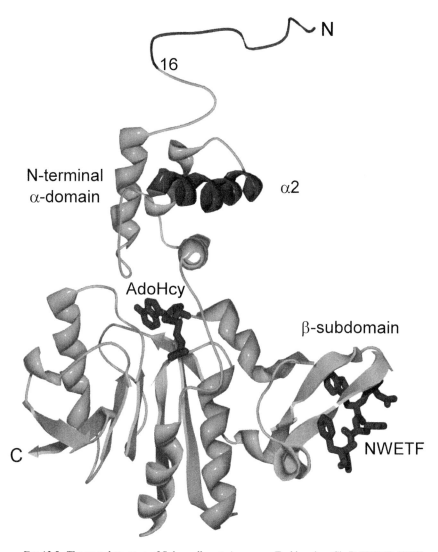

FIG. 12.5. The crystal structure of *Salmonella enterica* serovar Typhimurium CheR (PDB ID 1BC5). The bound reaction product *S*-adenosyl-L-homocysteine (AdoHcy) and the receptor pentapeptide (ac-NWETF) are rendered as stick structures. Sixteen amino acids at the N-terminus that were not resolved in the crystal structure are depicted as (an unstructured) random coil. Alpha helix two (α2, darkened) of the N-terminal α-domain contains conserved lysine and arginine residues that are implicated in methylation site sequence recognition. Based on Figure 1A of Perez et al. [151].

and has the characteristic features of a nucleotide binding fold [55]. The *Salmonella* CheR also has a β subdomain that appears to be common to all two-domain methyl-transferases involved in chemotaxis [178]. In *E. coli* and *Salmonella* CheR, this domain is known to bind the C-terminal pentapeptide found in Tar, Tsr, and Tcp (NWE[T/S]F), but not Trg and Tap [56, 219], and this binding interaction is alone responsible for localizing CheR to receptor signaling complexes [178]. Together, the two domains possess the catalytic and molecular recognition properties that generate specificity for the MCP substrate.

2.   CheR Phylogenetic Distribution and Occurrences in Multidomain Proteins

A survey of bacterial genomes, now numbering in the hundreds, has shown that chemotaxis systems based on MCPs and two-component (CheA) signaling are found throughout the *Archaea* and *Bacteria*. CheR is associated with the majority of these systems.[4] Curated sequence databases permit us to look beyond *E. coli* and *Salmonella* to determine how widely the two-domain organization of CheR is distributed [18, 114, 129]. For example, the Pfam database [18] facilitates a determination of the prevalence of the two-domain organization of CheR, in that N- and C-domains are given separate designations (PF03705 and PF01739, respectively).[5] In nearly all instances (>99%), the N-domain is associated with a C-domain. Conversely, C-domains have an associated N-domain in more than 95% of the cases.[6] These statistics imply that recognition of the MCP as a substrate requires the two-domain organization.

As intact proteins (and predicted proteins), the two-domain organization shown in Figure 12.5 accounts for 85% of the instances in the genome database. The remaining 15% constitute small, but interesting, groups. A few examples are depicted in Figure 12.6. These CheR-containing proteins are often found in organisms where more than one protein in the genome is predicted to contain the CheR domain. Two of the motifs in Figure 12.6 (*M. xanthus* FrzF and *B. burgdorferi* CheW-3) suggest different mechanisms for localizing CheR in receptor signaling complexes. *M. xanthus* FrzF is representative of a domain organization found in ~4% of the CheR-containing proteins, where a C-terminal extension contains tetratrico peptide repeats (TPRs). This is a motif known to facilitate protein-protein interaction and the assembly of multi-protein complexes [51]. CheW-CheR fusion proteins are observed in species of *Borrelia* and *Treponema*. (Currently, these examples represent a small fraction, ~1%, of the predicted proteins in which the CheR domain is detected.) Because CheW facilitates the activation of CheA

---

4. *Helicobacter pylori*, a bacterium in which chemotaxis augments virulence [199], has neither *cheR* nor *cheB* [153]. The mechanism of CheR- and CheB-independent adaptation is not known.

5. The NCBI Conserved Domain Architecture Retrieval Tool [129] enables similar analyses.

6. Most of these N-domain-negative cases are ambiguous because (1) sufficient polypeptide is found before the C-domain to accommodate the N-domain and (2) no others domains are identified in the region.

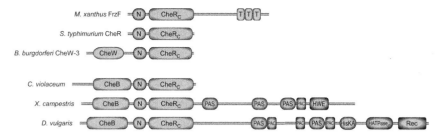

FIG. 12.6. Domain organization of CheR-containing proteins. Domain organizations of proteins containing CheR N-terminal (N) and catalytic domain (CheR$_C$) retrieved from the PFAM [18] and SMART [114] databases, indicating the species (and the GenBank accession number): *M. xanthus* FrzF (AAA25398) [130], *S. typhimurium* CheR (AAL20834) [55], *B. burgdorferi* CheW-3 (AAB81649) [69], *Chromobacterium violaceum* (AAQ59256) [38], *Xanthomonas campestris* (AAM42956) [156], and *Desulfovibrio vulgaris* (AAS94932) [84]. Additional domain abbreviations: T, tetratricopeptide repeats [51]; CheW, chemotaxis SH3-like protein [25, 78]; PAS domain [141, 230]; PAC, PAS domain S-box [154, 230]; HWE [92]; HisKA, histidine kinase phosphate-accepting dimerization domain [203]; HATPase$_C$, histidine kinase ATP binding and catalytic domain [26]; and Rec, response regulator receiver domain [49, 192].

through the formation of a ternary (MCP-CheW-CheA) complex [73, 168], an N-terminal CheW domain found in proteins like *B. burgdorferi* CheW-3 might tether CheR to the receptor complex. The relevance of tethering CheR near the receptor methylation sites is discussed in Section IV.C.1.

Domain structures for three more CheR-containing proteins are also depicted in Figure 12.6. These all have in common the catalytic domain of the methylesterase CheB. These CheB-CheR fusions (presumably) have the ester formation and hydrolysis activities in a single protein. The *E. coli* and *Salmonella* chemotaxis system uses phosphorylation of an N-terminal regulatory (receiver) domain to generate esterase activity (Figure 12.3) [6]. In contrast, this regulatory domain is not present in the CheB-CheR fusions, suggesting that other mechanisms are involved in regulating MCP methylation and demethylation. The CheB-CheR fusion ensures that the two enzyme activities colocalize, but other considerably more complex, fusion proteins are predicted (Figure 12.6), which also contain PAS/PAC domains [230] and two-component kinase and receiver domains [92, 216].

If these fusion proteins represent the mature forms, the enzymatic activities normally associated with both the excitation and adaptation branches of the signaling pathway will also colocalize. Why is this necessary? The possibility of unintended MCP-signaling protein interactions does not exist in the *E. coli* system, where just one copy of each gene in the pathway is present and the cell has one receptor array composed of the five MCPs, but a survey of the prokaryotic genomes demonstrates that as often as not multiple copies of *cheR* and *cheB* are present in the genome of an organism. Thus, it seems that a requirement for a correctly functioning pathway is the appropriate partnering of the signaling components, which these fusion proteins can facilitate.

1. Structure and Organization of the Receptor Dimer

A significantly more detailed picture for the structure and organization of *E. coli* and *Salmonella* MCPs has developed since the crystal structure for the cytoplasmic domain of *E. coli* Tsr has become available [97, 98]. The MCP is organized as a dimer (Figure 12.7), and its structure and the structure-function relationships are among the most intensively studied aspects of the transmembrane signaling in bacteria [63, 64]. The structure of the cytoplasmic domain has (1) provided added evidence that MCPs organize as dimers and (2) revealed a molecular basis for interactions among dimers as a trimer-of-dimers interaction (Figure 12.7, *left*).

The dimeric organization of MCPs was firmly established once the crystal structure of the *Salmonella* Tar ligand binding domain was solved, with and without aspartate bound [136, 225]. In these structures, two symmetry-related binding sites were identified (situated at the interface between the two subunits), and the bound aspartate has direct interactions with both subunits. When Kim, Yokota, and Kim solved the structure of the Tsr cytoplasmic domain fragment [97], they also constructed an atomic model for the intact receptor, guided by (1) the crystal structures of the periplasmic and cytoplasmic fragments, (2) an assumed helical structure for the intervening polypeptide (transmembrane and HAMP domains), and (3) refinement of the structure with computational energy minimization, subject to a twofold symmetry constraint. This model is shown in Figure 12.7, modified to include the C-terminal CheR docking site of Tar, Tsr, and Tcp [219] through the addition of amino acids at the C-terminus to generate a polypeptide 551 amino acids in length (corresponding to the number of residues in *E. coli* Tsr). The last 35 amino acids are depicted in a random coil conformation, because the evidence suggests that this region functions like a tether to localize CheR near the sites of methylation [219].

As Figure 12.7 shows, the dimer is nearly all helical; large regions of uninterrupted helix generated a model structure that is very long relative to its girth [97]. As a consequence of this organization, the functional domains of the receptor map onto the structure in a linear manner (Figure 12.7, *right*). The representations of the MCP dimer also illustrate that the change resulting from ligand binding must be transmitted over a hundred ångstroms to the methylation sites, and farther still to the site of interaction with CheW and CheA.

This information is transmitted through the HAMP domain (amino residues 216 through 268 in Tsr), located just inside the membrane. The HAMP domain

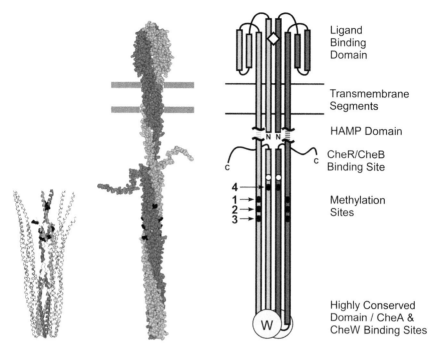

Ligand
Binding
Domain

Transmembrane
Segments

HAMP Domain

CheR/CheB
Binding Site

Methylation
Sites

Highly Conserved
Domain / CheA &
CheW Binding Sites

FIG. 12.7. MCP structure models. *Left*: The X-ray crystal structure of Tsr cytoplasmic fragment (PDB ID 1QU7) depicting the trimer-of-dimers organization. The two subunits in the front-facing dimer of the trimer are rendered as ribbons in different shades of gray. The polypeptide backbones of other two dimers are depicted in stick mode. The side chains of methyl-accepting residues are shown with a black space-filling representation. Atomic (*middle*) and schematic (*right*) models of the full-length MCP dimer. The space-filling representation of an atomic model of an *E. coli* MCP dimer, based on the model generated by Kim et al. [97]. The subunits are shaded differently and the sites of methylation are blackened. This representation has been modified to depict the last 35 residues as an unstructured polypeptide segment, which serves to tether CheR *via* a specific binding interaction with the last five amino acids (e.g., NWETF). The HAMP domain is depicted as an uninterrupted α-helix in the atomic model, but because detailed structure is not known it is depicted as a dashed region in the schematic diagram of the receptor dimer shown at the *right*. The *left* and *center* portions of the figure are adapted from Figures 4a and 5 of Kim et al. [97], respectively, and the *right* portion is based on Figure 1 of Chalah and Weis [44].

is known primarily as a conserved sequence motif (~50 amino acid residues in length), named for the prokaryotic signaling proteins (Histidine kinases, Adenylate cyclases, MCPs, Phosphatases) in which it is found [10]. No high-resolution structure information for the HAMP domain is yet available, which is why the schematic representation of the MCP dimer (Figure 12.7, *right*) depicts this region as a dashed segment. A site-directed cysteine and disulfide scanning study of the HAMP region in Tar [40] and electron microscopic images of membranous Tsr assemblies [212] provide evidence consistent with an organization of the HAMP domain that is more compact than the atomic model depicts.

The methylation sites, which are found in two clusters separated from one another in the primary structure (Figure 12.2), are brought together in the tertiary structure as a result of the anti-parallel coiled-coil organization in the methyl-accepting (MA) domain. This proximity is apparent in Figure 12.7. The MA domain, which is conserved evolutionarily [109], is the defining feature of MCPs, and to date, the Tsr MA domain provides a representative structure. The *E. coli* Tsr MA domain is annotated in the SMART database as a 248-amino-acid domain (residues 269 through 516), an assignment that maps well onto the structural model of the dimer. The MA domain follows immediately after the HAMP domain (residues 216 through 268) in the Tsr sequence and terminates in the CheR dock-ing tether (residues 517 through 551). The MA domain is made of two helical segments of equal length (~ residues 269 through 389 and 396 through 516), which are engaged in the coiled-coil interaction, and a hairpin formed around a six amino acid segment [GEQGRG (residues 390 through 395)], located between the two helical segments. MA domain dimerization occurs through a second coiled-coil interaction between the helices of the two participating domains. This is reflected in the double heptad repeat found in the sequence [97].

The degree of conservation varies within the MA domain, with the highest conservation found in the region near the hairpin including the trimer-of-dimers interaction site, and the CheA and CheW binding sites. This region (~ residues 360 to 425 in *E. coli* Tsr), sometimes referred to as the highly conserved domain (HCD), is postulated to result from the numerous protein-protein interactions that must be satisfied, including (1) MA domain dimerization, (2) the trimer-of-dimers interaction, and (3) CheA and CheW binding.

In regions farther from the hairpin, the homology is lower but still significant, as the coiled-coil repeat indicates. Two reasons can be cited for this: (1) the number and location of methylation sites vary among MCPs and (2) the presence of "indels," which are insertions/deletions that occur in both coiled-coil segments in multiples of the coiled-coil repeat (seven residues) and that have the effect of shortening (or lengthening) the MA domain. The indel pattern within a set of 29 MA domains ana-lyzed by Le Moual and Koshland in 1996 led to identification of three classes [109]. Now, with over 10^3 MA domains sequences available for analysis, more classes have been identified and a picture for the evolution of the MA domain is developing.[7]

2.   Formation of Receptor Arrays and Subcellular Localization

The first study of the cellular localization of MCPs and signaling proteins in *E. coli* was conducted by Maddock and Shapiro [127]. This study produced a

---

7. Roger Alexander and Igor Zhulin, manuscript in preparation.

striking result that was contrary to the expectation at the time: MCPs were clustered in the membrane and the clusters were located preferentially at the poles of the cell. In addition, CheA and CheW promoted more extensive cluster formation and polar localization. Binding interactions between the MCPs, CheA, and CheW were obvious agents of cluster formation, but it was the trimer-of-dimer interaction found in the Tsr MA domain crystal structure that provided insights into the molecular mechanism. Armed with this information, a model of a "core" complex consisting of MA domains, CheW, and CheA was built from the structures of the individual proteins, which was used to generate a receptor signaling array with hexagonal symmetry [175].

The features of this model (based on MCP-MCP trimer-of-dimers interaction and bridging interactions between MCPs, CheW, and CheA) were generally compatible with the results of the cellular localization studies. These subsequent studies, which have used either antibody-staining or fluorescent fusion proteins, affirmed the polar localization of signaling proteins [13, 41, 185] and identified factors responsible for cluster formation [123, 126]. The core complex also has binding sites for CheR (on some MCPs) and CheB (on some MCPs and CheA) that leads to their colocalization. Disrupting these interactions affects chemotactic function in a manner consistent with their roles in adaptation [9, 13]. Evidence supporting the existence of MCP interactions at the trimer-of-dimers site *in vivo* was obtained in a site-directed mutagenesis study that targeted the trimer-of-dimers interaction site, where a strong correlation between chemotactic ability and polar complex formation was noted [5]. Moreover, MCP trimers could be efficiently trapped in the cell when a trivalent sulfhyrdryl-specific cross-linking agent (TMEA) was used in combination with engineered MCPs that contained a single cysteine residue located near the trimer-of-dimers interaction site [197].

Remarkably, when two different MCPs (e.g., Tsr and Tar) with the cysteine mutation were expressed together mixed trimers were trapped in proportions that approximately reflected statistical mixing (i.e. $Tsr_3:2Tsr_2Tar:2TsrTar_2:Tar_3$, for a 1:1 mixture of Tsr and Tar). This observation provided supported for the formation of heterogeneous MCP dimer arrays making contacts through the trimer-of-dimers interaction site. As might be expected from the high degree of MA domain sequence conservation, polar cluster formation is conserved in *Archaea* and *Bacteria* [75]. Thus, considering the results of CheA regulation and these cellular localization studies together, the view has developed that MCPs function in cooperative manner to regulate kinase activity in a signaling array that forms through MCP-MCP interactions and interactions between MCPs, CheW, and CheA.

In an effort to obtain high-resolution structure information of a receptor-CheW-CheA complex, single-particle electron microscopic analysis was used to generate a 27-Å resolution reconstruction of a ~1.3 MDa kinase-active assembly, which was comprised of 24 Tar cytoplasmic fragment monomers, 6 CheW monomers, and 4 CheA monomers [67, 68]. The arrangement of subunits was compatible with the expected location of the MCP-CheW-CheA interaction near the hairpin of the MA domain and a dimeric organization for the MA domain, but interestingly the trimer-of-dimers organization observed in the Tsr MA domain crystal structure was not compatible with the electron density envelope [68]. Completely removing the organizing influence of the membrane (through the use of a soluble receptor fragment) can be expected to have a significant and unpredictable impact on the assembly, and thus the full significance of this arrangement remains to be established. Nonetheless, these data represent the first step in the quest to decipher the structural basis of kinase regulation and the role played by methylation. The single particle approach might prove to have greater relevance for investigations of receptor-CheW-CheA structures formed with the soluble MCPs that naturally lack membrane anchors (see Section IV.B.3).

Polymorphic assemblies of intact Tsr have been observed by electron microscopy, in inner membrane preparations and in the cell [113, 212, 227]. The assemblies were observed within cells in which Tsr was overexpressed in the absence of CheA and CheW, and were manifested as pronounced invaginations of the inner membrane [113]. In isolated inner membrane preparations, three types of assemblies were discerned: 2D crystalline arrays of limited extent and "rounded" assemblies and "zippers" [212]. Projection images of the rounded and zipper assemblies are shown in Figure 12.8 (*top* and *bottom*, respectively).

An inspection of these assemblies enabled the identification of the receptor dimer long axis, depicted with schematic representations of the Tsr dimer at the right in Figure 12.8. According to these interpretations, the interactions among MCP MA domains facilitate assembly formation, which can occur either between receptors that are nearby in the membrane or between clusters of MCPs in separate patches of membrane brought together by the interaction. Contacts among MA domains of nearby MCPs, consistent with the trimer-of-dimers interaction, are observed near the center rounded assemblies. The zipper assembly is stabilized by the interdigitation of MA domains made between apposing receptor-containing membrane patches.

The regularity in each type of assembly facilitated measurements of characteristic dimensions, radii in the rounded assemblies (r_p is an example in the center of Figure 12.8 *top*), and distances between parallels in the zipper assemblies. Estimates for the end-to-end length (~300 Å) and length of the cytoplasmic domain (~190 Å) proved to be ~20% shorter than the corresponding dimensions predicted from the atomic model of Figure 12.7. The discrepancy has been attributed to a more compact fold of the HAMP domain. However, these observations confirmed the general topology of the MCP dimer and provide additional evidence for interactions among receptors in the region of the MA domain.

Membrane samples containing receptor at elevated levels, such as those shown in Figure 12.8, find common use in experiments addressing questions of receptor function (e.g., methylation, demethylation, and CheA regulation).

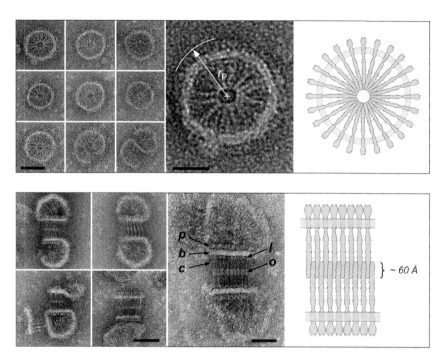

FIG. 12.8. Electron microscopic images of Tsr assemblies. Negatively stained detergent-treated inner membrane preparations isolated from *E. coli* cells overexpressing the serine receptor Tsr display characteristic "zipper" (*top*) and "wheel" assemblies (*bottom*). Scale bars for the images of the assemblies in the galleries are 50 nm and 25 nm for the enlarged images [212]. Cartoons depicting the packing arrangements of receptor dimers are shown at *right*. The enlarged image of the zipper assembly (*upper middle*) is labeled to indicate the location of protein domains and transmembrane segments: *p*, periplasmic domain; *b*, membrane bilayer; *c*, cytoplasmic domain; *l*, HAMP (linker) domain; and *o*, region of cytoplasmic domain overlap. The figure is based on Figures 4 and 5 of Weis et al. [212].

These receptor-containing membranes are complex and exhibit a considerable degree of polymorphism, which may influence quantitative measurements significantly. Therefore, the conclusions, which are based on experiments that use such samples, and require an exacting quantitative analysis to reach, should be interpreted with caution. Nevertheless, the qualitative trends displayed in such samples often compare favorably to those prepared at reduced (cellular) expression levels and/or show a substantial degree of agreement with the analogous properties measured in cells.

The extensive receptor interactions evident in Figure 12.8 are consistent with cooperative regulation of CheA observed both *in vitro* (as in Figure 12.4) and in living cells [186, 187]. In addition, CheR-mediated receptor transmethylation is required both for effective methylation in the test tube [118] and effective

CheR-mediated adaptation in the cell [9]. Thus, qualitative or semi-quantitative conclusions based on data obtained in such experiments are more likely to be valid indicators of the behavior in the cell, but must have support from such comparisons.

3. Unorthodox MCPs: "There Is An Exception to Every Generalization"

There are more than 2,000 open reading frames identified in the genome sequencing projects (completed and in progress) of *Bacteria* and *Archaea* that code for proteins containing an MA domain. The number is growing without an end in sight. Nonetheless, some general features are emerging.

  (i) MCPs have at least one N-terminal sensing domain and one C-terminal MA domain.
 (ii) The N-terminal sensing domain is membrane associated.
(iii) MA domain-containing proteins are reversibly methylated.

These three points are valid in most cases, but exceptions (and they do occur) represent interesting variations that can prove informative.

MCPs are not always composed of an N-terminal sensing domain and one C-terminal MA domain. In some cases, the predicted protein does not have an N-terminal sensing domain. These proteins may (1) interact with receptors that contain sensing domains but lack the MA-domain (e.g., the Htr-sensory rhodopsin interaction [86]), (2) play a structural (scaffolding) role in a receptor array (no known examples), or (3) have a sensing function contained within the MA-domain. There is some support for case 3. The MA domain of *B. subtilis* McpC is reported to sense carbohydrates, which postulated to occur through a direct interaction between the MA domain and a PTS protein [104]. If this postulate holds up to further experimentation, it will represent a new mode by which MA domains receive signal inputs.

In a small number of instances, MA-domain-containing proteins have sensor-like domains located after the MA domain. For example, DcrH is one of two MCPs in *Desulfovibrio vulgaris*, which is predicted to have a membrane organization similar to that of "orthodox" MCPs [84] and contains two hemerythrin domains in tandem after the MA domain [222]. Finally, and infrequently, more than one MA domain is predicted to be present within a single polypeptide. For example, the *Vibrio cholerae* genome is predicted to code for a protein 641 amino acids in length (AAF94560) [82], in which two MA-domains (~275 residues apiece) are joined in succession. The protein lacks a recognizable sensor domain, and its role in signaling is not known. What function is served by linking two MA domains in a single protein? The answer is not known, but such a protein may facilitate the assembly of signaling complexes, as do MA domains deliberately engineered to dimerize through the introduction of leucine zippers [50, 125].

Most MCPs are membrane associated, but some important exceptions have been identified. Car, the *Halobacterium salinarium* cytoplasmic arginine receptor, is a transducer without transmembrane segments that fractionates with the cell cytoplasm [195]. HemAT, first identified in *H. salinarium* and *B. subtilis*, is a soluble oxygen sensor consisting of a heme-containing globin domain and the MA domain [89]. It has been suggested that the HemAT dimer may incorporate into the membrane array of chemotaxis receptors via interactions between MA domains, which according to this scenario would mean that the globin domain is positioned near the inner membrane surface to detect oxygen [228]. The taxis-like proteins (Tlps) of *Rhodobacter sphaeroides* are a third example of MA-containing proteins that are not membrane associated. TlpC localizes to a discrete electron dense region within the cytoplasm of *R. sphaeroides* [205].

MA domains are not always involved in reversible methylation. The *E. coli* receptors were anointed MCPs after the methyl-accepting property was identified. It is now apparent that not all proteins with MA-like domains are substrates of CheR. As pointed out previously, whether or not any particular MCP is truly a substrate for CheR can be determined with sequence analysis combined with appropriate tests of methyl-accepting function. The *E. coli* Aer protein is a case in point, and is an example of an MA-domain-containing protein that is not a substrate of the transferase [24]. It is interesting to note that an engineered *E. coli* Aer-Tar chimera protein is an efficient methyl-accepting aerotaxis receptor [24], which is a strategy that seems to have been arrived at naturally. Both a putative Aer protein from *Burkholderia pseudomallei* (CAH34699) and a putative Aer-like protein in *Vibrio cholerae* (NP_229757) have CheR docking sites, and identifiable sites of methylation. Unlike *E. coli* Aer, these Aer orthologs are expected to be CheR substrates.

C.   Modes of Interactions and Factors Influencing Receptor Methylation

Much of what is known of CheR-mediated receptor methylation is based on studies of the *E. coli* (and *Salmonella*) MCPs. Recent attention has focused on determining the molecular details of enzyme substrate interactions: the structure of the active site, how the substrates are arranged in the active site, and the factors that influence the reaction rate (i.e., the effects of ligands).

1.   CheR Localization: The Ball and Chain

a.   **Intersubunit Methylation**

A distinguishing feature of CheR-mediated methylation is the multiple sites of modification, which are spatially colocalized as consequence of the MA domain hairpin structure. For this reason, and because receptors are clustered in the membrane (as Figure 12.3 depicts), the local concentration of methylation sites is large. It is now recognized that the *E. coli* and *Salmonella* enzymes are localized near the sites of methylation through a specific interaction between the enzyme and the C-terminus of some MCPs, an interaction distinct from and in addition to the active-site/methylation-site interaction. An early observation in support of a requirement for the Tar C-terminus in the process of adaptation (methylation), but not in generating responses to ligand (kinase regulation), was made with a form of Tar truncated by 35 amino acids [164].

A significant advance in understanding the molecular basis for this requirement was made when it was determined that CheR bound with micromolar affinity to the last five amino acids (NWETF) in the Tar sequence (and Tsr, and NWESF in Tcp) [219, 226], and that this binding interaction was required for efficient receptor methylation [219]. Soon after, the co-crystal structure of *Salmonella* CheR and the N-acetylated NWETF revealed the specific nature of the interaction [56], in which the pentapeptide was docked onto the β subdomain of CheR in β-strand conformation (Figure 12.5). From this structure, it was possible to identify specific interactions between the side chains of the peptide and CheR. The interactions with tryptophan (W) and phenylalanine (F) were found to be especially important [56, 177].

As a result of finding a CheR binding site at the C-terminus of Tar, the process of intersubunit methylation was postulated (Figure 12.9). According to this mechanism, CheR binds to the receptor primarily through the NWETF/β-subdomain interaction, and methyl group transfer takes place through comparatively low affinity active-site/methylation-site interactions. The model also suggested that CheR bound to one MCP dimer via the tethering interaction could catalyze methyl group transfer on an adjacent dimer (i.e., in trans). This feature was based on previous observations, including (1) MCPs formed only homodimers [139]; (2) *E. coli* Tar and Tsr (and *Salmonella* Tcp) possessed the CheR docking site, but Trg and Tap did not; and (3) *E. coli* cells lacking MCPs with the CheR docking site (Tar and Tsr) were able to generate responses to Trg-mediated attractants (were able to regulate CheA activity) but were deficient in adaptation toward these same attractants (were unable to catalyze effective receptor methylation) [81]. (Additional lines of evidence are cited in [219].) Thus, the receptor dimers that on their own were unable to bind to CheR at the docking site (the C-terminal peptide) required assistance from receptor dimers with docking sites.

Subsequent investigations have generally supported and extended the intersubunit methylation model. First, defects in adaptation, chemotaxis, and methylation due to C-terminal truncations of MCPs could be corrected by the presence of an MCP containing NWETF at the C-terminus [9, 110, 118]. At least one MCP with the docking motif at the C-terminus was sufficient for efficient receptor methylation irrespective of the ligand binding specificities. Second, the methylation rates and adaptation efficiencies of Trg-containing membranes and Trg-expressing cells, respectively, could be increased significantly by fusing the C-terminus of Tsr to Trg [14, 66]. A similar, but somewhat less successful, resurrection of Tap via Tap-Tar fusion proteins has also been observed [209]. Third, evidence for the specificity of the CheR/docking-site interaction has been collected through the introduction of point mutations and deletions in the pentapeptide, as well as the addition of amino acid residues to the C-terminus [106, 176, 177].

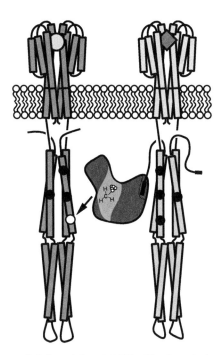

FIG. 12.9. Interdimer methylation of *E. coli* MCPs. The C-terminal pentapeptide of the high-abundance MCPs Tar and Tsr (NWETF) indicated by the thickened line segment binds to the β subdomain of CheR. A polypeptide tether localizes CheR in the vicinity to promote more frequent interaction between the CheR active site and the methylation sites on the receptor. The tethering interaction facilitates methylation of adjacent MCPs dimers. This is Figure 1 of Antommattei et al. [9].

Does CheR binding to an MCP at this site simply facilitate localization or is CheR regulated by NWETF binding? A binding study suggests that the latter is not the case. The affinity of CheR for AdoHcy was unaffected by pentapeptide, present in sufficient amounts to saturate NWETF/β-subdomain interaction [226]. Because the binding reactions were independent, it was concluded that CheR activity was probably not regulated allosterically through the docking interaction. Thus, the CheR-NWETF interaction appears to be simply one of a tethering interaction (a ball on a chain), which localizes CheR in the vicinity of the methylation sites.

The activity of the methylesterase CheB is, in contrast, regulated by phosphorylation. Activity increases substantially (~100-fold) as a result of phosphorylation on the N-terminal receiver domain [7]. Like CheR, efficient demethylation depends at least in part[8] on an interaction between CheB and the C-terminus of

---

8. CheB binds to the P2 domain of CheA with micromolar affinity [119], which contributes to the colocalization of CheB in MCP/CheW/CheA complexes [13].

MCPs such as Tar localizing CheB to receptor complexes [13, 15]. The CheB-pentapeptide interaction is significantly weaker. The K_D is about 150 μM [17], versus 5 μM for the corresponding interaction with CheR. Although the key residues on CheB implicated in the interaction have been identified [16], the structural basis is yet to be determined. The recent data of Li and Hazelbauer provide support for an intersubunit mechanism of receptor demethylation [122]. It thus seems that CheR and CheB have significant similarities in the mechanism of localization and interaction with the sites of modification.

b. Consequences of Bivalent CheR-MCP Interactions

The effect of bivalency on the dynamics and affinity of the CheR/MCP interaction has also been explored through simulation. Levin, Shimizu, and Bray developed a general theoretical framework to simulate the binding of a bivalent ligand to, and diffusion on, a 2D lattice of sites [115], and applied it to the CheR/MCP system. As might be anticipated from the general properties of multivalent interactions [128, 158], bivalency increased the affinity for (and residence time on) the lattice and led to a surface diffusion process that resulted from the stochastic nature of binding and rebinding at the two sites.

Levin et al. explored the situation in which the two sites are equal in affinity ($K_D \sim 2$ μM), as well as the consequences of a 10^2- to 10^3-fold weaker interaction at one site, which is more likely to correspond to the situation with CheR. Weaker binding at one site increased the number of lattice sites visited by CheR. This property favors efficient methylation. The quantitative immunoblotting study of Li and Hazelbauer provides the most recent evidence that CheR and CheB are far less abundant in the cell than Tar and Tsr (by $\sim$100-fold and $\sim$75-fold, respectively) [121], which suggests that CheR (relatively few in number) must effectively explore the entire receptor array.

The simulations suggest that instead of remaining anchored in one place, CheR, via diffusion, can explore the receptor array by a brachiating mechanism [115], but also (and perhaps mainly) by unbinding and rebinding through the pentapeptide/β-subdomain interaction. As Tar and Tsr are expressed most abundantly in the cell, accounting for $\sim$90% of all MCPs present [121], the CheR rebinding probability is significant. Meanwhile, the low affinity active-site/methylation-site interaction allows tethered CheR to efficiently explore all sites within reach.[9]

---

9. A study of receptor methylation (and demethylation) by Li and Hazelbauer estimated a neighborhood of seven MCP dimers around a (within reach of) tethered CheR (and for CheB, five MCP dimers) [122]. To generate this estimate, Trg methylation rates were measured as a function of the amount of Tar also present in the membrane, which was included to provide CheR tethering sites. The rate of Trg methylation increased to a maximum level as the Tar content increased from a Tar/Trg molar ratio of 0 to 2, and neighborhood estimates were generated by modeling. The reason for the increase in rate could, as the model assumed, result solely from an increasing proportion of Trg dimers in the vicinity of Tar but could equally well depend on an increase in the fraction of CheR bound to Tar, which would also increase with the Tar content. The model did not account for this, nor was the fraction of CheR bound to Tar measured. Thus, the meaning of the neighborhood estimates is unclear.

c. Properties and Distribution of the Tethering Moiety

With all available evidence suggesting that NWETF-mediated tethering inter-action is a requirement for efficient methylation (and demethylation) in *E. coli* chemotaxis, it might be asked how widespread is tethering as a mechanism for protein localization in bacteria chemotaxis systems? What other protein pairs engage in tethering interactions, and how prevalent is the CheR-MCP tethering interaction in other bacterial species?

Tethering interactions involving other chemotaxis proteins seem to be utilized in at least two other situations in *E. coli*: (1) between the phosphatase CheZ and the short form of CheA [102,146], and (2) between CheY (the response regulator) and CheZ [181]. An assessment of the CheR-MCP tethering interaction in different bacterial species was generated through an analysis of the MA-containing proteins in completed prokaryotic genome sequences. To determine MCPs with tethering segments, an operational definition for the CheR tether was adopted based on two main criteria. First, the docking site is a XHXXH pentapeptide located at or near the C-terminus, where X residues typically have either polar or charged side chains and H residues have hydrophobic side chains. The arrangement found most often was XWXXF, but other motifs were included (XLXXF, XFXXF, XFXXY).

The second criterion identifies the putative tether sequence as a polypeptide segment that (1) immediately follows the MA domain in the primary sequence, (2) is without significant homology to a known domain fold, and (3) is often identified as a region of intrinsic disorder [124]. These two criteria were used to accommodate reasonable variations on the *E. coli* theme (the NWETF pentapep-tide and a tether length of ~30 residues). The dimer model (Figure 12.7) served as a point reference in developing the criteria. Understandably, the criteria may not register tethers that depart significantly from this paradigm.[10]

From this analysis, it is clear that the tethering moiety is not common to all prokaryotes with MCPs or CheR. Examples of the *E. coli*-like interaction have been found mainly in the *Proteobacteria* (all subdivisions), to a lesser degree in the *Spirochaetes*, and sporadically among the *Clostridia* (phylum *Firmicutes*). The moiety seems to be absent from *Archaea* and many phyla in *Bacteria*. Table 12.1 summarizes the analysis of species with completed genome sequences that have at least one MCP with a tethering moiety at the C-terminus. Table 12.1 also lists the total number of predicted proteins containing MA domains identified in the genome, the number of MCPs with a CheR docking segment (or a putative docking segment), and the numbers of predicted proteins with CheR and CheB domains. From an inspection of Table 12.1, it is clear (and in hindsight fortuitous) that *E. coli* represents one of the simplest chemotaxis systems to decipher, with just five MCPs, one CheR, and one CheB.

---

10. Generally excluded from consideration are cases that appear to be protein fragments (e.g., with-out sensing domains and proteins without well-defined MA domains).

TABLE 12.1

DISTRIBUTION OF SIGNALING PROTEINS CONTAINING THE MCP-SIGNAL (MA) DOMAINS, CHER DOMAINS, CHEB CATALYTIC DOMAINS, AND MCPS WITH CHER DOCKING SITES.[a]

Species	Proteins with MA domains	NWETF-like MCPs	CheR	CheB	Ref.
α-Proteobacteria					
Agrobacterium tumefaciens C58	20	9	1	1	[217]
Caulobacter crescentus CB15	18	5	3	2	[144]
Gluconobacter oxydans 621H	3	1	1	1	[155]
Sinorhizobium meliloti 1021	9	4	3	4	[70]
Zymomonas mobilis subsp. *Mobilis* ZM4	3	1	1	2	[171]
β-Proteobacteria					
Bordetella bronchiseptica RB50	8	1	1	1	[148]
Burkholderia mallei ATCC 23344	17	3	2	1	[143]
Burkholderia pseudomallei K96243	21	5	2	2	[87]
Chromobacterium violaceum ATCC 12472	42	4	4	5	[36]
Dechloromonas aromatica RCB	25	3	3	2	b
Nitrosomonas europaea ATCC 19718	3	2	1	1	[42]
Ralstonia solanacearum GMI1000	21	4	2	1	[165]
Ralstonia eutrophia JMP134	21	5	4	3	b
δ-Proteobacteria					
Desulfovibrio vulgaris subsp. *vulgaris* Hildenborough	28	2	3	3	[84]
Geobacter metallireducens GS-15	17	4	7	8	c
Geobacter sulfurreducens PCA	32	6	5	3	[135]
Bdellovibrio bacteriovorus HD100	20	5	2	2	[162]
ε-Proteobacteria					
Wolinella succinogenes DSM 1740	32	1	1	1	[12]
γ-Proteobacteria					
Shewanella oneidensis MR-1	26	2	3	3	[83]
Vibrio cholerae O1 bv. *eltor* N16961	45	2	3	2	[82]
Vibrio vulnificus YJ016	51	1	3	3	[46]
Vibrio vulnificus CMCP6	53	1	3	3	[99]
Erwinia carotovora subsp. Atroseptica SCRI1043	36	19	2	1	[19]
Escherichia coli K12	5	2	1	1	[27]
Escherichia coli CFT073	4	2	1	1	[214]
Escherichia coli O157:H7	5	2	1	1	[80]
Escherichia coli O157:H7 EDL933	5	2	1	1	[152]
Salmonella typhimurium LT2	9	4	1	1	[132]
Salmonella typhi CT18	6	2	1	1	[147]
Salmonella enterica sv. Choleraesuis SC-B67	9	4	1	1	[48]
Salmonella enterica sv. Paratyphi A str. ATCC 9150	6	3	1	1	[131]
Salmonella enterica sv. Typhi Ty2	6	2	1	1	[54]
Shigella flexneri 2a 2457T	4	2	1	1	[210]
Shigella flexneri 2a 301	4	2	1	1	[90]
Photorhabdus luminescens subsp. *laumondii* TTO1	2	1	1	1	[59]
Yersinia pestis KIM	7	2	1	1	[53]

Continued

TABLE 12.1

DISTRIBUTION OF SIGNALING PROTEINS CONTAINING THE MCP-SIGNAL (MA) DOMAINS, CHER DOMAINS, CHEB CATALYTIC DOMAINS, AND MCPS WITH CHER DOCKING SITES.[a]—cont'd

Species	Proteins with MA domains	NWETF-like MCPs	CheR	CheB	Ref.
Yersinia pestis CO92	6	2	1	1	[149]
Yersinia pestis bv. Medievalis 91001	7	2	1	1	[183]
Yersinia pseudotuberculosis IP 32953	8	3	1	1	[43]
Pseudomonas aeruginosa PA01	26	1	4	4	[196]
Pseudomonas syringae pv. *phaseolicola* 1448A	49	1	4	4	[91]
Pseudomonas syringae pv. *tomato* DC3000	49	1	4	4	[39]
Pseudomonas syringae pv. *syringae* B728a	51	1	4	4	[65]
Xanthomonas campestris pv. *campestris* 8004	20	9	4	6	[156]
Xanthomonas campestris pv. *campestris* ATCC 33913	20	9	4	6	[52]
Xanthomonas axonopodis pv. *citri* 306	22	11	4	3	[52]
Xanthomonas oryzae pv. *oryzae* KACC10331	12	8	3	3	[112]
Spirochetes					
Borrelia burgdorferi B31	4	1	3	2	[69]
Borrelia garinii PBi	4	1	3	2	[76]
Leptospira interrogans sv. *Copenhageni*	12	1	2	3	[142]
Leptospira interrogans sv. *Lai*	13	1	2	3	[161]
Firmicutes					
Clostridium acetobutylicum	38	1	3	1	[145]

[a]Domain prevalence was determined using blastp [4] in completed genomes posted at the NCBI *Entrez* Genome Project web site [208]. The cytoplasmic signaling (MA) domain of *E. coli* K12 Tsr (amino acid residues 269 through 516), CheR (1 through 286), and CheB (159 through 349) were used to search for MA domains, CheR domains, and CheB domains, respectively.
[b]Genome completed at the DOE Joint Genome Institute.
[c]Draft Assembly completed at the DOE Joint Genome Institute.

One of the major challenges in deciphering signal transduction pathways in bacteria with multiple pathways will be to determine when and how the components of the different pathways remain segregated. Some entries in Table 12.1 are more "*E. coli*-like" than others. The fraction of MCPs in *E. coli* with a CheR tether is ~40%. A few entries (e.g., *Agrobacterium tumefactions*, *Erwinia carotovora*, and the *Xanthomonas* sp.) in Table 12.1 resemble *E. coli* in this respect, which suggests that CheR localization by the recognized tethering interaction may be the predominant mechanism, especially in those species that have a single *cheR* locus. For others (e.g., *Pseudomonas* and *Vibrio* sp.), the fraction of MCPs predicted to have tethers is small (<5%), and there is more than one protein predicted to contain the CheR domain. In these species, it seems that other mechanisms for CheR localization will also be used.

Table 12.2 illustrates the length distribution of selected putative tethering moieties. The MA domains of MCPs and putative MCPs were aligned with

TABLE 12.2

LENGTH DISTRIBUTION OF SELECTED C-TERMINAL TETHERING SEGMENTS.[a]

Gsp	Accession No	MA	Tether (length)	MCP
Rso	CAD14857	248	FRLDPTMSAVRSVVAKAEPTIAPPVPKAAEAGPATPVPVAAPVAAPAQAPRRE PVKPSPEAARPKLQPRKAPVLAKRAARPDTAPATVPAASPKPLHQPLVAAGDDADWETF (110)	629
Bbu	NP_212814	248	FKIKDSKIENPENDDYDFRLIDCPENSFKDENQNLKSNGISTNASGHNNYSLD IESESSVRTINKRVDPKKAIDIADKDLNFDDDFSEF (88)	753
Atu	AAL41754	248	FRMGGTRAAVAGAGGYSAPRQSSAKPAFQPAPAAPVRKAPVKSASAAARPVA SPARALGQSLARAFGGAAEAPAAKDQDWTEF (81)	673
Gme	EAM80584	248	FKLADGGRADTGAAPKTERKPAARVAPSARNTPAAKAAPKAKAANGYHKPA APEHHEAELPKAVGYDDDWKEF (71)	587
Bba	NP_968363	248	LNTIILGGGSVMVDAPAASPAKTSKKVPTFAKPTSKKSNVIAMKATPARKESQD MIPFDEDDRAKVGTTDGF (70)	532
Wsu	NP_906362	248	IAKIVGLAMDNLFTTHTKAAPTKPSSSISPKLKQIPHAKTELKRSVPSFPKKERRS KEDEIFPLDADDLKEF (70)	664
Bps	CAH34699	248	FRVNARVPARHDDARAPRAGAAAAMRERASVARPRPAMRAAPAPSLALAN ASTAQAPARATANADADWQTF (69)	583
Gsu	AAR34013	234	FRTDDRGASSRSAARRPVAKKKAAISHLGHGMSNGYHTEPATSRKVAVGGGVD LNLDTDHLDDQFEKF (66)	568
Xor	YP_200108	234	FRLGNGGRGAASGSSRPGPRRIAAHAPAAVSSAPPRRTNVRQISAVAATATATADAIDESQFASF (63)	722
Vch	NP_229757	248	FRLPDQDTSAPSLLKAVNKRPQSAPVTRHPASHIAKTPAKITAKASSRAQPVMQVAHDEEWESF (62)	659
Gme	EAM78143	234	FRLKNEGSRNPASKLAKTRNRVQVGHIAADNGHSGKLAPGTAKGYAFEMENSDAMDAEFEKF (60)	1046
Neu	CAD85774	248	FKLEHGMAAAHTDNPPSVERRSPNRATNVERLPQTRNRKKAATRSSVASVSAGTEGGWEEF (59)	776
Xor	YP_201486	248	FKVAAHAPVARRLSAIASAPSNARFAPQPAARVTPAAAAPRRSVGSSAASHADSGNWQDF (58)	753
Zmo	AAV88826	248	FKLGNGNTRQFASTPAVTSAPKKPAARPASRSTAKSAPAKAAAKPNPAPKPMADDDWSEF (58)	630
Dar	ZP_00149818	248	FAISKNQPTARRTASTDRAPQSSARMALPHREPVRLPEPSAKSSAKLVATKGDDEWEEF (57)	778
Xca	NP_637240	248	FKLAAQAPVARQLSAIASAPGNARSAARPAARTVAAAAPRPVRTAPATTADQSDWQEF (57)	699
Vvu	NP_763073	248	FSTDSSITTLETSPSAAPSAAKVTAMPTRGNISEMKKPKVANSGFRTEESDEWEEF (54)	807
Dvu	AAS96438	234	FRIPGGGTRPRSAPFIAGQEKPGQTRVEVDVPKAPATPRLRLDLSDDEDGDFERF (54)	720
Vch	NP_233472	248	FVTENNVTTFERRPKGPSTPPKTKPVVSMHKAPINQAVHKMPARAAEEGDEWEEF (52)	678
Xax	NP_642763	234	FRLGGQGAPTAPPKRGKQVAARPARAAGTRGWRSNAAPAYAMSEGPDEGQFARF (52)	552

Gsp	Accession No.	MA	Tether (length)	MCP
Psy	NP_790751	234	FTLDSSPKPTTSGSSIDHRSSPSRQPPRTVQPPARKAFAHSMASAPDESEFTRF (52)	551
Xax	NP_642219	248	FKTDSNDASRAASRRPAATPALAAKAVAAGRAPAPRLRAVVAAPGNDSSWQEF (51)	687
Lin	YP_001482	234	FKISDGDKRLTNGGGLLHRDFSGTGIIKATPTTRKDLPKGISPITEKFERY (49)	976
Son	AAN55170	248	FKVDEETRSARHTTELKKIPQKIPTLARVTPKPKAMTPKLNKADQDEWEEF (49)	654
Dar	ZP_00149817	248	FKVANAGGMPRLEAPRTSQRASVPQAPRGERIGARKVQALPSSLDDEWEEF (49)	945
Pae	NP_248866	248	FRLDTPPSVVQLASARPSAPRPSAPAPLARSGMARASKARKEDGWEEF (46)	679
Cac	NP_346765	234	FKLKGKGGFNYGENYGMEHAATKGGSKTYEMSNMDMYGSDFGKY (42)	555
Gox	YP_191946	248	FRLKSLHDAEKPLFESEMPASDLPSSQWGNESFQTSDEGWEEFR (41)	491
Eca	CAG74191	248	FKINQAVAQEHRAASASSLAALPKSLLPKPTSAGSSNANWETF (41)	556
Eco	MCP2 (Tar)	248	FRLAASPLTNKPQTPSRPASEQPPAQPRLRIAEQDPNWETF (39)	553
Son	AAN55357	234	FKLAISSERPHQGKSTHMTASVRKTAKAKQHVERADFERF (38)	646
Eco	MCP1 (Tsr)	248	FRIQQQRETSAVVKTVTPAAPRKMAVADSEENWETF (35)	551
Cvi	AAQ58687	234	FHLGSLHAASPAPAAVHRPQAALPAAADHHGFMQF (33)	536
Sty	MCPC	248	FRIQKQPRREASPTTLSKGLTPQPAAEQANWESF (32)	547
Eco	MCP3 (Trg)	248	FRLHKHSVSAEPRGAGEPVSFATV	546
Eco	Aer	248	LH	506

[a]**Gsp** (genus-species): Atu, *Agrobacterium tumefaciens*; Bba, *Bdellovibrio bacteriovorus*; Bbu, *Borrelia burgdorferi*; Bps, *Burkholderia pseudomallei*; Cac, *Clostridium acetobutylicum*; Cvi, *Chromobacterium violaceum*; Dar, *Dechloromonas aromatica*; Dvu, *Desulfovibrio vulgaris*; Eca, *Erwinia carotovora*; Eco, *Escherichia coli* K12; Gme, *Geobacter metallireducens*; Gsu, *G. sulfurreducens*; Gox, *Gluconobacter oxydans*; Lin, *Leptospira interrogans* sv. Lai; Neu, *Nitrosomonas europaea*; Pae, *Pseudomonas aeruginosa*; Psy, *P. syringae* pv. tomato; Vch, *Vibrio cholerae*; Vvu, *V. vulnificus* CMPC6; Son, *Shewanella oneidensis*; Sty, *Salmonella typhimurium* LT2; Wsu, *Wolinella succinogenes*; Xax, *Xanthomonas axonopodis*; Xor, *X. oryzae*; Zmo, *Zymomonas mobilis*. **Accession No.**: Genbank #; **MA**: MA domain size as defined in the text; **Tether (length)**: tether (and length) as defined in the text; **MCP**: The length of the MCP in amino acid residues.

ClustalW [47]. The displayed sequences show the last two amino acids of the MA domain at the left.[11] Table 12.2 also lists (1) the estimated MA domain lengths, (2) the tether lengths, and (3) the number of amino acids in the MCPs. The tethering moieties in Table 12.2 exhibit a remarkable range in length, from ~30 to over 100 amino-acid residues. Except for the putative tether in *Clostridium acetylbutylicum* (*Cac* NP_346765), every tether has at least one proline, and often several.[12]

The MA domain rarely has a proline residue, and thus the difference is often striking. The incorporation frequency of proline varies significantly from one tether to the next [e.g., one in five residues for the tether in the *Ralstonia solanacearum* MCP listed in Table 12.2 (*Rso* CAD14857), to one in thirty for the *B. burgdorferi* MCP tether (*Bbu* NP_212814)]. The variation in proline (and glycine) content might reflect differing requirements for conformational dynamics in the tether. The pentapeptides (in boldface type) are invariably found at the very C-terminus of the MCP, although rare exceptions occur (*Gox* YP_191946, Table 12.2).[13] The pentapeptides frequently conform to the X(W/F)XXF motif used as a selection criterion, but variations do occur, and it is reasonable to assume that changes in the pentapeptide sequence will be mirrored by amino acid changes in the β subdomain [56] to generate binding specificity.

Figure 12.10 illustrates the domain arrangement of a few MCPs. The examples provide a (small) indication of the diversity in the domain organization of the N-terminal portion of the protein, and also depict the comparative regularity in the position of the MA domain (MA). These MCPs were chosen to illustrate tethering moieties at the C-terminus, except in two cases (*E. coli* Aer and *B. subtilis* McpB). Aer helps to define the end of MA domain in the more closely related MCPs, which was used in generating estimates of tether lengths. Two MCPs (one from *G. metallireducens* and *B. subtilis* McpB) have MA domains that differ in length from the others as a result of indels in the coiled coil segments [109]. The different domain lengths may determine the specificity of interactions among receptor dimers, as discussed in material following, but this feature is yet to be demonstrated experimentally.

The MA domain lengths reported in Table 12.2 either correspond to those identified in the SMART database [114] or were determined through multiple sequence alignment analysis with the *E. coli* and *Salmonella* MCPs. *E. coli* and

---

11. The end of the MA domain is operationally defined as the last residue in the *E. coli* Aer protein, at the bottom of Table 12.2.

12. The *Clostridium acetylbutylicum* (*Cac* NP_346765) tether appears to utilize a great number of glycine residues (approximately one in five) to increase tether flexibility.

13. The docking segment in the MCP from *Desulfovibrio gigas*, GFERFGADT$_{649}$, is a second example of an emedded pentapeptide [180].

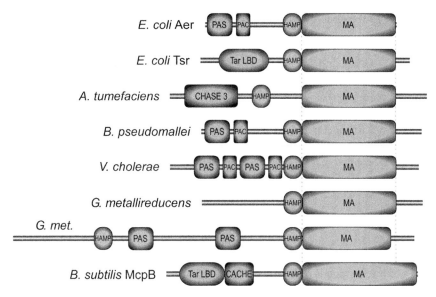

Fig. 12.10. Organization of MA domain-containing proteins. Identity (top to bottom) of proteins containing MA domains (with the accession number) retrieved from the PFAM [18] and SMART [114] databases: *E. coli* Aer (NP_417543) [23, 160], *E. coli* Tsr (NP_418775) [163], *Agrobacterium tumefaciens* Mcp (AAL41754) [217], *Burkholderia pseudomallei* putative Aer (CAH34699), *Vibrio cholerae* Mcp (NP_229757) [82], *Geobacter metallireducens* Mcp (EAM80584), *G. metallireducens* Mcp (EAM78143), and *B. subtilis* McpB (AAA20554) [79]. Domain abbreviations (not found in the legend to Figure 12.6): Tar LBD, aspartate receptor ligand binding domain [136, 225]; CHASE3 domain, cyclase/histidine kinase-associated sensing extracellular domain 3 [229]; and CACHE2, Ca^{2+} channel and chemotaxis receptor domain 2 [8].

Salmonella MCPs are represented by a single of MA domains (class I, according to Le Moual and Koshland [109]) that accounts for the majority of MCPs in Table 12.2, although several have smaller MA domains (less seven residues in each coiled-coil segment). In addition, it is frequently observed that the MCP genes in a genome are distributed among two or more different classes, which according to the MCP dimer model in Figure 12.7 should result in MA domains that differ in length.

As an example, four pairs of MCPs are listed in Table 12.2 from four different species (*Geobacter metallireducens*, *Gme*; *Shewanella oneidensis*, *Son*; *Xanthomonas oryzae*, *Xor*; *X. axonopodis*, *Xax*). One MCP in each pair contains a 248-amino-acid residue MA domain (class I). The other contains a 234-residue domain. In these examples, both MCPs are predicted to have CheR tethering moieties, although in general this is not the case. These examples were chosen to illustrate the insights that might be gained about the mechanisms for forming class-specific receptor complexes, and the role that specific recruitment of CheR may play.

Investigations of the *E. coli* system have provided the evidence that clusters of receptor dimers are generated by the trimer-of-dimers interaction, which can be formed between dimers of different ligand specificity [150]. One plausible requirement for trimerization and signal complex formation is membership in the same MA domain class (same length). The distance from the membrane surface to the trimer-of-dimers interaction site (and to the CheW and CheA binding sites) will vary between the Ma domain classes because the lengths of the coiled-coil segments are different. These differences could prevent effective interactions among MA domains in different classes, and as a result promote the formation of class-specific receptor signaling complexes.

Does the formation of class-specific signal complexes also lead to class-specific CheR-receptor interactions, and if so why? The "why" is a consequence of a need for specificity in signal regulation. Class-specific signaling complexes are likely to respond to different inputs (stimuli) for different purposes, which may have different timing requirements for response and adaptation. Compatible with this postulate is the observation that the pentapeptide amino acid compositions appear to be class specific. This property is evident in the four pairs of MCPs cited previously. Tethers attached to MA domains 234 residues in length have XxxPheXxxPosPhe pentapeptides, and those attached to MA domains 248 residues in length have XxxTrpXxxNegPhe pentapeptides (where Xxx are residues with polar side chains, Pos is either Lys or Arg, and Neg is either Glu or Asp). The differences may dictate binding specificity to the different predicted CheR homologs for these bacteria (Table 12.1). The two factors (MA domain length and CheR binding specificity) will work together to generate class-specific receptor attributes for the process of methylation (and demethylation).

Are there other mechanisms of CheR localization that are yet to be discovered? It appears so. The genomes in which tethering moieties are identified represent a fraction of all prokaryotic genomes, and the phylogenetic distribution as noted previously (and in Table 12.1), are sporadic. What are the alternatives? First, there could be other docking sequences that bear little similarity to the known and proposed examples. Second, different tethering mechanisms exist. The analysis of CheR domain organization provided hints that other mechanisms may exist, ones in which the tethering moiety is a part of the CheR protein rather than the MCP. The two CheR fusion proteins shown in Figure 12.6 possess domains that can localize CheR (i.e., the C-terminal TPRs in *M. xanthus* FrzF and the N-terminal CheW domain in the CheW-CheR fusion protein, *B. burgdorferi* CheW-3 [178]). Third, localization does not rely on tethering. *Bacillus subtilis* might be an organism that functions this way. None of the *B. subtilis* MCPs have a C-terminus that resembles a tether, and there is a single CheR in the genome [105].[14] In addition, possibly relevant is the fact that the modification sites in *B. subtilis* have specific and different roles in adaptation [157, 198], whereas in *E. coli* (and related systems) all sites appear to contribute to adaptation in the way. These other mechanisms of CheR localization require further investigation.

2. Methylation Site: Active Site Interactions

CheR facilitates AdoMet and MCPs direct methyl group transfer by nucleophilic attack of the γ-carboxylate oxygen in the methyl-accepting glutamate residue [218]. The interaction with the MA domain must be specific to direct the transfer to just one of two glutamate residues found within the consensus sequence, but the interaction should also be dynamic to facilitate methylation at the different sites.

These requirements are consistent with a specific molecular recognition event, but one of comparatively low affinity. Measurements of the methylation rates at different sites in the MCPs, made prior to the availability of the CheR and MA domain structures, resulted in two general conclusions: (1) that better agreement with the methylation site consensus sequence, EEXXA(T/S), led to more rapid methylation at that site [44, 163, 201, 202] and (2) that the modification state of the adjacent site had an effect on the rate [172, 173].

The atomic-level details of methylation-site/active-site interaction are coming to light through structure-function studies [151, 178], now that both the CheR and the MA domain structures are available. In broad strokes, the N-terminal domain is known to be important, the second α helix of the N-terminal domain has been identified as the most influential moiety (Figure 12.5), and the important residues within helix 2 are being pinpointed.

The first 19 residues at the N-terminus of CheR are relatively *un*important. These residues were not well resolved in the crystal structure [55, 56], and deleting them resulted in a form of CheR that retained 85% of the wild-type transferase activity [151], but deletion of 89 residues, which eliminated the entire N-terminal domain, resulting in a nonchemotactic phenotype in the swarm assay [178]. Several lysine and arginine residues are located in α-helix 2 (Figure 12.5), and the net positive charge of helix 2 is complementary to the net negative charge of the methylation helices [55, 151]. Consequently, electrostatic interactions between CheR and the methylation helices are thought to be important in promoting an effective arrangement for methyl group transfer. Site-directed disulfide

---

14. *B. subtilis* CheR has the β subdomain, as do probably all MCP methyltransferases. It has been suggested that the number of amino acids residues in the loops of the β subdomain may determine the ability to bind a pentapeptide. The implication is that *B. subtilis* CheR does not bind the C-terminus of an MCP [178].

bond formation, used to scan the reactivity of cysteine residues introduced into helix 2 of the CheR N-terminal domain with a cysteine substituted for glutamic acid 308 in Tar (adjacent to the third methyl-accepting glutamate residue, 309), revealed a pattern of reactivity consistent with close association between helix 2 and methylation helix 1 (MH1) of Tar [178].

In addition, mutations in helix 2 can have a strong effect in reducing the methylation rate of Tar-containing membranes [151]. One highly conserved residue, arginine 53 (*Salmonella* CheR numbering), was identified as especially important in promoting recognition. An R53A mutant displayed significantly reduced methyltransferase activity at all the major sites of Tar [151], yet a GFP-CheR protein bearing this mutation exhibited polar localization [178], which ruled out the possibility that the R53A mutation interfered with the tethering interaction. Based in part on these results, a structure of the CheR-MA domain complex was generated by computational modeling, with an approach that used electrostatic and distance constraints [151].

In this model (Figure 12.11a), the MA domain dimer (blue) is docked into a cleft formed by the CheR N and C domains (green) such that α-helix 2 in the N-domain (α2, in yellow) is arranged in an antiparallel manner to methylation helix 1 (MH1, red). The positively charged residues of α2 thus have roles in facilitating the antiparallel arrangement and registration along the axis to position AdoMet near a methyl-accepting glutamate. The enlarged view of α2 and the methylation region in Figure 12.11b depicts these interactions. α2 residues K46, R53, and R59 are seen to be in close proximity to E310, E303, and E296, respectively, which are adjacent to the methyl-accepting glutamates in sites 3, 2, and 1, respectively. The model predicts these interactions are distributed across more than one methylation site, which is consistent with the observation that the modification state of methylation adjacent sites can influence the rate [172, 173].

The model brings into sharper focus some aspects of the CheR-methylation helix interaction that were only suspected previously. The complex depicted in Figure 12.11 involves an MA-domain dimer. The requirement for dimerization, if true, may explain why soluble cytoplasmic receptor fragments, which often do not dimerize, display poor activity as CheR substrates. In addition, as a consequence of the antiparallel orientation between α2 and the methylation helix CheR is expected to dock onto the MA domain in two different orientations. One orientation is adopted when CheR methylates residues in MH1 (as Figure 12.11 depicts), and a second (in which CheR is rotated 180 degrees relative to the first) is adopted when MH2 is being methylated. Figure 12.12 depicts these two arrangements schematically. The CheR-MCP tethering interaction can plausibly allow for sufficient freedom of motion to allow CheR to accommodate these two orientations.

In the dual recognition process between MCPs and CheR, it is generally accepted that the active-site/methylation-site interaction is a substantially weaker interaction.

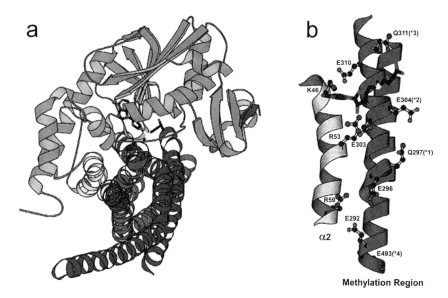

Methylation Region

FIG. 12.11. A transferase/receptor methylation site complex generated by computational modeling [151]. (a) A MCP cytoplasmic domain dimer, depicted in blue and viewed along the helical axes, is docked in a cleft formed by the N-terminal α-helical domain and the SAM binding domain of CheR (green). The methylation region of one cytoplasmic domain (depicted in red) is closely apposed to α-helix two (α2, yellow) of the CheR helical domain. (b) A close-up of the α2-methylation region interaction illustrating the charge complementary between α2 (lysine-46, arginine-53, arginine-59) and residues in the methylation helix (glutamate-296, glutamate-303, and glutatamate-310). The major methylation sites are noted by numbers in parentheses adjacent to the residue. In this model structure, SAM is positioned in proximity to methylation site 2 (glutamate-304). This is Figure 6 from Perez et al. [151].

The localization of GFP-CheR fusion proteins in the cell provides evidence that the tethering interaction is the primary determinant of colocalization with the MCPs. The polar localization observed with the GFP-wild-type-CheR fusion protein is abrogated by mutations that disrupt the tethering interaction, either (1) by deleting the docking segment from the MCP or (2) through a double mutation of key residues (H192A/R197A) in the β subdomain [178]. There are no direct measurements of the strength of the active-site/methylation-site interaction, possibly because of the weakness of the interaction. The original study that identified the pentapeptide-β-subdomain interaction [219] utilized soluble cytoplasmic fragments truncated at the C-terminus to provide evidence for the importance of the binding interaction, and thereby the relative weakness of the active-site/methylation-site interaction.

   The model of the CheR/MA-domain complex (Figure 12.11) suggests that the MA domain dimer is required for the interaction, and monomeric cytoplasmic

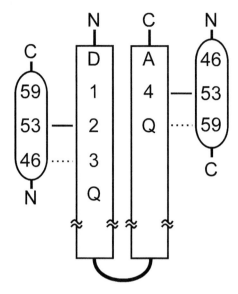

FIG. 12.12. A diagram of the directional interaction between α–helix 2 of CheR and the methylation helices. α2 (oval) is shown in duplicate to depict its orientation in two separate interactions, either with methylation site 2 (*left*) *or* methylation site 4 (*right*), which are antiparallel interactions with the methylation helices (squares). The positively charged amino acid residues on α2 are indicated by the numbers that correspond to their location in the primary sequence. The numbers on the MA domain represent the major methylation sites, which are spaced seven residues apart. Flanking residues seven residues away from the nearest site are indicated by one-letter amino acid abbreviations. Arg 53 on α2 interacts has an essential interaction close to the methylation site (—), whereas Arg59 and Lys46 are postulated to be involved in secondary interactions with adjacent sites (····). Note that α2 is rotated through 180 degrees between the site 2 and site 4 interaction. This is an adaptation of Figure 6 in Perez et al. [151].

fragments (CFs) would not satisfy this requirement. However, the CheR binding affinities measured in complexes involving the soluble CF [219] and Tsr-containing membranes [120] are equivalent and indicates either that dimerization of the MA domain is either not important or not the sole determinant in promoting the interaction. This other determinant may be prebinding of AdoMet because the CheR crystal structures show contributions from the N-domain, the linker, and the C-domain to the AdoHcy binding [55].

Accordingly, AdoMet binding may be needed to maintain a conformation of CheR that promotes the interaction with the methylation helices. Detectable influences of receptor modification and attractant binding, which were absent in the isothermal titration calorimetry experiments involving CheR and Tsr (conducted without a co-substrate analog present) [120], may become measurable when CheR is presaturated with an AdoMet analog (e.g., AdoHcy or singefungin). The issue

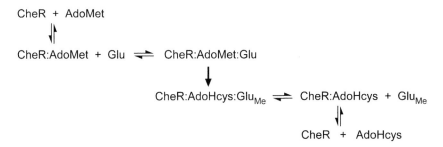

FIG. 12.13. A scheme of the elementary steps during the methyl group transfer process with receptor-tethered CheR. The steps are the binding of AdoMet and the methylation helix (Glu), methyl transfer, and the release of methylated products (Glu_{Me}) and AdoHcy.

merits further investigation, as it may serve to elucidate the mechanism by which the binding of ligand changes the rate of receptor methylation.

If AdoMet binding is required to stabilize a CheR conformation that can effectively interact with the methylation helices, or if the pathway for binding AdoMet and the release of AdoHcy is blocked by bound methylation helices, an ordered substrate binding would be necessary for methyl group transfer (Figure 12.13). A previous analysis of the enzymatic mechanism of CheR led to the conclusion that the order is random [182], but the parameters determined in that kinetic analysis reflected AdoMet binding and the tethering interaction, which have been proven to be independent [226]. An argument can be made that the tethering of CheR is more essential for facilitating multiple turnover events in an ordered AdoMet-methylation helix binding process than random binding, and therefore it may be of interest to examine this issue further.

Where binding experiments have yet to provide direct estimates for the strength (or rather weakness) of the active-site/methylation-site interaction, CheR cellular localization studies [178] and measurements of receptor methylation [9, 14] do give results consistent with a weak interaction under the conditions that AdoMet is bound to CheR. Defects in chemotaxis that result from compromised CheR localization *in vivo* can be overcome by the CheR overexpression to compensate for inefficient localization through mass action [177]. Figure 12.14 is a plot of initial rates of methylation measured as a function of the Tsr concentration [9]. Michaelis-Menten kinetic analysis generates estimates for K_M of ~5 µM when an MCP is present in the membrane with a CheR tethering site (closed symbols), in this case Tar. When no MCP with a tethering moiety is present (open symbols), the data are consistent with a much larger value for K_M (>100 µM).

3.   Effects of Ligands

The effects of ligand on the rates of receptor methylation and demethylation are qualitatively depicted in the two-state model for kinase regulation of Figure 12.3a.

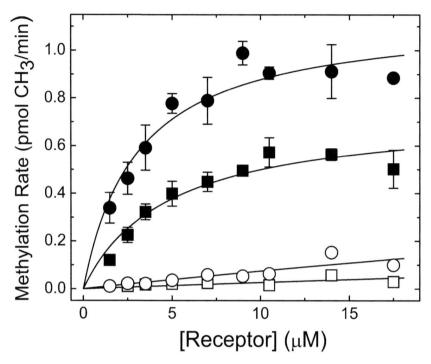

Fig. 12.14. Transmethylation in trTsr/Tar mixtures. Initial rates of methylation with membrane samples of coexpressed Tsr & Tar (1-to-1 molar ratio) were measured as a function of the receptor concentration and are plotted versus trTsr$_{QQEQ}$ concentration: trTsr$_{QQEQ}$, C-terminal truncated Tsr with the third site available; Tar$_{QQQQ}$, full-length Tar with four major sites amidated (no major sites available); and trTar$_{QQQQ}$, C-terminal truncated Tar$_{QQQQ}$. Legend: trTsr$_{QQEQ}$/Tar$_{QQQQ}$ (■); trTsr$_{QQEQ}$/Tar$_{QQQQ}$ plus 1 mM serine (●); trTsr$_{QQEQ}$/trTar$_{QQQQ}$ (□); trTsr$_{QQEQ}$/trTar$_{QQQQ}$ with 1 mM serine (○). This is Figure 6 of Antommattei et al. [9].

According to this picture, methyl group addition to kinase-inhibiting (inactive) receptors occurs more rapidly than to receptors in a kinase-activating state. *In vitro* measurements of methylation demonstrate that saturating concentrations of attractants binding typically increases methylation rates two- to threefold [9, 44, 138, 215].

The continued emphasis devoted to understanding ligand-mediated effects can be posed as two interrelated questions. What is the structural change in the cytoplasmic domain that results from ligand binding? To what extent is the change confined to the receptor dimer that binds ligand (versus being spread via dimer-dimer interactions)? Most of the effort has been directed toward elucidating changes within the MCP dimer, but as evidence accumulates in support of a role for receptor clusters in signaling interdimer interactions are also being tested.

Within the dimer, the effect of aspartate binding on structure is subtle, but several lines of evidence have been developed in support of a binding event that induces a piston-like displacement of transmembrane segment 2 toward the cytoplasm relative to transmembrane segment 1 (~1 to 2 Å) [63]. Recent experiments that mimic this displacement via site-directed mutagenesis generate forms of Tar that are methylated more rapidly and inhibit CheA [58, 138] thus providing evidence compatible with the involvement of a piston motion in altering the methylation rate (and in regulating kinase activity).

In addition, ligand binding seems to influence interactions among dimers. Two lines of evidence provide support for an aspartate-induced decrease in the interaction among Tar dimers. First, interdimer disulfide bond formation is reduced in a Tar double-cysteine mutant by aspartate [88]. Second, the polar localization of MCPs was reduced immediately after large attractant stimuli were delivered to E. coli and B. subtilis [108]. Moreover, the receptor distribution re-polarized when the cells were allowed to adapt to these large stimuli, suggesting that the receptor methylation promotes clustering. Previous observations, made with soluble and template-assembled signaling complexes, are consistent with this idea. These complexes show a significantly greater tendency to form when the receptor fragments have a high level of covalent modification [125, 179]. The effect of the ligands on the receptor structure that leads to these changes is yet to be established.

To achieve specific adaptation in a cluster of dimers that are heterogeneous with respect to binding specificity, an argument can be made that the effect of binding ligand should be localized in the dimer. On the other hand, the cooperative mechanisms of kinase regulation invoke interactions among several receptor dimers [34, 35, 133]. To determine the extent to which attractant-stimulated increases in methylation are localized to the dimer binding ligand, Tsr-Tar mixtures were engineered to promote methylation exclusively by an interdimer route (Figure 12.9), in which Tar dimers could tether CheR but were unable to accept methyl groups (Tar_{QQQQ}, Figure 12.9 right) and Tsr dimers were able to accept methyl groups at the third site but were unable to tether CheR ($trTsr_{QQEQ}$, Figure 12.9 left).

The results supported dimer-localized effects because only serine generated an increase in the methylation rate. Aspartate binding to Tar had no effect on methylation rates, indicating that the effect produced by aspartate binding was not communicated to the methylation regions of Tsr [9]. The methylation rate decreased significantly (~10-fold) when both Tar and Tsr were truncated ($trTar_{QQQQ}$ and $trTsr_{QQEQ}$), yet serine (and not aspartate) increased the methylation rate in the same proportion [9]. Evidently, attractant-induced increases in the rate of methylation do not require the tethering of CheR [111, 118].

It has been likewise postulated in models of robust adaptation that CheB acts preferentially on activated receptors [3], but substantial CheB activation (potentially 100-fold) is a result of phosphorylation [7], and the extent to which this enzyme-level activation is directed toward kinase-activating (attractant-free) receptor dimers is not known. Perhaps the cluster of receptor dimers that activate the CheA dimer(s) are the collective targets of the enhanced CheB action, which are mixtures of receptor dimers that differ in ligand specificity. The issue of global versus local targeting of CheB action has been investigated previously [95, 166], where evidence consistent with global regulation has been observed. The matter will undoubtedly receive further attention in light of the better understanding for receptor organization and the mechanism of CheB localization.

The major sites of methylation are modified at different rates [44, 202], and increases in rate generated by attractants vary from one site to the next. A recent study of *E. coli* Tsr methylation examined the methylation of four engineered forms of Tsr with one major methylation site available for methylation and the others blocked by amidation (EQQQ, QEQQ, QQEQ, QQQE) [44]. Site 4 was found to methylate most rapidly in the absence of ligand (and not site 3, as in Tar [202]). Tsr sites 1 through 3 (in Tsr_{EQQQ}, Tsr_{QEQQ}, and Tsr_{QQEQ}) were methylated more rapidly when serine was present, but no effect of serine was observed at site 4 [44].

It was proposed that the influence of serine binding was observable at the first three sites and not the fourth, because sites 1 through 3 reside in MH1, which is connected to transmembrane helix 2 through the HAMP domain (Figure 12.7). The effect of CheW was also studied. CheW was also viewed as a ligand that bound to the highly conserved domain of the receptor. CheW binding generated significant increases in the methylation rate at all four sites. The largest proportional effect was observed at site 1 (residue 297), where a sixfold increase in rate was produced with CheW (compared to a threefold increase produced by serine) [44]. In addition, CheW "amplified" the increase produced by serine. At site 1, the increase produced by serine was four times larger when CheW was present. The results demonstrated that CheW had a substantial impact on the methylation rate, and were interpreted to suggest that the arrangement between receptor dimers was altered by CheW in a way that increased the accessibility of CheR to the methylation sites [44]. In addition, if MCP-CheW interactions are dynamic (binding and unbinding) on the time scale of adaptation, CheW may potentially contribute to the regulation of receptor methylation.

The methylation sites are clustered nearby in the adaptation subdomain of the cytoplasmic domain (Figure 12.7). What are the physical properties of this domain that are likely to play a role in shifting the kinase-activating/kinase-inhibiting equilibrium, and generating changes in the methylation rate? Two features are receiving attention in this respect. First, a reduction in the net negative charge on the methylation helices that result from the formation of glutamyl esters may drive a conformational change that regulates kinase activity [190].

Second, ligand-mediated changes in the dynamics of the cytoplasmic domains are proposed to be involved in kinase regulation [98]. Such a change may also influence the methylation rate. The cytoplasmic domain is exceptionally dynamic as a soluble fragment [140, 169, 220], and evidence suggests that this dynamic behavior is retained in the intact receptor [215]. Additional investigation is required to elucidate the extent and manner in which these factors play a role in signal regulation.

V. Conclusions

The determination of the Tsr cytoplasmic fragment structure has had a significant impact on understanding receptor methylation, which reaches far beyond *E. coli*, given the widespread distribution of the MA domain and the high degree of sequence conservation. In the *E. coli/Salmonella* system, the MA domain structure has provided a detailed structure template for probing MA-domain/CheR interactions. More structure information of receptor/CheW/CheA complexes and the receptor/CheR complex is needed, and in particular structures that trap the kinase-activating (methylation inactive) and kinase-inhibiting (methylation active) states.

E. coli and *Salmonella* currently provide the best access to in-depth structure-function relationships, where a more complete understanding of receptor cluster structure and dynamics is being developed. Yet, the *E. coli* system will prove to be too simple to develop a complete picture for the roles of multiple chemotaxis pathways (e.g., their localization, specificity of interaction, and function). This picture is being developed for the chemotaxis (and chemotaxis-like) pathways in *Rhodobacter sphaeroides* (for example), where the evidence indicates that specific localization is critical for function [206].

There are more lessons to be learned from analyses of signaling in different species of *Bacteria* and *Archaea*, as the contrasts in signaling logic between *E. coli* and *B. subtilis* demonstrate [72, 157, 198]. From these studies we can hope to obtain a better view of the remarkably diverse manner in which signaling pathways are assembled and function in microorganisms. Strong evidence for the regulation of gene expression based on an MCP-like signaling pathway has been found [100].

It is probable more examples will be found. Methylation in these other systems, when it occurs, can be plausibly expected to have a role in generating temporal comparisons if the *E. coli* paradigm is followed, but the kinetics of methylation and demethylation will be suitably adapted to the time scale of the process involved. Finally, the amassing database of genomes is now significant, and continued advances in the data-mining tools of bioinformatics will draw together and add significance to the biochemical and structural investigations made in various organisms. In doing so, substantial progress toward understanding the full scope of receptor methylation will be made.

REFERENCES

1. Albert, R., Chiu, Y. W., and Othmer, H. G. (2004). Dynamic receptor team formation can explain the high signal transduction gain in *Escherichia coli*. *Biophys J* 86:2650–2659.
2. Alexandre, G., Greer-Phillips, S., and Zhulin, I. B. (2004). Ecological role of energy taxis in microorganisms. *FEMS Microbiol Rev* 28:113–126.
3. Alon, U., Surette, M. G., Barkai, N., and Leibler, S. (1999). Robustness in bacterial chemotaxis. *Nature* 397:168–171.
4. Altschul, S. F., Madden, T. L., Schaffer, A. A., Zhang, J., Zhang, Z., Miller, W., and Lipman, D. J. (1997). Gapped BLAST and PSI-BLAST: A new generation of protein database search programs. *Nucleic Acids Res* 25:3389–3402.
5. Ames, P., Studdert, C. A., Reiser, R. H., and Parkinson, J. S. (2002). Collaborative signaling by mixed chemoreceptor teams in *Escherichia coli*. *Proc Natl Acad Sci USA* 99:7060–7065.
6. Anand, G. S., Goudreau, P. N., and Stock, A. M. (1998). Activation of methylesterase CheB: evidence of a dual role for the regulatory domain. *Biochemistry* 37:14038–14047.
7. Anand, G. S., and Stock, A. M. (2002). Kinetic basis for the stimulatory effect of phosphorylation on the methylesterase activity of CheB. *Biochemistry* 41:6752–6760.
8. Anantharaman, V., and Aravind, L. (2000). Cache – a signaling domain common to animal Ca(2+)-channel subunits and a class of prokaryotic chemotaxis receptors. *Trends Biochem Sci* 25:535–537.
9. Antommattei, F. M., Munzner, J. B., and Weis, R. M. (2004). Ligand-specific activation of *Escherichia coli* chemoreceptor transmethylation. *J Bacteriol* 186:7556–7563.
10. Aravind, L., and Ponting, C. P. (1999). The cytoplasmic helical linker domain of receptor histidine kinase and methyl-accepting proteins is common to many prokaryotic signaling proteins. *FEMS Microbiol Lett* 176:111–116.
11. Aswad, D. W., and Koshland, D. E. Jr. (1975). Evidence for an S-adenosylmethionine requirement in the chemotactic behavior of *Salmonella typhimurium*. *J Mol Biol* 97:207–223.
12. Baar, C., Eppinger, M., Raddatz, G., Simon, J., Lanz, C., Klimmek, O., Nandakumar, R., Gross, R., Rosinus, A., Keller, H., Jagtap, P., Linke, B., Meyer, F., Lederer, H., and Schuster, S. C. (2003). Complete genome sequence and analysis of *Wolinella succinogenes*. *Proc Natl Acad Sci USA* 100:11690–11695.
13. Banno, S., Shiomi, D., Homma, M., and Kawagishi, I. (2004). Targeting of the chemotaxis methylesterase/deamidase CheB to the polar receptor-kinase cluster in an *Escherichia coli* cell. *Mol Microbiol* 53:1051–1063.
14. Barnakov, A. N., Barnakova, L. A., and Hazelbauer, G. L. (1998). Comparison in vitro of a high- and a low-abundance chemoreceptor of *Escherichia coli*: Similar kinase activation but different methyl-accepting activities. *J Bacteriol* 180:6713–6718.
15. Barnakov, A. N., Barnakova, L. A., and Hazelbauer, G. L. (1999). Efficient adaptational demethylation of chemoreceptors requires the same enzyme–docking site as efficient methylation. *Proc Natl Acad Sci USA* 96:10667–10672.
16. Barnakov, A. N., Barnakova, L. A., and Hazelbauer, G. L. (2001). Location of the receptor-interaction site on CheB, the methylesterase response regulator of bacterial chemotaxis. *J Biol Chem* 276:32984–32989.
17. Barnakov, A. N., Barnakova, L. A., and Hazelbauer, G. L. (2002). Allosteric enhancement of adaptational demethylation by a carboxyl-terminal sequence on chemoreceptors. *J Biol Chem* 277:42151–42156.

18. Bateman, A., Coin, L., Durbin, R., Finn, R. D., Hollich, V., Griffiths-Jones, S., Khanna, A., Marshall, M., Moxon, S., Sonnhammer, E. L., Studholme, D. J., Yeats, C., and Eddy, S. R. (2004). The Pfam protein families database. *Nucleic Acids Res* 32:D138–141.

19. Bell, K. S., Sebaihia, M., Pritchard, L., Holden, M. T., Hyman, L. J., Holeva, M. C., Thomson, N. R., Bentley, S. D., Churcher, L. J., Mungall, K., Atkin, R., Bason, N., Brooks, K., Chillingworth, T., Clark, K., Doggett, J., Fraser, A., Hance, Z., Hauser, H., Jagels, K., Moule, S., Norbertczak, H., Ormond, D., Price, C., Quail, M. A., Sanders, M., Walker, D., Whitehead, S., Salmond, G. P., Birch, P. R., Parkhill, J., and Toth, I. K. (2004). Genome sequence of the enterobacterial phytopathogen *Erwinia carotovora* subsp. Atroseptica and characterization of virulence factors. *Proc Natl Acad Sci USA* 101:11105–11110.

20. Berg, H. C. (2000). Motile behavior of bacteria. *Physics Today* 53:24–29.

21. Berg, H. C. (2004). The bacterial rotory motor, in D.D. Hackney and F. Tamanoi (eds.), *The Enzymes*, (vol. 23), pp. 143–202, Amsterdam: Elsevier (Academic Press).

22. Berg, H. C., and Purcell, E. M. (1977). Physics of chemoreception. *Biophysical Journal* 20:193–219.

23. Bibikov, S. I., Biran, R., Rudd, K. E., and Parkinson, J. S. (1997). A signal transducer for aerotaxis in Escherichia coli. *J Bacteriol* 179:4075–4079.

24. Bibikov, S. I., Miller, A. C., Gosink, K. K., and Parkinson, J. S. (2004). Methylation-independent aerotaxis mediated by the *Escherichia coli* Aer protein. *J Bacteriol* 186:3730–3737.

25. Bilwes, A. M., Alex, L. A., Crane, B. R., and Simon, M. I. (1999). Structure of CheA, a signal-transducing histidine kinase. *Cell* 96:131–141.

26. Bilwes, A. M., Quezada, C. M., Croal, L. R., Crane, B. R., and Simon, M. I. (2001). Nucleotide binding by the histidine kinase CheA. *Nat Struct Biol* 8:353–360.

27. Blattner, F. R., Plunkett, G., 3rd, Bloch, C. A., *et al.* (1997). The complete genome sequence of *Escherichia coli* K-12. *Science* 277:1453–1474.

28. Bollinger, J., Park, C., Harayama, S., and Hazelbauer, G. L. (1984). Structure of the Trg protein: Homologies with and differences from other sensory transducers of *Escherichia coli. Proc Natl Acad Sci USA* 81:3287–3291.

29. Borkovich, K. A., Alex, L. A., and Simon, M. I. (1992). Attenuation of sensory receptor signaling by covalent modification. *Proc Natl Acad Sci USA* 89:6756–6760.

30. Bornhorst, J. A., and Falke, J. J. (2000). Attractant regulation of the aspartate receptor-kinase complex: limited cooperative interactions between receptors and effects of the receptor modification state. *Biochemistry* 39:9486–9493.

31. Bornhorst, J. A., and Falke, J. J. (2003). Quantitative analysis of aspartate receptor signaling complex reveals that the homogeneous two-state model is inadequate: development of a heterogeneous two-state model. *J Mol Biol* 326:1597–1614.

32. Bourret, R. B., and Stock, A. M. (2002). Molecular information processing: lessons from bacterial chemotaxis. *J Biol Chem* 277:9625–9628.

33. Boyd, A., and Simon, M. I. (1980). Multiple electrophoretic forms of methyl-accepting chemotaxis proteins generated by stimulus-elicited methylation in *Escherichia coli. J Bacteriol* 143:809–815.

34. Bray, D., and Duke, T. (2004). Conformational spread: The propagation of allosteric states in large multiprotein complexes. *Annu Rev Biophys Biomol Struct* 33:53–73.

35. Bray, D., Levin, M. D., and Morton-Firth, C. J. (1998). Receptor clustering as a cellular mechanism to control sensitivity. *Nature* 393:85–88.

36. Brazilian National Genome Project Consortium (2003). The complete genome sequence of *Chromobacterium violaceum* reveals remarkable and exploitable bacterial adaptability. *Proc Natl Acad Sci USA* 100:11660–11665.

37. Bren, A., and Eisenbach, M. (2000). How signals are heard during bacterial chemotaxis: protein-protein interactions in sensory signal propagation. *J Bacteriol* 182:6865–6873.

38. Brito, C. F., Carvalho, C. B., Santos, F., Gazzinelli, R. T., Oliveira, S. C., Azevedo, V., and Teixeira, S. M. (2004). *Chromobacterium violaceum* genome: molecular mechanisms associated with pathogenicity. *Genet Mol Res* 3:148–161.

39. Buell, C. R., Joardar, V., Lindeberg, M., Selengut, J., Paulsen, I. T., Gwinn, M. L., Dodson, R. J., Deboy, R. T., Durkin, A. S., Kolonay, J. F., Madupu, R., Daugherty, S., Brinkac, L., Beanan, M. J., Haft, D. H., Nelson, W. C., Davidsen, T., Zafar, N., Zhou, L., Liu, J., Yuan, Q., Khouri, H., Fedorova, N., Tran, B., Russell, D., Berry, K., Utterback, T., Van Aken, S. E., Feldblyum, T. V., D'Ascenzo, M., Deng, W. L., Ramos, A. R., Alfano, J. R., Cartinhour, S., Chatterjee, A. K., Delaney, T. P., Lazarowitz, S. G., Martin, G. B., Schneider, D. J., Tang, X., Bender, C. L., White, O., Fraser, C. M., and Collmer, A. (2003). The complete genome sequence of the *Arabidopsis* and tomato pathogen *Pseudomonas syringae* pv. *tomato* DC3000. *Proc Natl Acad Sci USA* 100:10181–10186.

40. Butler, S. L., and Falke, J. J. (1998). Cysteine and disulfide scanning reveals two amphiphilic helices in the linker region of the aspartate chemoreceptor. *Biochemistry* 37:10746–10756.

41. Cantwell, B. J., Draheim, R. R., Weart, R. B., Nguyen, C., Stewart, R. C., and Manson, M. D. (2003). CheZ phosphatase localizes to chemoreceptor patches via CheA-short. *J Bacteriol* 185:2354–2361.

42. Chain, P., Lamerdin, J., Larimer, F., Regala, W., Lao, V., Land, M., Hauser, L., Hooper, A., Klotz, M., Norton, J., Sayavedra-Soto, L., Arciero, D., Hommes, N., Whittaker, M., and Arp, D. (2003). Complete genome sequence of the ammonia-oxidizing bacterium and obligate chemolithoautotroph *Nitrosomonas europaea*. *J Bacteriol* 185:2759–2773.

43. Chain, P. S., Carniel, E., Larimer, F. W., Lamerdin, J., Stoutland, P. O., Regala, W. M., Georgescu, A. M., Vergez, L. M., Land, M. L., Motin, V. L., Brubaker, R. R., Fowler, J., Hinnebusch, J., Marceau, M., Medigue, C., Simonet, M., Chenal-Francisque, V., Souza, B., Dacheux, D., Elliott, J. M., Derbise, A., Hauser, L. J., and Garcia, E. (2004). Insights into the evolution of *Yersinia pestis* through whole-genome comparison with *Yersinia pseudotuberculosis*. *Proc Natl Acad Sci USA* 101:13826–13831.

44. Chalah, A., and Weis, R. M. (2005). Site-specific and synergistic stimulation of methylation on the bacterial chemotaxis receptor Tsr by serine and CheW. *BMC Microbiol* 5:12.

45. Chelsky, D., and Dahlquist, F. W. (1981). Multiple sites of methylation in the methyl accepting chemotaxis proteins of *Escherichia coli*. *Prog Clin Biol Res* 63:371–381.

46. Chen, C. Y., Wu, K. M., Chang, Y. C., Chang, C. H., Tsai, H. C., Liao, T. L., Liu, Y. M., Chen, H. J., Shen, A. B., Li, J. C., Su, T. L., Shao, C. P., Lee, C. T., Hor, L. I., and Tsai, S. F. (2003). Comparative genome analysis of *Vibrio vulnificus*, a marine pathogen. *Genome Res* 13:2577–2587.

47. Chenna, R., Sugawara, H., Koike, T., Lopez, R., Gibson, T. J., Higgins, D. G., and Thompson, J. D. (2003). Multiple sequence alignment with the Clustal series of programs. *Nucleic Acids Res* 31:3497–3500.

48. Chiu, C. H., Tang, P., Chu, C., Hu, S., Bao, Q., Yu, J., Chou, Y. Y., Wang, H. S., and Lee, Y. S. (2005). The genome sequence of *Salmonella enterica* serovar Choleraesuis, a highly invasive and resistant zoonotic pathogen. *Nucleic Acids Res* 33:1690–1698.

49. Cho, H. S., Pelton, J. G., Yan, D., Kustu, S., and Wemmer, D. E. (2001). Phosphoaspartates in bacterial signal transduction. *Curr Opin Struct Biol* 11:679–684.

50. Cochran, A. G., and Kim, P. S. (1996). Imitation of Escherichia coli aspartate receptor signaling in engineered dimers of the cytoplasmic domain. *Science* 271:1113–1116.

51. D'Andrea, L. D., and Regan, L. (2003). TPR proteins: The versatile helix. *Trends Biochem Sci* 28:655–662.

52. Monteiro-Vitorello, C. B., Van Sluys, M. A., Almeida, N. F., Alves, L. M., do Amaral, A. M., Bertolini, M. C., Camargo, L. E., Camarotte, G., Cannavan, F., Cardozo, J., Chambergo, F., Ciapina, L. P., Cicarelli, R. M., Coutinho, L. L., Cursino-Santos, J. R., El-Dorry, H., Faria, J. B., Ferreira, A. J., Ferreira, R. C., Ferro, M. I., Formighieri, E. F., Franco, M. C., Greggio, C. C., Gruber, A., Katsuyama, A. M., Kishi, L. T., Leite, R. P., Lemos, E. G., Lemos, M. V., Locali, E. C., Machado, M. A., Madeira, A. M., Martinez-Rossi, N. M., Martins, E. C., Meidanis, J., Menck, C. F., Miyaki, C. Y., Moon, D. H., Moreira, L. M., Novo, M. T., Okura, V. K.,

Oliveira, M. C., Oliveira, V. R., Pereira, H. A., Rossi, A., Sena, J. A., Silva, C., de Souza, R. F., Spinola, L. A., Takita, M. A., Tamura, R. E., Teixeira, E. C., Tezza, R. I., Trindade dos Santos, M., Truffi, D., Tsai, S. M., White, F. F., Setubal, J. C., and Kitajima, J. P. (2002). Comparison of the genomes of two *Xanthomonas* pathogens with differing host specificities. *Nature* 417:459–463.

53. Deng, W., Burland, V., Plunkett, G., 3rd, Boutin, A., Mayhew, G. F., Liss, P., Perna, N. T., Rose, D. J., Mau, B., Zhou, S., Schwartz, D. C., Fetherston, J. D., Lindler, L. E., Brubaker, R. R., Plano, G. V., Straley, S. C., McDonough, K. A., Nilles, M. L., Matson, J. S., Blattner, F. R., and Perry, R. D. (2002). Genome sequence of *Yersinia pestis* KIM. *J Bacteriol* 184:4601–4611.

54. Deng, W., Liou, S. R., Plunkett, G. III, Mayhew, G. F., Rose, D. J., Burland, V., Kodoyianni, V., Schwartz, D. C., and Blattner, F. R. (2003). Comparative genomics of *Salmonella enterica* serovar Typhi strains Ty2 and CT18. *J Bacteriol* 185:2330–2337.

55. Djordjevic, S., and Stock, A. M. (1997). Crystal structure of the chemotaxis receptor methyltransferase CheR suggests a conserved structural motif for binding S-adenosylmethionine. *Structure* 5:545–558.

56. Djordjevic, S., and Stock, A. M. (1998). Chemotaxis receptor recognition by protein methyltransferase CheR. *Nat Struct Biol* 5:446–450.

57. Djordjevic, S., Stock, A. M., Chen, Y., and Stock, J. B. (1999). Protein methyltransferases involved in signal transduction, in X. Cheng and R.M. Blumenthal (eds.), *S-adenosylmethionine-Dependent Methyltransferases: Structures and Functions*, pp. 149–183, Singapore: World Scientific Publishing.

58. Draheim, R. R., Bormans, A. F., Lai, R. Z., and Manson, M. D. (2005). Tryptophan residues flanking the second transmembrane helix (TM2) set the signaling state of the Tar chemoreceptor. *Biochemistry* 44:1268–1277.

59. Duchaud, E., Rusniok, C., Frangeul, L., Buchrieser, C., Givaudan, A., Taourit, S., Bocs, S., Boursaux-Eude, C., Chandler, M., Charles, J. F., Dassa, E., Derose, R., Derzelle, S., Freyssinet, G., Gaudriault, S., Medigue, C., Lanois, A., Powell, K., Siguier, P., Vincent, R., Wingate, V., Zouine, M., Glaser, P., Boemare, N., Danchin, A., and Kunst, F. (2003). The genome sequence of the entomopathogenic bacterium *Photorhabdus luminescens*. *Nat Biotechnol* 21:1307–1313.

60. Dunten, P., and Koshland, D. E. Jr. (1991). Tuning the responsiveness of a sensory receptor via covalent modification. *J Biol Chem* 266:1491–1496.

61. Engstrom, P., and Hazelbauer, G. L. (1980). Multiple methylation of methyl-accepting chemotaxis proteins during adaptation of *E. coli* to chemical stimuli. *Cell* 20:165–171.

62. Falke, J. J., Bass, R. B., Butler, S. L., Chervitz, S. A., and Danielson, M. A. (1997). The two-component signaling pathway of bacterial chemotaxis: a molecular view of signal transduction by receptors, kinases, and adaptation enzymes. *Annu Rev Cell Dev Biol* 13:457–512.

63. Falke, J. J., and Hazelbauer, G. L. (2001). Transmembrane signaling in bacterial chemoreceptors. *Trends Biochem Sci* 26:257–265.

64. Falke, J. J., and Kim, S. H. (2000). Structure of a conserved receptor domain that regulates kinase activity: the cytoplasmic domain of bacterial taxis receptors. *Curr Opin Struct Biol* 10:462–469.

65. Feil, H., Feil, W. S., Chain, P., Larimer, F., DiBartolo, G., Copeland, A., Lykidis, A., Trong, S., Nolan, M., Goltsman, E., Thiel, J., Malfatti, S., Loper, J. E., Lapidus, A., Detter, J. C., Land, M., Richardson, P. M., Kyrpides, N. C., Ivanova, N., and Lindow, S. E. (2005). Comparison of the complete genome sequences of *Pseudomonas syringae* pv. *syringae* B728a and pv. *tomato* DC3000. *Proc Natl Acad Sci USA* 102:11064–11069.

66. Feng, X., Lilly, A. A., and Hazelbauer, G. L. (1999). Enhanced function conferred on low-abundance chemoreceptor Trg by a methyltransferase-docking site. *J Bacteriol* 181:3164–3171.

67. Francis, N. R., Levit, M. N., Shaikh, T. R., Melanson, L. A., Stock, J. B., and DeRosier, D. J. (2002). Subunit organization in a soluble complex of tar, CheW, and CheA by electron microscopy. *J Biol Chem* 277:36755–36759.

68. Francis, N. R., Wolanin, P. M., Stock, J. B., Derosier, D. J., and Thomas, D. R. (2004). Three-dimensional structure and organization of a receptor/signaling complex. *Proc Natl Acad Sci USA* 101:17480–17485.
69. Fraser, C. M., Casjens, S., Huang, W. M., Sutton, G. G., Clayton, R., Lathigra, R., White, O., Ketchum, K. A., Dodson, R., Hickey, E. K., Gwinn, M., Dougherty, B., Tomb, J. F., Fleischmann, R. D., Richardson, D., Peterson, J., Kerlavage, A. R., Quackenbush, J., Salzberg, S., Hanson, M., van Vugt, R., Palmer, N., Adams, M. D., Gocayne, J., Weidman, J., Utterback, T., Watthey, L., McDonald, L., Artiach, P., Bowman, C., Garland, S., Fuji, C., Cotton, M. D., Horst, K., Roberts, K., Hatch, B., Smith, H. O., and Venter, J. C. (1997). Genomic sequence of a Lyme disease spirochaete, *Borrelia burgdorferi*. *Nature* 390:580–586.
70. Galibert, F., Finan, T. M., Long, S. R., Puhler, A., Abola, P., Ampe, F., Barloy-Hubler, F., Barnett, M. J., Becker, A., Boistard, P., Bothe, G., Boutry, M., Bowser, L., Buhrmester, J., Cadieu, E., Capela, D., Chain, P., Cowie, A., Davis, R. W., Dreano, S., Federspiel, N. A., Fisher, R. F., Gloux, S., Godrie, T., Goffeau, A., Golding, B., Gouzy, J., Gurjal, M., Hernandez-Lucas, I., Hong, A., Huizar, L., Hyman, R. W., Jones, T., Kahn, D., Kahn, M. L., Kalman, S., Keating, D. H., Kiss, E., Komp, C., Lelaure, V., Masuy, D., Palm, C., Peck, M. C., Pohl, T. M., Portetelle, D., Purnelle, B., Ramsperger, U., Surzycki, R., Thebault, P., Vandenbol, M., Vorholter, F. J., Weidner, S., Wells, D. H., Wong, K., Yeh, K. C., and Batut, J. (2001). The composite genome of the legume symbiont *Sinorhizobium meliloti*. *Science* 293:668–672.
71. Galperin, M. Y. (2005). A census of membrane-bound and intracellular signal transduction proteins in bacteria: bacterial IQ, extroverts and introverts. *BMC Microbiol* 5:35.
72. Garrity, L. F., and Ordal, G. W. (1997). Activation of the CheA kinase by asparagine in *Bacillus subtilis* chemotaxis. *Microbiology* 143:2945–2951.
73. Gegner, J. A., Graham, D. R., Roth, A. F., and Dahlquist, F. W. (1992). Assembly of an MCP receptor, CheW, and kinase CheA complex in the bacterial chemotaxis signal transduction pathway. *Cell* 70:975–982.
74. Gestwicki, J. E., and Kiessling, L. L. (2002). Inter-receptor communication through arrays of bacterial chemoreceptors. *Nature* 415:81–84.
75. Gestwicki, J. E., Lamanna, A. C., Harshey, R. M., McCarter, L. L., Kiessling, L. L., and Adler, J. (2000). Evolutionary conservation of methyl-accepting chemotaxis protein location in *Bacteria* and *Archaea*. *J Bacteriol* 182:6499–6502.
76. Glockner, G., Lehmann, R., Romualdi, A., Pradella, S., Schulte-Spechtel, U., Schilhabel, M., Wilske, B., Suhnel, J., and Platzer, M. (2004). Comparative analysis of the *Borrelia garinii* genome. *Nucleic Acids Res* 32:6038–6046.
77. Grebe, T. W., and Stock, J. B. (1999). The histidine protein kinase superfamily. *Adv Microb Physiol* 41:139–227.
78. Griswold, I. J., Zhou, H., Matison, M., Swanson, R. V., McIntosh, L. P., Simon, M. I., and Dahlquist, F. W. (2002). The solution structure and interactions of CheW from *Thermotoga maritima*. *Nat Struct Biol* 9:121–125.
79. Hanlon, D. W., and Ordal, G. W. (1994). Cloning and characterization of genes encoding methyl-accepting chemotaxis proteins in *Bacillus subtilis*. *J Biol Chem* 269:14038–14046.
80. Hayashi, T., Makino, K., Ohnishi, M., Kurokawa, K., Ishii, K., Yokoyama, K., Han, C. G., Ohtsubo, E., Nakayama, K., Murata, T., Tanaka, M., Tobe, T., Iida, T., Takami, H., Honda, T., Sasakawa, C., Ogasawara, N., Yasunaga, T., Kuhara, S., Shiba, T., Hattori, M., and Shinagawa, H. (2001). Complete genome sequence of enterohemorrhagic *Escherichia coli* O157:H7 and genomic comparison with a laboratory strain K-12. *DNA Res* 8:11–22.
81. Hazelbauer, G. L., and Engstrom, P. (1980). Parallel pathways for transduction of chemotactic signals in *Escherichia coli*. *Nature* 283:98–100.
82. Heidelberg, J. F., Eisen, J. A., Nelson, W. C., Clayton, R. A., Gwinn, M. L., Dodson, R. J., Haft, D. H., Hickey, E. K., Peterson, J. D., Umayam, L., Gill, S. R., Nelson, K. E., Read, T. D., Tettelin, H., Richardson, D., Ermolaeva, M. D., Vamathevan, J., Bass, S., Qin, H., Dragoi, I.,

Sellers, P., McDonald, L., Utterback, T., Fleishmann, R. D., Nierman, W. C., White, O., Salzberg, S. L., Smith, H. O., Colwell, R. R., Mekalanos, J. J., Venter, J. C., and Fraser, C. M. (2000). DNA sequence of both chromosomes of the cholera pathogen *Vibrio cholerae*. *Nature* 406:477–483.

83. Heidelberg, J. F., Paulsen, I. T., Nelson, K. E., Gaidos, E. J., Nelson, W. C., Read, T. D., Eisen, J. A., Seshadri, R., Ward, N., Methe, B., Clayton, R. A., Meyer, T., Tsapin, A., Scott, J., Beanan, M., Brinkac, L., Daugherty, S., DeBoy, R. T., Dodson, R. J., Durkin, A. S., Haft, D. H., Kolonay, J. F., Madupu, R., Peterson, J. D., Umayam, L. A., White, O., Wolf, A. M., Vamathevan, J., Weidman, J., Impraim, M., Lee, K., Berry, K., Lee, C., Mueller, J., Khouri, H., Gill, J., Utterback, T. R., McDonald, L. A., Feldblyum, T. V., Smith, H. O., Venter, J. C., Nealson, K. H., and Fraser, C. M. (2002). Genome sequence of the dissimilatory metal ion-reducing bacterium *Shewanella oneidensis*. *Nat Biotechnol* 20:1118–1123.

84. Heidelberg, J. F., Seshadri, R., Haveman, S. A., Hemme, C. L., Paulsen, I. T., Kolonay, J. F., Eisen, J. A., Ward, N., Methe, B., Brinkac, L. M., Daugherty, S. C., Deboy, R. T., Dodson, R. J., Durkin, A. S., Madupu, R., Nelson, W. C., Sullivan, S. A., Fouts, D., Haft, D. H., Selengut, J., Peterson, J. D., Davidsen, T. M., Zafar, N., Zhou, L., Radune, D., Dimitrov, G., Hance, M., Tran, K., Khouri, H., Gill, J., Utterback, T. R., Feldblyum, T. V., Wall, J. D., Voordouw, G., and Fraser, C. M. (2004). The genome sequence of the anaerobic, sulfate-reducing bacterium *Desulfovibrio vulgaris* Hildenborough. *Nat Biotechnol* 22:554–559.

85. Hess, J. F., Oosawa, K., Matsumura, P., and Simon, M. I. (1987). Protein phosphorylation is involved in bacterial chemotaxis. *Proc Natl Acad Sci USA* 84:7609–7613.

86. Hoff, W. D., Jung, K. H., and Spudich, J. L. (1997). Molecular mechanism of photosignaling by archaeal sensory rhodopsins. *Annu Rev Biophys Biomol Struct* 26:223–258.

87. Holden, M. T., Titball, R. W., Peacock, S. J., Cerdeno-Tarraga, A. M., Atkins, T., Crossman, L. C., Pitt, T., Churcher, C., Mungall, K., Bentley, S. D., Sebaihia, M., Thomson, N. R., Bason, N., Beacham, I. R., Brooks, K., Brown, K. A., Brown, N. F., Challis, G. L., Cherevach, I., Chillingworth, T., Cronin, A., Crossett, B., Davis, P., DeShazer, D., Feltwell, T., Fraser, A., Hance, Z., Hauser, H., Holroyd, S., Jagels, K., Keith, K. E., Maddison, M., Moule, S., Price, C., Quail, M. A., Rabbinowitsch, E., Rutherford, K., Sanders, M., Simmonds, M., Songsivilai, S., Stevens, K., Tumapa, S., Vesaratchavest, M., Whitehead, S., Yeats, C., Barrell, B. G., Oyston, P. C., and Parkhill, J. (2004). Genomic plasticity of the causative agent of melioidosis, *Burkholderia pseudomallei*. *Proc Natl Acad Sci USA* 101:14240–14245.

88. Homma, M., Shiomi, D., Homma, M., and Kawagishi, I. (2004). Attractant binding alters arrangement of chemoreceptor dimers within its cluster at a cell pole. *Proc Natl Acad Sci USA* 101:3462–3467.

89. Hou, S., Larsen, R. W., Boudko, D., Riley, C. W., Karatan, E., Zimmer, M., Ordal, G. W., and Alam, M. (2000). Myoglobin-like aerotaxis transducers in *Archaea* and *Bacteria*. *Nature* 403:540–544.

90. Jin, Q., Yuan, Z., Xu, J., Wang, Y., Shen, Y., Lu, W., Wang, J., Liu, H., Yang, J., Yang, F., Zhang, X., Zhang, J., Yang, G., Wu, H., Qu, D., Dong, J., Sun, L., Xue, Y., Zhao, A., Gao, Y., Zhu, J., Kan, B., Ding, K., Chen, S., Cheng, H., Yao, Z., He, B., Chen, R., Ma, D., Qiang, B., Wen, Y., Hou, Y., and Yu, J. (2002). Genome sequence of *Shigella flexneri* 2a: insights into pathogenicity through comparison with genomes of *Escherichia coli* K12 and O157. *Nucleic Acids Res* 30:4432–4441.

91. Joardar, V., Lindeberg, M., Jackson, R. W., Selengut, J., Dodson, R., Brinkac, L. M., Daugherty, S. C., Deboy, R., Durkin, A. S., Giglio, M. G., Madupu, R., Nelson, W. C., Rosovitz, M. J., Sullivan, S., Crabtree, J., Creasy, T., Davidsen, T., Haft, D. H., Zafar, N., Zhou, L., Halpin, R., Holley, T., Khouri, H., Feldblyum, T., White, O., Fraser, C. M., Chatterjee, A. K., Cartinhour, S., Schneider, D. J., Mansfield, J., Collmer, A., and Buell, C. R. (2005). Whole-genome sequence analysis of *Pseudomonas syringae* pv. *phaseolicola* 1448A reveals divergence among pathovars in genes involved in virulence and transposition. *J Bacteriol* 187:6488–6498.

92. Karniol, B., and Vierstra, R. D. (2004). The HWE histidine kinases, a new family of bacterial two-component sensor kinases with potentially diverse roles in environmental signaling. *J Bacteriol* 186:445–453.

93. Kehry, M. R., Bond, M. W., Hunkapiller, M. W., and Dahlquist, F. W. (1983). Enzymatic deamidation of methyl-accepting chemotaxis proteins in *Escherichia coli* catalyzed by the *cheB* gene product. *Proc Natl Acad Sci USA* 80:3599–3603.

94. Kehry, M. R., and Dahlquist, F. W. (1982). The methyl-accepting chemotaxis proteins of *Escherichia coli*. Identification of the multiple methylation sites on methyl-accepting chemotaxis protein I. *J Biol Chem* 257:10378–10386.

95. Kehry, M. R., Doak, T. G., and Dahlquist, F. W. (1985). Sensory adaptation in bacterial chemotaxis: regulation of demethylation. *J Bacteriol* 163:983–990.

96. Kehry, M. R., Engstrom, P., Dahlquist, F. W., and Hazelbauer, G. L. (1983). Multiple covalent modifications of Trg, a sensory transducer of *Escherichia coli*. *J Biol Chem* 258:5050–5055.

97. Kim, K. K., Yokota, H., and Kim, S. H. (1999). Four-helical-bundle structure of the cytoplasmic domain of a serine chemotaxis receptor. *Nature* 400:787–792.

98. Kim, S. H., Wang, W., and Kim, K. K. (2002). Dynamic and clustering model of bacterial chemotaxis receptors: structural basis for signaling and high sensitivity. *Proc Natl Acad Sci USA* 99:11611–11615.

99. Kim, Y. R., Lee, S. E., Kim, C. M., Kim, S. Y., Shin, E. K., Shin, D. H., Chung, S. S., Choy, H. E., Progulske-Fox, A., Hillman, J. D., Handfield, M., and Rhee, J. H. (2003). Characterization and pathogenic significance of *Vibrio vulnificus* antigens preferentially expressed in septicemic patients. *Infect Immun* 71:5461–5471.

100. Kirby, J. R., and Zusman, D. R. (2003). Chemosensory regulation of developmental gene expression in *Myxococcus xanthus*. *Proc Natl Acad Sci USA* 100:2008–2013.

101. Kort, E. N., Goy, M. F., Larsen, S. H., and Adler, J. (1975). Methylation of a membrane protein involved in bacterial chemotaxis. *Proc Natl Acad Sci USA* 72:3939–3943.

102. Kott, L., Braswell, E. H., Shrout, A. L., and Weis, R. M. (2004). Distributed subunit interactions in CheA contribute to dimer stability: a sedimentation equilibrium study. *Biochim Biophys Acta* 1696:131–140.

103. Krikos, A., Mutoh, N., Boyd, A., and Simon, M. I. (1983). Sensory transducers of *E. coli* are composed of discrete structural and functional domains. *Cell* 33:615–622.

104. Kristich, C. J., Glekas, G. D., and Ordal, G. W. (2003). The conserved cytoplasmic module of the transmembrane chemoreceptor McpC mediates carbohydrate chemotaxis in *Bacillus subtilis*. *Mol Microbiol* 47:1353–1366.

105. Kunst, F., Ogasawara, N., Moszer, I., Albertini, A. M., Alloni, G., Azevedo, V., Bertero, M. G., Bessieres, P., Bolotin, A., Borchert, S., Borriss, R., Boursier, L., Brans, A., Braun, M., Brignell, S. C., Bron, S., Brouillet, S., Bruschi, C. V., Caldwell, B., Capuano, V., Carter, N. M., Choi, S. K., Codani, J. J., Connerton, I. F., Cummings, N. J., Daniel, R. A., Denizot, F., Devine, K. M., Düsterhöft, A., Ehrlich, S. D., Emmerson, P. T., Entian, K. D., Errington, J. , Fabret, C., Ferrari, E., Foulger, D., Fritz, C., Fujita, M., Fujita, Y., Fuma, S., Galizzi, A., Galleron, N., Ghim, S.-Y., Glaser, P., Goffeau, A., Golightly, E. J., Grandi, G., Guiseppi, G., Guy, B. J., Haga, K., Haiech, J., Harwood, C. R., Hénaut, A., Hilbert, H., Holsappel, S., Hosono, S., Hullo, M.-F., Itaya, M., Jones, L., Joris, B., Karamata, D., Kasahara, Y., Klaerr-Blanchard, M., Klein, C., Kobayashi, Y., Koetter, P., Konigstein, G., Krogh, S., Kumano, M., Kurita, K., Lapidus, A., Lardinois, S., Lauber, J., Lazarevic, V., Lee, S.-M., Levine, A., Liu, H., Masuda, S., Mauël, C., Médigue, C., Medina, N., Mellado, R. P., Mizuno, M., Moestl, D., Nakai, S., Noback, M., Noone, D., O'Reilly, M., Ogawa, K., Ogiwara, A., Oudega, B., Park, S.-H., Parro, V., Pohl, T. M., Portetelle, D., Porowollik, S., Prescott, A. M., Presecan, E., Pujic, P., Purnelle, B., Rapoport, G., Rey, M., Reynolds, S., Rieger, M., Rivolta, C., Rocha, E., Roche, B., Rose, M., Sadaie, Y., Sato, T., Scanlan, E., Schleich, S., Schroeter, R., Scoffone, F., Sekiguchi, J., Sekowska, A., Seror, S. J., Serror, P., Shin, B.-S., Soldo, B., Sorokin, A.,

Tacconi, E., Takagi, T., Takahashi, H., Takemaru, K., Takeuchi, M., Tamakoshi, A., Tanaka, T., Terpstra, P., Tognoni, A., Tosato, V., Uchiyama, S., Vandenbol, M., Vannier, F., Vassarotti, A., Viari, A., Wambutt, R., Wedler, E., Wedler, H., Weitzenegger, T., Winters, P., Wipat, A., Yamamoto, H., Yamane, K., Yasumoto, K., Yata, K., Yoshida, K., Yoshikawa, H.-F., Zumstein, E., Yoshikawa, H., and Danchin, A. (1997). The complete genome sequence of the gram-positive bacterium *Bacillus subtilis*. *Nature* 390:249–256.

106. Lai, W. C., and Hazelbauer, G. L. (2005). Carboxyl-terminal extensions beyond the conserved pentapeptide reduce rates of chemoreceptor adaptational modification. *J Bacteriol* 187:5115–5121.

107. Lamanna, A. C., Gestwicki, J. E., Strong, L. E., Borchardt, S. L., Owen, R. M., and Kiessling, L. L. (2002). Conserved amplification of chemotactic responses through chemoreceptor interactions. *J Bacteriol* 184:4981–4987.

108. Lamanna, A. C., Ordal, G. W., and Kiessling, L. L. (2005). Large increases in attractant concentration disrupt the polar localization of bacterial chemoreceptors. *Mol Microbiol* 57:774–785.

109. Le Moual, H., and Koshland, D. E. Jr. (1996). Molecular evolution of the C-terminal cytoplasmic domain of a superfamily of bacterial receptors involved in taxis. *J Mol Biol* 261:568–585.

110. Le Moual, H., Quang, T., and Koshland, D. E. Jr. (1997). Methylation of the *Escherichia coli* chemotaxis receptors: intra- and interdimer mechanisms. *Biochemistry* 36:13441–13448.

111. Le Moual, H., Quang, T., and Koshland, D. E., Jr. (1998). Conformational changes in the cytoplasmic domain of the *Escherichia coli* aspartate receptor upon adaptive methylation. *Biochemistry* 37:14852–14859.

112. Lee, B. M., Park, Y. J., Park, D. S., *et al.* (2005). The genome sequence of *Xanthomonas oryzae* pathovar *oryzae* KACC10331, the bacterial blight pathogen of rice. *Nucleic Acids Res* 33:577–586.

113. Lefman, J., Zhang, P., Hirai, T., Weis, R. M., Juliani, J., Bliss, D., Kessel, M., Bos, E., Peters, P. J., and Subramaniam, S. (2004). Three-dimensional electron microscopic imaging of membrane invaginations in *Escherichia coli* overproducing the chemotaxis receptor Tsr. *J Bacteriol* 186:5052–5061.

114. Letunic, I., Copley, R. R., Schmidt, S., Ciccarelli, F. D., Doerks, T., Schultz, J., Ponting, C. P., and Bork, P. (2004). SMART 4.0: towards genomic data integration. *Nucleic Acids Res* 32:D142–D144.

115. Levin, M. D., Shimizu, T. S., and Bray, D. (2002). Binding and diffusion of CheR molecules within a cluster of membrane receptors. *Biophys J* 82:1809–1817.

116. Levit, M. N., and Stock, J. B. (2002). Receptor methylation controls the magnitude of stimulus-response coupling in bacterial chemotaxis. *J Biol Chem* 277:36760–36765.

117. Li, G., and Weis, R. M. (2000). Covalent modification regulates ligand binding to receptor complexes in the chemosensory system of *Escherichia coli*. *Cell* 100:357–365.

118. Li, J., Li, G., and Weis, R. M. (1997). The serine chemoreceptor from *Escherichia coli* is methylated through an interdimer process. *Biochemistry* 36:11851–11857.

119. Li, J., Swanson, R. V., Simon, M. I., and Weis, R. M. (1995). The response regulators CheB and CheY exhibit competitive binding to the kinase CheA. *Biochemistry* 34:14626–14636.

120. Li, J., and Weis, R. M. (1996). Measurements of protein-protein interaction by isothermal titration calorimetry with applications to the bacterial chemotaxis system, in D.R. Marshak (ed.), *Techniques in Protein Chemistry* (vol. 7), pp. 34–47, Boston: Academic Press (Elsevier).

121. Li, M., and Hazelbauer, G. L. (2004). Cellular stoichiometry of the components of the chemotaxis signaling complex. *J Bacteriol* 186:3687–3694.

122. Li, M., and Hazelbauer, G. L. (2005). Adaptational assistance in clusters of bacterial chemoreceptors. *Mol Microbiol* 56:1617–1626.

123. Liberman, L., Berg, H. C., and Sourjik, V. (2004). Effect of chemoreceptor modification on assembly and activity of the receptor-kinase complex in *Escherichia coli*. *J Bacteriol* 186:6643–6646.

124. Linding, R., Jensen, L. J., Diella, F., Bork, P., Gibson, T. J., and Russell, R. B. (2003). Protein disorder prediction: implications for structural proteomics. *Structure (Camb)* 11:1453–1459.

125. Liu, Y., Levit, M., Lurz, R., Surette, M. G., and Stock, J. B. (1997). Receptor-mediated protein kinase activation and the mechanism of transmembrane signaling in bacterial chemotaxis. *Embo J* 16:7231–7240.

126. Lybarger, S. R., Nair, U., Lilly, A. A., Hazelbauer, G. L., and Maddock, J. R. (2005). Clustering requires modified methyl-accepting sites in low-abundance but not high-abundance chemoreceptors of *Escherichia coli*. *Mol Microbiol* 56:1078–1086.

127. Maddock, J. R., and Shapiro, L. (1993). Polar location of the chemoreceptor complex in the *Escherichia coli* cell. *Science* 259:1717–1723.

128. Mammen, M., Choi, S. K., and Whitesides, G. M. (1998). Polyvalent interactions in biological systems: Implications for design and use of multivalent ligands and inhibitors. *Angewandte Chemie-International Edition* 37:2755–2794.

129. Marchler-Bauer, A., and Bryant, S. H. (2004). CD-Search: protein domain annotations on the fly. *Nucleic Acids Res* 32:W327–331.

130. McCleary, W. R., McBride, M. J., and Zusman, D. R. (1990). Developmental sensory transduction in *Myxococcus xanthus* involves methylation and demethylation of FrzCD. *J Bacteriol* 172:4877–4887.

131. McClelland, M., Sanderson, K. E., Clifton, S. W., Latreille, P., Porwollik, S., Sabo, A., Meyer, R., Bieri, T., Ozersky, P., McLellan, M., Harkins, C. R., Wang, C., Nguyen, C., Berghoff, A., Elliott, G., Kohlberg, S., Strong, C., Du, F., Carter, J., Kremizki, C., Layman, D., Leonard, S., Sun, H., Fulton, L., Nash, W., Miner, T., Minx, P., Delehaunty, K., Fronick, C., Magrini, V., Nhan, M., Warren, W., Florea, L., Spieth, J., and Wilson, R. K. (2004). Comparison of genome degradation in Paratyphi A and Typhi, human-restricted serovars of *Salmonella enterica* that cause typhoid. *Nat Genet* 36:1268–1274.

132. McClelland, M., Sanderson, K. E., Spieth, J., Clifton, S. W., Latreille, P., Courtney, L., Porwollik, S., Ali, J., Dante, M., Du, F., Hou, S., Layman, D., Leonard, S., Nguyen, C., Scott, K., Holmes, A., Grewal, N., Mulvaney, E., Ryan, E., Sun, H., Florea, L., Miller, W., Stoneking, T., Nhan, M., Waterston, R., and Wilson, R. K. (2001). Complete genome sequence of *Salmonella enterica* serovar Typhimurium LT2. *Nature* 413:852–856.

133. Mello, B. A., Shaw, L., and Tu, Y. (2004). Effects of receptor interaction in bacterial chemotaxis. *Biophys J* 87:1578–1595.

134. Mello, B. A., and Tu, Y. (2003). Quantitative modeling of sensitivity in bacterial chemotaxis: The role of coupling among different chemoreceptor species. *Proc Natl Acad Sci USA* 100:8223–8228.

135. Methe, B. A., Nelson, K. E., Eisen, J. A., Paulsen, I. T., Nelson, W., Heidelberg, J. F., Wu, D., Wu, M., Ward, N., Beanan, M. J., Dodson, R. J., Madupu, R., Brinkac, L. M., Daugherty, S. C., DeBoy, R. T., Durkin, A. S., Gwinn, M., Kolonay, J. F., Sullivan, S. A., Haft, D. H., Selengut, J., Davidsen, T. M., Zafar, N., White, O., Tran, B., Romero, C., Forberger, H. A., Weidman, J., Khouri, H., Feldblyum, T. V., Utterback, T. R., Van Aken, S. E., Lovley, D. R., and Fraser, C. M. (2003). Genome of *Geobacter sulfurreducens*: metal reduction in subsurface environments. *Science* 302:1967–1969.

136. Milburn, M. V., Prive, G. G., Milligan, D. L., Scott, W. G., Yeh, J., Jancarik, J., Koshland, D. E., Jr., and Kim, S. H. (1991). Three-dimensional structures of the ligand-binding domain of the bacterial aspartate receptor with and without a ligand. *Science* 254:1342–1347.

137. Miller, A. F., and Falke, J. J. (2004). Chemotaxis receptors and signaling. *Adv Protein Chem* 68:393–444.

138. Miller, A. S., and Falke, J. J. (2004). Side chains at the membrane-water interface modulate the signaling state of a transmembrane receptor. *Biochemistry* 43:1763–1770.

139. Milligan, D. L., and Koshland, D. E. Jr. (1988). Site-directed cross-linking. Establishing the dimeric structure of the aspartate receptor of bacterial chemotaxis. *J Biol Chem* 263:6268–6275.

140. Murphy, O. J., 3rd, Yi, X., Weis, R. M., and Thompson, L. K. (2001). Hydrogen exchange reveals a stable and expandable core within the aspartate receptor cytoplasmic domain. *J Biol Chem* 276:43262–43269.

141. Nambu, J. R., Lewis, J. O., Wharton, K. A., Jr., and Crews, S. T. (1991). The *Drosophila* single-minded gene encodes a helix-loop-helix protein that acts as a master regulator of CNS midline development. *Cell* 67:1157–1167.

142. Nascimento, A. L., Ko, A. I., Martins, E. A., Monteiro-Vitorello, C. B., Ho, P. L., Haake, D. A., Verjovski-Almeida, S., Hartskeerl, R. A., Marques, M. V., Oliveira, M. C., Menck, C. F., Leite, L. C., Carrer, H., Coutinho, L. L., Degrave, W. M., Dellagostin, O. A., El-Dorry, H., Ferro, E. S., Ferro, M. I., Furlan, L. R., Gamberini, M., Giglioti, E. A., Goes-Neto, A., Goldman, G. H., Goldman, M. H., Harakava, R., Jeronimo, S. M., Junqueira-de-Azevedo, I. L., Kimura, E. T., Kuramae, E. E., Lemos, E. G., Lemos, M. V., Marino, C. L., Nunes, L. R., de Oliveira, R. C., Pereira, G. G., Reis, M. S., Schriefer, A., Siqueira, W. J., Sommer, P., Tsai, S. M., Simpson, A. J., Ferro, J. A., Camargo, L. E., Kitajima, J. P., Setubal, J. C., and Van Sluys, M. A. (2004). Comparative genomics of two *Leptospira interrogans* serovars reveals novel insights into physiology and pathogenesis. *J Bacteriol* 186:2164–2172.

143. Nierman, W. C., DeShazer, D., Kim, H. S., Tettelin, H., Nelson, K. E., Feldblyum, T., Ulrich, R. L., Ronning, C. M., Brinkac, L. M., Daugherty, S. C., Davidsen, T. D., Deboy, R. T., Dimitrov, G., Dodson, R. J., Durkin, A. S., Gwinn, M. L., Haft, D. H., Khouri, H., Kolonay, J. F., Madupu, R., Mohammoud, Y., Nelson, W. C., Radune, D., Romero, C. M., Sarria, S., Selengut, J., Shamblin, C., Sullivan, S. A., White, O., Yu, Y., Zafar, N., Zhou, L., and Fraser, C. M. (2004). Structural flexibility in the *Burkholderia mallei* genome. *Proc Natl Acad Sci USA* 101:14246–14251.

144. Nierman, W. C., Feldblyum, T. V., Laub, M. T., Paulsen, I. T., Nelson, K. E., Eisen, J. A., Heidelberg, J. F., Alley, M. R., Ohta, N., Maddock, J. R., Potocka, I., Nelson, W. C., Newton, A., Stephens, C., Phadke, N. D., Ely, B., DeBoy, R. T., Dodson, R. J., Durkin, A. S., Gwinn, M. L., Haft, D. H., Kolonay, J. F., Smit, J., Craven, M. B., Khouri, H., Shetty, J., Berry, K., Utterback, T., Tran, K., Wolf, A., Vamathevan, J., Ermolaeva, M., White, O., Salzberg, S. L., Venter, J. C., Shapiro, L., and Fraser, C. M. (2001). Complete genome sequence of *Caulobacter crescentus*. *Proc Natl Acad Sci USA* 98:4136–4141.

145. Nolling, J., Breton, G., Omelchenko, M. V., Makarova, K. S., Zeng, Q., Gibson, R., Lee, H. M., Dubois, J., Qiu, D., Hitti, J., Wolf, Y. I., Tatusov, R. L., Sabathe, F., Doucette-Stamm, L., Soucaille, P., Daly, M. J., Bennett, G. N., Koonin, E. V., and Smith, D. R. (2001). Genome sequence and comparative analysis of the solvent-producing bacterium *Clostridium acetobutylicum*. *J Bacteriol* 183:4823–4838.

146. O'Connor, C., and Matsumura, P. (2004). The accessibility of cys-120 in CheA(S) is important for the binding of CheZ and enhancement of CheZ phosphatase activity. *Biochemistry* 43:6909–6916.

147. Parkhill, J., Dougan, G., James, K. D., Thomson, N. R., Pickard, D., Wain, J., Churcher, C., Mungall, K. L., Bentley, S. D., Holden, M. T., Sebaihia, M., Baker, S., Basham, D., Brooks, K., Chillingworth, T., Connerton, P., Cronin, A., Davis, P., Davies, R. M., Dowd, L., White, N., Farrar, J., Feltwell, T., Hamlin, N., Haque, A., Hien, T. T., Holroyd, S., Jagels, K., Krogh, A., Larsen, T. S., Leather, S., Moule, S., O'Gaora, P., Parry, C., Quail, M., Rutherford, K., Simmonds, M., Skelton, J., Stevens, K., Whitehead, S., and Barrell, B. G. (2001). Complete genome sequence of a multiple drug resistant *Salmonella enterica* serovar Typhi CT18. *Nature* 413:848–852.

148. Parkhill, J., Sebaihia, M., Preston, A., Murphy, L. D., Thomson, N., Harris, D. E., Holden, M. T., Churcher, C. M., Bentley, S. D., Mungall, K. L., Cerdeno-Tarraga, A. M., Temple, L., James, K., Harris, B., Quail, M. A., Achtman, M., Atkin, R., Baker, S., Basham, D., Bason, N., Cherevach, I., Chillingworth, T., Collins, M., Cronin, A., Davis, P., Doggett, J., Feltwell, T., Goble, A., Hamlin, N., Hauser, H., Holroyd, S., Jagels, K., Leather, S., Moule, S., Norberczak, H., O'Neil, S., Ormond, D., Price, C., Rabbinowitsch, E., Rutter, S., Sanders, M., Saunders, D., Seeger, K., Sharp, S., Simmonds, M., Skelton, J., Squares, R., Squares, S., Stevens, K., Unwin, L., Whitehead, S., Barrell, B. G., and Maskell, D. J. (2003). Comparative analysis of the genome sequences of *Bordetella pertussis, Bordetella parapertussis* and *Bordetella bronchiseptica. Nat Genet* 35:32–40.

149. Parkhill, J., Wren, B. W., Thomson, N. R., Titball, R. W., Holden, M. T., Prentice, M. B., Sebaihia, M., James, K. D., Churcher, C., Mungall, K. L., Baker, S., Basham, D., Bentley, S. D., Brooks, K., Cerdeno-Tarraga, A. M., Chillingworth, T., Cronin, A., Davies, R. M., Davis, P., Dougan, G., Feltwell, T., Hamlin, N., Holroyd, S., Jagels, K., Karlyshev, A. V., Leather, S., Moule, S., Oyston, P. C., Quail, M., Rutherford, K., Simmonds, M., Skelton, J., Stevens, K., Whitehead, S., and Barrell, B. G. (2001). Genome sequence of *Yersinia pestis,* the causative agent of plague. *Nature* 413:523–527.

150. Parkinson, J. S., Ames, P., and Studdert, C. A. (2005). Collaborative signaling by bacterial chemoreceptors. *Curr Opin Microbiol* 8:116–121.

151. Perez, E., West, A. H., Stock, A. M., and Djordjevic, S. (2004). Discrimination between different methylation states of chemotaxis receptor Tar by receptor methyltransferase CheR. *Biochemistry* 43:953–961.

152. Perna, N. T., Plunkett, G., 3rd, Burland, V., Mau, B., Glasner, J. D., Rose, D. J., Mayhew, G. F., Evans, P. S., Gregor, J., Kirkpatrick, H. A., Posfai, G., Hackett, J., Klink, S., Boutin, A., Shao, Y., Miller, L., Grotbeck, E. J., Davis, N. W., Lim, A., Dimalanta, E. T., Potamousis, K. D., Apodaca, J., Anantharaman, T. S., Lin, J., Yen, G., Schwartz, D. C., Welch, R. A., and Blattner, F. R. (2001). Genome sequence of enterohaemorrhagic *Escherichia coli* O157:H7. *Nature* 409:529–533.

153. Pittman, M. S., Goodwin, M., and Kelly, D. J. (2001). Chemotaxis in the human gastric pathogen *Helicobacter pylori:* different roles for CheW and the three CheV paralogues, and evidence for CheV2 phosphorylation. *Microbiology* 147:2493–2504.

154. Ponting, C. P., and Aravind, L. (1997). PAS: a multifunctional domain family comes to light. *Curr Biol* 7:R674–677.

155. Prust, C., Hoffmeister, M., Liesegang, H., Wiezer, A., Fricke, W. F., Ehrenreich, A., Gottschalk, G., and Deppenmeier, U. (2005). Complete genome sequence of the acetic acid bacterium *Gluconobacter oxydans. Nat Biotechnol* 23:195–200.

156. Qian, W., Jia, Y., Ren, S. X., He, Y. Q., Feng, J. X., Lu, L. F., Sun, Q., Ying, G., Tang, D. J., Tang, H., Wu, W., Hao, P., Wang, L., Jiang, B. L., Zeng, S., Gu, W. Y., Lu, G., Rong, L., Tian, Y., Yao, Z., Fu, G., Chen, B., Fang, R., Qiang, B., Chen, Z., Zhao, G. P., Tang, J. L., and He, C. (2005). Comparative and functional genomic analyses of the pathogenicity of phytopathogen *Xanthomonas campestris* pv. *campestris. Genome Res* 15:757–767.

157. Rao, C. V., Kirby, J. R., and Arkin, A. P. (2004). Design and diversity in bacterial chemotaxis: a comparative study in *Escherichia coli* and *Bacillus subtilis. PLoS Biol* 2:E49.

158. Rao, J. H., Lahiri, J., Weis, R. M., and Whitesides, G. M. (2000). Design, synthesis, and characterization of a high-affinity trivalent system derived from vancomycin and L-Lys-D-Ala-D-Ala. *Journal of the American Chemical Society* 122:2698–2710.

159. Reader, R. W., Tso, W. W., Springer, M. S., Goy, M. F., and Adler, J. (1979). Pleiotropic aspartate taxis and serine taxis mutants of *Escherichia coli. J Gen Microbiol* 111:363–374.

160. Rebbapragada, A., Johnson, M. S., Harding, G. P., Zuccarelli, A. J., Fletcher, H. M., Zhulin, I. B., and Taylor, B. L. (1997). The Aer protein and the serine chemoreceptor Tsr independently

sense intracellular energy levels and transduce oxygen, redox, and energy signals for *Escherichia coli* behavior. *Proc Natl Acad Sci USA* 94:10541–10546.

161. Ren, S. X., Fu, G., Jiang, X. G., Zeng, R., Miao, Y. G., Xu, H., Zhang, Y. X., Xiong, H., Lu, G., Lu, L. F., Jiang, H. Q., Jia, J., Tu, Y. F., Jiang, J. X., Gu, W. Y., Zhang, Y. Q., Cai, Z., Sheng, H. H., Yin, H. F., Zhang, Y., Zhu, G. F., Wan, M., Huang, H. L., Qian, Z., Wang, S. Y., Ma, W., Yao, Z. J., Shen, Y., Qiang, B. Q., Xia, Q. C., Guo, X. K., Danchin, A., Saint Girons, I., Somerville, R. L., Wen, Y. M., Shi, M. H., Chen, Z., Xu, J. G., and Zhao, G. P. (2003). Unique physiological and pathogenic features of *Leptospira interrogans* revealed by whole-genome sequencing. *Nature* 422:888–893.

162. Rendulic, S., Jagtap, P., Rosinus, A., Eppinger, M., Baar, C., Lanz, C., Keller, H., Lambert, C., Evans, K. J., Goesmann, A., Meyer, F., Sockett, R. E., and Schuster, S. C. (2004). A predator unmasked: life cycle of *Bdellovibrio bacteriovorus* from a genomic perspective. *Science* 303:689–692.

163. Rice, M. S., and Dahlquist, F. W. (1991). Sites of deamidation and methylation in Tsr, a bacterial chemotaxis sensory transducer. *J Biol Chem* 266:9746–9753.

164. Russo, A. F., and Koshland, D. E. Jr. (1983). Separation of signal transduction and adaptation functions of the aspartate receptor in bacterial sensing. *Science* 220:1016–1020.

165. Salanoubat, M., Genin, S., Artiguenave, F., Gouzy, J., Mangenot, S., Arlat, M., Billault, A., Brottier, P., Camus, J. C., Cattolico, L., Chandler, M., Choisne, N., Claudel-Renard, C., Cunnac, S., Demange, N., Gaspin, C., Lavie, M., Moisan, A., Robert, C., Saurin, W., Schiex, T., Siguier, P., Thebault, P., Whalen, M., Wincker, P., Levy, M., Weissenbach, J., and Boucher, C. A. (2002). Genome sequence of the plant pathogen *Ralstonia solanacearum*. *Nature* 415:497–502.

166. Sanders, D. A., and Koshland, D. E. Jr. (1988). Receptor interactions through phosphorylation and methylation pathways in bacterial chemotaxis. *Proc Natl Acad Sci USA* 85:8425–8429.

167. Schubert, H. L., Blumenthal, R. M., and Cheng, X. (2003). Many paths to methyltransfer: a chronicle of convergence. *Trends Biochem Sci* 28:329–335.

168. Schuster, S. C., Swanson, R. V., Alex, L. A., Bourret, R. B., and Simon, M. I. (1993). Assembly and function of a quaternary signal transduction complex monitored by surface plasmon resonance. *Nature* 365:343–347.

169. Seeley, S. K., Weis, R. M., and Thompson, L. K. (1996). The cytoplasmic fragment of the aspartate receptor displays globally dynamic behavior. *Biochemistry* 35:5199–5206.

170. Segall, J. E., Block, S. M., and Berg, H. C. (1986). Temporal comparisons in bacterial chemotaxis. *Proc Natl Acad Sci USA* 83:8987–8991.

171. Seo, J. S., Chong, H., Park, H. S., Yoon, K. O., Jung, C., Kim, J. J., Hong, J. H., Kim, H., Kim, J. H., Kil, J. I., Park, C. J., Oh, H. M., Lee, J. S., Jin, S. J., Um, H. W., Lee, H. J., Oh, S. J., Kim, J. Y., Kang, H. L., Lee, S. Y., Lee, K. J., and Kang, H. S. (2005). The genome sequence of the ethanologenic bacterium *Zymomonas mobilis* ZM4. *Nat Biotechnol* 23:63–68.

172. Shapiro, M. J., and Koshland, D. E. Jr. (1994). Mutagenic studies of the interaction between the aspartate receptor and methyltransferase from *Escherichia coli*. *J Biol Chem* 269:11054–11059.

173. Shapiro, M. J., Panomitros, D., and Koshland, D. E. Jr. (1995). Interactions between the methylation sites of the *Escherichia coli* aspartate receptor mediated by the methyltransferase. *J Biol Chem* 270:751–755.

174. Shimizu, T. S., Aksenov, S. V., and Bray, D. (2003). A spatially extended stochastic model of the bacterial chemotaxis signalling pathway. *J Mol Biol* 329:291–309.

175. Shimizu, T. S., Le Novere, N., Levin, M. D., Beavil, A. J., Sutton, B. J., and Bray, D. (2000). Molecular model of a lattice of signalling proteins involved in bacterial chemotaxis. *Nat Cell Biol* 2:792–796.

176. Shiomi, D., Homma, M., and Kawagishi, I. (2002). Intragenic suppressors of a mutation in the aspartate chemoreceptor gene that abolishes binding of the receptor to methyltransferase. *Microbiology* 148:3265–3275.

177. Shiomi, D., Okumura, H., Homma, M., and Kawagishi, I. (2000). The aspartate chemoreceptor Tar is effectively methylated by binding to the methyltransferase mainly through hydrophobic interaction. *Mol Microbiol* 36:132–140.

178. Shiomi, D., Zhulin, I. B., Homma, M., and Kawagishi, I. (2002). Dual recognition of the bacterial chemoreceptor by chemotaxis-specific domains of the CheR methyltransferase. *J Biol Chem* 277:42325–42333.

179. Shrout, A. L., Montefusco, D. J., and Weis, R. M. (2003). Template-directed assembly of receptor signaling complexes. *Biochemistry* 42:13379–13385.

180. Silva, G., Oliveira, S., Gomes, C. M., Pacheco, I., Liu, M. Y., Xavier, A. V., Teixeira, M., Legall, J., and Rodrigues-pousada, C. (1999). *Desulfovibrio gigas* neelaredoxin. A novel superoxide dismutase integrated in a putative oxygen sensory operon of an anaerobe. *Eur J Biochem* 259:235–243.

181. Silversmith, R. E. (2005). High mobility of carboxyl-terminal region of bacterial chemotaxis phosphatase CheZ is diminished upon binding divalent cation or CheY-P substrate. *Biochemistry* 44:7768–7776.

182. Simms, S. A., and Subbaramaiah, K. (1991). The kinetic mechanism of S-adenosyl-L-methionine: glutamylmethyltransferase from *Salmonella typhimurium*. *J Biol Chem* 266:12741–12746.

183. Song, Y., Tong, Z., Wang, J., Wang, L., Guo, Z., Han, Y., Zhang, J., Pei, D., Zhou, D., Qin, H., Pang, X., Han, Y., Zhai, J., Li, M., Cui, B., Qi, Z., Jin, L., Dai, R., Chen, F., Li, S., Ye, C., Du, Z., Lin, W., Wang, J., Yu, J., Yang, H., Wang, J., Huang, P., and Yang, R. (2004). Complete genome sequence of *Yersinia pestis* strain 91001, an isolate avirulent to humans. *DNA Res* 11:179–197.

184. Sourjik, V. (2004). Receptor clustering and signal processing in *E. coli* chemotaxis. *Trends Microbiol* 12:569–576.

185. Sourjik, V., and Berg, H. C. (2000). Localization of components of the chemotaxis machinery of *Escherichia coli* using fluorescent protein fusions. *Mol Microbiol* 37:740–751.

186. Sourjik, V., and Berg, H. C. (2002). Receptor sensitivity in bacterial chemotaxis. *Proc Natl Acad Sci USA* 99:123–127.

187. Sourjik, V., and Berg, H. C. (2004). Functional interactions between receptors in bacterial chemotaxis. *Nature* 428:437–441.

188. Springer, M. S., Goy, M. F., and Adler, J. (1979). Protein methylation in behavioural control mechanisms and in signal transduction. *Nature* 280:279–284.

189. Springer, W. R., and Koshland, D. E. Jr. (1977). Identification of a protein methyltransferase as the *cheR* gene product in the bacterial sensing system. *Proc Natl Acad Sci USA* 74:533–537.

190. Starrett, D. J., and Falke, J. J. (2005). Adaptation mechanism of the aspartate receptor: electrostatics of the adaptation subdomain plays a key role in modulating kinase activity. *Biochemistry* 44:1550–1560.

191. Stewart, R. C., and Dahlquist, F. W. (1987). Molecular-components of bacterial chemotaxis. *Chemical Reviews* 87:997–1025.

192. Stock, A. M., Mottonen, J. M., Stock, J. B., and Schutt, C. E. (1989). Three-dimensional structure of CheY, the response regulator of bacterial chemotaxis. *Nature* 337:745–749.

193. Stock, A. M., Robinson, V. L., and Goudreau, P. N. (2000). Two-component signal transduction. *Annu Rev Biochem* 69:183–215.

194. Stock, J., and Simms, S. (1988). Methylation, demethylation, and deamidation at glutamate residues in membrane chemoreceptor proteins. *Adv Exp Med Biol* 231:201–212.

195. Storch, K. F., Rudolph, J., and Oesterhelt, D. (1999). Car: a cytoplasmic sensor responsible for arginine chemotaxis in the archaeon *Halobacterium salinarum*. *Embo Journal* 18:1146–1158.

196. Stover, C. K., Pham, X. Q., Erwin, A. L., Mizoguchi, S. D., Warrener, P., Hickey, M. J., Brinkman, F. S., Hufnagle, W. O., Kowalik, D. J., Lagrou, M., Garber, R. L., Goltry, L., Tolentino, E., Westbrock-Wadman, S., Yuan, Y., Brody, L. L., Coulter, S. N., Folger, K. R., Kas, A., Larbig, K., Lim, R., Smith, K., Spencer, D., Wong, G. K., Wu, Z., Paulsen, I. T.,

Reizer, J., Saier, M. H., Hancock, R. E., Lory, S., and Olson, M. V. (2000). Complete genome sequence of *Pseudomonas aeruginosa* PA01, an opportunistic pathogen. *Nature* 406:959–964.

197. Studdert, C. A., and Parkinson, J. S. (2004). Crosslinking snapshots of bacterial chemoreceptor squads. *Proc Natl Acad Sci USA* 101:2117–2122.

198. Szurmant, H., and Ordal, G. W. (2004). Diversity in chemotaxis mechanisms among the *Bacteria* and *Archaea*. *Microbiol Mol Biol Rev* 68:301–319.

199. Terry, K., Williams, S. M., Connolly, L., and Ottemann, K. M. (2005). Chemotaxis plays multiple roles during *Helicobacter pylori* animal infection. *Infect Immun* 73:803–811.

200. Terwilliger, T. C., Bogonez, E., Wang, E. A., and Koshland, D. E. Jr. (1983). Sites of methyl esterification on the aspartate receptor involved in bacterial chemotaxis. *J Biol Chem* 258:9608–9611.

201. Terwilliger, T. C., and Koshland, D. E. Jr. (1984). Sites of methyl esterification and deamination on the aspartate receptor involved in chemotaxis. *J Biol Chem* 259:7719–7725.

202. Terwilliger, T. C., Wang, J. Y., and Koshland, D. E. Jr. (1986). Surface structure recognized for covalent modification of the aspartate receptor in chemotaxis. *Proc Natl Acad Sci USA* 83:6707–6710.

203. Tomomori, C., Tanaka, T., Dutta, R., Park, H., Saha, S. K., Zhu, Y., Ishima, R., Liu, D., Tong, K. I., Kurokawa, H., Qian, H., Inouye, M., and Ikura, M. (1999). Solution structure of the homodimeric core domain of *Escherichia coli* histidine kinase EnvZ. *Nat Struct Biol* 6:729–734.

204. Wadhams, G. H., and Armitage, J. P. (2004). Making sense of it all: bacterial chemotaxis. *Nat Rev Mol Cell Biol* 5:1024–1037.

205. Wadhams, G. H., Martin, A. C., Porter, S. L., Maddock, J. R., Mantotta, J. C., King, H. M., and Armitage, J. P. (2002). TlpC, a novel chemotaxis protein in *Rhodobacter sphaeroides*, localizes to a discrete region in the cytoplasm. *Mol Microbiol* 46:1211–1221.

206. Wadhams, G. H., Martin, A. C., Warren, A. V., and Armitage, J. P. (2005). Requirements for chemotaxis protein localization in *Rhodobacter sphaeroides*. *Mol Microbiol* 58:895–902.

207. Webre, D. J., Wolanin, P. M., and Stock, J. B. (2004). Modulated receptor interactions in bacterial transmembrane signaling. *Trends Cell Biol* 14:478–482.

208. NCBI Genome Project Website. *http://www.ncbi.nlm.nih.gov/entrez/query.fcgi?CMD= search&DB=genomeprj.*

209. Weerasuriya, S., Schneider, B. M., and Manson, M. D. (1998). Chimeric chemoreceptors in *Escherichia coli*: signaling properties of Tar-Tap and Tap-Tar hybrids. *J Bacteriol* 180:914–920.

210. Wei, J., Goldberg, M. B., Burland, V., Venkatesan, M. M., Deng, W., Fournier, G., Mayhew, G. F., Plunkett, G., 3rd, Rose, D. J., Darling, A., Mau, B., Perna, N. T., Payne, S. M., Runyen-Janecky, L. J., Zhou, S., Schwartz, D. C., and Blattner, F. R. (2003). Complete genome sequence and comparative genomics of *Shigella flexneri* serotype 2a strain 2457T. *Infect Immun* 71:2775–2786.

211. Weis, R. M., Chasalow, S., and Koshland, D. E. Jr. (1990). The role of methylation in chemotaxis. An explanation of outstanding anomalies. *J Biol Chem* 265:6817–6826.

212. Weis, R. M., Hirai, T., Chalah, A., Kessel, M., Peters, P. J., and Subramaniam, S. (2003). Electron microscopic analysis of membrane assemblies formed by the bacterial chemotaxis receptor Tsr. *J Bacteriol* 185:3636–3643.

213. Weis, R. M., and Koshland, D. E., Jr. (1988). Reversible receptor methylation is essential for normal chemotaxis of *Escherichia coli* in gradients of aspartic acid. *Proc Natl Acad Sci USA* 85:83–87.

214. Welch, R. A., Burland, V., Plunkett, G., 3rd, Redford, P., Roesch, P., Rasko, D., Buckles, E. L., Liou, S. R., Boutin, A., Hackett, J., Stroud, D., Mayhew, G. F., Rose, D. J., Zhou, S., Schwartz, D. C., Perna, N. T., Mobley, H. L., Donnenberg, M. S., and Blattner, F. R. (2002). Extensive mosaic structure revealed by the complete genome sequence of uropathogenic *Escherichia coli*. *Proc Natl Acad Sci USA* 99:17020–17024.

215. Winston, S. E., Mehan, R., and Falke, J. J. (2005). Evidence that the adaptation region of the aspartate receptor is a dynamic four-helix bundle: cysteine and disulfide scanning studies. *Biochemistry* 44:12655–12666.

216. Wolanin, P. M., Thomason, P. A., and Stock, J. B. (2002). Histidine protein kinases: key signal transducers outside the animal kingdom. *Genome Biol* 3:Reviews 3013.

217. Wood, D. W., Setubal, J. C., Kaul, R., Monks, D. E., Kitajima, J. P., Okura, V. K., Zhou, Y., Chen, L., Wood, G. E., Almeida, N. F., Jr., Woo, L., Chen, Y., Paulsen, I. T., Eisen, J. A., Karp, P. D., Bovee, D., Sr., Chapman, P., Clendenning, J., Deatherage, G., Gillet, W., Grant, C., Kutyavin, T., Levy, R., Li, M. J., McClelland, E., Palmieri, A., Raymond, C., Rouse, G., Saenphimmachak, C., Wu, Z., Romero, P., Gordon, D., Zhang, S., Yoo, H., Tao, Y., Biddle, P., Jung, M., Krespan, W., Perry, M., Gordon-Kamm, B., Liao, L., Kim, S., Hendrick, C., Zhao, Z. Y., Dolan, M., Chumley, F., Tingey, S. V., Tomb, J. F., Gordon, M. P., Olson, M. V., and Nester, E. W. (2001). The genome of the natural genetic engineer *Agrobacterium tumefaciens* C58. *Science* 294:2317–2323.

218. Woodard, R. W., Tsai, M. D., Floss, H. G., Crooks, P. A., and Coward, J. K. (1980). Stereochemical course of the transmethylation catalyzed by catechol O-methyltransferase. *J Biol Chem* 255:9124–9127.

219. Wu, J., Li, J., Li, G., Long, D. G., and Weis, R. M. (1996). The receptor binding site for the methyltransferase of bacterial chemotaxis is distinct from the sites of methylation. *Biochemistry* 35:4984–4993.

220. Wu, J., Long, D. G., and Weis, R. M. (1995). Reversible dissociation and unfolding of the *Escherichia coli* aspartate receptor cytoplasmic fragment. *Biochemistry* 34:3056–3065.

221. Wylie, D., Stock, A., Wong, C. Y., and Stock, J. (1988). Sensory transduction in bacterial chemotaxis involves phosphotransfer between Che proteins. *Biochem Biophys Res Commun* 151:891–896.

222. Xiong, J., Kurtz, D. M., Jr., Ai, J., and Sanders-Loehr, J. (2000). A hemerythrin-like domain in a bacterial chemotaxis protein. *Biochemistry* 39:5117–5125.

223. Yamamoto, K., and Imae, Y. (1993). Cloning and characterization of the *Salmonella typhimurium*-specific chemoreceptor Tcp for taxis to citrate and from phenol. *Proc Natl Acad Sci USA* 90:217–221.

224. Yao, V. J., and Spudich, J. L. (1992). Primary structure of an archaebacterial transducer, a methyl-accepting protein associated with sensory rhodopsin I. *Proc Natl Acad Sci USA* 89:11915–11919.

225. Yeh, J. I., Biemann, H. P., Pandit, J., Koshland, D. E., and Kim, S. H. (1993). The three-dimensional structure of the ligand-binding domain of a wild-type bacterial chemotaxis receptor. Structural comparison to the cross-linked mutant forms and conformational changes upon ligand binding. *J Biol Chem* 268:9787–9792.

226. Yi, X., and Weis, R. M. (2002). The receptor docking segment and s-adenosyl-L-homocysteine bind independently to the methyltransferase of bacterial chemotaxis. *Biochim Biophys Acta* 1596:28–35.

227. Zhang, P., Bos, E., Heymann, J., Gnaegi, H., Kessel, M., Peters, P. J., and Subramaniam, S. (2004). Direct visualization of receptor arrays in frozen-hydrated sections and plunge-frozen specimens of *E. coli* engineered to overproduce the chemotaxis receptor Tsr. *J Microsc* 216:76–83.

228. Zhang, W., and Phillips, G. N. Jr. (2003). Structure of the oxygen sensor in *Bacillus subtilis*: signal transduction of chemotaxis by control of symmetry. *Structure (Camb)* 11:1097–1110.

229. Zhulin, I. B., Nikolskaya, A. N., and Galperin, M. Y. (2003). Common extracellular sensory domains in transmembrane receptors for diverse signal transduction pathways in *Bacteria* and *Archaea*. *J Bacteriol* 185:285–294.

230. Zhulin, I. B., Taylor, B. L., and Dixon, R. (1997). PAS domain S-boxes in *Archaea*, *Bacteria* and sensors for oxygen and redox. *Trends Biochem Sci* 22:331–333.

Part IV

Recognition of Damaged Proteins in Aging by Protein Methyltransferases

13

Protein L-Isoaspartyl, D-Aspartyl O-Methyltransferases: Catalysts for Protein Repair

CLARE M. O'CONNOR

Biology Department
Boston College
140 Commonwealth Avenue
Chestnut Hill, MA 02467, USA

I. Abstract

Protein L-isoaspartyl, D-aspartyl *O*-methyltransferases (PIMTs) are ancient enzymes distributed through all phylogenetic domains. PIMTs catalyze the methylation of L-isoaspartyl, and to a lesser extent D-aspartyl, residues arising from the spontaneous deamidation and isomerization of protein asparaginyl and aspartyl residues. PIMTs catalyze the methylation of isoaspartyl residues in a large number of primary sequence configurations, which accounts for the broad specificity of the enzyme for protein substrates both *in vitro* and *in vivo*.

PIMT-catalyzed methylation of isoaspartyl substrates initiates the repair of the polypeptide backbone in its damaged substrates by a spontaneous mechanism that involves a succinimidyl intermediate. The repair process catalyzed by PIMTs is not completely efficient, however, leaving open the possibility that unidentified enzymatic activities cooperate with PIMT in the repair process. Structurally, PIMTs are members of the class I family of AdoMet-dependent methyltransferases. PIMTs have a unique topological arrangement of strands in the central β sheet that provides

THE ENZYMES, Vol. XXIV

a signature for this class of enzymes. The regulation and physiological significance of PIMT has been studied in several model organisms.

PIMTs are constitutively synthesized by cells, but they can be upregulated in response to conditions that are potentially damaging to protein structures, or when proteins are stored for prolonged periods of time. Disruption of PIMT genes in bacteria and simple eukaryotes produces subtle phenotypes that are apparent only under stress. Loss of PIMT function in transgenic mice leads to fatal epilepsy, suggesting that PIMT function is particularly important to neurons in mammals.

II. Introduction

Due to the inherent chemical reactivities of amino acid side chains, proteins are subject to a variety of spontaneous modifications that can negatively affect the functionality of a protein. These modifications include oxidative damage [1, 2], deamidation [3], and racemization [4] reactions. If not metabolized, the abnormal products of these reactions can accumulate over time in long-lived proteins. Therefore, cells use both proteolytic systems and repair enzymes to reduce the burden of structurally damaged proteins. Much of the damage incurred by proteins is irreparable, and the destruction of damaged proteins by proteolytic systems is well established.

By contrast, only a very small number of protein repair enzymes have been identified. The focus of this chapter is one of these repair enzymes, the protein L-isoaspartyl/D-aspartyl methyltransferase (PIMT[1]: E.C.2.1.1.77) that catalyzes the S-adenosyl-L-methionine (AdoMet)-dependent methylation of L-isoaspartyl (L-isoAsp), and D-aspartyl (D-Asp) residues in age-damaged proteins. The goal of this chapter is to review the biochemistry of PIMT-catalyzed methylation, structural studies of PIMT, the distribution and regulation of PIMT activities in living organisms, and insights into the biological significance of protein isoAsp methylation gained from transgenic models. In addition to articles that focus on PIMT, the review includes information obtained from high-throughput genomic studies. Because of the large number of published manuscripts, it has not been possible to include all of the studies whose results have contributed to our current understanding of PIMT function. Many questions remain unanswered with respect to PIMT function, and this chapter attempts to point out questions in need of further investigation.

PIMT was serendipitously discovered in 1965 by Axelrod and Daly [5] while analyzing the metabolism of [^{14}C-methyl]AdoMet in extracts prepared from

---

1. Abbreviations: PIMT (protein L-isoaspartyl, D-aspartyl O-methyltransferase), AdoMet (S-adenosyl-L-methionine), AdoHcy (S-adenosyl-L-homocysteine), CaM (calmodulin), NLS (nuclear localization signal), and ROS (reactive oxygen species).

bovine pituitaries. The specificity of the reaction was not recognized at the time. Instead, the activity was described as a "methanol-forming enzyme," based on the presence of radioactive methanol in the extracts, which we now know was derived from the spontaneous hydrolysis of protein carboxyl [^{14}C-methyl] esters formed by PIMT. A few years later, the enzyme was correctly identified as a protein carboxylmethyltransferase [6, 7] and purified to homogeneity from calf thymus [8]. It was another 10 years before D-Asp [9, 10] and L-isoAsp [11, 12] residues were identified as the substrate residues modified by PIMT.

The complete amino acid sequences of the human erythrocyte and bovine brain PIMTs were described in 1989 [13, 14], paving the way for the first molecular cloning of the gene [15], molecular analyses of PIMT expression [16, 17], and the development of transgenic models. The first 3D structure of PIMT was solved for the *Thermatoga* enzyme [18] and followed by additional crystallographic studies exploring the catalytic mechanism of PIMT [19, 20]. Physiological roles for PIMT continue to be explored in model organisms ranging widely in their biological complexity [21–24]. Results from these studies, discussed in more detail in material following, indicate that PIMT is an ancient enzyme that has been strongly conserved during evolution. PIMT activity is expressed constitutively in most cells, but its function is not essential under many conditions. Evidence suggests, however, that PIMT plays an important role in the maintenance of functional protein structures under stress conditions and during long-term storage.

III. Biochemistry of PIMT-Catalyzed Reactions

A. PHYSICAL PROPERTIES OF PIMTs

PIMT activities were first characterized and purified from mammalian sources [8], but have subsequently been shown to be widely distributed throughout all domains of life (see section V). In mammals, PIMT activity can be detected in all tissues, although the measured specific activities vary considerably between tissues [25–27]. The enzyme has been purified to homogeneity from several tissues, including spleen [8], brain [14], and erythrocytes [13, 28]. With few exceptions [29], PIMT fractionates as a cytosolic enzyme with a calculated molecular weight of 24,000 to 27,000. The chromatographic behavior of PIMT on gel filtration columns is consistent with that of a globular monomeric protein [28, 30]. Two major isozymes of PIMT have been purified from bovine brain [31] and human erythrocytes [30, 32], which differ only in their C-terminal sequences [33, 34].

Analysis of cDNA clones indicates that the isozymes arise by differential splicing [33, 35]. Substitution of an -RDEL for an -RWK terminus changes the surface charge of the native protein sufficiently that the two isozymes can be resolved by anion exchange chromatography. The significance of the different C-termini is unknown. Although the -RDEL terminus of the more acidic isozyme is similar to that of an endoplasmic retention signal, both isozymes are found in the cytosolic fraction and neither form is associated with microsomes [34]. Biochemically, there are no significant differences in the substrate preferences or the kinetic parameters of the reactions catalyzed by the isozymes [30–32, 36], suggesting that they play similar roles in cells. Supporting this conclusion, both isozymes also function similarly in transgenic models [37].

PIMTs are ubiquitously distributed in plants as well (see section VIII.B), but the enzymes have been less thoroughly studied. A PIMT activity has been purified to homogeneity from wheat germ [27], which displays chromatographic properties consistent with a slightly acidic monomeric enzyme with a molecular weight of 25,000. Measured K_ms for the wheat enzyme with isoAsp substrates are almost two orders of magnitude higher than those of human PIMT with the same isoAsp substrates, suggesting that the active sites of mammalian and plant PIMTs have undergone considerable divergence. From genomic information, it is clear that higher plants have duplicated *PIMT* genes and it is likely that two closely related PIMTs are expressed (see section VIII.B.2). In this respect, two different PIMTs have been overexpressed from *Arabidopsis* cDNAs, but the biochemical properties of the second PIMT have not been investigated in detail [38, 39].

Most of the biochemical information on PIMTs from lower eukaryotes and microbes has been obtained using recombinant proteins. The physical properties of these recombinant enzymes are similar to those previously isolated from mammalian and plant tissues. With the exception of two PIMTs from hyperthermophiles (see section IV.C), the enzymes are approximately the same size as those from higher eukaryotes, and they catalyze the methylation of both protein and peptide substrates. Table 13.1 lists PIMTs that have been studied in some detail, together with additional sources of information about the enzymes, including relevant accession numbers for online databases.

B.   PIMT CATALYZES THE FORMATION OF PROTEIN CARBOXYL METHYL ESTERS

PIMT catalyzes the transfer of a methyl group from AdoMet to either the α-carboxyl group of isoAsp residues or the β-carboxyl group of D-Asp residues in peptide and protein substrates (Figure 13.1). The evidence supporting the unusual nature of these substrate residues and their spontaneous generation during aging is discussed in section VII. Stereochemically, D-Asp and L-isoAsp residues are surprisingly similar in the distribution of functional groups [11, 40], accounting for PIMT's ability to recognize both classes of substrates. The products of the reaction are a protein carboxyl methyl ester and a molecule of S-adenosyl-L-homocysteine (AdoHcy), which acts as a potent end-product inhibitor of the reaction [41, 42].

TABLE 13.1

ACCESSION INFORMATION FOR PIMTs IN ONLINE DATABASES.*

Species	Gene Name	TrEMBL/Swiss Protein	Database Identifier	PDB Accession	Literature References
E. coli	*pcm*	P24206	b2743		[195]
P. furiosus	*pcm*	Q8TZR3	PF1922	1JG1-1JG4	[19]
T. maritima	*pcm*	Q56308	TM0704	1DL5	[18]
S. tododaii	*pcm*	Q972K9	ST1123	1VBF	[93]
S. pombe	*pcm2*	Q9URZ1	SPAC869.08		
C. elegans	*pcm-1*	Q27873	C10F3.5		[45]
D. melanogaster	*Pcmt*	Q27869	CG2152	1R18	[20]
A. thaliana	*PIMT1*	Q42539	At3g48330		[38]
A. thaliana	*PIMT2*	Q8GXQ4	At5g50240		[39]
T. aestivum	*PCM*	Q43209			[27]
M. musculus	*Pcmt1*	P23506	MGI:97502		[16]
H. sapiens	*PCMT1*	P22061	HGNC:8728	1I1N, 1KR5	[43, 94, 164]

*Accession numbers and gene names were obtained from various online databases. There are numerous links between the databases, and key URLs follow. Detailed information for bacterial species and higher plants can be obtained via The Institute for Genomic Research (*www.tigr.org*). Information for human, mouse, and *S. pombe* can be accessed through the Sanger Center (*www.sanger.ac.uk*) or the National Center for Biotechnology Information (*www.ncbi.nlm.nih.gov*). Species-specific databases include information for *Drosophila* (*flybase.bio.indiana.edu*), *C. elegans* (*www.wormbase.org*), and *Arabidopsis* (*www.arabidopsis.org*). Other useful URLs are those for TrEMBL/Swiss Protein (*us.expasy.org/sprot*) and the Protein Data Bank (*www.rcsb.org*).

Kinetics studies have suggested that PIMT works by a rapid-equilibrium random sequential bi-bi mechanism [41, 42], but structural studies are more consistent with an ordered sequential mechanism (see section IV.B) in which AdoMet binding precedes peptide binding and the release of methylated peptide precedes AdoHcy release [20, 43]. The turnover number of PIMT calculated from kinetic data is always very low, on the order of 0.1 to 1.0 mol min^{-1}, although V_{max} varies with different isoAsp substrates [20, 44, 45]. In general, the turnover number with peptide substrates is significantly higher than that with protein substrates, suggesting that the greater conformational flexibility of peptide substrates facilitates product formation.

The protein carboxyl methyl esters formed by PIMT are unstable, turning over with half-times ranging from minutes to hours at physiological pH [46, 47]. The processing of protein carboxyl methyl esters appears to be completely nonenzymatic. At the present time, there is no compelling evidence for protein demethylases or any other enzymatic activities that might participate in processing esters [48]. Studies with synthetic peptide substrates indicate that the immediate product of demethylation is a succinimide structure (Figure 13.1), which forms spontaneously as methanol is released [11, 49].

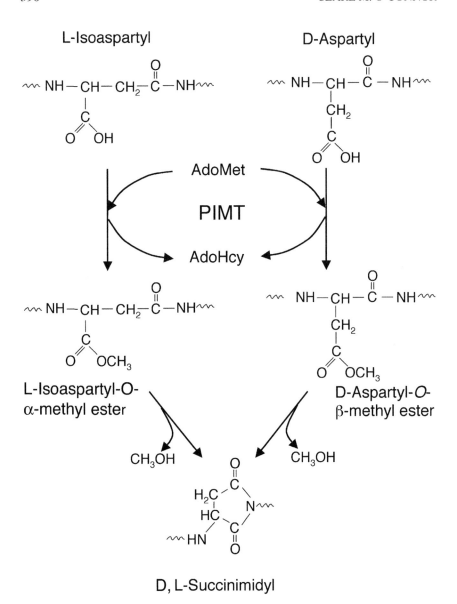

FIG. 13.1. Enzymatic reaction. PIMT catalyzes the transfer of methyl groups from AdoMet to protein L-isoaspartyl and D-aspartyl residues. Methyl esters spontaneously hydrolyze to form a succinimidyl structure.

that for Asn deamidation, which has been assigned a value of 1.0 because this is the rate-limiting reaction in cells. Generalizing from this model, L-isoAsp-containing peptides are predicted to comprise the major products of deamidation reactions. By contrast, D-Asp residues are the least frequent products of deamidation reactions, because the rate of succinimide hydrolysis is an order of magnitude faster than that of succinimide racemization. In addition, cleavage of succinimide rings is biased, producing about three times more isoAsp than Asp isomer. Thus, protein deamidation is likely to generate isoAsp-containing proteins with high frequency and proteins with racemized Asp residues with a much lower frequency.

The sequence requirements surrounding the L-isoAsp methyl-accepting site have been explored by using synthetic peptides as PIMT substrates. These experiments indicate that PIMT recognizes L-isoAsp residues in a wide range of primary sequences with limited commonality. In one comparative study [44], the affinity of human erythrocyte PIMT for 35 different L-isoAsp-containing peptides varied over several orders of magnitude. Measured K_ms for the most active substrates were in the 0.1- to 1.0-μM range, whereas PIMT recognized less active substrates with K_ms approaching and even exceeding 1 mM. High-affinity substrates had at least one amino acid N-terminal and two amino acids C-terminal to the modified isoAsp residue, suggesting that neighboring residues were important in positioning substrates within the active site. The wide range of peptide substrates modified by PIMT is consistent with the broad specificity of the enzyme observed in intact cells (see the section VI.A.1).

Considering that D-Asp β-[³H]methyl esters have been isolated from intact erythrocytes incubated with L-[³H-methyl]-L-methionine [10, 55, 56], it is surprising that peptides with D-Asp residues are generally not methylated or are poorly methylated by PIMT. Both human [56] and *Pyrococcus* [40] PIMTs modify D-Asp residues in synthetic peptides, but the measured K_ms are two to three orders of magnitude higher than the K_ms for the corresponding L-isoAsp peptides. The peptides may not provide good models for physiological methyl-accepting sites, in that purified PIMT catalyzes the formation of D-Asp β-[³H]methyl esters using erythrocyte membranes as a substrate *in vitro* [36]. Furthermore, the production of D-Asp β-methyl esters in the purified system was reduced in the presence of L-isoAsp peptides, suggesting that D-Asp and L-isoAsp substrates compete for the PIMT active site. Taken together, the biochemical data is best explained by a dual role for PIMT in recognizing both L-isoAsp and D-Asp substrates.

D.   METHYL-ACCEPTING SITES IN PROTEIN SUBSTRATES

From the very first identification of a protein carboxylmethyltransferase (then termed a carboxymethylase) activity in calf spleen [6], it has been clear that PIMT recognizes a broad range of protein substrates. Endogenous methyl-accepting proteins can be detected in extracts prepared from all mammalian tissues [25, 57] and across phyla [58–60]. The identification of these endogenous methyl-accepting proteins in cellular extracts is complicated by the lability of protein methyl esters under mildly basic conditions [6, 25, 46], thus requiring the use of acidic gel systems for their resolution. Under the appropriate separation conditions, methyl-accepting activity is always found to be distributed among a heterogeneous group of proteins characteristic of a particular cell or tissue [46, 61–63]. This broad specificity for protein substrates has proven useful for biochemical analyses of PIMT. A variety of commercially available inexpensive proteins, including most notably ovalbumin and gelatin, have been useful for monitoring the purification of PIMT activities and characterizing its enzymatic properties [6, 31].

The broad specificity demonstrated by PIMT is readily explained by the unusual nature of protein D-Asp and L-isoAsp residues, which arise spontaneously as proteins age. Because all Asn and Asp residues in a protein are theoretically subject to the types of spontaneous damage that give rise to D-Asp and L-isoAsp residues (Figure 13.2), cells would be expected to contain many different methyl-accepting proteins. Detailed studies with individual proteins indicate, however, that structural features of a protein strongly influence the actual rate of protein isomerization.

In general, isoAsp residues arise at higher rates in flexible regions of polypeptides [64–66], most probably because of steric difficulties associated with succinimide formation in more structured regions of polypeptides [67]. In this respect, methyl-accepting sites have been identified near the N-termini of several proteins, including serine hydroxymethyltransferase [68], α-globin [69], and protein kinase A [70]. Calmodulin (CaM) has provided a particularly useful model for exploring the effects of protein conformation on the generation of isoAsp residues, because of the profound conformational changes associated with Ca^{2+}binding. In the presence of Ca^{2+}, CaM is highly structured and demonstrates very low methyl-accepting activity at sites near the N-terminus and in the flexible central helix [71]. In the absence of Ca^{2+}, CaM is very flexible and isoAsp residues form rapidly under physiological conditions at multiple sites in the protein, but most significantly in the third and fourth EF hands responsible for high-affinity Ca^{2+}binding [64, 65, 72]. Methyl-accepting sites in the EF-hands are derived from both Asn and Asp residues, altering CaM's Ca^{2+}binding properties [73] and contributing to a severe reduction in enzymatic activity [72].

Additional evidence for the formation of protein isoAsp residues on a physiologically significant time scale has come from the biotechnology industry. In a growing number of cases, the appearance of an isoAsp residue is reported to adversely affect the stability of a recombinant protein designed for therapeutic use. Examples of recombinant proteins negatively affected by isoAsp formation include human growth hormone [74], epidermal growth factor [75, 76], interleukin-1β [77], the thrombin inhibitor hirudin [78], tissue plasminogen activator [79], stem cell factor [80], and recombinant antibodies [81, 82]. Sequence analyses indicate that isoAsp residues do not arise randomly in protein sequences. Instead, there are clearly hot spots for isoAsp generation in the sequences. Comparison of the primary sequences surrounding isoAsp residues reveals a strong tendency for a glycine residue to be situated at the *N+1* position, consistent with an increased propensity for succinimide formation at these sites [50, 83, 84].

Protein methyl-accepting activity has been widely used to assess the isoAsp content of proteins and has even been developed into a commercial kit, but the presence of an isoAsp residue is not sufficient to constitute a methyl-accepting site. This has been shown most clearly with a naturally occurring isoAsp variant of bovine ribonuclease that has been purified to homogeneity and studied in considerable detail. The isoAsp residue, which arises from deamidation of Asn-67 [85, 86], is located in the crystal structure in a highly structured surface loop [87] (where it is refractory to modification by PIMT [88]). Structural constraints in the intact protein are responsible for this poor methyl-accepting activity, because a tryptic fragment containing isoAsp-67 is readily methylated by PIMT [88]. This result provides a cautionary note to investigators using protein methyl-accepting activity to assess the isoAsp content of proteins [89], which is difficult to measure with chemical methods.

E.   PIMT-CATALYZED METHYLATION INITIATES THE REPAIR OF POLYPEPTIDE BACKBONES

The appearance of an isoAsp residue in a protein introduces an additional carbon into the polypeptide backbone, profoundly affecting local structure. Biochemical evidence indicates that PIMT-catalyzed methylation initiates the repair of the polypeptide backbone. The mechanism underlying protein repair has been studied extensively with peptide substrates [49, 88, 90, 91]. As discussed in section III.B, methyl esters formed by PIMT spontaneously hydrolyze to form an internal succinimide (Figure 13.1), which hydrolyzes to a mixture of isoAsp and Asp peptides (Figure 13.2).

L-Asp products of the hydrolysis can be considered repaired, in that the normal polypeptide backbone configuration is restored whereas L-isoAsp hydrolysis products can be methylated again by PIMT. Thus, by multiple cycles of methylation and demethylation isoAsp residues in synthetic peptide substrates are converted to L-Asp residues in high yields [88, 90, 91]. A minor fraction of synthetic peptide substrates, representing 10 to 20% of the initial substrate, is not repaired following incubation by PIMT because the succinimide has racemized. The D-isoAsp products generated by succinimide hydrolysis are considered irreparable, because D-Asp products are generally poor substrates for PIMT (see section III.C).

Because of their greater structural complexity, it has been more difficult to demonstrate the structural repair of isoaspartyl-containing proteins by PIMT. Fortunately, several proteins can be converted to isomerized forms in sufficient quantity to provide useful models for potential repair, and PIMT has been shown to restore impressive amounts of enzymatic activity to isoAsp-containing forms of both CaM [72] and the bacterial HPr phosphocarrier protein [92]. Significantly, PIMT was unable to completely restore activity to either CaM or HPr even after exhaustive methylation, probably because irreparable isomers were generated during succinimide hydrolysis.

The inefficiency of the protein repair pathway outlined in Figure 13.2 is striking in that it lacks the stereochemical specificity and efficiency characteristic of most cellular processes. In addition to the fact that the products of the processing pathway initiated by PIMT are heterogeneous, no mechanism has been identified for converting isoAsp residues to Asn residues, even though Asn are the likely origin of many isoAsp residues in proteins. It is an open, but important, question whether other proteins participate in and help to guide the repair process *in vivo*.

IV.   Crystallographic Studies of PIMTs

A.   CLASS I METHYLTRANSFERASES WITH A SIGNATURE TOPOLOGY

The first crystal structure of a PIMT was solved by Skinner et al. [18] for the enzyme from *Thermatoga maritima*, a hyperthermophilic eubacterium. The structure predicted from their model is a modified Rossman fold consisting of a central seven-stranded β sheet flanked by an α helix on either side. Similar $\alpha\beta\alpha$ structures have been subsequently observed in the orthologous PIMTs from *Pyrococcus furiosus* [19], *Sulfolobus tokodaii* [93], *Drosophila melanogaster* [20], and humans [43, 94], consistent with strong evolutionary conservation. Figure 13.3 shows the topology diagram and 3D structure of fly PIMT as an example. PIMTs consist of a PIMT-specific N-terminal subdomain (lightly shaded in Figure 13.3) and a central methyltransferase domain (dark shading), which is also common to class I AdoMet-dependent methyltransferases with other substrate specificities [95, 96]. The β sheet contains seven strands, one of which is antiparallel to the other six strands. In PIMTs, the central sheet has the topological arrangement of $3\uparrow2\uparrow1\uparrow4\uparrow5\uparrow6\downarrow7\uparrow$, which is a defining characteristic of PIMTs. By contrast, the strands in the central β sheets in other class I AdoMet-dependent MTases are arranged in a $3\uparrow2\uparrow1\uparrow4\uparrow5\uparrow7\downarrow6\uparrow$ topology [95, 96]. As might be expected, strongly conserved amino acid side chains in the conserved methyltransferase domain are involved in binding AdoMet and AdoHcy. Figure 13.3 shows a molecule of AdoHcy bound to PIMT, with an asterisk indicating the position of the sulfur atom.

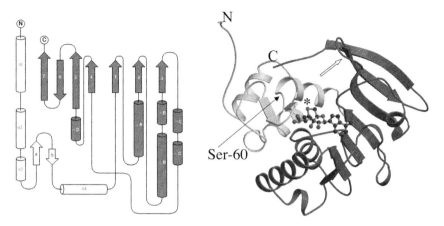

Fɪɢ. 13.3. Topology diagram and 3D structure of *Drosophila* PIMT (PDB accession number 1R18). The methyltransferase fold common to Class I AdoMet-dependent is shown in dark shading, and the N-terminal subdomain characteristic of PIMTs is shown in light shading [20]. An asterisk shows the position of the sulfur atom in AdoHcy. The open arrow indicates a loop whose position is significantly different in AdoMet and AdoHcy co-crystals [19]. The position of Ser-60 essential for catalysis is indicated.

B.   Sᴜʙsᴛʀᴀᴛᴇ Bɪɴᴅɪɴɢ ᴀɴᴅ Cᴀᴛᴀʟʏsɪs

Two highly conserved regions that are unique to PIMTs form the sides of the substrate binding cleft and provide functional groups important in binding isoAsp substrates [18, 19]. One side of the substrate binding cleft is contributed by helix α4 and a small β hairpin, consisting of strands βa and βb, located at the end of the N-terminal subdomain. The short β hairpin extends over the AdoHcy binding cleft in the crystal structure (open arrow in Figure 13.3). The other side of the substrate binding cleft consists of a short hydrophobic coil near the C-terminus, extending from the end of strand β7 in the central sheet. From a co-crystal of *P. furiosus* PIMT (adenosine) and an isoAsp-containing hexapeptide substrate (VYP(iso)DHA), it is possible to identify interactions important in substrate binding [19]. The crystallographic model shows that peptides bind to the enzyme in an extended conformation suggestive of local denaturation.

Many stabilizing interactions arise from hydrogen bonds between the substrate and the polypeptide backbone of PIMT. The carboxylate group of the bound isoaspartyl residue forms hydrogen bonds with residues on both sides of the substrate binding cleft. One hydrogen bond involves an invariant serine in the N-terminal subdomain. In fly PIMT, this Ser-60 is located in a short loop connecting the βb strand with the α4 helix (Figure 13.3). Mutagenesis has confirmed the importance of Ser-60 in catalysis. Substitution of Ser-60 with either Thr or Gln progressively reduces catalytic activity, and a S60A mutant is completely inactive [20]. The isoAsp carboxylate group also forms hydrogen bonds with the backbone nitrogen of Val-219 in the conserved C-terminal sequence on the other side of the substrate binding cleft.

The orientation of the C-terminus in fly PIMT is rotated approximately 90 degrees relative to the C-termini in other PIMT crystals. Effectively, this displacement generates a more open conformation in the substrate binding cleft that allows greater solvent access to AdoHcy. In the homologous structures from *P. furiosus* and humans, the C-terminus folds back over the bound AdoHcy, limiting solvent access to 1.4% and 0.5% of the respective structures. By contrast, 9% of the AdoHcy in the *D. melanogaster* structure is accessible to solvent. The differences between PIMT structures suggests that PIMT alternates between open and closed conformations. These conformational changes at the C-terminus could be important in the exchange of substrates and products [97]. The open conformation would facilitate the exchange of AdoHcy and AdoMet, whereas the closed conformation would facilitate the binding of isoAsp-containing substrates, while dehydrating the active site. Considering all of the structural data together, including the relative placements of bound cofactors and peptide substrates, the data are consistent with an ordered sequential mechanism for PIMT-catalyzed reactions. The structural data predicts the AdoMet binding would precede peptide binding and that methylated peptide release would precede AdoHcy release (see section III.B).

C.   ADDITIONAL C-TERMINAL DOMAINS IN PIMTS FROM SOME HYPERTHERMOPHILES

The αβα domain described previously is common to all PIMTs and represents the entire structures of PIMTs from humans, flies, and *P. furiosus*. Additional species-specific C-terminal domains are present, however, in *T. maritima* and *S. tokodaii* PIMTs, *T. maritima* PIMT contains an additional irregularly structured C-terminal domain consisting of 103 amino acids [18]. The structure of the domain is unlike other known folds and the sequence appears to be unique to *T. maritima*. Nonetheless, the domain is required for enzymatic activity, because a recombinant protein lacking the C-terminal domain displays much lower levels of activity than the full-length protein [98].

An entirely different C-terminal domain and domain structure has been identified in PIMT from the archaeon, *S. tokodaii*. The additional C-terminal domain in *S. tokodaii* PIMT consists of approximately 30 amino acids arranged in a single loop and a short α-helix [93]. The C-terminal domain in *S. tokodaii* PIMT is postulated to promote oligomerization. In the crystal structure, C-terminal domains interact with one another to form a coiled coil structure at the center of a hexamer. The hexamer is thought to be assembled from three homodimers held together by disulfide linkages involving Cys-149, a residue that is notably lacking in orthologous PIMT sequences. Cys-149 is located in a surface helix connecting strands of the central sheet and is required for dimerization. Mutagenesis indicates that the C-terminal domains are not required for hexamer formation, but they help to stabilize the hexamer, as shown by differential scanning calorimetry. It is not clear if oligomerization is required for PIMT activity, because the enzymatic properties of *S. tokodaii* PIMT have not been studied. Interestingly, the *S. tokodaii* genome contains a coding sequence for another PIMT without the additional C-terminal domain, and thus it is not clear if both forms are physiologically important.

V. Phylogenetic Distribution of PIMT Activities Deduced from Whole Genome Sequences

As the data continues to accumulate from whole-genome sequencing projects, it is clear PIMT orthologs are distributed throughout all phylogenetic kingdoms and domains. PIMT can therefore be classified as an ancient enzyme, but not as one whose activity is essential for life because orthologs are missing in some species. Information about PIMT gene names, locus identification, and database accession numbers is give in Table 13.1 for the PIMTs that have been studied in detail. More information about the evolution of PIMTs is available for prokaryotic sequences than for eukaryotic sequences, because of the larger number of completed genome sequences. Prokaryotic orthologs of PIMT, known as *pcm* sequences, form the Cluster of Orthologous Group (COG) 2518 [99]. The phylogenetic distribution of *pcm* sequences within the prokaryotes is unlike that of any other COG, and thus PIMT function appears to have evolved independently of other metabolic enzymes. Supporting this conclusion, *pcm* sequences are rarely observed in gene fusions.

Within the prokaryotes, *pcm* orthologs are much more widely distributed in the archaea than in the eubacteria, perhaps reflecting the ability of archaea to live under extreme conditions that might be expected to cause structural damage to proteins. Multiple *pcm* genes are present in the genomes of some archaea, such as *Archaeoglobus fulgidis* and *S. tokodaii*, although it is not clear if both genes encode functional proteins. *A. fulgidis* has multiple gene duplications, including *pcm*, which show differential codon usage and are therefore thought to have arisen from lateral gene transfer [100]. Lateral gene transfer of an archaeal *pcm* sequence is also the most likely origin of the *pcm* gene in *T. maritima*, a hyperthermophilic eubacterium [101]. Sequence comparisons using the BLAST algorithm [102] indicate that the closest orthologs to the *T. maritima* sequence are those from *A. fulgidis* and other archaea, followed by orthologs from the eubacteria.

PIMT-encoding sequences are less widely distributed within the eubacteria than in other domains, and few generalities can be made about their distribution within the eubacteria. Potential PIMT-encoding sequences are present in about half of the eubacterial genomes sequences to date, including *E. coli*, *Rhizobium meliloti*, and many opportunistic pathogens. Orthologs of *pcm* are generally not detected in mycoplasma and other obligate pathogens, such as *Treponema* and *Borrelia*. The absence of PIMT in obligate pathogens is not surprising, in that these organisms rely on their hosts for many essential functions. One exception is the ulcer bacterium, *Helicobacter pylori*, an obligate parasite of the human gut. *H. pylori* may have retained PIMT function because its ecological niche is an acidic environment containing high concentrations of urea, which could reduce the structural stability of its proteins.

PIMT-encoding sequences are nearly universally distributed in the eukaryotes. The only exceptions identified to date are *Encephalitozoon cuniculi*, a parasitic microsporidium, and yeasts from the order *Saccharomyces*. PIMT orthologs are readily identified in genomes from all other orders of yeast, suggesting that the ancestral yeast PIMT-coding sequence was lost before the genome duplication and subsequent specialization that accompanied the evolution of the Saccharomycetes [103, 104]. PIMT-encoding sequences have been maintained in the genomes of all multicellular eukaryotes studied to date. Higher plants are unique in possessing duplicated genes that appear capable of encoding PIMT activities (see section VIII.B.2). Animal genomes contain a single copy of a PIMT-encoding sequence, but alternative transcript splicing is used in mammalian species to generate multiple isozymes (see section VIII.A.2).

Figure 13.4 shows a sequence comparison of the PIMT-encoding sequences from several model organisms and other species in which PIMTs have been studied in detail. The phylogenetic tree derived from the alignment is shown in Figure 13.5. The most highly conserved regions of the PIMT sequence include a core methyltransferase fold common to many AdoMet-dependent methyltransferases [96] and additional PIMT-specific regions implicated in the binding of isoaspartyl substrates [18, 19].

VI. Biochemistry of Protein Isoaspartyl Methylation *In Vivo*

The analysis of protein isoaspartyl methylation reactions in intact cells is a complex undertaking, due to the poor permeability of cells to AdoMet as well as the many different types of methylated molecules in cells. Consequently, the most complete information about isoAsp methylation in intact cells has been obtained from a few cell types, such as human erythrocytes and *Xenopus* oocytes, which offer unique advantages for studies of protein isoAsp methylation.

A. HUMAN ERYTHROCYTES

Erythrocytes were a fortuitous choice for methylation studies for several reasons. First, the incorporation of methyl groups into protein methyl esters from

```
H. sapiens       --MAWK-SGGASHSELIHN LRKNGIIKTDKVFEVMLATD RSHYAK----CNPYMDSPQS IGFQ  56
D. melanogaster  --MAWR-SVGANNEDLIRQ LKDHGVIASDAVAQAMKETD RKHYSP----RNPYMDAPQP IGGG  56
C. elegans       --MAWR-SSGSTNSELIDN LRNNRVFASQRAYDAMKSVD RGDFAP----RAPYEDAPQR IGYN  56
S. pombe         --MFWS-FNLSSNAALVQH LVESKFLTNQRAIKAMNATS RSFYCP----LSPYMDSPQS IGYG  56
A. thaliana      MKQFWSPSSINKNKAMVEN LQNHGIVTSDEVAKAMEAVD RGVFVTD--RSSAYVDSPMS IGYN  61
P. furiosus      --MMDEKELYEKWMRTVEM LKAEGIIRSKEVERAFLKYP RYLFVEDKYKKYAHIDEPLP IPAG  61
E. coli          --MVSRR-----VQALLDQ LRAQGIQ-DEQVLNALAAVP REKFVDEAFEQKAWDNIALP IGQG  55


H. sapiens       ATISAPHMHAYALELLFDQ LHEGAKALDVGSGSGILTACFARMVGCTG----KVIGIDHIKE 114
D. melanogaster  VTISAPHMHAFALEYLRDH LKPGARILDVGSGSGYLTACFYRYIKAGVDADTRIVGIEHQAE 119
C. elegans       ATVSAPHMHAAALDYLQNH LVAGAKALDVGSGSGYLTVCMAMMVGRNG----TVVGIEHMPQ 114
S. pombe         VTISAPHMHATALQELEPV LQPGCSALDIGSGSGYIVAAMARMVAPNG----TVKGIEHIPQ 114
A. thaliana      VTISAPHMHAMCLQLLEKH LKPGMRVLDVGSGTGYLTACFAVMVGTEG----RAIGVEHIPE 119
P. furiosus      QTVSAPHMVAIMLEIAN-- LKPGNNILEVGTGSGWNAALISEIVKT------DVYTIERIPE 115
E. coli          QTISQPYWVARMTELLE-- ITPQSRVLEIGTGSGYQTAILAHLVQ------HVCSVERIKG 108


H. sapiens       IVDDSVNNVRKDD------PTLLSSGRVQLVV GDGRMGYAEEAPYDA IHVGAAAPVVPQALID 171
D. melanogaster  IVRRSKANLNTDD------RSMLDSGQLLIVE GDGRKGYPPNAPYNA IHVGAAAPDTPTELIN 176
C. elegans       LVELSEKNIRKHH------SEQLERGNVIIIE GDGRQGFAEKAPYNA IHVGAASKGVPKALTD 171
S. pombe         LVETSKKNLLKDINHDEVLMEMYKEKRLQINV GDGRMGTSEDEKFDA IHVGASASELPQKLVD 177
A. thaliana      LVASSVKNIEASAA-----SPFLKERSLAVHV GDGRQGWAEFAPYDA IHVGAAAPEIPEALID 177
P. furiosus      LVEFAKRNLERAG------------VKNVHVIL GDGSKGFPPKAPYDV IIVTAGAPKIPEPLIE 167
E. coli          LQWQARRRLKNLD------------LHNVSTRH GDGWQGWQARAPFDA IIVTAAPPEIPTALMT 160


H. sapiens       QLKPGGRLILPVGPAGGN QMLEQYDKLQDGSIMKPLMG VIYVPLTDKEKQWSRWK 227
D. melanogaster  QLASGGRLIVPVGPDGGS QYMQQYDKDANGKVEMTRLMG VMYVPLTDLRS------ 226
C. elegans       QLAEGGRMMIPVEQVDGN QVFMQIDKIN-GKIEQKIVEH VIYVPLTSREEQWNRN- 225
S. pombe         QIKSPGKILIPIG--TYS QNIYLIEKNEQGKISKRTLFP VRYVPLTDSPDDSSDY- 230
A. thaliana      QLKPGGRLVIPVG--NIF QDLQVVDKNSDGSVSIKDETS VRYVPLTSREAQLRGD- 230
P. furiosus      QIKIGGKLIIPVGSYHLW QELLEVRKTK-DGIKIKNHGG VAFVPLIG-EYGWKE-- 219
E. coli          QIDEGGILVLPVGEEH-- QYLKRVRRRG-GEFIIDTVEA VRFVPLVKGELA----- 208
```

FIG. 13.4. Alignment of PIMT sequences using the Clustal W algorithm from the European Bioinformatics Institute (*http://www.ebi.ac.uk*). Invariant residues are shown by white letters on a black background. Strongly conserved regions are shaded.

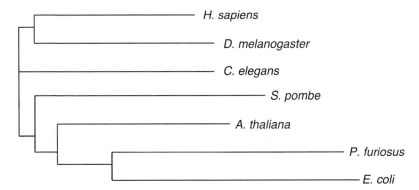

FIG. 13.5. Phylogenetic tree for PIMTs constructed from the alignment data in Figure 13.4 using the phylip algorithm from the European Bioinformatics Institute (*http://www.ebi.ac.uk*).

L-[methyl-^{3}H]methionine is not obscured by the incorporation of radioactivity into newly synthesized proteins or methylated nucleic acids. Second, PIMT is the major methyltransferase in erythrocyte cytoplasm [28, 55], ensuring that a large fraction of the radioactivity is incorporated into protein [^{3}H]methyl esters [47]. In addition, the stoichiometries of individual methylation reactions can be calculated because the biochemical composition of the red cell membrane had already been well defined [105]. Many of the first studies that established the characteristics of PIMT-catalyzed reactions were carried out in Steve Clarke's laboratory at about the same time that his group identified the unusual nature of PIMT substrates.

1. Characteristics of Protein IsoAsp Methylation in Intact Cells

A heterogeneous group of endogenous methyl-accepting proteins is observed when human erythrocytes are incubated with high concentrations of L-[methyl-^{3}H]methionine, the permeable precursor to [^{3}H]AdoMet in erythrocytes [106]. Both membrane and cytosolic proteins are methylated by PIMT [47, 51, 52, 107]. Figure 13.6 shows a stained gel (left) and fluorogram (right) of cytosolic (lane A) and membrane (lane B) proteins obtained from a typical labeling experiment [108]. The membranes (lane B) in this experiment are "pink ghosts," which retain some of the loosely associated hemoglobin (Hb). This membrane-bound Hb can be released from the membrane with either 100 mM NaCl (lane C) or a nonionic detergent (Lane D).

In the cytosolic fraction, protein methyl esters are readily detected with both α-globin and β-globin and with an unidentified 35,000-kDa protein. Cytosolic Hb has a very low methyl-accepting activity, but its sheer predominance obscures the detection of other cytosolic methyl-accepting proteins, which include CaM [109, 110] and many other proteins [47]. In the membrane fraction (lane B), both

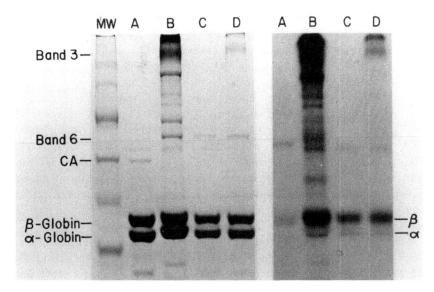

FIG. 13.6. Methylation of membrane and cytosolic proteins in human erythrocytes. Stained gel (*left*) and fluorogram (*right*) of proteins prepared from erythrocytes incubated with L-[methyl-^{3}H]-methionine and separated by SDS-PAGE at pH 2.4. Cytosolic proteins (lane A), membrane proteins (lane B), and proteins extracted from membranes with 100 mM NaCl (lane C) or 0.5% Nonidet P-40 (lane D). Reproduced from Figure 1 in O'Connor and Yutzey [108].

integral and peripheral proteins have been identified as methyl acceptors. Some of the highest levels of methyl-accepting activity are associated with the cytoskeletal proteins, ankyrin and band 4.1 [51, 52]. Band 3, the anion transporter protein, is also methylated in intact cells [111, 112].

The heavy exposure of the fluorogram in Figure 13.6 makes it difficult to resolve band 3 methylation in the membrane fraction (lane B), but this is apparent in the detergent extract of membranes (lane D). Proteolytic mapping indicates that the physiological methylation sites in band 3 are distributed throughout its sequence [111]. A number of other unidentified methyl-accepting accepting proteins are also detected in the membrane fraction. Interestingly, the small fraction of Hb that associates with membranes is more highly methylated than cytosolic Hb (see material following).

The calculated stoichiometries for individual methylation reactions vary over several orders of magnitude, but all of the reactions are markedly substoichiometric. Bands 2.1 (ankyrin) and 4.1 represent some of the most active methyl-accepting proteins in red cells, yet only about 1% of band 4.1 polypeptides are methylated at steady state [52]. Methylation stoichiometries are much lower for other erythrocyte proteins. Only one in every thousand band 3 molecules and one in every

300,000 Hb molecules is estimated to be methylated at steady state [47]. These low stoichiometries are partly explained by the unusual nature of the methylated residues. The spontaneous processes responsible for the generation of D-Asp and L-isoAsp residues (Figure 13.2) are relatively slow, and isomerized variants would be expected to comprise only a small fraction of any one polypeptide.

Another factor underlying the low stoichiometries of methylation detected in intact cells is the efficient metabolism of isoAsp residues by PIMT. Biosynthetic labeling studies indicate that erythrocyte protein methyl esters are metabolically labile [47, 51]. When cells are incubated continuously with L-[methyl-^{3}H]methionine, the rate of [^{3}H]methanol production from ester turnover is close to an order or magnitude higher than the rate of radioactive label incorporation into protein methyl esters, indicating that the average life time of a protein methyl ester is very short. Pulse-chase analysis confirms that individual protein methyl esters turn over with half-times ranging from minutes to hours.

Other lines of evidence also indicate that isoAsp-containing proteins are maintained at very low levels in erythrocytes due to their active metabolism by PIMT [47, 113]. When erythrocyte membranes are used as the substrate for PIMT and [^{3}H]AdoMet *in vitro*, methyl groups are not incorporated into the physiological methyl-acceptors unless the cells had been preincubated with methylation inhibitors for several hours prior to membrane isolation [55]. In addition, the methyl-accepting activity of erythrocytes in PIMT-knockout mice is significantly higher than that of normal mice [23, 114].

2.   Age-dependent Increases in Protein Methylation

The protein repair process may become less efficient as cells age, and this decline may contribute to the higher levels of methyl-accepting activity observed in older cells [52, 113, 115]. Circulating erythrocytes have a narrowly-defined lifespan of 120 days. As erythrocytes age, membrane blebbing reduces the surface area and disrupts the normal process of membrane transport, causing cells to become more dense.

This property can be exploited to separate cells into rough age fractions by centrifugation on density gradients. Fractions enriched in older cells have three to four times higher levels of methylated proteins than younger cells in biosynthetic labeling experiments [52, 113, 115]. The spectrum of methylated proteins does not change with age, however with individual proteins demonstrating fairly uniform increases in methyl-accepting activity. At the same time the number of protein methyl esters is increasing in older cells, the specific activity of PIMT is declining, because erythrocytes are unable to synthesize replacements for denatured enzymes *de novo*. Thus, PIMT should become increasingly saturated with substrates in older erythrocytes, which could potentially lead to less effective protein repair.

The reduced resistance of older erythrocytes to oxidative damage may contribute to higher levels of methyl-accepting proteins in older cells. It is well established that levels of oxidant defense enzymes decline in older erythrocytes. Failure to resist oxidative damage is manifest in Heinz bodies, aggregates of oxidatively-damaged Hb that associate with the membrane and are hypothesized to play a role in the removal of aged blood cells from the circulation [116]. As shown in Figure 13.6, the 1 to 2% of Hb that associates with erythrocyte membranes is also significantly enriched in methyl-accepting sites. The specific methyl-accepting activity of membrane Hb is about 10-fold higher than that of cytosolic Hb [108].

Reactive oxygen species (ROS) may contribute to this higher methyl-accepting activity because even a short incubation of Hb with acetylphenylhydrazine produces a dramatic increase in Hb's methyl-accepting capacity [108]. Other experiments with intact erythrocytes have also implicated ROS in the generation of methyl-accepting sites. Erythrocytes exposed to the ROS generators, *t*-butyl peroxide or hydrogen peroxide, have elevated rates of methylation relative to control cells [117]. Even larger increases in methyl-accepting activity are observed in cells with an inherited deficiency in glucose-6-phosphate dehydrogenase that sensitizes erythrocytes to oxidative stress [118]. The mechanism by which ROS increase methyl-accepting sites is unclear. Chemically, ROS are not expected to react directly with Asn or Asp residues to generate isoAsp residues. The production of isoAsp residues is more likely to be a secondary effect of chemical damage to other amino acids, which alters the 3D structure of the protein.

B.   NONERYTHROID MAMMALIAN CELLS

Biosynthetic labeling studies with L-[methyl-^{3}H]methionine have established that the basic features of carboxyl methylation reactions are similar in erythrocytes and nonerythroid cells, although fewer biochemical details are available in nonerythroid cells. In these experiments, the base lability of methyl esters has been used to differentiate carboxyl methylation reactions from the background of incorporation of radioactivity from L-[methyl-^{3}H]methionine into polypeptide backbones and into base-stable linkages with other amino acids. In both brain tissue slices [119, 120] and cultured cells [121] [122], the substrates for carboxyl methylation are heterogeneous.

In platelets, methyl esters are associated with more than 30 different proteins resolved by SDS-PAGE [121]. Protein methyl esters in platelets are metabolically labile, as shown by the rapid and continuing evolution of [^{3}H]methanol during culture. Similarly, a diverse set of methyl-accepting proteins has been identified in neuroblastoma cells by SDS-PAGE [122]. Individual substrates for PIMT have not been identified in nonerythroid cells, because additional purification steps would be required to separate methyl-accepting proteins from other proteins with similar electrophoretic mobilities. Another complicating factor arises from the fact that simple base treatment does not allow one to distinguish protein Asp methyl esters from esters formed by other enzymes, such as those formed during the C-terminal processing of CaaX-box proteins [123] and C-terminal esters in protein phosphatase 2A [124].

C. XENOPUS OOCYTES

Erythrocytes are highly specialized cells that have lost their ability to synthesize new protein as well as their intracellular organelles. To understand isoAsp methylation reactions in more typical cells, protein isoAsp methylation has been analyzed in *Xenopus laevis* oocytes, which offer some unique advantages for these studies. Fully grown oocytes are roughly 1.2 mm in diameter and are easily microinjected with impermeable molecules and macromolecules. Oocytes have an exceptionally large nucleus, the germinal vesicle, which can be manually isolated with minimal contamination [125], and other organelles can be purified by standard differential centrifugation techniques. Finally, oocytes have been extensively used to study quantitative aspects of protein synthesis and translational regulation [126], making it possible to appreciate the importance of protein carboxyl methylation in the context of overall protein metabolism.

1. Kinetic Analysis of Protein Methylation

Xenopus oocytes can be easily injected with impermeable substrates for PIMT, including [³H]AdoMet, isoAsp-containing peptides, and protein substrates. Direct injection of [³H]AdoMet allows the investigator to monitor methylation reactions withount the background incorporation into proteins encountered using L-[³H-methyl] methionine as the metabolic precursor. Kinetic analyses of [³H]AdoMet utilization in oocytes established that protein carboxyl methylation reactions represent approximately one-third of the oocyte AdoMet utilization [127].

The calculated rate of protein carboxylmethylation is very similar to that of protein synthesis for both the aggregate of oocyte proteins [127] and for an individual oocyte protein, CaM [128]. Together, the results indicate that PIMT-catalyzed reactions are metabolically significant in oocytes, even though cells are able to replace damaged proteins by *de novo* synthesis.

2. Subcellular Compartmentation of PIMT and Methyl-accepting Proteins

As in other cells, the substrates for PIMT in oocytes are heterogeneous [127]. Diverse groups of methyl-accepting proteins are found in both the nucleus and cytoplasm, which are quite distinct from each other. Methylation of nuclear proteins could occur in either the nucleus or cytoplasm, because the specific activities of PIMT in the nucleus and cytoplasm are identical. It is likely that PIMT diffuses freely through the nuclear pores, because it is smaller than the exclusion limit for nuclear pores in oocytes [129] and lacks a nuclear localization signal (NLS). Unfortunately, it has not been possible to identify the individual methyl-accepting proteins in oocytes, because the proteins cannot be labeled to a sufficiently high specific activity with a single injection of [³H]AdoMet.

Some isoAsp-containing proteins in oocytes are not methylated by PIMT because they are sequestered in cytoplasmic organelles. The largest organelles in oocytes are yolk platelets, which undergo a massive increase in size and number during oogenesis [125]. The process of oogenesis occurs over several months, during which time yolk protein is accumulated by receptor-mediated endocytosis and processed into semi-crystalline arrays within yolk platelets, where they are inaccessible to cytoplasmic PIMT.

Figure 13.7 shows an experiment in which yolk platelets were isolated from oocytes ranging in size from 0.6 to 1.2 mm in diameter and used as substrates for PIMT *in vitro* [62]. Both of the major yolk proteins, phosvitin and lipovitellin, are methylated by PIMT as well as other minor yolk proteins. The specific methyl-accepting of yolk proteins increases about fourfold between the 0.6-mm (lane 1) and 1.2-mm (lane 6) oocyte. The data in Figure 13.7 seriously underestimate the real rate at which methyl-accepting sites accumulate in yolk proteins, however, because newly synthesized protein is continually added to yolk plates by endocytosis. Nonetheless, the data indicate that methyl-accepting sites can accumulate rapidly in a physiological compartment where PIMT is not present. By contrast, the specific methyl-accepting activity of cytosolic proteins does not increase over the same physiological time span, consistent with a repair function for PIMT in oocytes.

3. Alternative Processing of IsoAsp Proteins

The large size of *Xenopus* oocytes has facilitated their use as a microinjection model for analyzing the metabolism of defined isoAsp-containing substrates for

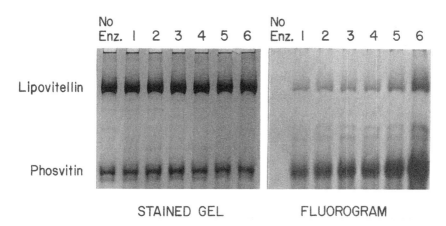

Fig. 13.7. Methyl-accepting activity of yolk proteins obtained from oocytes at different stages of oogenesis. Yolk proteins were isolated from vitellogenic oocytes ranging in diameter from 0.6 to 0.7 mm (lane 1), 0.7 to 0.8 mm (lane 2), 0.8 to 0.925 mm (lane 3), 0.925 to 1.075 mm (lane 4), 1.075 to 1.2 mm (lane 5), and 1.2 to 1.25 mm (lane 6) and used as the substrate for PIMT and [³H]AdoMet. Methylated proteins were separated by SDS-PAGE at pH 2.4 and methylated proteins were visualized by fluorography. Reproduced from Figure 5 in O'Connor [62].

PIMT *in vivo*. The oocyte PIMT is not saturated with endogenous substrates and demonstrates considerable reserve capacity in microinjection experiments. Rates of methylation increase by up to an order of magnitude following the injection of isoAsp peptides or proteins [128, 130, 131].

At the same time, however, a significant fraction of injected isoAsp-containing peptide and protein substrates are proteolytically degraded, suggesting that carboxyl methylation and proteolysis represent alternative pathways for the metabolism of isoAsp-containing proteins. This last conclusion was strengthened by the demonstration that an isomerized variant of CaM was protected from proteolytic digestion by incubation with PIMT and AdoMet prior to injection [131]. The proteolytic activity responsible for the degradation of isomerized CaM was identified as the proteasome, based on its sensitivity to lactacystin [73]. Interestingly, degradation of isomerized CaM does not appear to require ubiquitination, because purified proteasomes are able to degrade isomerized CaM directly.

VII. Accumulation of Isoaspartyl and Racemized Proteins During Aging

An implication of the repair hypothesis is that isoAsp and D-aspartyl sequences will accumulate over time in protein sequences that cannot be modified by PIMT *in vivo*. Accordingly, one would expect that methyl-accepting sites would accumulate rapidly in extracellular protein sequences that are not normally accessible to PIMT. Only small amounts of PIMT have been detected in plasma, presumably originating in damaged cells [132]. Methyl-accepting sites in intracellular proteins would also be expected to increase if PIMT activities became unable to cope with the burden of damaged proteins, due either to aging or enhanced rates of protein damage. Considerable experimental evidence supports both of these predictions, as discussed in material following. Although correlative, the data suggest that the absence of PIMT-catalyzed repair could have pathological implications for age-related diseases.

A. COLLAGEN

Collagen is the major extracellular matrix protein in vertebrates and might therefore be expected to accumulate isoAsp residues over time. Interestingly, the first description of isoAsp generation via a succinimide intermediate involved the $\alpha 1$ chain of collagen I [133]. More recent experiments point to a complex chemistry that generates potential substrates for PIMT. The Asp residue in the C-terminal telopeptide $A_{1209}HDGGR_{1214}$ of collagen $\alpha 1$ [134] is derivatized to a mixture of D-Asp and (D/L)-isoAsp variants (Figure 13.2), which can be distinguished using isoform-specific antibodies. Due to the normal turnover of bone tissue, the quantities of the various peptides isoforms can be monitored in urine samples obtained from patients. Comparing the isoform profile of individuals ranging in age from 0.2 to 74 years, investigators observed a steady decline in the normal isoform and a corresponding increase in damaged isoforms with donor age [134].

In another experiment, samples obtained directly from cortical bone were found to have higher proportions of isomerized variants than samples from trabecular bone, which turns over more slowly [135], suggesting that older cells are the source of the isomerized variants. Based on these results, it has been suggested that the collagen isoform profile could provide a useful clinical indicator of bone growth and turnover.

A rat model has been used to assess the functional consequences of isoAsp formation in collagen. The specific methyl-accepting activity of rat tail collagen increases dramatically between three and 20 weeks of age [136]. Collagen isolated from the tails of 20-week old rats had about one-third the ability of collagen from three-week old rats to support the motility of fibroblasts in a cell migration assay. Exhaustive methylation of the 20-week collagen with PIMT prior to the migration assay restored much of the lost activity, suggesting that isoaspartyl formation and/or racemization can disrupt the extracellular matrix and interfere with normal remodeling processes.

B. CRYSTALLINS

The mammalian eye lens has provided a useful physiological model for studying age-related changes in protein repair. The oldest fiber cells at the center of the lens nucleus are as old as the organism, and they are surrounded by layers of progressively younger fiber cells in the lens cortex. The predominant lens proteins are the crystallins, a group of developmentally regulated proteins responsible for lens transparency. Crystallins are remarkably stable proteins that accumulate over time a large number of chemical modifications caused by oxidation, cross-linking, deamidation, racemization, and other processes [137]. At least some of the D-Asp and L-isoAsp residues that have been detected chemically in crystallins are expected to be substrates for PIMT *in vivo*, because PIMT catalyzes the methylation of all crystallin classes *in vitro* [138, 139].

Detailed chemical information is available for one potential methyl-accepting site, Asp-151 of αA crystallin [140], which is highly prone to both racemization and isomerization. In this experiment, tryptic peptides corresponding to residues 146 through 157 were isolated from the lens tissues of newborns and adults aged 30, 60, and 80 years, and the proportion of each isoform was determined using standard chemical methods. The investigators found that the fraction of normal peptide decreased steadily over time, representing only 42% of the total at 80 years, whereas the D-isoAsp isomer showed the greatest increase with age, representing 35% of the total at 80 years. The D-isoAsp isoform was found to be particularly concentrated in the lens nucleus [141] by immunochemical localization with an isoform-specific antibody.

The pattern of crystallin isoforms and the changes observed during aging are in good agreement with transformations outlined in Figure 13.2 if protein repair becomes less efficient during aging. Measurements of PIMT and endogenous methyl-accepting proteins in lens tissues are, in fact, consistent with this possibility. In a comparative study of human lenses varying in age from 3 to 89 years, a negative correlation was observed between PIMT and lens protein methyl-accepting activities. PIMT levels declined steadily with age, whereas endogenous methyl-accepting activity showed the opposite trend [138, 139]. At any age, PIMT specific activity was significantly lower in the lens nucleus than in the cortex, whereas the opposite relationship was observed for methyl-accepting activity. The functional consequences of declining PIMT are unclear. Exceptionally low levels of PIMT have been observed in cataractous lenses [139, 142], but any relationship with cataract formation is strictly correlative.

C. AMYLOID

Extracellular L-isoAsp residues and D-Asp residues have also been identified in the amyloid deposits from patients with Alzheimer's disease [143-147, 148] and may play a role in the formation of these peptide aggregates. The major component of amyloid plaques is the β-amyloid peptide Aβ derived from the processing of the Alzheimer's precursor protein (APP). Various types of evidence suggest that racemization and isomerization occur at all three Asp residues in Aβ, located at positions 1, 7, and 23.

Structural analysis of the Aβ peptides isolated from neuritic plaques and the parenchymal vasculature reveals mixtures of racemized and isomerized Asp residues at both positions 1 and 7 [143, 144, 146], suggesting that the residues arise through the succinimide chemistry shown in Figure 13.2. The presence of the Aβ(isoAsp7) variant in both neuritic plaques and the cerebrovascular amyloid has been confirmed cytologically using isoform-specific antibodies [147-149]. Isomerized variants of Asp-23 have not been detected biochemically in amyloid, but this could reflect the fact that only a portion of the amyloid deposits can be solubilized sufficiently for biochemical characterization. Immunochemical experiments with peptide-specific antibodies indicate that Aβ(isoAsp23) is present in both neuritic plaques and the brain microvasculature [148, 149]. Staining is particularly intense in the cores of amyloid deposits, suggesting that Asp-23 variants of Aβ are highly insoluble.

Potential roles for isomerized and racemized Asp residues in amyloid formation have been addressed using synthetic peptides corresponding to the various Aβ variants. Modifications to Asp-23 appear to be the most deleterious. Synthetic Aβ1-42(isoAsp23) showed a much greater tendency to aggregate than either

Aβ1-42(isoAsp7) or normal Aβ1-42. Aggregation in this experiment was also highly correlated with neurotoxicity with cultured neurons [150]. Racemization of Asp23 may also increase its aggregation properties.

In a second study, the D-Asp23 variant of Aβ1-35 was much more prone to aggregation and fibril formation than either Aβ1-35(D-Asp7) or unmodified Aβ1-35 [145]. Interestingly, Asp-23 is the site of the Iowa D23N mutation associated with inherited early onset dementia, and the substitution of an Asn residue for Asp-27 increases the propensity of synthetic Aβ peptides to undergo isomerization [149]. It is not clear that PIMT could play a role in preventing Alzheimer's disease, in that the isomerization reactions that promote amyloid formation most likely occur in the mature protein when the relevant amino acids are not accessible to PIMT.

VIII. Physiological Studies of PIMT Function

PIMT activities are widely distributed across phylogenetic domains (see section V.), suggesting that PIMT plays a fundamental role in cellular protein metabolism. Nonetheless, cellular requirements for PIMT activity may vary widely. Significant variations in PIMT concentrations have been observed between different cell types within an organism and in a single cell type responding to physiological and environmental stimuli. This section reviews our current understanding of PIMT regulation in several species. The regulatory elements responsible for physiological changes in PIMT activities remain largely unknown. For the most part, PIMT appears to be constitutively produced in cells, although the data suggest that PIMT concentrations may increase in response to the burden of structurally damaged proteins.

A. MAMMALS

1. Cellular and Subcellular Distribution

PIMT is ubiquitously expressed in mammalian tissues, although PIMT activities vary widely among tissues and between cell types. In rodents, PIMT specific activities are particularly high in brain and testis, whereas much lower levels of enzyme activity are present in liver and kidney [25]. Immunohistochemical analyses of brain slices indicate that PIMT concentrations are significantly higher in neuronal cells than in glial cells, although oligodendrocytes and astroglial cells both stain positively with anti-PIMT antisera [151-153].

Biochemical measurements are consistent with the immunochemical results, because PIMT specific activities in the developing rat brain parallel its population by mature neurons [152]. PIMT activity may be important for neuronal function.

Brain tissue contains many different substrates for PIMT, including important neuronal proteins such as synapsin [57, 154, 155] and tubulin [156]. The loss of PIMT function in transgenic mice (see material following) results in neuronal death and the development of fatal epileptic seizures, whereas overexpression of PIMT in primary cortical neurons is able to protect cells against Bax-induced apoptosis [157].

Within the testis, PIMT activities are present at significantly higher levels in germ cells than in somatic cells. Rodent testes depleted of germ cells as a consequence of either mutation or X-ray irradiation have reduced levels of PIMT [158, 159], and during prepuberal development PIMT-specific activities increase in parallel with the population of the seminiferous tubules with spermatogenic cells and mature sperm [158–161]. The changes in PIMT activity during spermatogenesis have been followed using staged cell populations isolated from dissociated seminiferous tubules [158, 159].

These experiments indicate that PIMT is upregulated during the final haploid phase of spermatogenesis. It is tempting to speculate that high concentrations of PIMT in mature sperm may have important implications for fertility, because mature sperm are translationally inactive cells that may be stored in the epididymis and vas deferens for several weeks prior to ejaculation. Sperm isolated from both the caput (proximal) and caudal (distal) epididymides have been shown to possess both PIMT and methyl-accepting proteins [162, 163]. Unfortunately, it has not been possible to directly address the functional importance of PIMT to male fertility in transgenic mice, because PIMT-deficient mice die before they are fully fertile [24].

2. Gene Structure and Transcriptional Regulation

A single gene encoding PIMT is present in mammalian genomes. The first cloned *Pcmt* gene was isolated from a rat brain cDNA library by Sato et al. [15], who also reported the presence of multiple transcripts in rat brain mRNA. Subsequent experiments established that the mouse *Pcmt1* [16] and human *PCMT1* [164] genes are organized into eight exons that are transcribed and spliced into at least three transcripts of varying lengths [15]. An alternative splice site has been identified in the seventh exon [33, 35, 165], which is used to produce the coding sequences for isozymes with distinct C-termini (see section III.A). The 5′-flanking sequences of the mouse [16] and human [164] genes have features that are typical of constitutively expressed genes.

Both sequences are GC-rich, with CpG islands that start approximately 400 bp upstream from the transcription start site and extend through exon 1 and intron 1. The methylation status of the CpG islands is not known, but their presence suggests that the genes may be subject to silencing by DNA methylation. The genes lack conventional TATA boxes, and transcription begins at multiple start sites [35, 164]. Most of the consensus transcription factor binding sites that can be identified (using stringent criteria) are binding sites for widely expressed transcription factors such as Sp1, ETF, and AP1.

Interestingly, the sequences also contain potential antioxidant [166] and unfolded protein [167] response elements, but there is no evidence that *Pcmt1* transcription responds to these stimuli (O'Connor, unpublished data). A minimal sequence for the mouse promoter has been functionally defined by monitoring reporter gene expression in cultured NIH/3T3 cells [35]. This 407 bp sequence extends from the transcription start sites upstream through the GC-island, including the transcription factor binding sites. Overall, the features of the *Pcmt1* promoter are consistent with the wide expression of PIMT in mammalian tissues. Additional tissue-specific enhancers undoubtedly exist and contribute to the tissue-specific differences in *Pcmt1* regulation, but these factors have not yet been identified.

The regulation of PIMT levels appears to occur primarily at the transcriptional level in mammals, in that transcript levels closely parallel measured PIMT specific activities. High levels of *Pcmt1* transcripts are detected in mouse brain and testis, whereas much lower transcript levels are detected in preparations of liver RNA [16, 161]. Brain and testicular tissues also show parallel increases in PIMT enzymatic activities and *Pcmt1* transcript levels during early development, coincident with increased numbers of neurons and spermatogenic cells, respectively [158, 161].

The size distribution of *Pcmt1* transcripts is similar in all tissues except the testis. The major testicular transcript is a smaller 900-bp transcript with a unique 3'-untranslated region derived from alternative splicing in intron 7 [24, 35]. This transcript is positively regulated during the differentiation of pachytene spermatocytes to round spermatids. It is possible that this unique 3'-untranslated region is important for translational regulation during the late stages of spermiogenesis when the spermatid nucleus has condensed and become transcriptionally inactive. Like the protamine and testis-specific actin mRNAs that are translated in late spermatids [168], *Pcmt* transcripts are distributed in both the polysomal and non-polysomal fractions obtained by ultracentrifugation [24].

3. Functional Analysis in Transgenic Models

Loss of PIMT function in mice produces phenotypes that appear to primarily affect the nervous system. Two different strains of mice generated by homologous recombination demonstrate virtually identical phenotypes [23, 114]. PIMT-deficient mice appear fairly normal at birth, although the animals are slightly smaller than wild-type. Histologically, most tissues appear normal and the overall neuroanatomy is normal. Reproducible abnormalities are detected, however, in some pyramidal neurons in layer V of the precentral cortex, granule cells in dentate gyrus, and some astrocytes in the hippocampus [114]. A few weeks after birth, brains of knockout mice become enlarged and brains continue to grow at higher rates than in wild-type mice [114, 153]. The underlying biochemical disturbances accompanying these morphological changes may involve alterations in insulin signaling through PI3K/Akt pathway. Western blots and immunochemical data indicate that the insulin and IGF-1 receptors are upregulated in the knockout hippocampus, and components of the PI3K/Akt signaling pathway are more highly phosphorylated [153].

The biochemical changes in knockout mice become manifest in several behavioral disorders. PIMT-deficient mice fail to mate and they experience epileptic seizures that increase in severity before mice finally die from fatal seizures between 22 and 60 days after birth [23, 114, 169]. Mutant animals also display exaggerated sensitivity to convulsive agents, reduced memory and altered behaviors in several quantitative assays [170, 171]. Electrophysiological measurements with isolated brain slices indicate that synaptic transmission within the hippocampus is abnormal [170], possibly accounting for the behavioral disorders displayed by the mice.

Neurons appear to be the cell type most affected by the loss of PIMT function, because the epileptic phenotype is partially rescued by PIMT transgenes in which the neuron-specific *enolase* promoter controls *Pcmt* expression [172]. Rescued mice can live up to five times longer than knockout mice, and they accumulate fewer unmodified isoAsp-containing proteins in their brain tissue. More complete rescue is observed when either of the PIMT isozymes is expressed in knockout mice under the control of the stronger but less specific *prion* promoter [155]. The epileptic phenotype is also partially rescued when an adenovirus containing the PIMT1 coding sequence is injected into the ventricles of embryonic mice [37]. In all of the rescue experiments, there is a good correlation between the degree of phenotype rescue and protein repair.

Non-neural tissues appear to be less affected by the loss of PIMT activity, even though methyl-accepting proteins also accumulate in these tissues. No histological differences are apparent in the non-neural tissues of PIMT-deficient mice and no consistent phenotypes have been observed. Potential testicular defects have been examined in considerable detail [24], because of the high PIMT concentrations normally detected in testis (see previous material). Molecular analyses indicate that spermatogenesis occurs normally in the absence of PIMT activity. Sperm-specific markers appear at the normal times during development and acrosome reaction-competent sperm can be recovered from the epididymides of knockout mice [24].

It has not been possible to perform quantitative analyses on the fertilization-competence of PIMT-deficient sperm, in that the animals die from seizures about the same time that they reach sexual maturity. In addition, PIMT-deficient mice fail to mate. Knockout mice rescued with a neuron-specific *enolase:Pcmt* transgene are partially fertile [172], suggesting that behavioral factors contribute to the infertility of knockout mice. Additional experiments are required to exclude direct effects of PIMT depletion on sperm function, because the neuron-specific *enolase* promoter has been reported to be partially active in testis [202].

Overall, there is a strong correlation between the severity of mouse phenotypes and the accumulation of unprocessed isoAsp-containing proteins [23, 114, 172].

As predicted from the scheme in Figure 13.2, PIMT-deficient mice also have higher levels of racemized proteins [201], which are also unrepaired. Lack of protein repair may not be sufficient to totally account for the phenotypes observed in PIMT-deficient mice, however, because the ratio of AdoMet to AdoHcy is unexpectedly elevated in the brains of PIMT-deficient mice [173]. AdoMet is known to have pharmacological properties [174] and it is also involved in the synthesis of other neurotransmitters [175], raising the possibility that epigenetic factors contribute to the phenotypes displayed by PIMT-deficient mice.

B.   PLANTS

1.   *Distribution of PIMT Activities in Higher Plants*

Data from several studies have established that PIMT activities are widely distributed among all the major divisions of the plant kingdom [27, 40, 58, 59]. In a comparative study [59], functional PIMT activity was detected in crude extracts prepared from 45 different species, representing 23 families of both seedless and seed-bearing plants. Fairly similar levels of PIMT activity were observed in seeds obtained from coniferous species and from both monocotyledonous and dicotyledonous angiosperms.

Overall, lower levels of PIMT activity were detected in extracts prepared from mosses and ferns than in extracts prepared from seed-bearing plants. PIMT activity was not detected in several algal species, but this could reflect the limitations of using enzymatic assays rather than genomic information to identify sources of PIMTs. Biochemical measurements are probably best construed as providing only rough estimates of PIMT activity because some plant extracts contain substances that can interfere with biochemical measurements of PIMT activity [58, 59].

In higher plants, PIMT specific activities are often concentrated in seeds. It has been suggested that PIMT activity is important for maintaining the structural integrity of embryonic proteins during long-term storage in the seed [27]. Consistent with this hypothesis, PIMT is positively regulated during seed development [17] and is more concentrated in embryonic tissue than in the endosperm [59]. Depending on the species, seeds can be stored for considerable periods of time, retaining the ability to germinate and produce progeny. An extreme example is the sacred lotus plant, whose seeds can retain the ability to germinate for centuries.

PIMT appears to be very stable in lotus seeds, as shown by the recovery of robust PIMT activity from seeds estimated to be 95 and 416 years old [176]. In a more typical example, PIMT activity was detected in barley seeds aged for 17 years, even though the seeds were no longer viable. Higher levels of PIMT activity were detected in younger seeds stored for 5 or 11 years, which still maintained partial viability [59]. The older seeds also demonstrated higher methyl-accepting activity than younger seeds, suggesting that the seed PIMT may become progressively more saturated with potential substrates over time.

PIMT is not restricted to the seed tissue of higher plants. Enzymatic activity is also detected in extracts prepared from nonreproductive tissues. Considerable variability is observed, however, in the tissue distribution of PIMT between plant species. Relatively high levels of PIMT are present in corn stems and leaves and in carrot roots and leaves. By contrast, rice and wheat have only low levels of activity in the same tissues [177]. In *Arabidopsis*, PIMT enzymatic activity was not detected in tissues other than seeds, although *PIMT* transcripts were readily identified in RNA samples prepared from these tissues [38]. Additional studies will be needed to determine if this apparent contradiction reflects posttranscriptional regulation of PIMT levels in nonreproductive tissue or if the difference is caused by the presence of substances in tissue extracts that interfere with PIMT assays.

2.   Gene Structure and Transcriptional Regulation

The regulation of PIMT production in plants is expected to be at least as complex as in animals, because higher plants may contain more than one gene that encodes PIMT. Large scale genome sequencing and microarray projects involving *Arabidopsis* and cereal grains have already produced a wealth of information about *PIMT* gene organization and transcript processing that will need to be reconciled with biochemical information in the future. Information from the sequencing and microarray projects is available online in databases maintained by the Institute for Genomic Research (*www.tigr.org*) and the Arabidopsis Information Resource (*www.arabidopsis.org*).

These projects have identified paralogous *PIMT* sequences on two different chromosomes in the genomes of *Arabidopsis*, rice, and wheat. Like many other plant genes, the two *PIMT* genes most likely arose by ancestral gene duplication. The duplicated genes have undergone considerable sequence divergence, yet the predicted proteins within a species are more than 85% similar. There is good evidence that both genes are transcribed. Full-length cDNA and EST clones have been reported for both *PIMT* genes in *Arabidopsis* and rice. EST sequences have also been reported for both wheat genes, although only one full-length cDNA sequence has been isolated [27].

PIMT genes in plants are organized into four exons, which may be used for alternative splicing. The most detailed information about *PIMT* transcription is available for *Arabidopsis*. *PIMT1* transcripts are more abundant than *PIMT2* transcripts, but *PIMT2* may be subject to more complex regulation. Alternative splicing between exons 1 and 2 of *PIMT2* primary transcripts generates transcripts for two distinct proteins, each of which is predicted to have a NLS [39]. The NLS is functional in *Arabidopsis* cells, but their significance is unclear, in that PIMTs are expected to diffuse freely through nuclear pores [62]. NLS sequences are not present in other *PIMT2* sequences reported in the public databases or in transcripts of the *PIMT1* gene. It is not clear if alternative processing occurs during *PIMT1* transcription. Two transcripts hybridize with *PIMT1* probes on northern blots [38], but further analysis is needed to determine if this is the result of alternative processing of *PIMT1* transcripts or cross-hybridization of *PIMT2* transcripts with the *PIMT1* probe.

There is good evidence for the transcriptional regulation of PIMT activities in higher plants during development and in response to environmental stress. In wheat, *PIMT* transcription increases during caryopsis development coincident with increases in PIMT activity. PIMT transcription also increases during salt stress and desiccation [17]. The phytohormone, abscisic acid (ABA), is hypothesized to be the common regulator in these transitions. ABA is known to promote seed development, and it has been implicated as well in plant salt responses. Consistent with this hypothesis, ABA treatment of 4-day-old wheat seedlings induces *PIMT* transcription. In all cases, there is good correspondence between data obtained from northern blots and enzymatic assays, suggesting that regulation of PIMT levels in wheat occurs primarily at the transcriptional level.

ABA has also been implicated in regulation of *PIMT* transcription in *Arabidopsis*. The 5′-flanking sequences of both the *PIMT1* [38] and *PIMT2* [39] genes contain putative ABA response elements. Transcription of both genes increases during seed development and in response to ABA treatment. Some differences are observed, however, in the responses of the two genes to environmental stress. Transcription of *PIMT2* increases in response to both salt stress and dehydration, whereas *PIMT1* is minimally affected by the stresses.

3. Functional Importance of PIMT in Plants

Functional analyses of PIMT in plants are strictly correlative at present. Based on the distribution of PIMT activities and the positive regulation of *PIMT* transcript by ABA, it is likely that PIMT is important for maintaining protein function during long-term storage or under stress conditions. Potential roles for PIMT in plant survival can be tested in transgenic plants, because transgenic *Arabidopsis* lines deficient in single *PIMT* genes are already available in stock collections. The development of transgenic models will be complicated by the fact that *PIMT* genes are duplicated in higher plants, and it may be necessary to inactivate both genes in order to analyze PIMT function.

C. DROSOPHILA

1. Distribution of PIMT Activities in Drosophila

The small size and short lifespan of *Drosophila* have made it possible to analyze PIMT levels throughout the complete lifespan. PIMT can be detected biochemically in the early embryo before embryonic transcription is initiated, suggesting that maternal transcripts are translated in the oocyte and early embryo [178]. PIMT levels then remain essentially constant during embryogenesis before dropping to their lowest levels in larvae, and subsequently increasing during pupal development. The highest levels of PIMT are found in the adult fly. PIMT specific activity increases sharply during the first day after eclosion to the adult level, where it is maintained at an essentially constant level for the duration of the lifespan [179, 180].

2. Gene Structure and Regulation of PIMT Activity

The *Drosophila Pcmt* gene consists of four exons and is located in region 83B of the third chromosome. There is no evidence for alternative splicing during the processing of *Pcmt* transcripts, in that a single 1.6-kb transcript is detected on northern blots of *Drosophila* RNA [179]. The abundance of this transcript is always proportionate to the biochemical PIMT activity, suggesting that PIMT is primarily regulated at the transcriptional level in flies. The regulatory factors that control *Pcmt* transcription are unknown.

The upstream flanking region of the *Pcmt* gene possesses few consensus binding sites for known transcriptional regulators in *Drosophila*. Transcription initiates approximately 200 bp upstream from the initiation codon in the proximity of a TATA box and an arthropod capsite consensus element [181]. The absence of other binding sites for tissue- and stage-specific transcription factors seem to be consistent with constitutive regulation of the gene.

The only changes that have been noted in *Pcmt* expression occur in response to heat. Flies respond to a series of five 15-min heat shocks at 34°C by elevating PIMT levels approximately 50% [180]. In addition, the PIMT specific activity is slightly higher in flies raised at 29°C than in flies raised at 25°C (O'Connor, data not shown). These small temperature-dependent increases in PIMT expression do not appear to be mediated by the well-characerized *Drosopohila* heat-shock transcription factor [182], however, because no binding sites can be detected in the *Pcmt* flanking sequence.

3. Functional Analysis in Transgenic Models

No loss-of-function *Pcmt* mutants are currently available in *Drosophila*, and high-throughput RNAi screens have not identified any phenotypes associated with *Pcmt* silencing. Functional studies have been restricted to overexpression experiments, which suggest that PIMT plays an important role in mitigating the aging process. When the binary GAL4-UAS system [183] is used to drive constitutive overexpression of PIMT activity in flies, the median lifespan is increased by 30 to 40% [180]. The longevity effect is temperature dependent.

Lifespan is extended when flies are raised at 29°C , but not at 25°C, suggesting that PIMT activity can become limiting under conditions where rates of protein isomerization are expected to increase. PIMT is one of very few proteins that extend the *Drosophila* lifespan when overexpressed *in vivo*. Other overexpressed proteins that increase the *Drosophila* lifespan include Cu/Zn superoxide dismutase [184–186], mitochondrial Mn-superoxide dismutase [187], heat shock proteins [188], and methionine sulfoxide reductase [189]. A common feature of all these proteins is that they are involved in preventing or repairing damage to protein structures.

D.   *CAENORHABDITIS ELEGANS*

1.   *PIMT Activities During C. elegans Development*

The nematode *C. elegans* is a popular model organism in which to study both early development and aging. The *pcm-1* gene in *C. elegans* encodes a PIMT of 225 amino acids. The enzyme has been overexpressed in *E. coli* and has been shown to catalyze the methylation of isoAsp peptide substrates, although much less effectively than human PIMT [45]. Unlike mammalian PIMTs, the *C. elegans* enzyme does not recognize ovalbumin as a substrate, suggesting that considerable divergence has occurred in the enzyme active site.

Biochemical analyses of PIMT activity during the *C. elegans* lifespan are difficult because of the organism's small size, but PIMT is reported to be enriched in dauer larvae, a long-lived quiescent state that worms can enter in response to overcrowding or nutritional deprivation. The PIMT specific activity measured in dauer larvae is approximately twice that of mixed-stage populations, in good agreement with transcriptional data (see material following). Dauer larvae can survive for periods as long as 70 days in this altered metabolic state, which perhaps increases the metabolic requirement for PIMT to maintain protein integrity.

2.   *Gene Structure and Transcriptional Regulation*

The *pcm-1* gene is located on chromosome 5 and organized in 7 exons [45]. It is not clear if alternative processing generates multiple transcripts in *C. elegans*, because data on *pcm-1* transcription have been obtained exclusively from genomewide microarray experiments. These data reveal only small fluctuations in the concentration of *pcm-1* transcripts during development. In a carefully synchronized study of early development [190], *pcm-1* transcript levels remained essentially constant from the oocyte through the larval stages. Although the *pcm-1* transcripts in oocytes are of maternal origin, embryonic transcription subsequently maintains *pcm-1* transcripts at a steady state. Transcript levels do not change significantly during the adult lifespan [191], but are approximately twofold higher in dauer larvae [192, 193], in good agreement with the biochemical measurements [45].

In microarray experiments, the transcription profile of *pcm-1* does not cluster with those of other genes. A broader view is required to begin to detect associations between *pcm-1* transcription and that of other genes. Such a compendium of gene expression has been constructed for *C. elegans* genes based on 553 microarray experiments from 30 different laboratories [194]. The experimental data used in the analysis included experiments involving different developmental stages, growth conditions, aging, stress responses, and mutant strains.

The outcome of the analysis was the identification of groups of genes, or "mountains," whose expression was highly correlated. Using this statistical data, *pcm-1* was placed in Mount 6, consisting of 909 genes that are transcribed at elevated levels in neuronal cells [194]. Mount 6 includes a broad spectrum of proteins with diverse functions, and much additional work will be required to establish functional relationships between them. The tendency for Mount 6 transcripts to be elevated in neuronal cells is particularly interesting in light of the elevated PIMT concentrations in mammalian neurons, suggesting that the functional importance of PIMT to neuronal function is conserved during evolution. Because of its comparative simplicity, *C. elegans* may provide a useful model in which to investigate physiological roles for PIMT in the nervous system.

3. Functional Analysis of PIMT

PIMT-deficient worms have been generated using transposon mutagenesis [22]. The mutant strains were isolated by screening for the spontaneous excision of a Tc1 transposon located in the fourth intron of the *pcm-1* gene. Imprecision excision of the transposon generated a deletion encompassing exons 2 through 5. Homozygous PIMT-deficient worms were viable and developed normally, but showed several subtle phenotypes. In mixed cultures with wild-type worms, the frequency of the null allele decreased slowly over time.

The difference between the wild-type and mutant worms was a slight one, however, because the null allele was still detected after 65 generations. A second phenotype was observed during dauer phase. The mean survival time for mutant worms in dauer phase was 24.5 days, compared to 27 days for the wild-type control. The results suggest that the fitness of mutant worms is slightly less than that of wild-type worms, but the differences are subtle. It will be important to extend these studies in the future to exclude any contributions of genetic drift or genetic background to the phenotypes.

E. BACTERIA

1. Protein IsoAsp Methylation in E. coli

The functional significance of bacterial PIMTs has been studied most thoroughly in *E. coli*, because of the techniques available for genetic manipulation. The *E. coli* PIMT sequence is 31% identical to the human PIMT sequence, and the enzyme catalyzes methylation of the same isoAsp substrates as human PIMT, but with reduced efficiency [195]. The K_ms of *E. coli* PIMT for isoAsp peptides and ovalbumin are significantly higher than those of mammalian PIMTs with the same substrates. Surprisingly, *E. coli* PIMT may also be less effective than mammalian PIMT at modifying bacterial methyl-accepting proteins.

This possibility was serendipitously discovered during the analysis of *E. coli* that had been genetically engineered to overexpress rat PIMT [196]. Multiple proteins in extracts prepared from both wild-type and overexpressing *E.coli* were able to act as substrates for purified mammalian PIMT, but the specific methyl-accepting activities of proteins from PIMT-overexpressing bacteria were significantly lower than those from wild-type bacteria, suggesting that overexpressed rat PIMT had previously modified isoAsp residues *in vivo*. The most prominent methyl-accepting protein was identified as ribosomal protein S11, which was methylated to a stoichiometry estimated at 0.5 mol/mol S11. The results suggest that isoAsp sites are inefficiently methylated in intact *E. coli*, in contrast to their efficient methylation in higher eukaryotes. In addition, the results raise the possibility that L-isoAsp residues in S11 are generated by a catalytic mechanism, both because of the speed with which they arise and because their fractional representation is greater than that predicted for a spontaneous mechanism of isoAsp generation (Figure 13.2).

2.   Gene Structure

The regulation of *pcm* transcription has not been studied in detail. Genome annotation predicts that the *pcm* gene is part of an operon that includes four other coding sequences. The predicted coding sequences partially overlap one other, such that the *pcm* coding sequence would be situated at the 3′-end of a polycistronic transcript. The proximal gene, *SurE*, encodes a protein with nucleotidase activity that has also been described as a survival protein. *SurE* has been previously shown to interact genetically with *pcm* [197], but the mechanism is obscure. The functions of the other proteins in the operon have not been identified. The *E. coli* operon organization has a restricted distribution within the bacteria. It will be interesting to determine if *pcm* is transcribed as part of a polycistronic mRNA and if the relationship between the proteins encoding by the operon is coincidental or functional.

3.   Functional Analysis of PIMT in Bacteria

The functional importance of bacterial PIMT has been tested in mutant strains. A strain lacking PIMT activity was constructed by replacing the endogenous *pcm* gene with a chloramphenicol resistance gene [198]. Gene replacement did not affect the growth or stress resistance of vegetative cells, but produced phenotypes that were manifest only under stationary phase culture conditions. Mutant cells showed a reduced resistance to several stresses that could conceivably destabilize protein structures, including methanol, paraquat, and high salt.

In all cases, the mutant phenotypes could be complemented by a wild-type *pcm* gene, indicating that the loss of PIMT function was responsible for the phenotypes.

The *pcm* mutation also conferred a competitive disadvantage to stationary phase mutants co-cultured with normal cells. Interestingly, the mutant phenotypes have been recently shown to be restricted to cultures maintained under alkaline conditions [199], which are known to promote the generation of protein isoAsp residues by the spontaneous mechanism shown in Figure 13.2. In agreement with this prediction, extracts prepared from cells grown at pH 8 or 9 had elevated methyl-accepting activity in direct proportion to the alkalinity of the culture medium.

Other evidence that *E. coli* PIMT might be involved in maintaining native protein structures was obtained in a genetic screen designed to identify multicopy suppressors of protein aggregation and inclusion body formation [200]. In this screen, the *E. coli* tester strain overexpressed a three-domain fusion protein, consisting of the preS2 and S′ domains of the hepatitis B surface antigen N-terminal to β-galactosidase. The fusion protein is soluble in *E. coli* cytoplasm at 37°C, but denatures and forms insoluble aggregates at 43°C.

Transformation of the tester strain with a genomic segment containing the *pcm* coding sequence suppressed fusion protein aggregation at 43°C. In addition, PIMT overexpression also protected β-galactosidase from thermal denaturation in extracts prepared from the transformed cells. The mechanism by which PIMT suppresses aggregation requires further investigation. The fusion protein is a highly nonphysiological substrate produced at high levels in the cytoplasm, and its aggregation has not actually been linked to isoAsp formation. It will also be important to determine if a direct link exists between the suppression of fusion protein aggregation and PIMT catalytic activity.

REFERENCES

1. Berlett, B.S., and Stadtman, E.R. (1997). Protein oxidation in aging, disease, and oxidative stress. *J Biol Chem* 272:20313–20316.
2. Stadtman, E.R., and Levine, R.L. (2000). Protein oxidation. *Ann NY Acad Sci* 899:191–208.
3. Robinson, N.E., and Robinson, A.B. (2001). Deamidation of human proteins. *Proc Natl Acad Sci USA* 98:12409–12413.
4. Bada, J.L. (1984). *In vivo* racemization of mammalian proteins. *Methods Enzymol* 106:98–115.
5. Axelrod, J., and Daly, J. (1965). Pituitary gland: Enzymatic formation of methanol from S-adenosylmethionine. *Science* 150:892–893.
6. Liss, M., Maxam, A.M., and Cuprak, L.J. (1969). Methylation of protein by calf spleen methylase. *J Biol Chem* 244:1617–1622.
7. Morin, A.M., and Liss, M. (1973). Evidence for a methylated protein intermediate in pituitary methanol formation. *Biochem Biophys Res Commun* 52:373–378.
8. Kim, S., and Paik, W.K. (1970). Purification and properties of protein methylase II. *J Biol Chem* 245:1806–1813.
9. Janson, C.A., and Clarke, S. (1980). Identification of aspartic acid as a site of methylation in human erythrocyte membrane proteins. *J Biol Chem* 255:11640–11643.

10. McFadden, P.N., and Clarke, S. (1982). Methylation at D-aspartyl residues in erythrocytes: Possible step in the repair of aged membrane proteins. *Proc Natl Acad Sci USA* 79:2460–2464.
11. Murray, E.D. Jr., and Clarke, S. (1984). Synthetic peptide substrates for the erythrocyte protein carboxyl methyltransferase. *J Biol Chem* 259:10722–10732.
12. Aswad, D.W. (1984). Stoichiometric methylation of porcine adrenocorticotropin by protein carboxyl methyltransferase requires deamidation of asparagine 25. *J Biol Chem* 259:10714–10721.
13. Ingrosso, D., Fowler, A.V., Bleibaum, J., and Clarke, S. (1989). Sequence of the D-aspartyl/ L-isoaspartyl protein methyltransferase from human erythrocytes. *J Biol Chem* 264:20131–20139.
14. Henzel, W.J., Stults, J.T., Hsu, C.-A., and Aswad, D.W. (1989). The primary structure of a protein carboxyl methyltransferase from bovine brain that selectively methylates L-isoaspartyl sites. *J Biol Chem* 264:15905–15911.
15. Sato, M., Yoshida, T., and Tuboi, S. (1989). Primary structure of rat brain protein carboxyl methyltransferase deduced from cDNA sequence. *Biochem Biophys Res Commun* 161:342–347.
16. Romanik, E.A., Ladino, C.A., Killoy, L.C., D'Ardenne, S.C., and O'Connor, C.M. (1992). Genomic organization and tissue expression of the murine gene encoding the protein β-aspartate methyltransferase. *Gene* 118:217–222.
17. Mudgett, M.B., and Clarke, S. (1994). Hormonal and environmental responsiveness of a developmentally regulated protein repair L–isoaspartyl methyltransferase in wheat. *J Biol Chem* 269:25605–25612.
18. Skinner, M.M., Puvathingal, J.M., Walter, R.L., and Friedman, A.M. (2000). Crystal structure of protein isoaspartyl methyltransferase: A catalyst for protein repair. *Structure* 8:1189–1201.
19. Griffith, S.C., Sawaya, M.R., Boutz, D.R., Thapar, N., Katz, J.E., Clarke, S., and Yeates, T.O. (2001). Crystal structure of a protein repair methyltransferase from *Pyrococcus furiosus* with its L-isoaspartyl peptide substrate. *J Mol Biol* 313:1103–1116.
20. Bennett, E.J., Bjerregaard, J., Knapp, J.E., Chavous, D.A., Friedman, A.M., Royer, W.E. Jr., and O'Connor, C.M. (2003). Catalytic implications from the *Drosophila* protein L-isoaspartyl methyltransferase structure and site-directed mutagenesis. *Biochemistry* 42:12844–12853.
21. Li, C., and Clarke, S. (1992). A protein methyltransferase specific for altered aspartyl residues is important in *Escherichia coli* stationary-phase survival and heat-shock resistance. *Proc Natl Acad Sci USA* 89:9885–9889.
22. Kagan, R.M., Niewmierzycka, A., and Clarke, S. (1997). Targeted gene disruption of the *Caenorhabditis elegans* L-isoaspartyl protein repair methyltransferase impairs survival of dauer stage nematodes. *Arch Biochem Biophys* 348:320–328.
23. Kim, E., Lowenson, J.D., MacLaren, D.C., Clarke, S., and Young, S.G. (1997). Deficiency of a protein-repair enzyme results in the accumulation of altered proteins, retardation of growth, and fatal seizures in mice. *Proc Natl Acad Sci USA* 94:6132–6137.
24. Chavous, D.A., Hake, L.E., Lynch, R.J., and O'Connor, C.M. (2000). Translation of a unique transcript for protein isoaspartyl methyltransferase in haploid spermatids: Implications for protein storage and repair. *Mol Reprod Devel* 56:139–144.
25. Diliberto, E.J. Jr., and Axelrod, J. (1976). Regional and subcellular distribution of protein carboxymethylase in brain and other tissues. *J Neurochem* 26:1159–1165.
26. Paik, W.K., and Kim, S. (1985) Protein methylation, in R.B. Freedman and C. H. Hawkins (eds.), *The Enzymology of Post-translational Modifications*, pp. 187–228, London: Academic Press.
27. Mudgett, M.B., and Clarke, S. (1993). Characterization of plant L-isoaspartyl methyltransferases that may be involved in seed survival: Purification, cloning and sequence analysis of the wheat germ enzyme. *Biochemistry* 32:11100–11111.
28. Kim, S. (1974). S-adenosylmethionine: Protein carboxyl methyltransferase from erythrocyte. *Arch Biochem Biophys* 161:652–657.
29. Boivin, D., Gingras, D., and Beliveau, R. (1993). Purification and characterization of a membrane-bound protein carboxyl methyltransferase from rat kidney cortex. *J Biol Chem* 268:2610–2615.

30. Gilbert, J.M., Fowler, A., Bleibaum, J., and Clarke, S. (1988). Purification of homologous protein carboxyl methyltransferase isozymes from human and bovine erythrocytes. *Biochemistry* 27:5227–5233.

31. Aswad, D.W., and Deight, E.A. (1983). Purification and characterization of two distinct isozymes of protein carboxylmethylase from bovine brain. *J Neurochem* 40:1718–1726.

32. Ingrosso, D., Kagan, R.M., and Clarke, S. (1991). Distinct C-terminal sequences of isozymes I and II of the human erythrocyte L-isoaspartyl/D-aspartyl protein methyltransferase. *Biochem Biophys Res Commun* 175:351–358.

33. MacLaren, D.C., Kagan, R.M., and Clarke, S. (1992). Alternative splicing of the human isoaspartyl protein carboxyl methyltransferase RNA leads to the generation of a C-terminal - RDEL sequence in isozyme II. *Biochem Biophys Res Commun* 185:277–283.

34. Potter, S.M., Johnson, B.A., Henschen, A., and Aswad, D.W. (1992). The type II isoform of bovine brain protein L-isoaspartyl methyltransferase has an endoplasmic reticulum retention signal (...RDEL) at its C-terminus. *Biochemistry* 31:6339–6347.

35. Galus, A., Lagos, A., Romanik, E.A., and O'Connor, C.M. (1994). Structural analysis of transcripts for the protein L-isoaspartyl methyltransferase reveals multiple transcription initiation sites and a distinct pattern of expression in mouse testis: Identification of a 5′-flanking sequence with promoter activity. *Arch Biochem Biophys* 312:524–533.

36. O'Connor, C.M., Aswad, D.W., and Clarke, S. (1984). Mammalian brain and erythrocyte carboxyl methyltransferases are similar enzymes that recognize both D-aspartyl and L-isoaspartyl residues in structurally altered protein substrates. *Proc Natl Acad Sci USA* 81:7757–7761.

37. Ogawara, M., Shimizu, T., Nakajima, M., Setoguchi, Y., and Shirasawa, T. (2002). Adenoviral expression of protein-L-isoaspartyl methyltransferase (PIMT) partially attenuates the biochemical changes in PIMT-deficient mice. *J Neurosci Res* 69:353–361.

38. Mudgett, M.B., and Clarke, S. (1996). A distinctly regulated protein repair L-isoaspartyl-methyltransferase from *Arabidopsis thaliana*. *Plant Mol Biol* 30:723–737.

39. Xu, Q., Belcastro, M.P., Villa, S.T., Dinkins, R.D., Clarke, S.G., and Downie, A.B. (2004). A second protein L-isoaspartyl methyltransferase gene in Arabidopsis produces two transcripts whose products are sequestered in the nucleus. *Plant Physiol* 136:2652–2564.

40. Thapar, N., Griffith, S.C., Yeates, T.O., and Clarke, S. (2002). Protein repair methyltransferase from the hyperthermophilic archaeon *Pyrococcus furiosus*. *J Biol Chem* 277:1058–1065.

41. Jamaluddin, M., Kim, S., and Paik, W.K. (1975). Studies on the kinetic mechanism of S-adenosylmethionine: protein O-methyltransferase of calf thymus. *Biochemistry* 14:694–698.

42. Johnson, B.A., and Aswad, D.W. (1993). Kinetic properties of bovine brain protein L-isoaspartyl methyltransferase determined using a synthetic isoaspartyl peptide substrate. *Neurochem Res* 18:87–94.

43. Smith, C.D., Carson, M., Friedman, A.M., Skinner, M.M., Delucas, L., Chantalat, L., Weise, L., Shirasawa, T., and Chattopadhyay, D. (2002). Crystal structure of human L-isoaspartyl-O-methyltransferase with S-adenosylhomocysteine at 1.6-A resolution and modeling of an isoaspartyl-containing peptide at the active site. *Protein Science* 11:625–635.

44. Lowenson, J.D., and Clarke, S. (1991). Structural elements affecting the recognition of L-isoaspartyl residues by the L-isoaspartyl/D-aspartyl protein methyltransferase. *J Biol Chem* 266:19396–19406.

45. Kagan, R.M., and Clarke, S. (1995). Protein L-isoaspartyl methyltransferase from the nematode *Caenorhabditis elegans*: Genomic structure and substrate specificity. *Biochemistry* 34: 10794–10806.

46. Terwilliger, T.C., and Clarke, S. (1981). Methylation of membrane proteins in human erythrocytes. *J Biol Chem* 256:3067–3076.

47. O'Connor, C.M., and Clarke, S. (1984). Carboxyl methylation of cytosolic proteins in intact human erythrocytes. *J Biol Chem* 259:2570–2578.

48. Barber, J.R., and Clarke, S. (1985). Demethylation of protein carboxyl methyl esters: A nonenzymatic process in human erythrocytes? *Biochemistry* 24:4867–4871.

49. Johnson, B.A., and Aswad, D.W. (1985). Enzymatic protein carboxyl methylation at physiological pH: Cyclic imide formation explains rapid methyl turnover. *Biochemistry* 24:581–2586.

50. Stephenson, R.C., and Clarke, S. (1989). Succinimide formation from aspartyl and asparaginyl peptides as a model for the spontaneous degradation of proteins. *J Biol Chem* 264:6164–6170.

51. Freitag, C., and Clarke, S. (1981). Reversible methylation of cytoskeletal and membrane proteins in intact human erythrocytes. *J Biol Chem* 256:6102–6108.

52. Barber, J.R., and Clarke, S. (1983). Membrane protein carboxyl methylation increases with human erythrocyte age. *J Biol Chem* 258:1189–1196.

53. Geiger, T., and Clarke, S. (1987). Deamidation, isomerization, and racemization at asparaginyl and aspartyl residues in peptides. *J Biol Chem* 262:785–794.

54. Kim, S., and Li, C.H. (1979). Enzymatic methyl esterification of specific glutamyl residue in coricotropin. *Proc Natl Acad Sci USA* 76:4255–4257.

55. O'Connor, C.M., and Clarke, S. (1983). Methylation of erythrocyte membrane proteins at extracellular and intracellular D-aspartyl sites *in vitro*. *J Biol Chem* 258:8485–8492.

56. Lowenson, J.D., and Clarke, S. (1992). Recognition of D-aspartyl residues in polypeptides by the erythrocyte L-isoaspartyl/D-aspartyl protein methyltransferase. *J Biol Chem* 267:5985–5995.

57. Aswad, D.W., and Deight, E.A. (1983). Endogenous substrates for protein carboxyl methyltransferase in cytosolic fractions of bovine brain. *J Neurochem* 41:1702–1709.

58. Johnson, B.A., Ngo, S.Q., and Aswad, D.W. (1991). Widespread phylogenetic distribution of a protein methyltransferase that modifies L-isoaspartyl residues. *Biochem Int* 24:841–847.

59. Mudgett, M.B., Lowenson, J.D., and Clarke, S. (1997). Protein repair L-isoaspartyl methyltransferase in plants. *Plant Physiol* 115:1481–1489.

60. Niewmierzycka, A., and Clarke, S. (1999). Do damaged proteins accumulate in *Caenorhabditis elegans* L-isoaspartate methyltransferase (*pcm-1*) deletion mutants? *Arch Biochem Biophys* 364:209–218.

61. Johnson, B.A., and Aswad, D.W. (1985). Identification and topography of substrates for protein carboxyl methyltransferase in synaptic membrane and myelin-enriched fractions of bovine and rat brain. *J Neurochem* 45:1119–1127.

62. O'Connor, C.M. (1987). Regulation and subcellular distribution of a protein methyltransferase and its damaged aspartyl substrate sites in developing *Xenopus* oocytes. *J Biol Chem* 262:10398–10403.

63. Johnson, B.A., Najbauer, J., and Aswad, D.W. (1993). Accumulation of substrates for protein L-isoaspartyl methyltransferase in adenosine dialdehyde-treated PC-12 cells. *J Biol Chem* 268:6174–6181.

64. Ota, I.M., and Clarke, S. (1989). Calcium affects the spontaneous degradation of aspartyl/asparaginyl residues in calmodulin. *Biochemistry* 28:4020–4027.

65. Potter, S.M., Henzel, W.J., and Aswad, D.W. (1993). *In vitro* aging of calmodulin generates isoaspartate at multiple Asn-Gly and Asp-Gly sites in calcium-binding domains II, III, and IV. *Protein Science* 2:1648–1663.

66. Aritomi, M., Kunishima, N., Inohara, N., Ishibashi, Y.O., S., and Morikawa, K. (1997). Crystal structure of rat Bcl-x$_L$: Implications for the function of the Bcl-2 protein family. *J Biol Chem* 272:27886–27892.

67. Clarke, S. (1987). Propensity for spontaneous succinimide formation from aspartyl and asparaginyl residues in cellular proteins. *Int J Peptide Protein Res* 30:808–821.

68. Artigues, A., Farrant, H., and Schirch, V. (1993). Cytosolic serine hydroxymethyltransferase. Deamidation of asparaginyl residues and degradation in *Xenopus laevis* oocytes. *J Biol Chem* 268:13784–13790.

69. Ladino, C.A., and O'Connor, C.M. (1991). Identification of a site for carboxyl methylation in human α-globin. *Biochem Biophys Res Commun* 180:742–747.

70. Kinzel, V., Konig, N., Pipkorn, R., Bossemeyer, D., and Lehmann, W.D. (2000). The amino terminus of PKA catalytic subunit: A site for introduction of posttranslational heterogeneities by deamidation: D-Asp2 and D-isoAsp2 containing isozymes. *Protein Science* 9:2269–2277.

71. Ota, I.M., and Clarke, S. (1989). Enzymatic methylation of L-isoaspartyl residues derived from aspartyl residues in affinity-purified calmodulin. *J Biol Chem* 264:54–60.

72. Johnson, B.A., Langmack, E.L., and Aswad, D.W. (1987). Partial repair of deamidation-damaged calmodulin by protein carboxyl methyltransferase. *J Biol Chem* 262:12283–12287.

73. Tarcsa, E., Szymanska, G., Lecker, S., O'Connor, C.M., and Goldberg, A.L. (2000). Ca^{+2}-free calmodulin and calmodulin damaged by *in vitro* aging are selectively degraded by 26S proteasomes without ubiquitination. *J Biol Chem* 275:20295–20301.

74. Johnson, B.A., Shirokawa, J.M., Hancock, W.S., Spellman, M.W., Basa, L.J., and Aswad, D.W. (1989). Formation of isoaspartate at two distinct sites during *in vitro* aging of human growth hormone. *J Biol Chem* 264:14262–14271.

75. Galletti, P., Iardino, P., Ingrosso, D., Manna, C., and Zappia, V. (1989). Enzymatic methyl esterification of a deamidated form of mouse epidermal growth factor. *Int J Peptide Protein Res* 33:397–402.

76. George-Nascimento, C., Lowenson, J., Borissenko, M., Calderón, M., Medina-Selby, A., Kuo, J., Clarke, S., and Randolph, A. (1990). Replacement of a labile aspartyl residue increases the stability of human epidermal growth factor. *Biochemistry* 29:9584–9591.

77. Daumy, G.O., Wilder, C.L., Merenda, J.M., McColl, A.S., Geoghegan, K.F., and Otterness, I.G. (1991). Reduction of biological activity of murine recombinant interleukin-1β by selective deamidation at asparagine-149. *FEBS Lett* 278:98–102.

78. Tuong, A., Maftouh, M., Ponthus, C., Whitechurch, O., Roitsch, C., and Picard, C. (1992). Characterization of the deamidated forms of recombinant hirudin. *Biochemistry* 31:8291–8299.

79. Paranandi, M.V., Guzzetta, A.W., Hancock, W.S., and Aswad, D.W. (1994). Deamidation and isoaspartate formation during *in vitro* aging of recombinant tissue plasminogen activator. *J Biol Chem* 269:243–253.

80. Hsu, Y.-R., Chang, W.-C., Mendiaz, E.A., Hara, S., Chow, D.T., Mann, M.B., Langley, K.E., and Lu, H.S. (1998). Selective deamidation of recombinant human stem cell factor during in vitro aging: Isolation and characterization of the aspartyl and isoaspartyl homodimers and heterodimers. *Biochemistry* 37:2251–2262.

81. Harris, R.J., Kabakoff, B., Macchi, F.D., Shen, F.J., Kwong, M., Andya, J.D., Shire, S.J., and Bjork, N. (2001). Identification of multiple sources of charge heterogeneity in a recombinant antibody. *J Chromatogr B Biomed Sci Appl* 752:233–245.

82. Zhang, W., and Czupryn, M.J. (2003). Analysis of isoaspartate in a recombinant monoclonal antibody and its charge isoforms. *J Pharm Biomed Anal* 30:1479–1490.

83. Teshima, G., Porter, J., Yim, K., Ling, V., and Guzzetta, A. (1991). Deamidation of soluble CD4 at asparagine-52 results in reduced binding capacity for the HIV-1 envelope glycoprotein gp120. *Biochemistry* 30:3916–3922.

84. Bischoff, R., Lepage, P., Jaquinod, M., Cauet, G., Acker-Klein, M., Clesse, D., Laporte, M., Bayol, A., Van Dorsselaer, A., and Roitsch, C. (1993). Sequence-specific deamidation: Isolation and biochemical characterization of succinimide intermediates of recombinant hirudin. *Biochemistry* 32:725–734.

85. DiDonato, A., Galletti, P., and D'Alessio, G. (1986). Selective deamidation and enzymatic methylation of seminal ribonuclease. *Biochemistry* 25:8361–8368.

86. DiDonato, A., Ciardiello, M.A., deNigris, M., Piccoli, R., Mazzarella, L., and D'Alessio, G. (1993). Selective deamidation of ribonuclease A. *J Biol Chem* 268:4745–4751.

87. Capasso, S., DiDonato, A., Esposito, L., Sica, F., Sorrentino, G., Vitagliano, L., Zagari, A., and Mazzarella, L. (1996). Deamidation in proteins: The crystal structure of bovine pancreatic ribonuclease with an isoaspartyl residue at position 67. *J Mol Biol* 257: 492–496.

88. Galletti, P., Ciardiello, A., Ingrosso, D., DiDonato, A., and D'Alessio, G. (1988). Repair of isopeptide bonds by protein carboxyl *O*-methyltransferase: Seminal ribonuclease as a model system. *Biochemistry* 27:1752–1757.

89. Johnson, B.A., and Aswad, D.W. (1991). Optimal conditions for the use of protein L-isoaspartyl methyltransferase in assessing the isoaspartate content of peptides and proteins. *Anal Biochem* 192:384–391.

90. Johnson, B.A., Murray, E.D. Jr., Clarke, S., Glass, D.B., and Aswad, D.W. (1987). Protein carboxyl methyltransferase facilitates conversion of atypical L-isoaspartyl peptides to normal L-aspartyl peptides. *J Biol Chem* 262:5622–5629.

91. McFadden, P.N., and Clarke, S. (1987). Conversion of isoaspartyl peptides to normal peptides: Implications for the cellular repair of damaged proteins. *Proc Natl Acad Sci USA* 84:2595–2599.

92. Brennan, T.V., Anderson, J.W., Jia, Z., Waygood, E.B., and Clarke, S. (1994). Repair of spontaneously deamidated HPr phosphocarrier protein catalyzed by the L-isoaspartate-(D-aspartate) *O*-methyltransferase. *J Biol Chem* 269:24586–24595.

93. Tanaka, Y., Tsumoto, K., Yasutake, Y., Umetsu, M., Yao, M., Fukada, H., Tanaka, I., and Kumagai, I. (2004). How oligomerization contributes to the thermostability of an archaeon protein. Protein L-isoaspartyl-O-methyltransferase from *Sulfolobus tokodaii*. *J Biol Chem* 279:32957–32967.

94. Ryttersgaard, C., Griffith, S.C., Sawaya, M.R., MacLaren, D.C., Clarke, S., and Yeates, T.O. (2002). Crystal structure of human L-isoaspartyl methyltransferase. *J Biol Chem* 277: 10642–10646.

95. Martin, J.L., and McMillan, F.M. (2002). SAM (dependent) I AM: The S-adenosylmethioinine-dependent methyltransferase fold. *Curr Opin Struct Biol* 12:783–793.

96. Schubert, H.L., Blumenthal, R.M., and Cheng, X. (2003). Many paths to methyltransfer: A chronicle of convergence. *Trends Biochem Sci* 28:329–335.

97. Smith, C.D., Barchue, J., Mentel, C., DeLucas, L., Shirasawa, T., and Chattopadhyay, D. (1997). Crystallization and preliminary cryogenic X-ray diffraction analyses of protein L-isoaspartyl *O*-methyltransferase from human fetal brain. *Proteins* 28:457–460.

98. Ichikawa, J.D., and Clarke, S. (1998). A highly active protein repair enzyme from an extreme thermophile: The L-isoaspartyl methyltransferase from *Thermotoga maritima*. *Arch Biochem Biophys* 358:222–231.

99. Tatusov, R.L., Fedorova, N.D., Jackson, J.D., Jacobs, A.R., Kiryutin, B., Koonin, E.V., Krylov, D.M., Mazumder, R., Mekhedov, S.L., Nikolskaya, A.N., Rao, S., Smirnov, S., Sverdlov, A.V., Vasudevan, S., Wolf, Y.I., Yin, J.J., and Natale, D.A. (2003). The COG database: An updated version includes eukaryotes. *Bio Med Central Bioinformatics* 4:41.

100. Klenk, H.P., Clayton, R.A., Tomb, J.F., White, O., Nelson, K.E., Ketchum, K.A., Dodson, R.J., Gwinn, M., Hickey, E.K., Peterson, J.D., Richardson, D.L., Kerlavage, A.R., Graham, D.E., Kyrpides, N.C., Fleischmann, R.D., Quackenbush, J., Lee, N.H., Sutton, G.G., Gill, S., Kirkness, E.F., Dougherty, B.A., McKenney, K., Adams, M.D., Loftus, B., Peterson, S., Reich, C.I., McNeil, L.K., Badger, J.H., Glodek, A., Zhou, L., Overbeek, R., Gocayne, J.D., Weidman, J.F., McDonald, L., Utterback, T., Cotton, M.D., Spriggs, T., Artiach, P., Kaine, B.P., Sykes, S.M., Sadow, P.W., D'Andrea, K.P., Bowman, C., Fujii, C., Garland, S.A., Mason, T.M., Olsen, G.J., Fraser, C.M., Smith, H.O., Woese, C.R., and Venter, J.C. (1997). The complete genome sequence of the hyperthermophilic, sulphate-reducing archaeon Archaeoglobus fulgidus. *Nature* 390:364–370.

101. Nelson, K.E., Clayton, R.A., Gill, S.R., Gwinn, M.L., Dodson, R.J., Haft, D.H., Hickey, E.K., Peterson, J.D., Nelson, W.C., Ketchum, K.A., McDonald, L., Utterback, T.R., Malek, J.A., Linher, K.D., Garrett, M.M., Stewart, A.M., Cotton, M.D., Pratt, M.S., Phillips, C.A., Richardson, D., Heidelberg, J., Sutton, G.G., Fleischmann, R.D., Eisen, J.A., White, O., Salzberg, S.L., Smith, H.O., Venter, J.C., and Fraser, C.M. (1999). Evidence for lateral gene transfer between Archaea and bacteria from genome sequence of *Thermotoga maritima*. *Nature* 399:323–329.

102. Altschul, S.F., Madden, T.L., Schäffer, A.A., Zhang, J., Zhang, Z., Miller, W., and Lipman, D.J. (1997). Gapped BLAST and PSI-BLAST: a new generation of protein database search programs. *Nucl Acids Res* 25:3389–3402.

103. Kellis, M., Borrem, B.W., and Lander, E.S. (2004). Proof and evolutionary analysis of ancient genome duplication in the yeast *Saccharomyces cerevisiae*. *Nature* 428:617–624.

104. Dietrich, F.S., Voegeli, S., Brachat, S., Lerch, A., Gates, K., Steiner, S., Mohr, C., Pohlmann, R., Luedi, P., Choi, S., Wing, R.A., Flavier, A., Gaffney, T.D., and Philippsen, P. (2004). The Ashbya gossypii genome as a tool for mapping the ancient Saccharomyces cerevisiae genome. *Science* 304:304–307.

105. Fairbanks, G., Steck, T.L., and Wallach, D.F. (1971). Electrophoretic analyses of the major polypeptides of the human erythrocyte membrane. *Biochemistry* 10:2606–2617.

106. Oden, K.L., and Clarke, S. (1983). S-adenosyl-L-methionine synthetase from human erythrocytes: role in the regulation of cellular S-adenosylmethionine levels. *Biochemistry* 22:2978–2986.

107. O'Connor, C.M., and Clarke, S. (1985). Analysis of erythrocyte protein methyl esters by two-dimensional gel electrophoresis under acidic separating conditions. *Anal Biochem* 148:79–85.

108. O'Connor, C.M., and Yutzey, K.E. (1988). Enhanced carboxyl methylation of membrane-associated hemoglobin in human erythrocytes. *J Biol Chem* 263:1386–1390.

109. Runte, L., Jürgensmeier, C.U., and Söling, H.D. (1982). Calmodulin carboxylmethyl ester formation in intact human red cells and modulation of this reaction by divalent cations *in vitro*. *FEBS Lett* 147:125–130.

110. Brunauer, L.S., and Clarke, S. (1986). Methylation of calmodulin at carboxylic acid residues in erythrocytes. *Biochem J* 236:811–820.

111. Lou, L.L., and Clarke, S. (1986). Carboxyl methylation of human erythrocyte band 3 in intact cells. *Biochem J* 235:183–187.

112. Lou, L.L., and Clarke, S. (1987). Enzymatic methylation of band 3 anion transporter in intact human erythrocytes. *Biochemistry* 26:52–59.

113. Ladino, C.A., and O'Connor, C. M. (1990). Protein carboxyl methylation and methyl ester turnover in density-fractionated human erythrocytes. *Mech Ageing Dev* 55:123–137.

114. Yamamoto, A., Takagi, H., Kitamura, D., Tatsuoka, H., Nakano, H., Kawano, H., Kuroyanagi, H., Yahagi, Y., Kobayashi, S., Koizumi, K., Sakai, T., Saito, K., Chiba, T., Kawamura, K., Suzuki, K., Watanabe, T., Mori, H., and Shirasawa, T. (1998). Deficiency in protein L-isoaspartyl methyltransferase results in a fatal progressive epilepsy. *J Neurosci* 18:2063–2074.

115. Galletti, P., Ingrosso, D., Nappi, A., Gragnaniello, V., Iolascon, A., and Pinto, L. (1983). Increased methyl esterification of membrane proteins in aged red-blood cells. Preferential esterification of ankyrin and band-4.1 cytoskeletal proteins. *Eur J Biochem* 135:25–31.

116. Low, P.S. (1991). Role of hemoglobin denaturation and band 3 clustering in initiating red cell removal. *Adv Exp Med Biol* 319:525–546.

117. Ingrosso, D., D'Angelo, S., Di Carlo, E., Perna, A.F., Zappia, V., and Galletti, P. (2000). Increased methyl esterification of altered aspartyl residues in erythrocyte membrane proteins in response to oxidative stress. *Eur J Biochem* 267:4397–4405.

118. Ingrosso, D., Cimmino, A., D'Angelo, S., Alfinito, F., Zappia, V., and Galletti, P. (2002). Protein methylation as a marker of aspartate damage in glucose-6-phosphate dehydrogenase-deficient erythrocytes: role of oxidative stress. *Eur J Biochem* 269:2032–2039.

119. Kloog, Y., and Saavadra, J.M. (1983). Protein carboxylmethylation in intact rat posterior pituitary lobes *in vitro*. *J Biol Chem* 258:7129–7133.

120. Ohta, K., Seo, N., Yoshida, T., Hiraga, K., and Tuboi, S. (1987). Tubulin and high molecular weight microtubule-associated proteins as endogenous substrates for carboxymethyltransferase in brain. *Biochimie* 69:1227–1234.

121. Macfarlane, D.E. (1984). Inhibitors of cyclic nucleotide phosphodiesterases inhibit protein carboxyl methylation in intact blood platelets. *J Biol Chem* 259:1357–1362.

122. Barten, D.M., and O'Dea, R.F. (1989). Protein carboxylmethyltransferase activity in intact, differentiated neuroblastoma cells: Quantitation by S-[³H]adenosylmethionine prelabeling. *J Neurochem* 53:1156–1165.

123. Clarke, S. (1992). Protein isoprenylation and methylation at carboxyl-terminal cysteine residues. *Annu Rev Biochem* 61:355–386.

124. DeBaere, I., Derua, R., Janssens, V., VanHoof, C., Waelkens, E., Merlevede, W., and Goris, J. (1999). Purification of porcine brain protein phosphatase 2A leucine carboxyl methyltransferase and cloning of the human homologue. *Biochemistry* 38:16539–16547.

125. Dumont, J.N. (1971). Oogenesis in *Xenopus laevis* (Daudin). I. Stages of oocyte development in laboratory maintained animals. *J Morphol* 136:153–180.

126. Smith, L.D., and Richter, J.D. (1985). in C. Metz and A. Monroy (eds.), *Biology of Fertilization*. pp. 141–187, New York: Academic Press.

127. O'Connor, C.M., and Germain, B.J. (1987). Kinetic and electrophoretic analysis of trans-methylation reactions in intact *Xenopus laevis* oocytes. *J Biol Chem* 262:10404–10411.

128. Desrosiers, R.R., Romanik, E.A., and O'Connor, C.M. (1990). Selective carboxyl methylation of structurally altered calmodulins in *Xenopus* oocytes. *J Biol Chem* 265:21368–21374.

129. Feldherr, C.M., and Ogburn, J.A. (1980). Mechanism for the selection of nuclear polypeptides in Xenopus oocytes. II. Two-dimensional gel analysis. *J Cell Biol* 87:589–593.

130. Romanik, E.A., and O'Connor, C.M. (1989). Methylation of microinjected isoaspartyl peptides in *Xenopus* oocytes. *J Biol Chem* 264:14050–14056.

131. Szymanska, G., Leszyk, J.D., and O'Connor, C.M. (1998). Carboxyl methylation of deamidated calmodulin increases its stability in *Xenopus* oocyte cytoplasm. *J Biol Chem* 273:28516–28523.

132. Weber, D.J., and McFadden, P.N. (1997). Detection and characterization of a protein isoaspartyl methyltransferase which becomes trapped in the extracellular space during blood vessel injury. *J Prot Chem* 16:257–267.

133. Bornstein, P. (1970). Structure of alpha-1-CB8, a large cyanogen bromide produced fragment from the alpha-1 chain of rat collagen. The nature of a hydroxylamine-sensitive bond and composition of tryptic peptides. *Biochemistry* 9:2408–2421.

134. Cloos, P.A., and Fledelius, C. (2000). Collagen fragments in urine derived from bone resorption are highly racemized and isomerized: a biological clock of protein aging with clinical potential. *Biochem J* 345:473–480.

135. Fledelius, C., Johnsen, A.H., Cloos, P.A., Bonde, M., and Ovist, P. (1997). Characterization of urinary degradation products derived from type I collagen. Identification of a beta-isomerized Asp-Gly sequence within the C-terminal telopeptide (alpha1) region. *J Biol Chem* 272:9755–9763.

136. Lanthier, J., and Desrosiers, R.R. (2004). Protein L-isoaspartyl methyltransferase repairs abnormal aspartyl residues accumulated *in vivo* in type-I collagen and restores cell migration. *Exp Cell Res* 293:96–105.

137. Bloemendal, H., de Jong, W., Jaenicke, R., Lubsen, N.H., Slingsby, C., and Tardieu, A. (2004). Ageing and vision: structure, stability and function of lens crystallins. *Prog Biophys Mol Biol* 86:407–485.

138. McFadden, P.N., Horwitz, J., and Clarke, S. (1983). Protein carboxyl methyltransferase from cow eye lens. *Biochem Biophys Res Commun* 113:418–424.

139. McFadden, P.N., and Clarke, S. (1986). Protein carboxyl methyltransferase and methyl acceptor proteins in aging and cataractous tissue of the human eye lens. *Mech Ageing Dev* 34:91–105.

140. Fujii, N., Takemoto, L.J., Momose, Y., Matsumoto, S., Hiroki, K., and Akaboshi, M. (1999). Formation of four isomers at the asp-151 residue of aged human alphaA-crystallin by natural aging. *Biochem Biophys Res Commun* 265:746–751.

141. Fujii, N., Shimo-Oka, T., Ogiso, M., Momose, Y., Kodama, T., Kodama, M., and Akaboshi, M. (2000). Localization of biologically uncommon D-beta-aspartate-containing alphaA-crystallin in human eye lens. *Molecular Vision* 6:1–5.

142. Kodama, T., Mizobuchi, M., Takeda, R., Torikai, H., Shinomiya, H., and Y., Ohashi. (1995). Hampered expression of isoaspartyl protein carboxyl methyltransferase gene in the human cataractous lens. *Biochim Biophys Acta* 1245:269–272.

143. Roher, A.E., Lowenson, J.D., Clarke, S., Wolkow, C., Wang, R., Cotter, R.J., Reardon, I.M., Zürcher-Neely, H.A., Heinrikson, R.L., Ball, M.J., and Greenberg, B.D. (1993). Structural alterations in the peptide backbone of β-amyloid core protein may account for its deposition and stability in Alzheimer's disease. *J Biol Chem* 268:3072–3083.

144. Roher, A.E., Lowenson, J.D., Clarke, S., Woods, A.S., Cotter, R.J., Gowing, E., and Ball, M.J. (1993). β-Amyloid-(1–42) is a major component of cerebrovascular amyloid deposits: Implications for the pathology of Alzheimer disease. *Proc Natl Acad Sci USA* 90:10836–10840.

145. Tomiyama, T., Asano, S., Furiya, Y., Shirasawa, T., Endo, N., and Mori, H. (1994). Racemization of Asp[23] residue affects the aggregation properties of Alzheimer amyloid β protein analogues. *J Biol Chem* 269:10205–10208.

146. Kuo, Y.M., Emmerling, M.R., Woods, A.S., Cotter, R.J., and Roher, A.E. (1997). Isolation, chemical characterization, and quantitation of A beta 3-pyroglutamyl peptide from neuritic plaques and vascular amyloid deposits. *Biochem Biophys Res Commun* 237:188–191.

147. Fonseca, M.I., Head, E., Velazquez, P., Cotman, C.W., and Tenner, A.J. (1999). The presence of isoaspartic acid in beta-amyloid plaques indicates plaque age. *Exp Neurol* 157:277–288.

148. Shimizu, T., Fukuda, H., Murayama, S., Izumiyama, N., and Shirasawa, T. (2002). Isoaspartate formation at position 23 of amyloid beta peptide enhanced fibril formation and deposited onto senile plaques and vascular amyloids in Alzheimer's disease. *J Neurosci Res* 70:451–461.

149. Shin, Y., Cho, H.S., Fukumoto, H., Shimizu, T., Shirasawa, T., Greenberg, S.M., and Rebeck, G.W. (2003). Abeta species, including IsoAsp23 Abeta, in Iowa-type familial cerebral amyloid angiopathy. *Acta Neuropathol* 105:252–258.

150. Fukuda, H., Shimizu, T., Nakajima, M., Mori, H., and Shirasawa, T. (1999). Synthesis, aggregation, and neurotoxicity of the Alzheimer's AB1-42 amyloid peptide and its isoaspartyl isomers. *Bioorg Med Chem Lett* 9:953–956.

151. Billingsley, M.L., and Kuhn, D.M. (1985). Immunohistochemical localization of protein-O-carboxylmethyltransferase in rat brain neurons. *Neuroscience* 15:159–171.

152. Shirasawa, T., Endoh, R., Zeng, Y.-X., Sakamoto, K., and Mori, H. (1995). Protein L-isoaspartyl methyltransferase: developmentally regulated gene expression and protein localization in the central nervous system of aged rat. *Neurosci Lett* 188:37–40.

153. Farrar, C., Houser, C.R., and Clarke, S. (2005). Activation of the PI3K/Akt signal transduction pathway and increased levels of insulin receptor in protein repair-deficient mice. *Aging Cell* 4:1–12.

154. Paranandi, M.V., and Aswad, D.W. (1995). Spontaneous alterations in the covalent structure of synapsin I during *in vitro* aging. *Biochem. Biophys. Res Commun* 212:442–448.

155. Shimizu, T., Ikegami, T., Ogawara, M., Suzuki, Y., Takahashi, M., Morio, H., and T., Shirasawa. (2002). Transgenic expression of the protein-L-isoaspartyl methyltransferase (PIMT) gene in the brain rescues mice from the fatal epilepsy of PIMT deficiency. *J Neurosci Res* 69:341–352.

156. Najbauer, J., Orpiszewski, J., and Aswad, D.W. (1996). Molecular aging of tubulin: Accumulation of isoasparatyl sites *in vitro* and *in vivo*. *Biochemistry* 35:5183–5190.

157. Huebscher, K.J., Lee, J., Rovelli, G., Ludin, B., Matus, A., Stauffer, D., and Furst, P. (1999). Protein isoaspartyl methyltransferase protects from Bax-induced apoptosis. *Gene* 240: 333–341.

158. Gagnon, C., Axelrod, J., Musto, N., Dym, M., and Bardin, C.W. (1979). Protein carboxyl-methylation in rat testes: A study of inherited and X-ray-induced seminiferous tubule failure. *Endocrinology* 105:1440–1445.

159. O'Connor, C.M., Germain, B.J., Guthrie, K.M., Aswad, D.W., and Millette, C.F. (1989). Protein carboxyl methyltransferase activity specific for age-modified aspartyl residues in mouse testes and ovaries: Evidence for translation during spermiogenesis. *Gamete Res* 22:307–319.

160. Cusan, L., Andersen, D., Tuen, E., and Hansson, V. (1982). Changes in protein carboxyl-methylase isoenzymes during testicular development in the rat. *Arch Androl* 8:285–292.

161. Mizobuchi, M., Murao, K., Takeda, R., and Kakimoto, Y. (1994). Tissue-specific expression of isoaspartyl protein carboxyl methyltransferase gene in rat brain and testis. *J Neurochem* 62:322–328.

162. Bouchard, P., Gagnon, C., Phillips, D.M., and Bardin, C.W. (1980). The localization of protein carboxyl-methylase in sperm tails. *J Cell Biol* 86:417–423.

163. Williams-Ashman, H.G., Hatch, R., and Harvey, S.E. (1985). Protein O-carboxylmethylation in relation to male gamete production and function, in G. Weber (ed.), *Advances in Enzyme Regulation* (vol. XXIII), pp. 389–416, Oxford: Pergamon Press.

164. DeVry, C.G., Tsai, W., and Clarke, S. (1996). Structure of the human gene encoding the protein repair L-isoaspartyl (D-aspartyl) *O*-methyltransferase. *Arch Biochem Biophys* 335:321–332.

165. Takeda, R., Mizobuchi, M., Murao, K., Sato, M., and Takahara, J. (1995). Characterization of three cDNAs encoding two isozymes of an isoaspartyl protein carboxyl methyltransferase from human erythroid leukemia cells. *J Biochem* 117:683–685.

166. Rushmore, T.H., Morton, M.R., and Pickett, C.B. (1991). The antioxidant response element. *J Biol Chem* 266:11632–11639.

167. Kohno, K., Normington, K., Sambrook, J., Gething, M.-J., and Mori, K. (1993). The promoter region of the yeast *KAR2* gene contains a regulatory domain that responds to the presence of unfolded proteins in the endoplasmic reticulum. *Mol Cell Biol* 13:877–890.

168. Braun, R.E. (1998). Post-transcriptional control of gene expression during spermatogenesis. *Semin Cell Devel Biol* 9:483–489.

169. Kim, E., Lowenson, J.D., Clarke, S., and Young, S.G. (1999). Phenotypic analysis of seizure-prone mice lacking L-isoaspartate (D-aspartate) *O*-methyltransferase. *J Biol Chem* 274:20671–20678.

170. Ikegaya, Y., Yamada, M., Fukuda, T., Kuroyanagi, H., Shirasawa, T., and Nishiyama, N. (2001). Aberrant synaptic transmission in the hippocampal CA3 region and cognitive deterioration in protein-repair enzyme-deficient mice. *Hippocampus* 11:287–298.

171. Vitali, R., and Clarke, S. (2004). Improved rotorod performance and hyperactivity in mice deficient in a protein repair methyltransferase. *Behav Brain Res* 153:129–141.

172. Lowenson, J.D., Kim, E., Young, S.G., and Clarke, S. (2001). Limited accumulation of damaged proteins in L-isoaspartyl (D-aspartyl) *O*-methyltransferase-deficient mice. *J Biol Chem* 276:20695–20702.

173. Farrar, C., and Clarke, S. (2002). Altered levels of S-adenosylmethionine and S-adenosylhomocysteine in the brains of L-isoaspartyl (D-Aspartyl) O-methyltransferase deficient mice. *J Biol Chem* 277:27856–27863.
174. Bottiglieri, T. (2002). S-Adenosyl-L-methionine (SAMe): From the bench to the bedside; molecular basis of a pleiotrophic molecule. *Am J Clin Nutr* 76:1151S–1157S.
175. Axelrod, J. (2003). Journey of a late blooming biochemical neuroscientist. *J Biol Chem* 278:1–13.
176. Shen-Miller, J., Mudgett, M.B., Schopf, J.W., Clarke, S., and Berger, R. (1995). Exceptional seed longevity and robust growth: Ancient sacred lotus from China. *Am J Botany* 82:1367–1380.
177. Thapar, N., Kim, A.-K., and Clarke, S. (2001). Distinct patterns of expression but similar biochemical properties of protein L-isoaspartyl methyltransferase in higher plants. *Plant Physiol* 125:1023–1035.
178. Chavous, D.A. (2001). Overexpression of protein carboxyl methyltransferase extends lifespan in *Drosophila melanogaster*. Ph.D. thesis, Biology Dept., Boston College, Chestnut Hill, MA.
179. O'Connor, M.B., Galus, A., Hartenstine, M., Magee, M., Jackson, F.R., and O'Connor, C.M. (1997). Structural organization and developmental expression of the protein isoaspartyl methyltransferase gene from *Drosophila melanogaster*. *Insect Biochem Mol Biol* 27:49–54.
180. Chavous, D.A., Jackson, F.R., and O'Connor, C.M. (2001). Extension of the *Drosophila* lifespan by overexpression of a protein repair methyltransferase. *Proc Natl Acad Sci USA* 98:14814–14818.
181. Cherbas, L., and Cherbas, P. (1993). The arthropod initiator: The capsite consensus plays an important role in transcription. *Insect Biochem Mol Biol* 23:81–90.
182. Shuey, D.J., and Parker, C.S. (1986). Binding of Drosophila heat-shock gene transcription factor to the hsp 70 promoter. Evidence for symmetric and dynamic interactions. *J Biol Chem* 261:7934–7940.
183. Brand, A.H., and Perrimon, N. (1993). Targeted gene expression as a means of altering cell fates and generating dominant phenotypes. *Development* 118:401–415.
184. Orr, W.C., and Sohal, R.S. (1994). Extension of life-span by overexpression of superoxide dismutase and catalase in *Drosophila melanogaster*. *Science* 263:1128–1130.
185. Parkes, T.L., Elia, A.J., Dickinson, D., Hilliker, A.J., Phillips, J.P., and Boulianne, G.L. (1998). Extension of *Drosophila* lifespan by overexpression of human *SOD1* in motorneurons. *Nature Genet* 19:171–174.
186. Sun, J., and Tower, J. (1999). FLP recombinase-mediated induction of Cu/Zn-superoxide dismutase transgene expression can extend the life span of adult *Drosophila melanogaster* flies. *Mol Cell Biol* 19:216–228.
187. Sun, J., Folk, D., Bradley, T.J., and Tower, J. (2002). Induced overexpression of mitochondrial Mn-superoxide dismutase extends the life span of adult *Drosophila melanogaster*. *Genetics* 161:661–672.
188. Tatar, M., Khazaeli, A.A., and Curtsinger, J.W. (1997). Chaperoning extended life. *Nature* 390:30.
189. Ruan, H., Tang, X.D., Chen, M.-L., Joiner, M.-L. A., Sun, G., Brot, N., Weissbach, H., Heinemann, S.H., Iverson, L., Wu, C.-F., and Hoshi, T. (2002). High-quality life extension by the enzyme peptide methionine sulfoxide reductase. *Proc Natl Acad Sci USA* 99:2748–2753.
190. Baugh, L.R., Hill, A.A., Slonin, D.K., Brown, E.L., and Hunter, C.P. (2003). Composition and dynamics of the *Caenorhabditis elegans* early embryonic transcriptome. *Development* 130:889–900.
191. Hill, A.A., Hunter, C.P., Tsung, B.T., Tucker-Kellogg, G., and Brown, E.L. (2000). Genomic analysis of gene expression in *C. elegans*. *Science* 290:809–812.
192. McElwee, J.J., Schuster, E., Blanc, E., Thomas, J.H., and Gems, D. (2004). Shared transcriptional signature in *Caenorhabditis elegans* dauer larvae and long-lived *daf-2* mutants implicates detoxification system in longevity assurance. *J Biol Chem* 279:44533–44543.

193. Wang, J., and Kim, S.K. (2003). Global analysis of dauer gene expression in *Caenorhabditis elegans*. *Development* 130:1621–1634.
194. Kim, S.K., Lund, J., Kiraly, M., Duke, K., Jiang, M., Stuart, J.M., Eizinger, A., Wylie, B.N., and Davidson, G.S. (2001). A gene expression map for *Caenorhabditis elegans*. *Science* 293:2087–2092.
195. Fu, J.C., Ding, L., and Clarke, S. (1991). Purification, gene cloning, and sequence analysis of an L-isoaspartyl protein carboxyl methyltransferase from *Escherichia coli*. *J Biol Chem* 266:14562–14572.
196. David, C.L., Keener, J., and Aswad, D.W. (1999). Isoaspartate in ribosomal protein S11 of *Escherichia coli*. *J Bacteriology* 181:2872–2877.
197. Visick, J.E., Ichikawa, J.K., and Clarke, S. (1998). Mutations in the *Escherichia coli surE* gene increase isoaspartyl accumulation in a strain lacking the *pcm* repair methyltransferase but suppress stress-survival phenotypes. *FEMS Ltrs* 167:19–25.
198. Visick, J.E., Cai, H., and Clarke, S. (1998). The L-isoaspartyl protein repair methyltransferase enhances survival of aging *Escherichia coli* subjected to secondary environmental stresses. *J Bacteriol* 180:2623–2629.
199. Hicks, W.M., Kotlajich, M.V., and Visick, J.E. (2005). Recovery from long-term stationary phase and stress survival in *Escherichia coli* require the L-isoaspartyl protein carboxyl methyltransferase at alkaline pH. *Microbiology* 151:2151–2158.
200. Kern, R., Malki, A., Abdallah, J., Liebart, J.-C., Dubucs, C., Yu, M.H., and Richarme, G. (2005). Protein isoaspartate methyltransferase is a multicopy suppressor of protein aggregation in *Escherichia coli*. *J Bacteriol* 187:1377–1383.
201. Young, G.W., Hoffring, S.A., Mamula, M.J., Doyle, H.A., Bunick, G.J., Hu, Y., Aswad, D.W. (2005). Protein L-isoaspartyl methyltransferase catalyzes *in vivo* racemization of aspartate-25 in mammalian histone H2B. *J Biol Chem* 280:26094–26098.
202. Castillo, M.B., Celio, M.R., Andressen, C., Gotzos, V., Rulicke, T., Berger, M.C., Weber, J., and Berchtold, M.W. (1995). Production and analysis of transgenic mice with ectopic expression of parvalbumin. *Arch Biochem Biophys* 317:292–298.

Part V

Modification of Proteins by Methylation of Glutamine and Asparagine Residues

14

Modification of Glutamine Residues in Proteins Involved in Translation

HEIDI L. SCHUBERT

Department of Biochemistry
University of Utah
15 North Medical Drive East
Salt Lake City, UT 84112, USA

I. Abstract

The only known protein glutamine methylations take place on proteins related to the ribosomal function. The modification is a mono-methylation on the side-chain amide nitrogen (N5). The glutamine residue within a Gly-Gly-Gln (GGQ) motif of the polypeptide release factor (RF) is methylated, and the modification is involved in polypeptide chain termination. A defect in RF methylation results in a dramatic growth defect and substantial read-through of stop codons during translation.

Ribosomal protein L3 is also methylated on a glutamine residue. This modification is only found in bacteria and is not essential, but may be involved in efficient assembly of the ribosome structure. Gln methyltransferases (MTase) adopt the canonical class I MTase structure and contain a [D/N]PPY motif to position the substrate and facilitate catalysis. This motif was previously only thought to function in DNA N-MTases, but has been shown to act more generally on neutral planar amide substrates.

THE ENZYMES, Vol. XXIV

II.   Introduction

The type of protein methylation discussed within the chapters of this book covers the act of posttranslational modification of protein residues by an S-adenosyl-L-methionine (AdoMet)-dependent methyltransferase (MTase). After ATP, AdoMet is the second most widely used cofactor, and its methyl group is transferred to a wide variety of substrates, including DNA, RNA, small molecules, and amino acids to generate S-adenosyl-L-homocysteine (AdoHcy) and a methylated product. Glutamine was not one of the residues identified as methylated when protein methylation was initially being described and target residues within proteins were being characterized [1]. Indeed, even studies focused on the methylation of ribosomal proteins, of which we now know contain methylated glutamine, did not find the modification [2, 3].

These initial studies utilized liquid and paper chromatography of acid hydrolyzed proteins. Whereas Gln and Asn side chains normally yield glutamic acid or aspartic acid and ammonium, their methylated counterparts would yield methylamine. Methylamine is a volatile and extremely basic compound, and consequently has dramatically different chromatographic properties [4]. Since its original discovery, only a few major sites of N5-methylglutamine (Figure 14.1) have been identified and characterized as physiologically relevant posttranslational modifications.

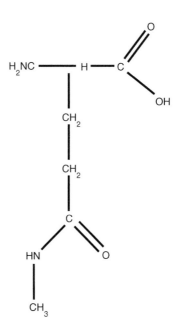

FIG. 14.1. Structure of N5-methylglutamine.

Gln methylation has more in common with methylations on neutrally charged amides in RNA and DNAs than with the charged nitrogens in Lys or Arg residues or at the N-terminus of proteins. A single methylation of the Gln side-chain amide nitrogen has been observed *in vivo*, whereas the charged amine side chains of arginine or lysine residues have the potential to carry multiple methylations. The modification alters the chemical properties of the amide side chain, interfering with hydrogen-bonding potential and significantly enhancing its hydrophobic character. Whereas some methylations have been shown to be reversible, such as the Glu methylations on chemoreceptors [5], the methylations on Gln appear to be permanent functional modifications.

Methyl groups have also been observed on a glutamine within the Methylcoenzyme M reductase structure [6]. The modification is not the typical N5-methylation, but rather a novel modification at the Cα carbon resulting in 2-(S)-methylglutamine. The methyl group does derive from AdoMet, but happens at a stage in protein synthesis prior to quaternary structure formation as the residue is buried within the core of the protein near the active site.

III. Ribosomal Protein Methylation

During protein synthesis, the growing polypeptide chain extends from the P-site tRNA on the ribosome and remains covalently attached until a stop codon in the mRNA enters the A site [7, 8]. To terminate transcription, the polypeptide must be cleaved from the P-site tRNA by hydrolysis. Ribosomal release factors (RFs) bind to the ribosome, recognize the stop codon, and trigger hydrolysis [9]. Ribosome recycling factors use GTP hydrolysis to disrupt the complexes between the RFs and the ribosome, and stimulate ribosome disassembly for future rounds of translation.

Release factors are a diverse family of proteins that show specificity for the individual stop codons [8-10]. In addition, the prokaryotic and eukaryotic RFs share almost no sequence similarity. Despite these differences, each RF contains two functional domains. One domain recognizes the stop codon and the other triggers peptide hydrolysis [11]. The only amino acids conserved between the diverse families of RFs are a short tri-peptide, GGQ. The CCA terminal trinucleotide of tRNAs is also invariant over evolution and hence it has been proposed that these two motifs interact during peptide termination. The GGQ may act more specifically by activating or positioning the water molecule involved in hydrolysis at the peptidyl transferase active site [12, 13].

The fact that the GGQ motif is conserved throughout all structurally diverse RFs from humans to bacteria highlights its universal importance in translation termination. Biochemical analysis revealed that mutation to either glycine residue resulted in a lack of function, loss of peptide termination, and stop codon read-through [12, 14, 15]. These mutations did not disrupt the physical interaction between the RF and the ribosome.

Study of the RFs was complicated by the inconsistent growth phenotypes associated with bacterial overexpression of the genes from *E. coli* versus genes from yeast [16-18]. Sequence data finally identified the residue number 246 as the cause of the phenotypic difference, four residues toward the N-terminus of the GGQ motif.

A threonine at this position, as seen in *E. coli* K12, causes reduced functionality of the release factor, whereas all other bacteria contain either an Ala or Ser at this position [19]. Additional steps were taken to check the sequence of the region surrounding residue 246 and the GGQ motif. MALDI-MS and MS/MS sequencing identified a posttranslational modification of the GGQ motif as an increase in 14 daltons, the addition of a methyl group [19]. Dincbas-Renqvist et al. then realized that during RF overexpression the MTase could not methylate all of the overexpressed protein. In combination with the reduced functionality of a Thr at residue 246, the methylation defect leads to a dominant negative phenotype in *E. coli* similar to that of a null mutant [19].

Two different groups used independent methods to identify the MTase responsible for both the RF modification and a similar Gln methylation on the ribosomal protein L3. Using genome analysis methods, Heurgué-Hamard et al. identified an MTase just upstream of the L3 ribosomal protein, designated as yfcB [20]. Ribosomes purified from cells deficient in this MTase could be methylated from cell extracts containing the MTase construct. Further analysis revealed that the *E. coli* gene hemK contained the highest similarity to yfcB and was a likely candidate for the only other known Gln MTase in *E. coli*, the RF MTase. Transfer of the radioactive methyl group (^{3}H-AdoMet) to overexpressed release factors (and hence under-methylated) was specific to hemK containing cell extracts [20].

The gene hemK was originally assigned a function in late pathway heme biosynthesis [21]. A hemK mutant *E. coli* exhibited a growth defect falsely attributed to a reduction in heme synthesis and a rise in the oxidative stress response. The researchers were further lead astray when the gene was located within a multifunctional operon containing the first heme biosynthetic enzyme, hemA. Nakahigashi et al. later identified suppressor mutations for their ΔhemK strain in the gene prfA encoding the *E. coli* RF1, suggesting that hemK was not involved in heme biosynthesis but instead involved in polypeptide termination [22, 23]. Remarkably, the suppressor mutation was identified at the same residue highlighted previously in *E. coli* RF2, Thr246. Methylation on Gln is significantly more crucial for RF activity in the Thr246 background than when Thr is mutated to Ala. The group went on to find a correlation between hemK function and polypeptide termination in the form of increased stop codon read-through [22].

In 2002, with the combination of sequence analysis and biochemical characterization hemK was finally correctly characterized as the MTase responsible for methylation of the glutamine residue of the GGQ motif contained within peptide release factors involved in peptide translation termination [20, 22]. The related protein, YfcB, was identified as the Gln MTase for ribosomal protein L3. YfcB is also known as PrmB, for protein ribosomal methylation. Due to the complication with the HemK nomenclature Heurgue-Hamard et al., went on to suggest a protein name change for HemK to PrmC, which will be used for the rest of this chapter. PrmA was identified at the same time as PrmB and is a N-MTase involved in methylating lysines and the N-terminus of ribosomal protein L11.

Prior to these functional studies, large scale sequence alignments correctly identified *E. coli* hemK as an AdoMet-dependent MTase [24]. The alignment correctly identified the GxGxG sequence within PrmC that comprises Motif I in all class I MTases, and structurally represents the turn after the first β strand in the MTase domain that is positioned underneath the SAM ligand [25]. Other motifs were identified in the PrmC sequence (including an [D/N]PPY sequence, known as motif IV, which was thought to be conserved only among DNA N-MTases). Consequently, early on hemk was incorrectly classified as an amino-targeted DNA MTase [24].

The [D/N]PPY motif has been shown to encircle the active site and adenosine substrate in the DNA-adenine-N6 MTase, TaqI [26]. Hydrogen bonds from the protein orient the amide nitrogen of the adenine such that the lone-pair electrons are directed toward the incoming methyl group [26–28]. The question was would Gln MTases use a similar mechanism on the Gln substrate. To understand this diverse use of the [D/N]PPY motif, the structure of PrmC was determined.

A. PrmC Structural Characterization

The crystal structure of *Thermotoga maritima* PrmC was determined to understand how the (N/D)PPY motif of PrmC facilitates methylation on the N5-nitrogen of glutamine [29]. This work was further corroborated by the structural characterization of the *E. coli* PrmC [30]. Fortuitously, the *T. maritima* structure contains a mixture of substrates (AdoMet and glutamine) and products (AdoHcy and a *N*5-methylglutamine bound to the MTase active site). To accommodate this crystallographically, the reactive methyl group was refined with an occupancy of 0.5 on both the AdoMet and MeGln molecules.

The structure of PrmC contains two major domains. Four α-helices constitute the N-terminal domain, and are connected to the canonical AdoMet-binding domain through a small β hairpin (Figure 14.2) [29]. The catalytic domain is similar to other class I MTase structures [31] and contains a seven-stranded mixed β sheet flanked by six α-helices. One loop on the structure, between β4 and αD, occupies multiple orientations in several independently solved structures. The loop is positioned near the active site and contains a few conserved residues in the loop, suggesting that it may be crucial for substrate recognition. The majority of the similarity to other MTases is contained within the AdoMet-binding domain, and the N-terminal domain of PrmC does not contain statistically significant similarity with any other MTase.

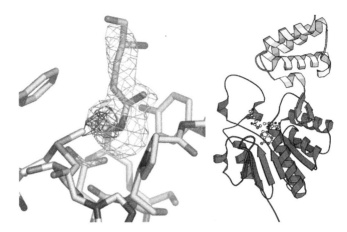

FIG. 14.2. Structure of PrmC. (right) Structure of the enzyme with the N-terminal helical domain on top, the β-hairpin subdomain in the middle, and the α/β catalytic domain below. The bound AdoMet (ball-and-stick, below) and methylglutamine (ball-and-stick, above) delineate the active site. In the three independently determined PrmC molecules, the β4-αD loop (highlighted by the dashed coil) adopts alternate conformations. Figures were prepared using the program Bobscript (Avatar Industries AB) and Pymol [41]. (left) The crystal structure of *T. maritima* PrmC contains a mixture of AdoMet and AdoHcy supported by 3σ Fo-Fc difference maps. The structure also contains a partially methylated glutamine residue as supported by both difference maps. *[Figure adapted with permission from* Biochemistry *2003, 42(19), 5592-5599).]*

B. ADOMET BINDING WITHIN THE ACTIVE SITE

In *T. maritima* PrmC, the substrate AdoMet is bound above the β sheet near the classic nucleotide-binding sequence, GxGxG, following the C-terminal end of β1 (Figures 14.2a and b). Several hydrophobic side chains of Ile128, Val152, Phe180, and Phe228 [*E. coli* numbering (ec) Leu116ec, Arg141ec, Trp168ec, and Asp216ec] help position against the AdoMet adenine ring. Hydrogen bonding interactions are formed between the hydroxyl groups of the ribose ring and the conserved side chain of Asp151 (Asp140ec). The methionine moiety of AdoMet interacts with Asn197 (Asn183ec), Gly129 (Gly117ec), and two water molecules.

The nonpolar side chain of Phe100 (Leu89ec) packs behind the positively charged sulfur atom of AdoMet and may function to destabilize the charged ligand, thereby enhancing the rate of methylation. It is interesting to note that like other class I MTase structures the AdoMet is completely buried within the core of the protein and turnover of this substrate will require substantial remodeling of the structure.

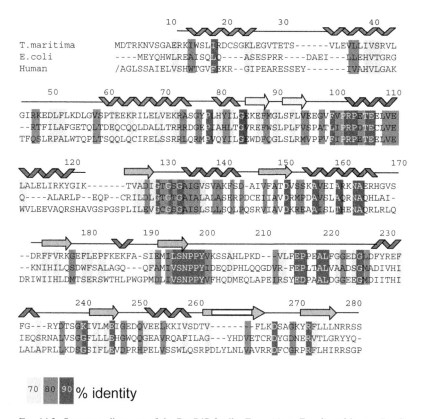

FIG. 14.3. Sequence alignment of the PrmB/C family. *T. maritima*, *E. coli*, and human PrmC are aligned underneath the *T. maritima* secondary structure with their conserved residues highlighted. *E. coli* PrmB is also aligned, and has slightly different conservation of residues (only invariant residues based on a seven sequence alignment are shown in black), highlighting its different evolution and substrate specificity. *[Figure adapted with permission from* Biochemistry 2003, 42(19), 5592-5599.*]*

An alignment of 14 members of the PrmC family revealed 15 invariant residues (Figure 14.3). These residues are concentrated at the active site cleft, a surface depression approximately 10 Å in diameter and 8 Å deep. The methyl group of the AdoMet substrate sits at the very bottom of the pocket [29]. Hydrophobic residues Tyr200 (Y186ec), Val201 (I187ec), Leu219 (L207ec), and Phe100 (L89c) sit on one side of the active site, and charged residues Arg103 (R29ec), Glu105 (D94ec), Glu246 (E237ec), Glu249 (W240ec), and Arg273 (R269ec) lie on the other. An overall electronegative charge is generated by the positioning of several additional negatively charged residues. This charge may complement the positive charges seen in the sequences surrounding the GGQ motif on RFs [12].

C.   A COMBINATION OF SUBSTRATES AND PRODUCTS IS BOUND TO THE ACTIVE SITE

A glutamine substrate is bound to the structure of *T. maritima* PrmC. Residual electron density remaining after refinement suggested the presence of a partially occupied methyl group on the amide nitrogen [29]. To accommodate the partial occupancy, the methyl substrate was also modeled as a mixture of AdoMet and AdoHcy (Figure 14.2b) [29]. The methyl group on the glutamine is positioned perpendicular to the glutamine amide bond plane. This position is suggestive of an intermediate stage of catalysis, immediately after methyl transfer, where the amide retains two protons, receives the methyl group, and adopts a partial positive charge. Proton removal, shift of the methyl group into the amide plane, and neutralization of the side chain would occur later to finish the reaction.

As observed in the adenosine MTase, TaqI, the [D/N]PPY motif in PrmC (NPPY) lies at the bottom of the active site and wraps around the glutamine binding site [29]. The proline residues facilitate a tight kink in the polypeptide chain such that the residues on either side point into the active site. The Asn197 (N183ec) side chain is available to hydrogen bond to the substrate, whereas the Tyr200 (Y186ec) sits at the back of the pocket and provides a planar hydrophobic framework to orient the planar substrate. The carbonyl oxygen of the first proline also points into the active site and hydrogen bonds with the substrate amide. The two hydrogen bonds on either side of the glutamine amide nitrogen restrict its location within the active site to a position optimal for the ensuing mechanistic chemistry. The hydrogen bonds from the glutamine to the [D/N]PPY motif lie within the plane of the sp2-hybridized nitrogen, and the lone-pair electrons from the nitrogen would project perpendicular to that plane, pointing toward the incoming methyl group (Figure 14.4a).

D.   MECHANISM

The symmetry of the methyl group is inverted during methyl transfer in a classic SN2 mechanism (Figure 14.5). The PrmC structure contains S-C-N angles of 156 degrees and 161 degrees, similar to the linear geometry of 180 degrees required for such a mechanism [29]. The half-occupancy of the methyl group refined in the crystal structure can be interpreted as a mixture of AdoMet/Gln and AdoHcy/MeGln. The distances between the AdoMet carbon and the Gln nitrogen are similar to that of the AdoHcy sulfur and the MgGln methyl group at 3.3 and 3.6 Å, respectively. The angles and locations of these atoms are all consistent with the mechanism of direct methyl transfer.

Because the methylated glutamine product is bound and constrained by the active site residues, it appears that the methyl group is not in the expected final product orientation but rather lies in an intermediate position perpendicular to the plane of the original glutamine amide bond [29]. It is likely that the nitrogen

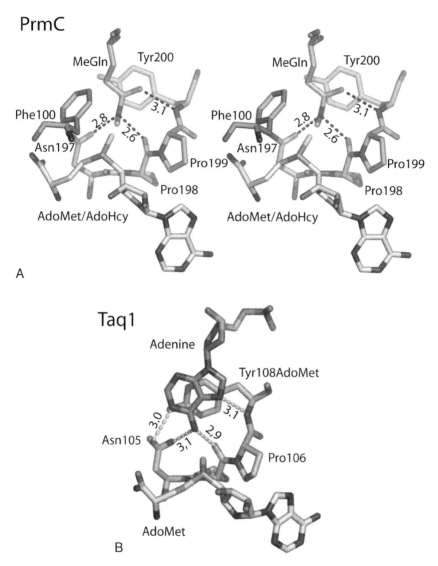

FIG. 14.4. The NPPY motif of Gln and adenosine MTases. (A) The active site of PrmC is formed by the invariant residues NPPY, which wrap around the substrate binding site and form hydrogen bonds to the ligand to specify its orientation and location. A large hydrophobic group is positioned behind the positively charged methyl group of AdoMet, enhancing the transfer of the methyl group and formation of the neutral AdoHcy product. (B) A similar architecture is seen in the Ade MTase TaqI (PDBcodes: 2ADM and 1G38).

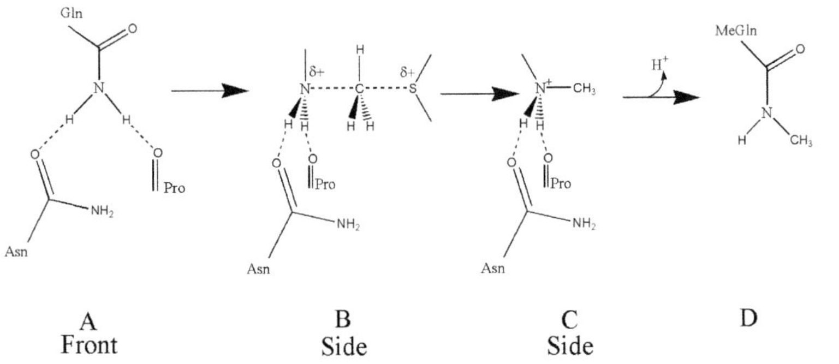

FIG. 14.5. The catalytic mechanism of Gln MTases. (A) The mechanism originates with the alignment of the sp2-hybridized ligand using hydrogen bonds to both the Asn side chain and one of the proline's hydroxyls. (B and C) As the incoming methyl group approaches the amide nitrogen, the hydrogens are pushed away eventually forming a sp3-hybridized nitrogen center using a traditional SN2-type reactions mechanism with inversion of symmetry. (D) During release from the active site a final proton is removed and the methyl group is positioned within the plane of the original amide bond. *[Figure adapted with permission from* Biochemistry *2003, 42(19), 5592-5599.]*

contains both protons and the partially occupied methyl group and that the nitrogen atom would contain a partial positive charge. It is possible that the atom could have only one proton and the methyl group attached and would thus be neutral. If neutral, the conformation is strained with respect to the expected final planar conformation [32]. Neither the DNA nor the Gln MTase (see Figure 14.4) structures provide a candidate for a catalytic general acid or general base and hence the proton is probably lost to solvent.

E.   COMPARISON WITH CLASS I METHYLTRANSFERASES

Both PrmC and TaqI bind the sp2-hybridized NH2 group of their substrate through hydrogen bonds with the (D/N)PPY motif. (Figure 14.4b) [26, 29]. In contrast, many other class I AdoMet-dependent MTases do not contain the [D/N]PPY motif and have evolved alternate methods of substrate binding, nucleophilic activation, and proton abstraction. The tertiary structures have been solved for several uncharacterized proteins, which are obviously MTases, and share high structural similarity to the structure of PrmC. Due to the high sequence similarity within the class I AdoMet-dependent MTases it would still be difficult to determine the true substrate of a new MTase sequence without additional biochemical characterization. Indeed, the PDB file 1DUS corresponds to *M. jannaschii* protein mj0882 and is the nearest structural neighbor characterized for *E. coli* PrmC. It has been discussed as a model of the RNA:(Guanine-N2) MTases RsmC/RsmD [33].

Structural comparison to the PDB database reveals several close structural neighbors, including *Methanococcus jannaschii* mj0882 (PDBcode:1DUS), *Mycobacterium tuberculosis* rv2118c (PDBcode:1I9G), the catechol O-MTase (PDBcode:1VID), and the glycine O-MTase (PDBcode:1XVA) [34]. The structural comparison served DALI-assigned Z scores in the range of 11 to 18 for these alignments despite only 10 to 21% sequence identity (Z score > 2 indicates significant structural similarity).

F.   COMPARISON WITH CLASS II METHYLTRANSFERASES

Since the first structure of PrmC (PDB code 1NV8), two additional unpublished structures of the *T. maritima* enzyme (PDB codes 1SG9 and 1VQ1) and the *E. coli* PrmC (PDB code 1T43) [30] have been solved. As expected, the *T. maritima* structures are very similar, with differences only in a flexible loop between β4 and αD. The r.m.s. deviation between the *T. maritima* and *E. coli* structures is 2.2 Å over 263 residues [34]. The two proteins share 28% sequence identity. The primary location for sequence conservation is within the C-terminal MTase domain.

The N-terminal domain retains some invariant residues in the hydrophobic core of the domain, but does not retain conservation along surface residues. This downplays the role of this domain in substrate recognition [30]. Perhaps the domain is not conserved, in that similar PrmC enzymes must recognize both eukaryotic RF1 and RF2 proteins as well as the drastically different prokaryotic eRF1 protein (see material following). This lack of specialization has rendered PrmC a uniquely nonspecific enzyme globally, but highly specific locally at the GGQ sequence. In fact, no substantial physical interactions have been observed between the *T. maritima* PrmC and RF1 as determined by size exclusion chromatography (Dong Hae Shin et al., personal communication).

IV.   Biological Significance: The Role of Methylated Release Factors

Human eRF1 and *E. coli* RF2 (both class I RFs) share little structural or sequence similarity aside from the GGQ motif (Figures 14.6a through d), which is always positioned on a surface loop [13, 35]. The eRF1 structure contains three α/β domains stretched out linearly such that domains 1 and 2 are only in contact through the central domain 3 (Figure 14.6d). In domain 1, the NIKS motif is responsible for recognition of the stop codon at the decoding center on the ribosome. The GGQ motif lies at the opposite end of the molecule in domain 2. The two motifs are separated by a distance of more than 75 Å [13].

Crystal structures of *E. coli* RF2 revealed a more compact molecule also containing three domains [35, 36]. The N-terminal domain contains a three-helix

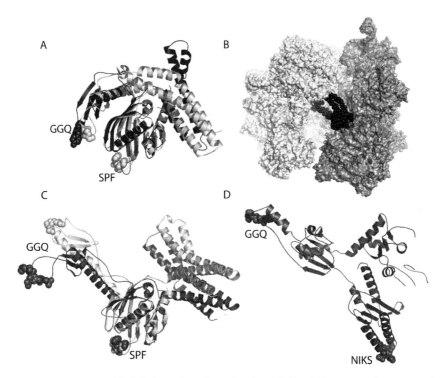

FIG. 14.6. Structural insight into release factor function. (A) Two different crystal structures of RF2 reveal a compact structure where the GGQ motif and the mRNA stop codon recognition site (SPF) are only 23 to 25 Å apart. *E. coli* RF2 (PDBcode: 1GQE) is shown in white [35] (PDBcode: 1GQE), and *T. maritima* RF2 (PDBcode: 1RQ0) is shown in black [36.] (B) Cryo-electron microscopy (EM) studies revealed the location of the RF on the ribosome spanning the decoding center and the peptidyl transferase center [37, 38] (figure based on PDBcodes 1MI6 and 1ML5). (C) Two groups modeled *E. coli* RF into the cryo-EM maps by allowing significant conformational changes in the regions between domains. The separation between the GGQ and SPF motifs in the new models is approximately 70 Å (1MI6: white, 1ML5: black). Figures 6a and 6c are similarly oriented based on alignment of their central core domain. (D) The human eRF1 structure contains significant topological differences to that of *E. coli* RF2 [13] (PDBcode: 1DT9). In the crystal structure the distance between the GGQ motif and the stop codon mRNA recognition sequence (NIKS) is similar to the span in tRNA and the models of the prokaryotic RF bound to the ribosome.

bundle that leads into the central β-sheet core domain followed by a small α/β domain containing the GGQ motif on an extended loop. The stop codon recognition sequence, SPF, lies on a loop within the core domain. The two motifs are separated by only 23 Å (Figure 14.6a) [35]. In both types of structures, the Gln side chain is packed against the surface of the protein. To facilitate Gln methylation, some conformational change (perhaps involving the flexible glycine residues) will be required. The region flanking the GGQ motif of both eRF1 and RF2 contains several basic residues, such as Arg, Lys, and His [12]. Although these residues may be evolutionarily conserved in order to make essential contacts with the ribosome, the MTase may also have adapted to make use of those residues when recognizing the extended loop as a substrate.

The peptidyl transferase center (PTC) on the 50S ribosome is positioned some 73 Å from the position of the mRNA decoding center on the 30S ribosome. The elongated eRF1 structure shares a remote resemblance to the structure of a tRNA and leads to the hypothesis that the RF would function similarly, with residues recognizing the stop codon 73 Å away from the GGQ motif and triggering peptide release at the PTC (Figure 14.6d) [13]. With the subsequent *E. coli* RF2 structure, this hypothesis was placed into question. Although the tRNA mimic idea was still attractive, substantial conformational changes were going to be necessary if the RF was going to recognize both ribosomal sites at once.

A more recent structure of RF1 from *T. maritima* partially supported this hypothesis, revealing the potential for rigid-body movements of the N- and C-terminal domains of RF1 with respect to the *E. coli* RF2 structure [36]. The N-terminal domain moves almost 90 degrees (Figure 14.6a), but only small changes are observed in the GGQ and codon recognition sites (which are crucial to the tRNA hypothesis of RF function). The two sites are still only 27Å apart. More dramatic conformational changes are supported by recent electron microscopy studies on whole ribosomes in the presence of release factors (Figure 14.6b).

In two separate experiments, the structure of the *E. coli* ribosome has been solved in the presence of RF2. RF2 has been modeled into electron density that spans the decoding center and PTC sites [37, 38] (Figure 14.6c). The models required rigid-body motions of the three separate domains of RF, assuming they are connected with flexible linkers between domains. Unfortunately, the resolution of the structure limits the information relating to position of the flexible loop containing the GGQ motif and specifically to the role of the methyl group.

The prediction that the GGQ motif plays a role in hydrolysis of the ester bond in polypeptide termination originated with mutational studies by Frolova [12]. A connection was made that the GGQ motif was just as highly conserved over evolution as the CCA tail of the tRNA. One hypothesis suggested that the GGQ motif might specifically recognize the CCA of the tRNA in the P site and position the release factor into the peptidyl transferase center [13]. This hypothesis was taken one step further to suggest that the Gln residue coordinates the nucleophilic water molecule involved in polypeptide-ester bond during hydrolysis [13]. At the time of these predictions, neither group realized that the Gln residue was methylated. The methylation has been shown to enhance the rate of peptide release [19].

Whereas the glutamine residue is invariant and its methylation seems crucial, the role of the modified residue in polypeptide hydrolysis is less than certain. Mutations of the Gln residue to Glu, Gly, Arg, Asn, Asp, and Lys are tolerated by the reaction *in vitro*, showing decreased rates, but not abolition of peptide termination [13, 19, 39]. Only mutation of Gln to Ile appears to abolish *in vitro* activity. The *in vivo* activity of Gln mutants has only been tested on the Gln→Ala, Gln→Glu, and Gln→Leu mutants. In all cases, the mutants do not complement thermosensitive RF mutants [40] or a yeast *E. coli* RF deletion strain [13]. Note that the Gln→Glu mutant retained activity *in vitro* but not *in vivo*, suggesting that these studies cannot fully be correlated.[1]

V.   The Extended Gln MTase Family

There are three families of genes currently ascribed to the Gln MTase family. These are designated PrmB/YfcB, PrmC/HemK, and hemk-rel-arch (Figure 14.7). The three share significant sequence identity, although the low similarity between the two proteins makes it difficult to discern the true function of their respective homologs without resorting to a full phylogenetic tree or genomic location analysis.

In most organisms, the enzyme and substrate genes prmC/hemK and prfA (RF1) lie in a tightly linked operon beginning with a third gene, the early heme biosynthetic enzyme hemA. This linkage provides additional information when determining the true prmC homologs from remote species. Of 79 genomes analyzed, eight did not show evidence of direct linkage and may be the result of lateral gene transfer [30]. PrmB/YfcB is the L3 Gln MTase. It is unknown which of the 13 glutamines in protein L3 is actually methylated. The *E. coli* enzyme has been characterized through ^{3}H-labeled AdoMet reactions on ribosomes from ΔyfcB mutant *E. coli* [20].

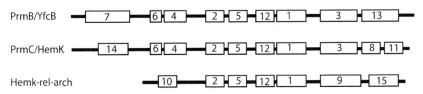

Fig. 14.7. The HemK/PrmC family contains three subclassifications (Interpro signatures IPR004556/7). Traditional PrmB – L3 Gln MTases, PrmB – RF Gln MTases, and the smaller Hemk-rel-arch family. They share most similarity in the traditional MTase domain and additional conserved regions specific to their subgroup. Regions of similarity have been defined (and numbered) using the MEME program [42]. *(Figure adapted from Heurgue-Hamard et al. [20].)*

1 Since the composition of this chapter, the structure of the complex between E. coli PrmC and RF1 has been solved (PDBcode 2B3T). Graille M, Heurgue-Hamard V, Champ S, Mora L, Scrima N, Ulryck N, van Tilbeurgh H, Buckingham RH. Mol Cell (2005) 20(6):917–927.

The two proteins PrmB and PrmC from *E. coli* share 32% identity (53% similarity). The term *hemk-rel-arch* defines a smaller subfamily of GlnMTases that lacks the N-terminal domain altogether and only retains similarity to the PrmC family via the MTase domain. In mammalian systems, there is a PrmC-like gene in the mitochondria that may act on the mitochondrial release factors (all of bacterial origin), whereas the nuclear genome contains the "hemk-rel-arch" gene and may function on the unique mammalian release factors.

VI. Conclusions

The newest members of the protein MTase family are that of the Gln MTase. Their catalytic function is quite similar to that of the DNA N-MTases, as they share the use of the [D/N]PPY motif for substrate recognition and catalysis. Two substrates are known to be methylated on glutamine. The Gln moiety on ribosomal protein L3 has no essential function and is only observed in bacteria. On the other hand, the methylation on the Gln residue within the GGQ motif in polypeptide release factors plays a crucial role in ribosomal translation termination. Future studies are still required to unearth the specific function of this methylation.

ACKNOWLEDGMENTS

HLS is funded through NIH grant GM56775.

REFERENCES

1. Paik, W.K., and Kim, S. (1975). Protein methylation: Chemical, enzymological and biological significance. *Adv Enzymol* 42:227–286.
2. Alix, J.H., and Hayes, D. (1974). Properties of ribosomes and RNA synthesized by *Escherichia coli* grown in the presence of ethionine III. Methylated proteins in 50S ribosomes of *E. coli* EA2. *J Mol Bio* 86:136–159.
3. Chang, C.N., and Chang, F.N. (1975). Methylation of the ribosomal proteins in *Escherichia coli*. Nature and stoichiometry of the methylated amino acids in 50S ribosomal proteins. *Biochem* 14:468–470.
4. Lhoest, J., and Colson, C. (1977). Genetics of ribosomal protein methylation in *Escherichia coli*. II. A mutant lacking a new type of methylated amino acid, N5-methylglutamine, in protein L3. *Mol Gen Genet* 154(2):175–180.
5. Levit, M.N., and Stock, J.B. (2002). Receptor methylation controls the magnitude of stimulus-response coupling in bacterial chemotaxis. *J Biol Chem* 277(39):36760–36765.
6. Selmer, T., Kahnt, J., Goubeaud, M., Shima, S., Grabarse, W., Ermler, U., and Thauer, R. K. (2000). The biosynthesis of methylated amino acids in the active site region of methyl-coenzyme M reductase. *J. Biol. Chem.* 275: 3755–3760.
7. Kisselev, L.L., and Buckingham, R.H. (2000). Translational termination comes of age. *Trends Biochem Sci* 25(11):561–566.

8. Ehrenberg, M., and Tenson, T. (2002). A new beginning of the end of translation. *Nat Struct Biol* 9(2):85–87.

9. Frolova, L., Le Geoff, X., Rasmussen, H. H., Cheperegin, S., Drugeon, G., Kress, M., Arman, I., Haenni, A. L., Celis, J. E., Philippe, M. & al., e. (1994). A highly conserved eukaryotic protein family possessing properties of polypeptide chain release factor. *Nature* 372:4065–4077.

10. Scolnick, E., et al. (1968). Release factors differing in specificity for termination codons. *PNAS* 62(2):768–774.

11. Frolova, L.Y., Merkulova, T.I., and Kisselev, L.L. (2000). Translation termination in eukaryotes: polypeptide release factor eRF1 is composed of functionally and structurally distinct domains. *RNA* 6(3):381–390.

12. Frolova, L. Y., Tsivkovskii, R. Y., Sivolobova, G. F., Oparina, N. Y., Serpinsky, O. I., Blinov, V. M., Tatkov, S. I., and Kisselev, L. L. (1999). Mutations in the highly conserved GGQ motif of class 1 polypeptide release factors abolish ability of human eRF1 to trigger peptidyl-tRNA hydrolysis. *RNA* 5:1014–1020.

13. Song, H., Mugnier, P., Das, A. K., Webb, H. M., Evans, D. R., Tuite, M. F., Hemmings, B. A., and Barford, D. (2000). The crystal structure of human eukaryotic release factor eRF1– mechanism of stop codon recognition and peptidyl-tRNA hydrolysis. *Cell* 100:311–321.

14. Seit-Nebi, A., Frolova, L., Ivanova, N., Poltaraus, A., and Kiselev, L. (2000). Mutation of a glutamine residue in the universal tripeptide GGQ in human eRF1 termination factor does not cause complete loss of its activity. *Mol. Biol. (Mosk)* 34:899–900.

15. Zavialov, A. V., Mora, L., Buckingham, R. H., and Ehrenberg, M. (2002). Release of peptide promoted by the GGQ motif of class 1 release factors regulates the GTPase activity of RF3. *Mol. Cell* 10:789–798.

16. Kawakami, K., Nakamura, Y., Mikuni, O., Kawakami, K., Nakamura, Y., Uno, M., Ito, K., and Nakamura, Y. (1990). Autogenous suppression of an opal mutation in the gene encoding peptide chain release factor 2. *Proc. Natl. Acad. Sci. USA* 87:8432–8436.

17. Mikuni, O., Kawakami, K., Nakamura, Y., Uno, M., Ito, K., and Nakamura, Y. (1991). Sequence and functional analysis of mutations in the gene encoding peptide-chain-release factor 2 of *Escherichia coli. Biochimie* 73, 1509–1516.

18. Uno, M., Ito, K., and Nakamura, Y. (1996). Functional specificity of amino acid at position 246 in the tRNA mimicry domain of bacterial release factor 2. *Biochimie* 78(11–12):935–943.

19. Dincbas-Renqvist, V., Engstrom, A., Mora, L., Heurgue-Hamard, V., Buckingham, R., and Ehrenberg, M. (2000). A posttranslational modification in the GGQ motif of RF2 from *Escherichia coli* stimulates termination of translation. *EMBO J.* 19:6900–6907.

20. Heurgue-Hamard, V., Champ, S., Engstrom, A., Ehrenberg, M., and Buckingham, R. H. (2002). The hemK gene in *Escherichia coli* encodes the N(5)-glutamine methyltransferase that modifies peptide release factors. *EMBO J.* 21:769–778.

21. Nakayashiki, T., Nishimura, K., and Inokuchi, H. (1995). Cloning and sequencing of a previously unidentified gene that is involved in the biosynthesis of heme in *Escherichia coli. Gene* 153(1):67–70.

22. Nakahigashi, K., et al. (2002). HemK, a class of protein methyl transferase with similarity to DNA methyl transferases, methylates polypeptide chain release factors, and hemK knockout induces defects in translational termination. *Proc Natl Acad Sci USA* 99(3):1473–1478.

23. Clarke, S. (2002). The methylator meets the terminator. *Proc Natl Acad Sci USA* 99(3): 1104–1106.

24. Bujnicki, J.M., and Radlinska, M. (1999). Is the HemK family of putative S-adenosylmethionine-dependent methyltransferases a "missing" zeta subfamily of adenine methyltransferases? A hypothesis. *IUBMB Life* 48(3):247–249.

25. Fauman, E.B., Blumenthal, R.M., and Cheng, X. (1998). *Structure and Evolution of AdoMet-Dependent Methyltransferases. S-adenosylmethionine-Dependent Methyltransferases: Structures and Functions*, X. Cheng and R.M. Blumenthal (eds.), pp. 1–38, New York: World Scientific.

26. Goedecke, K., Pignot, M., Goody, R. S., Scheidig, A. J., and Weinhold, E. (2001). Structure of the N6-adenine DNA methyltransferase M.TaqI in complex with DNA and a cofactor analog. *Nat. Struct. Biol.* 8:121–125.

27. Malone, T., Blumenthal, R.M., and Cheng, X. (1995). Structure-guided analysis reveals nine sequence motifs conserved among DNA amino-methyltransferases, and suggests a catalytic mechanism for these enzymes. *J Mol Biol* 253(4):618–632.

28. Newby, Z.E., Lau, E.Y., and Bruice, T.C. (2002). A theoretical examination of the factors controlling the catalytic efficiency of the DNA-(adenine-N6)-methyltransferase from Thermus aquaticus. *Proc Natl Acad Sci USA* 99(12):7922–7927.

29. Schubert, H.L., Phillips, J.D., and Hill, C.P. (2003). Structures along the catalytic pathway of PrmC/HemK, an N5-glutamine AdoMet-dependent methyltransferase. *Biochem* 42(19):5592–5599.

30. Yang, Z., Shipman, L., Zhang, M., Anton, B. P., Roberts, R. J., and Cheng, X. (2004). Structural characterization and comparative phylogenetic analysis of *Escherichia coli* HemK, a protein (N5)-glutamine methyltransferase. *J. Mol. Biol.* 340:695–706.

31. Schubert, H.L., Blumenthal, R.M., and Cheng, X. (2003). Many paths to methyltransfer: A chronic of convergence. *TIBS* 28(6):329–335.

32. Milner-White, E.J. (1997). The partial charge of the nitrogen atom in peptide bonds. *Prot Sci* 6(11):2477–82.

33. Bujnicki, J.M., and Rychlewski, L. (2002). RNA:(guanine-N2) methyltransferases RsmC/RsmD and their homologs revisited: Bioinformatic analysis and prediction of the active site based on the uncharacterized Mj0882 protein structure. *BMC Bioinformatics* 3(1):10.

34. Holm, L. and Sander, C. (1997). Dali/FSSP classification of three-dimensional protein folds. *Nucleic Acids Res* 25(1):231–234.

35. Vestergaard, B., Van, L. B., Andersen, G. R., Nyborg, J., Buckingham, R. H., and Kjeldgaard, M. (2001). Bacterial polypeptide release factor RF2 is structurally distinct from eukaryotic eRF1. *Mol Cell* 8:1375–1382.

36. Shin, D. H., Brandsen, J., Jancarik, J., Yokota, H., Kim, R., and Kim, S. H. (2004). Structural analyses of peptide release factor 1 from *Thermotoga maritima* reveal domain flexibility required for its interaction with the ribosome. *J. Mol. Biol.* 341:227–239.

37. Klaholz, B. P., Pape, T., Zavialov, A. V., Myasnikov, A. G., Orlova, E. V., Vestergaard, B., Ehrenberg, M., and van Heel, M. (2003). Structure of the *Escherichia coli* ribosomal termination complex with release factor 2. *Nature* 421:90–94.

38. Rawat, U. B., Zavialov, A. V., Sengupta, J., Valle, M., Grassucci, R. A., Linde, J., Vestergaard, B., Ehrenberg, M., and Frank, J. (2003). A cryo-electron microscopic study of ribosome-bound termination factor RF2. *Nature* 421:87–90.

39. Seit-Nebi, A., Frolova, L., Justesen, J., and Kisselev, L. (2001). Class-1 translation termination factors: invariant GGQ minidomain is essential for release activity and ribosome binding but not for stop codon recognition. *Nucleic Acids Res.* 29: 3982–3987.

40. Mora, L., Zavialov, A., Ehrenberg, M., and Buckingham, R. H. (2003). Stop codon recognition and interactions with peptide release factor RF3 of truncated and chimeric RF1 and RF2 from *Escherichia coli. Mol. Microbiol.* 50:1467–1476.

41. DeLano, W.L. (2002). *http://www.pymol.org.*

42. Bailey, T.L., and Elkan, C. (1994). *Fitting a mixture model by expectation maximization to discover motifs in biopolomers*, pp. 28–36, Menlow Park, CA: AAAI Press.

15

Modification of Phycobiliproteins at Asparagine Residues

ALAN V. KLOTZ

Lilly Research Laboratories
Eli Lilly and Company
Lilly Corporate Center
Indianapolis, IN 46285, USA

I. Abstract

Side-chain amide methylation of asparagine was described in a special complement of photosynthesis accessory pigment-protein complexes called phycobiliproteins nearly 20 years ago. Since that report, several investigations have assigned this posttranslational modification a functional role in tuning the spectroscopic properties of the phycobiliprotein chromophores. Asparagine methylation has not been reported in other systems and is restricted to the broader phycobiliprotein family. The methyltransferase responsible for this modification has been partially characterized but the structural gene has not been identified.

II. Introduction

Methylation on the side-chain amide of asparagine in proteins was first described nearly 20 years ago [1, 2], but a compelling physiological role has been somewhat elusive. More recently, with advances in the knowledge of protein glutaminyl methylation there has been a renaissance in the interest level for this type of stable protein methylation [3]. Glutaminyl side-chain methylation [4] was

THE ENZYMES, Vol. XXIV

reported prior to the recognition of asparagine methylation, and the two phenomena have traded insights and advances in the intervening period. The purpose of the current contribution is to summarize the status of asparagine methylation and to provide perspective on how the recent advances in glutaminyl methylation may pertain to asparagine methylation structure and function.

III. Occurrence

Asparagine methylation has been described in a special complement of photosynthesis accessory pigment-protein complexes called phycobiliproteins. These complexes are distributed throughout prokaryotic cyanobacteria, cryptomonads (Cryptophyta), and red algae (Rhodophyta), a range of taxonomic groups bridging the prokaryotic and eukaryotic phyla. Phycobiliproteins are unique pigment proteins that function as light-harvesting antennae. These proteins form macromolecular aggregates (7 to 15×10^6 daltons [5]) known as phycobilisomes, which electrostatically interact with the stromal side of the thylakoid membrane in the vicinity of photosystem II. They function to absorb excitation energy in the range of 450 to 650 nm, with energy transfer to the protein-chlorophyll complex of photosystem II in a process that occurs with >95% quantum efficiency [5]. Phycobiliproteins are composed of non-identical α and β subunits, each of which possesses linear tetrapyrrole (bilin) prosthetic groups covalently linked to the primary structure via cysteine thioether bonds [6].

Methylasparagine was uncovered in phycobiliproteins after extensive protein sequence analysis and X-ray crystallography work was accomplished. Table 15.1 provides a summary of phycobiliprotein sequences that have been confirmed to contain N^4-methylasparagine (NMA).[1] The persistence of NMA across the prokaryote/eukaryote boundary is an intriguing observation that implies a selective advantage to justify retention. One other noteworthy, but unexplained, phenomenon is the finding that some species possess NMA in one class of phycobiliproteins yet lack this modification at the homologous sequence position in a related phycobiliprotein [7, 8].

One reason NMA may have been overlooked in the past is the fact that Edman degradation for peptide sequence analysis can be subject to misinterpretation when no confirmatory analyses are performed. The authentic phenylthiohydantoin derivative of NMA in many chromatographic systems coelutes with the serine derivative and is accompanied by a minor component eluting near dimethylphenylthiourea [7-9]. Thus, NMA provides a chromatographic signature during Edman degradation if the sequence analyst evaluates the secondary peaks [9].

---

1. Early literature refers to this modification as N^5-methylasparagine, but we have adopted the recent convention of N^4-methylasparagine for consistency in this chapter. Alternative designations include γ-N-methylasparagine and NMA.

TABLE 15.1

SEQUENCE HOMOLOGIES AROUND THE N^4-METHYLASPARAGINE SITE IN PHYCOBILIPROTEIN B-SUBUNITS[a]

Organism	Protein	Homology Around β-72 Position	References
Synechococcus sp. PCC6301	Allophycocyanin β-subunit	...Thr-Arg-Pro-Gly-Gly-NMA-Met-Tyr...	[2]
Anabaena variabilis	Allophycocyanin β-subunit	...Thr-Arg-Pro-Gly-Gly-NMA-Met-Tyr...	[27]
Mastigocladus laminosus	Allophycocyanin β-subunit	...Thr-Arg-Pro-Gly-Gly-NMA-Met-Tyr	[7]
Synechococcus sp. PCC6301	C-Phycocyanin β-subunit	...Ile-Ala-Pro-Gly-Gly-NMA-Ala-Tyr...	[28]
Synechococcus sp. PCC7002	C-Phycocyanin β-subunit	...Ile-Ala-Pro-Gly-Gly-NMA-Ala-Tyr...	[2]
Mastigocladus laminosus	C-Phycocyanin β-subunit	...Ile-Ala-Pro-Gly-Gly-NMA-Ala-Tyr...	[7]
Mastigocladus laminosus	$\beta^{16.2}$ phycobiliprotein	...Ile-Arg-Pro-Gly-Gly-NMA-Ala-Tyr...	[7]
Porphyridium cruentum	B-Phycoerythrin β-subunit	...Ile-Ser-Pro-Gly-Gly-NMA-Cys-Tyr...	[29]
Gastroclonium coulteri	R-Phycoerythrin β-subunit	...Ile-Ala-Pro-Gly-Gly-NMA-Cys-Tyr...	[30]
Cryptomonad strain CBD	Phycoerythrin 566 β-subunit	...Ile-Ser-Pro-Ser-Gly-NMA-Cys-Tyr...	[8]

[a]The NMA structure is assigned by direct chemical identification or inference from sequence homology and liberation of methylamine upon acid hydrolysis. NMA: N^4-methylasparagine.

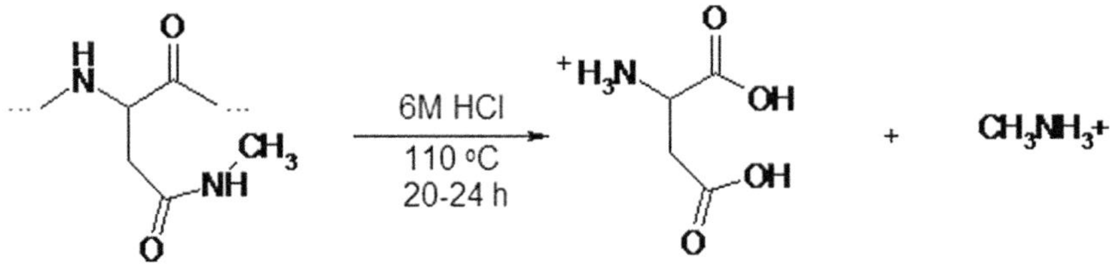

Fɪɢ. 15.1. Acid hydrolysis products of NMA. Methylamine is a characteristic acid hydrolysis product of NMA.

Retrospective analysis of original phycobiliprotein sequence analysis indicates that the misidentification was commonly serine or aspartate [1, 2, 7].

NMA and its sister, N-methylglutamine [4], can be tentatively identified by the unique release of methylamine upon acid hydrolysis (Figure 15.1). Methylamine is stable under acid hydrolysis conditions, and stoichiometric amounts can be recovered and quantified by amino acid analyzers employing cation exchange chromatography [1, 2]. Despite survey attempts by amino acid analysis, NMA has not been identified in other proteins, including bovine histones, porcine myelin basic protein, or the *S. typhimurium* aspartate chemoreceptor (which are known to contain posttranslational methylations) [2].

From the beginning it was recognized that primary sequence is not the sole determinant for asparagine methylation because the α subunits in phycobiliproteins lack asparagine methylation despite containing a potential Pro-Gly-Gly-Asn methylation site [1] with sequence and structural homologies to the β subunit. Further, DNA sequencing of the phycobiliprotein genes clearly indicates that asparagine is the encoded primary translation product [1, 2].

IV. Biochemistry

Formation of NMA has been linked to an enzymatic process requiring *S*-adenosyl-ʟ-methionine [2], and the cyanobacterial activity has been reconstituted *in vitro* [10]. Chemical mutagenesis has allowed the creation of mutants that lack the capability to methylate asparagines, but the responsible gene(s) have not been identified or sequenced. To date NMA has not been found to turn over during physiological stress of the cells by light deprivation or nitrogen starvation [2], and the methylation appears to be extremely stable.

Some information about the methyl transferase can be inferred from studies in which the asparagine site was substituted with glutamine [11]. In this case, the glutamine was partially methylated to a molar ratio of 0.27, implying that the structural fidelity for the activity is high but not absolute. The reconstitution of methyltransferase activity *in vitro* with apo-phycocyanin indicates that the methylation site in the full length protein (without covalently attached bilins) is accessible to the enzyme.

V.   Potential Functions

Potential functions for NMA have been sought in order to understand the intriguing evolutionary retention of this characteristic. Three hypotheses have been examined: the concept that asparagine methylation might diminish the rate of chemical deamidation at a structurally sensitive site, the idea that NMA may contribute to protein stability or protein-protein interaction, and the proposal that asparagine methylation improves the energy transfer of the phycobiliproteins under physiologically relevant light-harvesting conditions. The data and conclusions revolving around each of these proposed functions will be considered in turn.

VI.   Deamidation

Glutamine and asparagine deamidation contribute to protein aging, accelerate protein turnover, and may control some development functions. The mechanism for asparagine deamidation in polypeptides above pH 3 involves an intramolecular attack by the (n+1) amide nitrogen in the polypeptide backbone on the side-chain carboxamide. Asparagine deamidation has gained recognition as one of the major routes for chemical degradation of proteins [12–15], and is particularly relevant in long-lived proteins. A chemical rationale for asparagine methylation is the hypothesis that side-chain methylation is predicted to suppress the rate of deamidation. This has been demonstrated for both free amino acids [16] and peptides [17, 18] at pH 7.4, where the methyl group provided a >10-fold stabilization against deamidation. In peptides, the stabilization effect was 45-fold for the deamidation reaction. However, in both cases the diminished deamidation rate for N^4-methylasparagine resulted in a change in the predominant pathway to a side-chain nucleophilic attack on the α carbonyl (Figure 15.2).

In a peptide context, this reaction results in main chain cleavage for both …NMA-Ala… and …NMA-Gly… peptides, whereas the homologous peptide series containing Asn displays exclusively succinimide formation resulting in Asp/isoAsp creation. Therefore, the stabilizing effect of methylation against a potential deamidation pathway is counterbalanced by accentuating a potentially more deleterious peptide chain cleavage outcome.

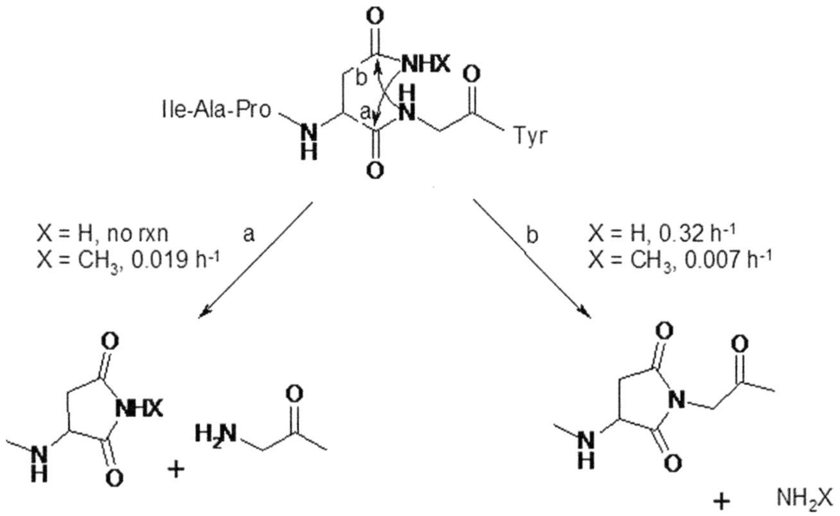

FIG. 15.2. Paths of degradation for Asn and NMA peptides. Peptides Ile-Ala-Pro-Asn/NMA-Gly-Tyr were studied at 60° C in 0.1M NaPO$_4$, pH 7.4 and the measured rate constants for peptide disappearance are provided [18]. The asparagine-containing peptide ($t_{0.5} = 2.17$ h) degrades exclusively by main chain amide attack on the amide side chain (path b), whereas the methylasparagine-containing peptide ($t_{0.5} = 26.6$ h) partitions between a side-chain attack on the peptide backbone resulting in chain scission (path a) and the expected main chain amide attack on the amide side chain (path b). The homologous Asn/NMA-Ala peptides have much longer half-lives ($t_{0.5} = 45$ and 84 h, respectively) due to steric effects, and consequently the degradants were not thoroughly characterized [17].

VII. Protein Stability

The best characterized phycobiliprotein system, C-phycocyanin, was examined by differential scanning calorimetry but no differences were found between fully methylated, under-methylated, and unmethylated forms [10]. Similarly, urea denaturation studies of C-phycocyanin containing NMA, Asp, or Gln at the β-72 position indicate no difference in protein stability [11]. These results demonstrate that methylation does not alter protein stability as a global property.

VIII. Specialized Energy Transfer

Soon after the elucidation of N^4-methylasparagine in phycobiliproteins, crystallographic analysis revealed that the β-72 methyl group was approximately 3.5 Å from the β-84 bilin, which serves as the terminal energy acceptor in the complex [19]. The bilins that absorb light energy and transfer the excitation energy to other bilins are called donors. The bilins that absorb excitation energy from incident light or by energy transfer are called acceptors and typically fluoresce in the isolated state [5]. Intramolecular energy transfer within phycobiliprotein trimers and hexamers is extremely fast and consequently the fluorescence emission spectrum is dominated by the acceptor bilin photophysics.

Studies with cyanobacteria strains deficient in asparagine methyltransferase have established that unmethylated phycobiliproteins are less efficient in directional energy transfer and that the phycobilisomes possess a lower emission quantum yield [10]. This observation has been confirmed by quantum yield measurements for an oxygen evolution endpoint [20]. Asparagine methylation does not impact the fluorescence properties of C-phycocyanin $\alpha\beta$ monomer, indicating that the effect is only expressed in higher-order phycobiliprotein aggregates.

Molecular dynamics calculations for C-phycocyanin indicate that β-72 NMA restricts hydrogen bonding networks in the protein [11]. The aggregate biophysical studies in the literature are consistent with β-72 NMA decreasing energy losses from the bilin excited state via nonradiative pathways of bilin deexcitation.

IX. Conclusions

The originally described δ-N-methylglutamine (N^5-methylglutamine) in ribosomal protein L3 is found in a site (...Gly-Gln-Asn-Gln*-Thr-Pro... [21]) that has no recognizable homology to the asparagine methylation sites in phycobiliproteins. Prokaryotic and eukaryotic peptidyl release factors 1 and 2 possess N^5-methylglutamine with an intriguing sequence homology (...Gly-Gly-Gln*...) at the methylation site [22–24], which is shared with the phycobiliprotein β-72 NMA recognition sequence (Figure 15.1).

The glutamine methylation site is contained in a turn connecting a β strand with an α helix [24]. This is structurally similar to the Pro-Gly-Gly-Asn sequence in phycobiliproteins, which forms a β turn between α helices. Mutational analysis of the GGQ sequence indicates that it is very sensitive to structural substitutions, and many of the mutations are lethal [22, 24]. Mutants that lack methylation capability for ribosomal protein L3 display a cold-sensitive phenotype and a diminished rate of ribosome assembly [25]. However, these ribosomes are fully active upon assembly, and it appears that glutamine methylation influences only the assembly process.

In contrast to this, although evolutionarily conserved, the NMA site in phycobiliproteins is not essential for protein function, and site-specific mutants lack a burden on growth rate in the laboratory. Presumably, NMA provides a competitive advantage under conditions of limited light quality or quantity, as may commonly be found in ecological niches such as the open ocean [26]. Asparagine methylation remains an uncommon posttranslational modification that has a functional utility restricted to specialized biophysical properties of protein chromophores.

REFERENCES

1. Klotz, A.V., Leary, J.A., and Glazer, A.N. (1986). Posttranslational methylation of asparaginyl residues: Identification of β-71 γ-N-methylasparagine in allophycocyanin. *J Biol Chem* 261:15891–15894.
2. Klotz, A.V., and Glazer, A.N. (1987). γ-N-Methylasparagine in phycobiliproteins. *J Biol Chem* 262:17350–17355.
3. Clarke, S. (2002). The methylator meets the terminator. *Proc Natl Acad Sci USA* 99:1104–1106.
4. Lhoest, J., and Colson, C. (1977). Genetics of ribosomal protein methylation in *Escherichia coli*. *Mol Gen Genet* 154:175–180.
5. Glazer, A.N. (1989). Light guides. *J Biol Chem* 264:1–4.
6. Glazer, A.N. (1984). Phycobilisome: A macromolecular complex optimized for light energy transfer. *Biochim Biophys Acta* 768:29–51.
7. Rümbeli, R., Suter, F., Wirth, M., Sidler, W., and Zuber, H. (1987). Isolation and localization of N^4-methylasparagine in phycobiliproteins from the cyanobacterium *Mastigocladus laminosus*. *Biol Chem Hoppe-Seyler* 368:1401–1406.
8. Wilbanks, S.M., Wedemeyer, G.J., and Glazer, A.N. (1989). Posttranslational modifications of the β-subunit of a cryptomonad phycoerythrin. *J Biol Chem* 264:17860–17867.
9. Klotz, A.V., Thomas, B.A., Glazer, A.N., and Blacher, R.W. (1990). Detection of methylated asparagine and glutamine residues in polypeptides. *Anal Biochem* 186:95–100.
10. Swanson, R.V., and Glazer, A.N. (1990). Phycobiliprotein methylation. *J Mol Biol* 214:787–796.
11. Thomas, B.A., McMahon, L.P., and Klotz, A.V. (1995). N^5-Methylasparagine and energy-transfer efficiency in C-phycocyanin. *Biochem* 34:3758–3770.
12. Ahern, T.J., and Klibanov, A.M. (1985). The mechanism of irreversible enzyme inactivation at 100°C. *Science* 228:1280–1284.
13. Aswad, D.W. (1984). Stoichiometric methylation of porcine adrenocorticotropin by protein carboxyl methyltransferase requires deamidation of asparagine 25. *J Biol Chem* 259:10714–10721.
14. Geiger, T., and Clarke, S. (1987). Deamidation, isomerization, and racemization at asparaginyl and aspartyl residues in peptides. *J Biol Chem* 262:785–794.
15. Voorter, C.E.M., de Haard-Hoekman, W.A., van den Oetelaar, P.J.M., Bloemendal, H., and de Jong, W.W. (1988). Spontaneous peptide bond cleavage in aging α-crystallin through a succinimide intermediate. *J Biol Chem* 263:19020–19023.
16. Klotz, A.V., and Higgins, B.M. (1991). Deamidation and succinimide formation by γ-N-methylasparagine: Potential pitfalls of amino acid analysis. *Arch Biochem Biophys* 291:113–120.
17. Klotz, A.V. (1993). The stability of γ-N-methylasparagine in peptides. *Bioorg Chem* 21:83–94.
18. Klotz, A.V., and Thomas, B.A. (1993). N^5-Methylasparagine and asparagine as nucleophiles in peptides: main-chain vs. side-chain amide cleavage. *J Org Chem* 58:6985–6989.
19. Duerring, M., Huber, R., and Bode, W. (1988). The structure of γ-N-methylasparagine in C-phycocyanin from *Mastigocladus laminosus* and *Agmenellum quadruplicatum*. *FEBS Lett* 236:167–170.
20. Thomas, B.A., Bricker, T.M., and Klotz, A.V. (1993). Posttranslational methylation of phycobilisomes and oxygen evolution efficiency in cyanobacteria. *Biochim Biophys Acta* 1143:104–108.
21. Muranova, T.A., Muranov, A.V., Markova, L.F., and Ovchinnikov, Y.A. (1978). The primary structure of ribosomal protein L3 from *Escherichia coli* 70S ribosomes. *FEBS Lett* 96:301–305.

22. Frolova, L.Y., Tsivkovskii, R.Y., Sivolobova, G.F., Oparina, N.Y., Serpinksy, O.I., Blinov, V.M., Tatkov, S.I., and Kisselev, L.L. (1999). Mutations in the highly conserved GGQ motif of class I polypeptide release factors abolish ability of human eRF1 to trigger peptidyl-tRNA hydrolysis. *RNA* 5:1014–1020.
23. Dinçbas-Renqvist, V., Engström, Å, Mora, L., Heurgué-Hamard, V., Buckingham, R., and Ehrenberg, M. (2000). A posttranslational modification in the GGQ motif of RF2 from *Escherichia coli* stimulates termination of translation. *EMBO J* 19:6900–6907.
24. Song, H., Mugnier, P., Das, P.K., Webb, H.M., Evans, D.R., Tuite, M.F., Hemmings, B.A., and Barford, D. (2000). The crystal structure of human eukaryotic release factor eRF1— mechanism of stop codon recognition and peptidyl-tRNA hydrolysis. *Cell* 100:311–321.
25. Lhoest, J., and Colson, C. (1981). Cold-sensitive ribosome assembly in an *Escherichia coli* mutant lacking a single methyl group in ribosomal protein L3. *Eur J Biochem* 121:33–37.
26. Ong, L.J., and Glazer, A.N. (1991). Phycoerythrins of marine unicellular cyanobacteria. *J Biol Chem* 266:9515–9527.
27. DeLange, R.J., Williams, L.C., and Glazer, A.N. (1981). The amino acid sequence of the β-subunit of allophycocyanin. *J Biol Chem* 256:9558–9566.
28. Freidenreich, P., Apell, G.S., and Glazer, A.N. (1978). Structural studies on phycobiliproteins. *J Biol Chem* 253:212–219.
29. Lundell, D.J., Glazer, A.N., DeLange, R.J., and Brown, D.M. (1984). Bilin attachment sites in the α and β-subunits of B-phycoerythrin. *J Biol Chem* 259:5472–5480.
30. Klotz, A.V., and Glazer, A.N. (1985). Characterization of the bilin attachment sites in R-phycoerythrin. *J Biol Chem* 260:4856–4863.

Part VI

Inhibition of Methyltransferases by Metabolites

16

Inhibition of Mammalian Protein Methyltransferases by 5′-Methylthioadenosine (MTA): A Mechanism of Action of Dietary SAMe?

STEVEN G. CLARKE

Department of Chemistry and Biochemistry and the Molecular Biology Institute
University of California, Los Angeles
405 Hilgard Avenue
Los Angeles, CA 90095, USA

I. Abstract

5′-deoxy-5′-methylthioadenosine (5′-methylthioadenosine, MTA) is a naturally occurring metabolite. As an experimental reagent, it has proved useful in providing investigators a window onto the role of protein methylation reactions in intact cells, although its mode of action is poorly understood in most cases. This chapter reevaluates its utility as a reagent. It appears now that MTA is at best a poor direct inhibitor of methyltransferases and that its effectiveness in intact cells may depend on its ability to inhibit S-adenosyl-L-homocysteine hydrolase.

This chapter reviews recent evidence that points to an important role for MTA as an intermediary in the beneficial pharmaceutical action of orally ingested

THE ENZYMES, Vol. XXIV

S-adenosyl-L-methionine (AdoMet, SAMe). These new results suggest that oral AdoMet may function not by enhancing the activity of cellular methyltransferases, as has been previously surmised, but by inhibiting their action. Such inhibition, particularly of protein methyltransferases involved in intracellular communication, may attenuate signal transduction pathways otherwise leading to inflammatory damage to tissues.

II. Introduction

5'-methylthioadenosine (5'-deoxy-5'-methylthioadenosine), or MTA, is a normal metabolite in polyamine and methionine metabolism, as well as a useful reagent for investigators studying protein methyltransferase action in intact cells [1, 2] (Figure 16.1). Recent evidence has been presented that this compound may also be a crucial mediator in the pharmaceutical action of SAMe, or S-adenosyl-L-methionine (AdoMet). These studies now suggest that the inhibition of protein methyltransferases and other cellular perturbations by MTA may underlie at least part of the beneficial action of SAMe, which in turn suggests a very different mode of action than previously ascribed to this widely used nutraceutical.

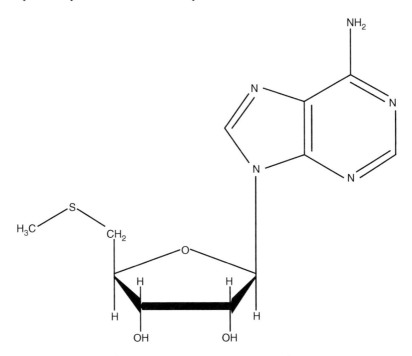

FIG. 16.1. Structure of 5'-methylthioadenosine (MTA, 5'-deoxy-5'-methylthioadenosine).

III. Formation and Metabolism of MTA in Mammalian Cells

The major endogenous production of MTA in mammalian cells occurs as a by-product of the enzymatic transfer of aminopropyl groups from decarboxylated S-adenosyl-L-methionine to form the polyamines spermine and spermidine [3, 4] (Figure 16.2). MTA can also be formed as a product of enzymatic reactions in which the alpha-aminobutyryl (3-amino-3-carboxypropyl) moiety of AdoMet is enzymatically transferred, as has been proposed in the biosynthetic pathways for diphthamide [5] and wybutosine [6]. However, MTA production from these

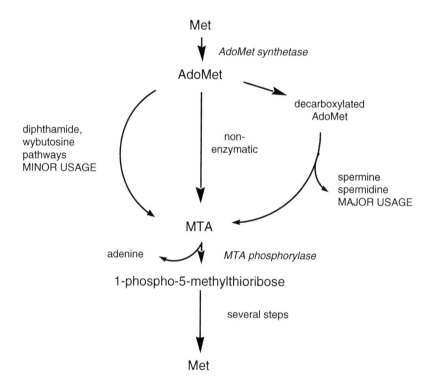

FIG. 16.2. Enzymatic and non-enzymatic pathways for the cellular formation and utilization of MTA. The major biosynthetic production of MTA occurs as a by-product of spermine and spermidine synthesis from decarboxylated S-adenosyl-L-methionine (AdoMet) (right-hand pathway). However, MTA can also readily form non-enzymatically from AdoMet itself (center pathway) and from enzymatic reactions where the aminobutryl group of AdoMet is transferred (left-hand pathways). The methylthioribose portion of MTA is converted to methionine in a series of steps, generally initiating with the phospholytic cleavage of the adenine base by MTA phosphorylase to give 1-phospho-5-methylthioribose (see text for more details).

pathways would be expected to be only a small fraction of that of the polyamine pathway. There have also been reports that MTA can be formed directly from AdoMet by an enzymatic activity present in yeast [7] and in mammalian cells [8]. However, this activity has not been characterized, and at least in mammalian cells does not appear to contribute significantly to MTA production [9].

Interestingly, MTA is also spontaneously produced from AdoMet in a non-enzymatic reaction (Figure 16.2). At the physiological conditions of pH 7.5 and 37° C, the half-time of MTA formation from AdoMet has been measured variously at 16 h by Creason et al. [10], 32 h by Wu et al. [11], and 42 h by Hoffman [12]. These results suggest that MTA can be readily generated from AdoMet, even in environments such as the extracellular space that lack the enzymes of polyamine formation and other biosynthetic reactions.

MTA is metabolized in mammalian cells in a series of enzymatic reactions that lead to the formation of adenine and methionine [13–15]. The sulfur atom and methyl group of MTA are conserved as the sulfur and methyl group of the methionine product. The ribose carbon atoms of MTA are also conserved as the backbone carbon atoms of the amino acid. The metabolic conversions of adenine to ATP and methionine to AdoMet then complete the cycle (Figure 16.2).

A common feature of many tumor cells is their deficiency in the first enzyme of the pathway that converts MTA to methionine, MTA phosphorylase [16]. It has been suggested that the increased levels of MTA in tumor cells may favor unregulated growth [17]. However, recent results have indicated that MTA phosphorylase deficiency may represent largely "collateral damage" from the co-deletion of its gene on chromosome 9p21 with the adjacent gene on the chomosome encoding the p16 protein, a cyclin-dependent kinase inhibitor [16]. The loss of the p16 protein itself and the resultant loss of control of the cell cycle may thus explain the apparent selection for MTA phosphorylase deficiency in cancer cells. However, it remains possible that MTA-dependent alterations to cells also contribute to tumor cell growth.

IV. Usefulness of MTA as a Reagent to Inhibit Protein Methylation in Intact Cells: Mechanism of Action

In mammalian cells, MTA has been very useful as a reagent to investigate the role of methylation reactions in intact cells, particularly protein methylation reactions (for recent examples see [18–21]). An important feature of MTA action is that it is rapidly transported across the plasma membrane of mammalian cells [22–25], allowing its use in intact cell systems. Transport occurs both by the nonspecific nucleoside transport system in the plasma membrane as well as by an apparent non-protein mediated, relatively efficient, diffusion mechanism [25]. MTA uptake from the medium is driven by its enzymatic conversion by phosphorolysis to adenine and 1-phospho-5-methylthioribose as the first step of the pathway for its conversion to methionine [23, 24] (Figure 16.2). In the absence of phosphorylase activity, often observed in tumor cells, MTA can be excreted from cells as well [17].

The assumption has often been made that MTA is simply a direct inhibitor of methyltransferases. This does not appear to be the case. MTA is generally a much better inhibitor of methylation reactions when incubated with intact cells than it is when incubated with purified methyltransferases or with cell-free extracts. For example, no inhibition of the protein isoprenylcysteine carboxyl methyltransferase (ICMT) is detected in *in vitro* assays at MTA concentrations of up to 12.8 mM [26, 27], whereas much lower concentrations (1.5 to 3 mM) are effective at inhibiting this enzyme in intact cells [28]. Similarly, no inhibition of MTA on protein lysine methyltransferases active on histones is found in nuclear preparations [29], whereas MTA is a good inhibitor of histone lysine methylation in intact cells [19, 20]. No inhibition of the bovine brain protein phosphatase 2A C-terminal leucine methyltransferase in cellular extracts is found at MTA concentrations of up to 8 mM [31]. These results suggest that MTA must be metabolized by cells in order to fully affect methyltransferase activity.

MTA does appear to have some direct inhibitory action on some methyltransferases. For example, 29% inhibition was seen with 0.5 mM MTA for the acetylserotonin methyltransferase and 20% inhibition at 1 mM MTA for the histamine methyltransferase [32]. For the calf brain protein L-isoaspartyl methyltransferase, MTA has been reported to directly inhibit the enzyme with a K_i of 41 μM [33]. Poorer inhibition is seen with the corresponding enzyme from an archaeon, with a K_i of 250 μM [34]. Variable results have been presented for the sensitivity of members of the protein arginine methyltransferase family to MTA. The major activity in chicken embryo fibroblasts was reported to be inhibited with a K_i of 396 μM [35], and that of mouse Krebs ascites cells was inhibited 30% by 100 μM MTA and 82% by 1 mM MTA [36]. On the other hand, no inhibition of the major recombinant protein arginine methyltransferase GST fusion proteins from yeast (Rmt1) or mammalian cells (PRMT1) was observed at 220 μM [37].

Probably the best description of the present situation is that given by Williams-Ashman and colleagues [1], who summarized the early literature with the observation that direct inhibition of methyltransferases by MTA is generally "feeble," especially compared to their inhibition by AdoHcy. In all of the cases studied, MTA is a much poorer inhibitor of methyltransferases than AdoHcy.

In summary, potent inhibition of methyltransferases directly by MTA is a possibility but has not been described yet for any enzyme. As described previously, in the cases where it has been measured, MTA is either a poor inhibitor or not an inhibitor of purified methyltransferases. In light of the important role of MTA in cell studies, it would be useful to reinvestigate the ability of this compound to inhibit purified preparations of a wider variety of various methyltransferases.

Unfortunately, much of the evidence on the direct action of MTA on methyltransferases comes from limited studies on only a few enzyme systems. Some of the differences cited previously in the direct inhibition of methyltransferases by MTA may possibly be due to species differences. It is also possible that impurities present in some preparations of MTA may have contributed to the inhibition observed in some cases.

So why then does the incubation of intact mammalian cells with MTA generally result in effective inhibition of methylation reactions, especially protein methylation reactions? A likely scenario of MTA's effective inhibitory action on methyltransferases in intact cells is that it inhibits AdoHcy hydrolase, the enzyme responsible for metabolizing AdoHcy [38] (Figure 16.3). Ferro et al. [38] demonstrated that incubation with MTA results in an apparent irreversible inhibition of the human erythrocyte AdoHcy hydrolase with a K_I value of 36 μM. Similar results for MTA inactivation of AdoHcy hydrolase from extracts of human erythrocytes, B-lymphoblasts, and T-lymphoblasts were obtained by Fox et al. [39].

The loss of AdoHcy hydrolase activity caused by MTA restricts its ability to convert methyltransferase-produced AdoHcy to adenosine and homocysteine, products that do not affect methyltransferase activity. However, the AdoHcy substrate is itself a strong product inhibitor of almost all AdoMet-dependent methyltransferases [40–42]. The MTA-dependent buildup of AdoHcy may then be responsible for the majority of the methyltransferase inhibition seen (Figure 16.3).

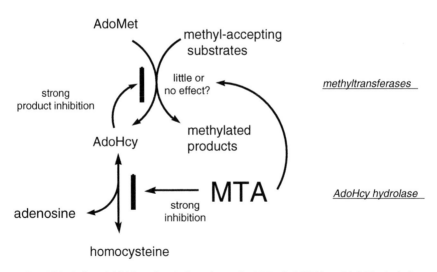

FIG. 16.3. Indirect inhibition of methyltransferases by MTA via inhibition of AdoHcy hydrolase, followed by AdoHcy buildup and the product inhibition of AdoHcy on methyltransferases. When analyzed directly, MTA has very little effect directly on methyltransferases (see text for details).

The complexity of MTA action as an inhibitor of protein methyltransferases is often not appreciated, as revealed by its characterization by various authors as a "general inhibitor of protein methyltransferases" [20], a "general methyltransferase blocker" [19], a "specific inhibitor of protein methyltransferases" [18, 43], or a "potent inhibitor of protein carboxylmethyltransferase" [44], all of which are incorrect! MTA inhibitory actions on methylation in cells may reflect contributions from some direct inhibition of some methyltransferases, from the inhibition of AdoHcy hydrolase resulting in a more general inhibition of methyltransferases according to their sensitivity of AdoHcy product inhibition [42], and perhaps from additional interactions not yet understood.

V. Pharmaceutical AdoMet (SAMe) as an Extracellular Time-Release Form of MTA, Resulting in the Inhibition of Intracellular Methyltransferases?

AdoMet has been available in the United States since 1999 as a nutraceutical with advertised benefits including enhanced mood, pain relief from osteoarthritis, and better liver function [45]. There is a large literature confirming these claims in controlled studies. Reviews for the effectiveness of SAMe as an antidepressant include those of Bressa [46], Mischoulon and Fava [47], Papakostas et al. [48], and Williams et al. [49]. The beneficial clinical action of SAMe in osteoarthritis has been reviewed by di Padova [50]. Finally, the effectiveness of SAMe for improving liver function has been reviewed by Lieber [51, 52].

However, the physiological sites of activity and the biochemical mechanisms for the beneficial actions of dietary AdoMet action have not been established [53]. It has often been assumed that its action derives in large part from increasing the intracellular cytosolic concentration of AdoMet to enhance the function of intracellular methyltransferases and other enzymes that utilize it in this space [45, 53]. However, two lines of evidence argue against such a role. In the first place, pharmacokinetic studies have revealed that whereas plasma [54, 55] and cerebrospinal fluid [56] AdoMet concentrations are increased with oral AdoMet ingestion, no significant increases in cytosolic levels occur intracellularly in mammalian cells [53, 57, 58]. Second, intracellular AdoMet levels are generally higher than the Km values for AdoMet of these enzymes [42], suggesting that even small changes in AdoMet concentration should have little or no effect on methyltransferase activity in any case.

It would appear that there is only one methyltransferase for which there is good evidence that extracellular AdoMet can serve as a methyl donor. This is the liver methyltransferase that catalyzes the formation of phosphatidylcholine from phosphatidylethanolamine. Although this enzyme is generally localized on the endoplasmic reticulum with the AdoMet binding site facing the cytosol, a small amount appears to be localized on the plasma membrane with the AdoMet binding site facing the extracellular space [59, 60]. Changes in plasma AdoMet levels might well affect the activity of this small fraction of the enzyme. However, this appears to be an exception to the rule that methyltransferase active sites are localized intracellularly and that only intracellular AdoMet can affect their activities.

The reason intracellular levels of AdoMet do not increase with oral SAMe ingestion appears to be that AdoMet is not transported across the plasma membrane of mammalian cells. There were initial reports of the transport of AdoMet across the plasma membrane of erythrocytes [61] and hepatocytes [62, 63]. However, further studies in erythrocytes have suggested that methionine and not AdoMet is a protein methyl donor in intact cells [64]. In hepatocytes, external AdoMet does not enter the intracellular AdoMet pool [59, 60]. A recent study of mammalian cultured cells and hepatocytes has concluded that uptake of AdoMet is restricted to what occurs by paracellular transport (i.e., between cells, not into cells) [65].

It is also possible that AdoMet can enter cells by pinocytosis along with other components of the extracellular medium in cells such as macrophage that undergo this process. However, even in this case the AdoMet would be delivered to the endosomes/lysosomes, and there is no evidence for the presence of methyltransferases within these organelles. Finally, the incorporation of radiolabel from the methyl group of intravenously administered AdoMet into creatinine has been taken as evidence for the uptake of AdoMet into hepatocytes [66]. However, given the evidence presented previously, this uptake may be alternatively accounted for by the uptake of MTA formed spontaneously from AdoMet and its subsequent conversion to methionine and AdoMet.

McMillan et al. [65] have suggested that the clinical effectiveness of SAMe might be improved by developing a more lipid-soluble derivative. However, this suggestion assumes that the sites of therapeutic action of AdoMet are intracellular, which does not appear to be the case.

Strengthening the argument that AdoMet is not transported across the plasma membrane of mammalian cells into the cytosol is the success of efforts to define transport systems for this molecule in other types of membranes where it does occur. It is clear that active transport of AdoMet does take place across the mitochondrial inner membrane. Its transporters have been characterized in both mammalian cells [67, 68] and in yeast [69]. Transporters have also been characterized in the yeast plasma membrane that catalyzes the direct uptake of AdoMet into these cells [70]. The fact that permeases for AdoMet have been readily characterized from mitochondria and from yeast plasma membrane, but not from mammalian plasma membrane, suggests that the failure to characterize mammalian plasma membrane counterparts for AdoMet transport may indeed reflect the absence of such permeases in these membranes.

Given the lack of transport of AdoMet into mammalian cells, one must then look for extracellular sites to explain the beneficial actions of dietary AdoMet. There are exceptional cases, such as the methylation of hepatic phospholipids described previously, where extracellular AdoMet can be utilized to some extent [59, 60]. Evidence has also been presented for an anti-oxidant role of AdoMet where transport into cells would presumably not be required [71].

However, this chapter seeks to explore evidence supporting another type of extracellular action of dietary AdoMet. As detailed previously, AdoMet in the plasma and cerebrospinal fluid is subject to a relatively robust spontaneous degradation reaction that produces MTA [10–12]. This reaction would be expected to result in the continuous formation of MTA from AdoMet in the extracellular environment, much in the same way as a controlled release drug. This MTA would then be expected to be transported into cells, where it might exert the beneficial effects ascribed to AdoMet, arguably by inhibiting methyltransferase action (Figure 16.4). This scenario suggests that instead of enhancing the activity

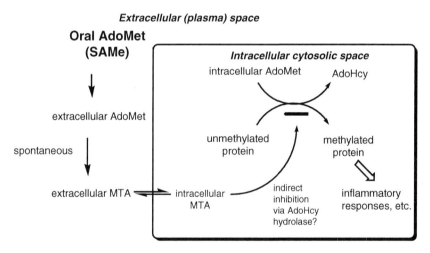

FIG. 16.4. A proposed mechanism for the beneficial effects of oral AdoMet administration to humans via MTA. In this model, increased extracellular AdoMet is slowly converted to MTA by a non-enzymatic mechanism. MTA is then transported into cells, inhibiting methyltransferases as depicted in Figure 16.3. Evidence has been presented that extracellular-supplied MTA and AdoMet can attenuate the inflammatory response. Because protein arginine methyltransferases have been implicated as mediators of the inflammatory response, it is possible that this potentiation is due directly to their inhibition by the effect of intracellular MTA on AdoHcy hydrolase. In this model, oral AdoMet can be viewed as a time-release form of the active species, MTA. There are upward of 300 distinct methyltransferases in mammalian cells [72], and MTA-derived inhibition of other types of these enzymes may also be involved.

of methyltransferases, dietary AdoMet may actually attenuate their activity and that the beneficial action results, at least in part, from such inhibition! The idea that MTA may be an active metabolite of oral AdoMet is not a new one. Feo and colleagues have suggested that the accumulation of MTA as an enzymatic and non-enzymatic product of AdoMet may be responsible for the chemopreventive effects observed for AdoMet [73, 74].

This chapter reviews the evidence for the action of MTA as an inhibitor of methyltransferases and for the correlation of the effects of extracellular AdoMet and MTA with such inhibition. The chapter also addresses the larger question of how increased levels of MTA may be responsible for the beneficial action of oral SAMe by other mechanisms. The question of whether MTA is the active or an active form of oral AdoMet is still open. Can orally administered SAMe in fact be considered a "time-release" form of MTA, and can MTA be responsible for part or all of its therapeutic actions? Is the modulation of protein methyltransferase activity a part of these processes?

VI. Effects of MTA on Cellular Functions and Correlation with Effects on Specific Protein Methyltransferases

This section reviews the evidence for the correlation of changes in cell function induced by MTA treatment with the activities of specific methylation reactions. It must be borne in mind, however, that MTA would be expected to affect the activities of many methyltransferases and ascribing physiological effects to any one enzyme is problematic. It remains to be determined how much of the effects of MTA on the systems described in the following are due to its action on one or a family of protein methyltransferases, on one or more other types of methyltransferases, or on one or more non-methyltransferase pathways. Much of the evidence linking protein methylation pathways and function is correlative and it is certainly possible that the effects seen are due to some combination of affected pathways. Nevertheless, the discussion following is arranged in terms of the enzyme systems that presently appear to be most closely linked to the physiological processes disrupted.

A. Protein Arginine Methyltransferases and Signaling Functions

In mammals, protein arginine residues are modified by a family of eight (and potentially nine) related enzymes designated PRMT1, PRMT2, and so on [75–78]. The physiological roles of these enzymes are under intense investigation at the present time, which include functions in RNA processing and signal transduction.

1. Lipopolysaccaride Receptor Signaling Pathways and Inflammation

Perhaps the best present evidence for the role of MTA (and AdoMet) comes from studies of inflammation. It seems clear that human responses to bacterial products such as lipopolysaccharide can give an exaggerated immune response that can be harmful to tissues. A role of dietary AdoMet as an anti-inflammatory has been proposed for some years [79]. It is also become increasingly clear that the effects of MTA are similar to those of AdoMet [80, 81].

Evidence obtained prior to the identification of protein arginine methyltransferase genes suggested a role for MTA-inhibited reactions in cellular signaling in the immune system. For instance, MTA inhibited the lipopolysaccharide-induced NF-kappaB activation of mouse 70Z/3 pre-B lymphocytes in a parallel fashion as the inhibition of protein methylation reactions that generated volatile radioactive products after base treatment [82]. The protein methyl groups were originally thought to be derived from methyl ester linkages on protein carboxyl groups, but it was later realized that base treatment of protein arginine residues can also generate volatile methylamine derivatives.

MTA was also found to inhibit the production of tumor necrosis factor in lipopolysaccharide-stimulated macrophages [83]. Anti-inflammatory effects of MTA have been recently confirmed and extended in two major studies [80, 81]. MTA was found to protect mice against lipopolysaccharide-induced toxic shock, preventing the accumulation of tumor necrosis factor alpha in serum [80]. MTA treatment was also found to enhance the serum levels of interleukin-10. Similar results were obtained in the mouse macrophage cell line RAW 264.7, with MTA attenuating tumor necrosis factor alpha and potentiating interleukin-10 production [80]. MTA also attenuated lipopolysaccharide-dependent inducible nitric oxide synthase induction and cyclooxygenase 2 induction.

Furthermore, MTA was found to decrease nitric oxide production in response to a mixture of pro-inflammatory cytokines [80]. Evidence was presented that these effects are not dependent on the conversion of MTA into AdoMet, nor are they related to effects of MTA on adenosine receptors or the cAMP-protein kinase A pathway [80]. AdoMet and MTA effects were also studied in rat hepatic macrophage (Kupffer cells), with similar results [81]. These results, taken in context of previous studies linking lipopolysaccharide-dependent signaling events to protein methylation [82], suggest that MTA may function at least in part by inhibiting the activity of these enzymes, perhaps in concert with parallel changes in protein phosphorylation pathways [80]. Furthermore, the authors suggest that some of the effects of AdoMet as an anti-inflammatory may be mediated by its transformation into MTA [80].

Clear evidence has been recently presented that AdoMet can modulate cytokine production stimulated by lipopolysaccharide, including attenuating tumor necrosis factor (TNF) production and stimulating interleukin 6 and 10

(IL-6 and IL-10) production in primary cultures of neocortex cells prepared from rat embyros [84]. It will be interesting to see if this effect is mediated by MTA.

Finally, evidence has been presented that another biological effect of lipopolysaccharide, the activation of the HuR RNA-binding protein, may be related to HuR methylation by the protein arginine methyltransferase 4 (PRMT4, or CARM1) [85]. HuR can stabilize the mRNA of species such as tumor necrosis factor alpha, and its activation by methylation of arginine residues can contribute to its activation in macrophage. The inhibition of PRMT4 by MTA, presumably via AdoHcy buildup, may thus contribute to its anti-inflammatory action.

2. Interferon Signaling Pathways, STAT Modification, and Transcription: Is Protein Arginine Methylation Involved?

Growth factor and cytokine stimulation of cells results in transcriptional events dependent on the phosphorylation of STAT (signal transducer and activator of transcription) proteins at serine and tyrosine residues. The type I receptor for the cytokine interferons of the alpha and beta family physically interacts with the PRMT1 protein arginine methyltransferase [86], suggesting a role for this type of methylation reaction in cytokine signaling. MTA at 0.3 mM has been shown to inhibit the STAT1-dependent responses to interferon alpha in HeLa cells [18]. These effects have been correlated with the methylation of STAT1 at a specific arginine residue (Arg-31) that was suggested to enhance STAT1 binding to DNA by blocking its interaction with its endogenous inhibitor PIAS1 [18]. This suggests that oral AdoMet treatment and subsequent generation of MTA may lead to reduced interferon responsiveness, perhaps lessening the inflammatory response.

However, this view has been recently challenged by Meissner et al., who failed to detect Arg-31 methylation on STAT1 [87]. Komyod [88] also failed to confirm the presence of methylated arginine residues on STAT1 and STAT3, although they did not directly measure methylation on the protein. Meissner et al. [87] show that MTA at 0.3 mM inhibits the interferon alpha-induced tyrosine phosphorylation of STAT1 in the human breast cancer cell line MCF-7, and suggest that this and not a methylation change is responsible for the reduced DNA binding and reduced transcriptional activation. Meissner et al. also reported that MTA inhibits the dephosphorylation of kinase p38 and the transcription of a STAT1-independent reporter gene [87]. Further work is clearly needed to establish whether protein arginine methylation reactions are involved in this signaling pathway, and whether their inhibition by MTA may be linked to the therapeutic action of oral AdoMet administration.

3. Interleukin Signaling Pathways

MTA has been shown to decrease cytokine gene expression in leukemic Jurkat T-cells [89]. This effect has now been linked to interleukin 2 (IL-2) signaling in these lymphocytes that is dependent on PRMT5 protein arginine methylation by using

RNA interference to target specific methyltransferases [89]. These authors have provided evidence that proteins containing symmetric dimethylarginine residues are bound to the IL-2 promoter when T-cells are activated and that it is the recruitment of these species that regulates gene expression in the MTA-sensitive step.

4. Fibroblast Growth Factor Signaling: Does MTA Affect Phosphorylation or Methylation?

Basic fibroblast growth factor (FGF-2) is a mitogen and signal for morphogenic changes in a variety of cells. It is recognized by the FGF receptor, whose activity is mediated by its protein tyrosine kinase activity. MTA is a potent inhibitor of FGF-2-induced tyrosine kinase activity [90, 91] and of its nuclear accumulation [92]. At a concentration of 3 mM, MTA is effective in almost completely abolishing protein methylation in mouse cell lines [90]. The question has been raised whether this inhibition is a direct effect of MTA on the receptor kinase or results from its inhibition of cellular kinase activity or activities [90, 91]. Indirect evidence that the methylation reaction is involved in the signaling system is seen in the similarity in the concentration dependence of MTA's effects on protein methylation, FGF-2-stimulated cell growth, and protein tyrosine phosphorylation [91]. MTA certainly does not appear to be a general inhibitor of tyrosine protein kinases because it has no effect on epidermal growth factor or platelet-derived growth factor [90].

The higher-molecular-weight subforms of FGF-2 are known to be modified on arginine residues, with five dimethylated arginine residues present at the N-terminus of the 22-kDa and 22.5-kDa forms, and seven to eight dimethyl arginine residues present in the N-terminus of the 24-kDa form [93, 94]. The suggestion has been made that the loss of FGF-2 subform methylation may restrict it from entering the nucleus and initiating its biological effects [92–94]. FGF-2 is an unusual growth factor in that it is present and acts intracellularly, and thus its methylation by endogenous protein arginine methyltransferases may be an important part of its signaling function. At this point, it is not clear whether the effect of MTA on FGF-2 signaling is due to the methylation of FGF-2 itself or due to methylation of another protein or small molecule, or whether it may be a direct effect on the tyrosine kinase activity itself [90, 92].

The action of MTA on basic fibroblast growth factor (FGF-2) dependent changes in morphology and gene expression in rat primary astrocytes has been studied recently [95]. FGF-2 induced a stellate morphology in these cells and inhibited the action of transforming growth factor-beta1 on inducing glial fibrillary acidic protein synthesis. Recent work has shown that FGF-2 modulates the activity of the ciliary neurotrophic factor and the STAT pathway in astrocytes by effects on chromatin remodeling, including changes in the methylation of lysine residues in histone H3 [19]. MTA may also affect protein lysine methylation, and this may be an additional player in this complex web (see section VI.B).

5. Nerve Growth Factor Signaling and Differentiation

Methylation inhibitors, including MTA, have been found to inhibit nerve-growth-factor-dependent neurite growth and regeneration in the rat PC-12 cell line, suggesting that a methylation event may be an early step in the signaling pathway [96]. MTA was also found to inhibit the primary response gene expression in rat PC-12 cells in response to nerve growth factor in parallel with the inhibition of protein arginine/methyl ester methylation [97]. The nature of the methylation events stimulated by nerve growth factor have not yet been established. Evidence has been presented for the involvement of small G-protein substrates of the isoprenylcysteine carboxyl methyltransferases [98]. It has been suggested that this effect of MTA may be more closely linked to its ability to directly inhibit protein kinase activity [99]. However, a link between the methylation of specific proteins separated on 2D gels and nerve-growth-factor-stimulated differentiation was also demonstrated with the AdoHcy hydrolase inhibitor dihydroxycyclopentenyl adenine [100], suggesting the importance of protein methylation reactions.

Interestingly, this inhibitor did not affect the nerve-growth-factor-mediated phosphorylation of Trk [100]. Recently, evidence has been presented for a central role of the protein arginine methyltransferase PRMT1 in nerve growth factor signaling [101], although it is still not possible to rule out the participation of other methyltransferases, including the recently characterized brain-specific PRMT8 [102].

Nevertheless, these studies indicate that nerve growth factor signaling pathways may be a target for an MTA-mediated effect of oral AdoMet in humans. It is unclear how the disruption of such pathways may contribute to the action of AdoMet as an antidepressant. MTA does block nerve-growth-factor-mediated survival of chick embryonic sympathetic neurons [103], suggesting the possibility of action via a differential regulation of neuronal activity.

6. Cell Transformation, Proliferation, and Tumor Formation

In an early study, the transformation of chicken embryo fibroblasts by Rous sarcoma virus to give foci formation was found to be inhibited by MTA and a variety of other compounds at similar concentrations required to inhibit protein arginine methylation activity [35]. These results provided an initial suggestion that protein arginine methylation was connected to the control of cell division.

In mammalian cells, proliferation in human peripheral lymphocyte cell cultures stimulated with mitogens or antigens is inhibited by MTA [104], suggesting again an effect on cell cycle control. This cytostatic action was also observed in fibroblasts [105]. Here, the failure to restore growth when spermidine was added was evidence that the effect of MTA was not on the enzymes of polyamine synthesis [105]. Inhibition of growth but not differentiation was also seen in the promyelocytic cell line HL-60 and human granulocytic precursor cells [106]. It is not at all clear how these effects of MTA may be related to protein arginine methylation or to methylation at all. In fact, the growth arrest by MTA on mouse S49 cells has been suggested to be due to the inhibition of cAMP phosphodiesterase [107].

7. Pre-mRNA Splicing

MTA treatment of human HeLa cells at a concentration of 0.75 mM has been shown to inhibit coilin methylation and cause the expression of spinal motor neuron protein "gems" [108]. The loss of methylation here has been associated with the disruption of the assembly of spliceosome complexes, possibly by defects in Sm protein and hnRNP protein arginine methylation [108]. How defects in mRNA splicing may relate to the beneficial effects of AdoMet is not clear at present.

B. PROTEIN LYSINE METHYLTRANSFERASES AND CHROMOSOME REMODELING FUNCTIONS

Recently, the role of protein lysine methyltransferases in chromatin remodeling has come to be appreciated [109, 110]. MTA in relatively low concentrations (0.3 mM) has been shown to effectively block the dimethyl lysine modification of histone H3 at the Lys4 site in E18 rat cortical cell lines [19] and in human primary lymphoblast cells [30]. In the latter study, MTA was found to limit histone H3 methylation and the transcription of a viral gene [30]. Similarly, MTA was found to attenuate the expression of endothelial nitric oxide synthase by trichostatin A in vascular smooth muscle cells and was correlated with the decrease in histone H3 Lys4 methylation [21].

C. PROTEIN ISOPRENYLCYSTEINE O-METHYLTRANSFERASE AND NUCLEAR LAMIN FUNCTION: EFFECTS OF MTA ON NUCLEAR ASSEMBLY AND APOPTOSIS

The methyl esterification of isoprenylated proteins was shown to be important for their cellular localization [111]. Moderate inhibition (30 to 60%) of lamin B methyl esterification has been found when Chinese hamster ovary (CHO) and PC13 mouse embryonal carcinoma cell lines were treated with 3 mM MTA [28]. MTA was slightly less effective at 1.5 mM, but lost much of its effect at lower concentrations. Inhibition of lamin B methylation by MTA was correlated by an inhibition of its assembly into the nuclear envelope [28]. Most interestingly, three compounds known to inhibit AdoHcy hydrolase (3-deazaadenosine, 2-deoxyadenosine, and D-deazaaristeromycin) were found not to inhibit lamin B methyl esterification, while having their expected effect on total protein N-methylation. The correlation of MTA inhibition of lamin B methylation and nuclear assembly was strengthened with the finding that none of these AdoHcy inhibitors affected nuclear assembly [28].

The mechanism of MTA action in this case is puzzling. MTA is at best a very poor inhibitor of the isoprenylcysteine carboxyl methyltransferase that methylesterifies lamin B. Little or no inhibition was found at concentrations up to 0.5 mM, whereas AdoHcy was found to be a good inhibitor with a Ki value of 9.2 µM [26]. In a later study, using dimethylformamide as a solvent for MTA, no inhibition was seen at MTA concentrations up to 12.8 mM in the presence of 10% dimethylformamide [27]. These results suggest that there is little or no direct effect of MTA on this enzyme. However, it is not clear that the inhibition seen in intact cells may arise from the intracellular accumulation of AdoHcy because compounds that would be expected to increase the intracellular concentration of AdoHcy had apparently little or no effect on lamin B methylation or its assembly into the nuclear envelope [28]. Further work would be useful to explore the mechanism of MTA action in this area.

Recent work has suggested that the MTA-dependent inhibition of lamin B methylation and its delayed assembly into the nuclear membrane can induce apoptosis in at least one cell line [44]. In the human leukemia U927 cell line, incubation with 0.5 to 1 mM MTA resulted in the initiation of cell death within one day with a mitotic-like nuclear fragmentation pattern and other features consistent with apoptosis [44].

Interestingly, in other cells AdoMet and MTA have been found to either inhibit apoptosis or to promote it. In the human HuH7 hepatoma cell line, both AdoMet and MTA were found to promote apoptosis induced by the protein phosphatase inhibitor okadaic acid [112], consistent with the results described previously for the human leukemia cell line [44]. However, a very different result was seen in cultured rat hepatocytes that had not been transformed. Here, both AdoMet and AdoHcy protected these non-cancerous cells from apoptosis [112]. Similar results have been found in the liver-cancer derived HepG2 cell line and primary hepatocytes with a proapoptotic role in only the tumor cells [113]. These effects appear to involve changes in protein phosphorylation and alternative splicing. It remains to be seen whether these effects may be modulated by changes in protein methylation.

These results are remarkable for two reasons. First, the common actions of AdoMet and MTA are consistent with an intermediary role of MTA in the pharmacological effects of oral AdoMet. Second, these results suggest that for normal cells AdoMet/MTA may exert some part of their beneficial effect by minimizing apoptosis in liver cells.

D.   PROTEIN L-ISOASPARTYL/D-ASPARTYL REPAIR O-METHYLTRANSFERASE

This methyltransferase recognizes age-damaged proteins in all mammalian tissues, initiating the conversion of "kinked" isoaspartyl residues to normal aspartyl residues [114]. MTA has been shown to be an effective inhibitor of protein carboxyl methylation in intact human erythrocytes, with only about 10% methylation occurring when cells are incubated in 5 mM MTA [115]. The major methylation reaction in these cells is protein carboxyl methylation catalyzed by the protein L-isoaspartyl (D-aspartyl) methyltransferase. Results have been presented that the bovine brain enzyme may be directly inhibited by MTA at relatively low concentrations, with a Ki measured at 41 μM [33], an apparent exception to the scenario that the effects of MTA on methyltransferases are generally directed via its effects on AdoHcy hydrolase. No physiological effects of MTA inhibition in red cells have been reported.

VII. Biochemical Actions of AdoMet/MTA Not Involving Methyltransferases

There have been a number of suggestions for rationalizing the beneficial effects of dietary AdoMet that do not involve methyltransferases, and evidence has been presented that even if MTA is a crucial intermediate in the therapeutic actions, its effects might also include those independent of methyltransferases. The suggestion raised here that MTA derived from AdoMet is an active component in the beneficial action of SAMe certainly does not preclude other mechanisms of action.

The previous discussion is focused on the actions of MTA in leading to the inhibition of cellular methyltransferases. However, MTA can also affect a number of other biochemical pathways. Thus, MTA produced non-enzymatically from dietary AdoMet can exert its beneficial actions by not only modulating protein methylation but also by affecting several other types of reactions [74]. The action of MTA derived from AdoMet on these reactions may certainly contribute to the therapeutic action of dietary AdoMet.

A.   MTA IS AN INHIBITOR OF POLYAMINE SYNTHESIS

MTA is an effective product inhibitor of the enzymes of polyamine synthesis, spermine synthase, and spermidine synthase [3, 4, 105, 116]. However, the inhibition of polyamine synthesis does not appear to underlie the cytostatic effects of MTA on cell growth because it is not reversed by polyamine addition [105].

B.   MTA INHIBITS cAMP PHOSPHODIESTERASES

MTA appears to be an inhibitor of one or more species of cAMP phosphodiesterases, with a Ki of 62 μM measured for the high-affinity enzyme from S49 cells [107]. Such inhibition would be expected to increase the level of cAMP in cells and lead to enhanced activity of the protein kinase A signaling pathway.

The effects of cAMP signaling are cell type dependent, and it is difficult to predict what the physiological outcome would be of any MTA inhibition of phosphodiesterases [117].

C. MTA INTERACTS WITH ADENOSINE RECEPTORS

MTA is a partial agonist of the adenosine A1 receptor and an antagonist of the A2 receptors [118, 119]. The relaxation of rabbit thoracic aorta strip preparations produced by MTA has been ascribed to MTA's binding to adenosine receptors, although it is not clear which subtype is involved [118, 120]. Recent studies have linked adenosine A1 and A2A receptors to depression in animal models [121, 122]. It will be interesting to now ask if part of the anti-depressive effect of oral AdoMet in humans may be related to increased concentrations of MTA and its effects on adenosine-mediated signal transduction pathways, including changes in potassium ion and calcium ion channels [119]. Because these pathways are linked to adenylate cyclase, the effects of MTA on cAMP phosphodiesterase (see previous material) may attenuate or potentiate physiological changes linked to adenosine receptors.

D. MTA INTERACTS WITH BENZODIAZEPINE RECEPTORS

Evidence has been presented that both AdoHcy and MTA can inhibit the binding of ligands such as the sedative flunitrazepam to rat brain benzodiazepine receptors [123]. This result raises the possibility that MTA and AdoHcy may be endogenous effectors for GABA receptors and provides an avenue to explore the biochemical basis of the antidepressant action of oral AdoMet.

E. MTA HAS ANTIOXIDANT ACTIVITY

MTA appears to have antioxidant activity as a substrate for enzymatic oxidation pathways and by its ability to autooxidize [124]. Administration of MTA to rats was effective (as was AdoMet) in minimizing oxidative liver injury induced by carbon tetrachloride [124]. Hence, MTA formed from AdoMet may contribute to the overall defense against oxidant damage in humans.

F. MTA HAS EFFECTS ON PROTEIN PHOSPHORYLATION AND DEPHOSPHORYLATION

MTA clearly inhibits protein phosphorylation cascades when incubated with intact cells [90]. However, as discussed previously it is unclear whether these effects represent direct inhibition of protein kinases or reflect a secondary effect of MTA's inhibition of methylation pathways. It has been reported that MTA can inhibit protein kinase activity in PC12 cell extracts with a Ki of 50 μM [99].

MTA may also affect protein phosphatase action, inhibiting the dephosphorylation of kinase p38 [87].

G.   MTA CAN INHIBIT DIPHTHAMIDE SYNTHESIS

MTA is thought to be the product of the biosynthetic step that adds the alpha-aminobutyryl group to the histidine residue that becomes the diphthamide residue in elongation factor-2, an essential component of the protein synthesis apparatus. Data has been presented that MTA in fact inhibits diphthamide synthesis in a mouse lymphoma cell line [125].

VIII.   Effects of Dietary AdoMet Not Mediated by Either AdoMet or MTA: Role of Adenosine?

A recent study has suggested that another metabolic product of AdoMet, adenosine, may be responsible for the stimulatory effects of AdoMet on IL-6 and IL-10 production in a monocyte cell line [84]. Importantly, these authors found that 3-deazaadenosine, an inhibitor of AdoHcy hydrolase, interfered with rather than stimulated the AdoMet-dependent IL-6 and IL-10 production, suggesting that a mechanism distinct from that suggested previously for MTA (also an AdoHcy hydrolase inhibitor) occurs here. Significantly, however, adenosine does not appear to be responsible for the effect on tumor necrosis factor alpha attenuation by AdoMet/MTA [84]. At this point, it is not clear how adenosine is formed from AdoMet if it is not transported into cells. One possible source is the release of the adenine base by MTA phosphorylase (Figure 16.2).

IX.   Conclusions

The suggestion here that part or all of the beneficial actions of oral AdoMet may be a function of the generation of MTA and the inhibition of protein methyltrans-ferases can now be tested directly. It will be important to ask whether plasma MTA levels are in fact increased in people taking SAMe, and whether protein methylation is in fact inhibited in animal models. Interestingly, it has been reported that human serum contains an MTA phosphorylase activity that might be expected to prevent any accumulation of MTA in the plasma and the cells [126]. However, this activity requires relatively high concentrations of thiols such as reduced glutathione, and in the normal oxidizing environment of the plasma, little or no activity might be expected [126]. It will also be interesting to see if urinary levels of MTA increase with dietary AdoMet supplementation. Small amounts of MTA are known to be excreted here, about 0.6 μmol/d (127), and dietary AdoMet may increase these levels.

One consequence of the idea that the active component of oral SAMe treatment is MTA is that this same compound is generated equally well from the *S, S*-stereochemical form of AdoMet produced enzymatically, as well as the racemized *R,S*-form [12]. Thus, it may make no difference in pharmaceutical preparations which form or ratio of forms is present, and suggests that synthetic AdoMet that exists as a mixture of the *S,S* and *R,S* forms may be a more economical alternative to the fermentative production of the largely *S,S* form. Certainly, therapeutic benefit has been demonstrated from SAMe preparations that are synthetically derived and contain approximately 55% of the *R,S* form and 45% of the *S,S* form [128], as well as those derived from fermentation that contain approximately 20% of the *R,S* form and 80% of the *S,S* form.

Finally, the question can be raised of whether MTA can be used therapeutically in addition to or instead of SAMe? Such administration would not have the "time-release" character of its spontaneous formation from AdoMet, but it may be possible to obtain higher concentrations that may enhance the beneficial effects [2]. These considerations also suggest that the development of more specific inhibitors of protein methyltransferases would also be useful. High-throughput screens have already identified one promising inhibitor of mammalian arginine protein methyltransferases as a symmetrical sulfonated urea designated AMI-1 [129].

ACKNOWLEDGMENTS

This work was supported by NIH grant GM026020. I thank my colleagues (especially Stephen Young, Sarah Villa, Jonathan Lowenson, Tara Gomez, and Catherine Clarke) for their helpful comments.

REFERENCES

1. Williams-Ashman, H.G., Seidenfeld, J., and Galletti, P. (1982). Trends in the biochemical pharmacology of 5′-deoxy-5′-methylthioadenosine. *Biochem Pharmacol* 31:277–288.
2. Avila, M.A., Garcia-Trevijano, E.R., Lu, S.C., Corrales, F.J., and Mato, J.M. (2004). Molecules in focus: Methylthioadenosine. *Internat J Biochem Cell Biol* 36:2125–2130.
3. Pegg, A.E. (1986) Recent advances in the biochemistry of polyamines in eukaryotes. *Biochem J* 232:249–262.
4. Tabor, C.W., and Tabor, H. (1984). Polyamines. *Annu Rev Biochem* 53:749–790.
5. Liu, S., Milne, G.T., Kuremsky, J.G., Fink, G.R., and Leppla, S.H. (2004). Identification of the proteins required for biosynthesis of diphthamide, the target of bacterial ADP-ribosylating toxins on translational elongation factor 2. *Mol Cell Biol* 24:9847–9897.
6. Kalhor, H.R., Penjwini, M., and Clarke, S. (2005). A novel methyltransferase required for the formation of the hypermodified nucleoside wybutosine in eukaryotic tRNA. *Biochem Biophys Res Commun* 334:433–440.
7. Mudd, S.H. (1959). Enzymatic cleavage of S-adenosylmethionine. *J Biol Chem* 234:87–92.

8. Swiatek, K.R., Simon, L.N., and Chao, K.-L. (1973). Nicotinamide methyltransferase and S-adenosylmethionine:5'-methylthioadenosine hydrolase: Control of transfer ribonucleic acid methylation. *Biochemistry* 12:4670–4674.

9. Eloranta, T.O., and Kajander, E.O. (1984). Catabolism and lability of S-adenosyl-L-methionine in rat liver extracts. *Biochem J* 224:137–144.

10. Creason, G.L., Madison, J.T., and Thompson, J.F. (1985). Soybeans and radish leaves contain only one of the sulfonium diasteromers of S-adenosylmethionine. *Phytochemistry* 24:1151–1155.

11. Wu, E.-E., Huskey, W.P., Borchardt, R.T., and Schowen, R.L. (1983). Chiral instability at sulfur of S-adenosylmethionine. *Biochemistry* 22:2828–2832.

12. Hoffman, J.L. (1986). Chromatographic analysis of the chiral and covalent instability of S-adenosyl-L-methionine. *Biochemistry* 25:4444–4449.

13. Backlund, P.S. Jr., and Smith, R.A. (1981). Methionine synthesis from 5'-methylthioadenosine in rat liver. *J Biol Chem* 256:1533–1535.

14. Backlund, P.S. Jr., and Smith, R.A. (1982). 5'-methylthioadenosine metabolism and methionine synthesis in mammalian cells grown in culture. *Biochem Biophys Res Commun* 108:687–695.

15. Trackman, P.C., and Abeles, R.H. (1983). Methionine synthesis from 5'-S-methylthioadenosine. Resolution of enzyme activities and identification of 1-phospho-5-S-methylthioribulose. *J Biol Chem* 258:6717–6720.

16. Nobori, T., Takabayashi, K., Tran, P., Orvis, L., Batova, A., Yu, A.L., and Carson, D.A. (1996). Genomic cloning of methylthioadenosine phosphorylase: A purine metabolic enzyme deficient in multiple different cancers. *Proc Natl Acad Sci USA* 93:6203–6208.

17. Kamatani, N., and Carson, D.A. (1980). Abnormal regulation of methylthioadenosine and polyamine metabolism in methylthioadenosine phosphorylase-deficient human leukemic cell lines. *Cancer Res* 40:4178–4182.

18. Mowen, K.A., Tang, J., Zhu, W., Schurter, B.T., Shuai, K., Herschman, H.R., and David, M. (2001). Arginine methylation of STAT1 modulates IFN alpha/beta-induced transcription. *Cell* 104:731–741.

19. Song, M.-R., and Ghosh, A. (2004). FGF2-induced chromatin remodeling regulations CNFT-mediated gene expression and astrocyte differentiation. *Nature Neurosci* 7:229–235.

20. Liang, X., Lu, Y., Wilkes, M., Neubert, T.A., and Resh, M.D. (2004). The N-terminal SH4 region of the Src family kinase Fyn is modified by methylation and heterogeneous fatty acylation. *J Biol Chem* 279:8133–8139.

21. Fish, J.E., Matouk, C.C., Rachlis, A., Lin, S., Tai, S.C., D'Abreo, C., and Marsden, P.A. (2005). The expression of endothelial nitric-oxide synthase is controlled by a cell-specific histone code. *J Biol Chem* 280:24824–24838.

22. Eloranta, T.O., Tuomi, K., and Raina, A.M. (1982). Uptake and utilization of 5'-methylthioadenosine by cultured baby-hamster kidney cells. *Biochem J* 204:803–807.

23. Carteni-Farina, M., Della-Ragione, F., Cacciapuoti, G., Porcelli, M., and Zappia, V. (1983). Transport and metabolism of 5'-methylthioadenosine in human erythrocytes. *Biochim Biophys Acta* 727:221–229.

24. Carson, D.A., Willis, E.H., and Kamatani, N. (1983). Metabolism to methionine and growth stimulation by 5'-methylthioadenosine and 5'-methylthioinosine in mammalian cells. *Biochem Biophys Res Commun* 112:391–397.

25. Stoeckler, J.D., and Li, S.-Y. (1987). Influx of 5'-deoxy-5'-methylthioadenosine in HL-60 human leukemia cells and erythrocytes. *J Biol Chem* 262:9542–9546.

26. Stephenson, R.C., and Clarke, S. (1990). Identification of a C-terminal protein carboxyl methyltransferase in rat liver membranes using a synthetic farnesyl cysteine-containing substrate. *J Biol Chem* 265:16248–16254.

27. Stephenson, R.C., and Clarke, S. (1992). Characterization of a rat liver protein carboxyl methyltransferase involved in the maturation of proteins with the -CXXX C-terminal sequence motif. *J Biol Chem* 267:13314–13319.

28. Chelsky, D., Sobotka, C., and O'Neill, C.L. (1989). Lamin B methylation and assembly into the nuclear envelope. *J Biol Chem* 264:7637–7643.
29. Duerre, J.A., Wallwork, J.C., Quick, D.P., and Ford, K.M. (1977). *In vitro* studies of the methylation of histones in rat brain nuclei. *J Biol Chem* 252:5981–5885.
30. Chau, C.M., and Lieberman, P.M. (2004). Dynamic chromatin boundaries delineate a latency control region of Epstein-Barr virus. *J Virol* 78:12308–12319.
31. Xie, H., and Clarke, S. (1993). Methyl esterification of C-terminal leucine residues in cytosolic 36-kDa polypeptides of bovine brain: A novel eucaryotic protein carboxyl methylation reaction. *J Biol Chem* 268:13364–13371.
32. Zappia, B., Zydek-Cwick, C.R., and Schlenk, F. (1969). The specificity of *S*-adenosylmethionine derivatives in methyl transfer reactions. *J Biol Chem* 244:4499–4509.
33. Oliva, A., Galletti, P., Zappia, V., Paik, W.K., and Kim, S. (1980). Studies on substrate specificity of *S*-adenosylmethionine:protein-carboxyl methyltransferase from calf brain. *Eur J Biochem* 104:595–602.
34. Thapar, N., Griffith, S.C., Yeates, T.O., and Clarke, S. (2002) Protein repair methyltransferase from the hyperthermophiilic archeaon *Pyrococcus furiosus*: Unusual methyl-accepting affinity for D-aspartyl and N-succinyl-containing peptides. *J Biol Chem* 277:1058–1065.
35. Enouf, J., Lawence, F., Tempete, C., Robert-Gero, M., and Lederer, E. (1979). Relationship between inhibition of protein methylase I and inhibition of Rous Sarcoma Virus-induced cell transformation. *Cancer Res* 39:4497–4502.
36. Casellas, P., and Jeanteur, P. (1978). Protein methylation in animal cells. II. Inhibition of *S*-adenosyl-L-methionine:protein(arginine) *N*-methyltransferase by analogs of *S*-adenosyl-L-homocysteine. *Biochem Biophys Acta* 519:255–258.
37. Gary, J. D., Lin, W.-J., Yang, M.C., Herschman, R.R., and Clarke, S. (1996). The predominant protein-arginine methyltransferase from *Saccharomyces cerevisiae*. *J Biol Chem* 271: 12585–12594.
38. Ferro, A.J., Vandenbark, A.A., and MacDonald, M.R. (1981). Inactivation of *S*-adenosylhomocysteine hydrolase by 5′-deoxy-5′-methylthioadenosine. *Biochem Biophys Res Commun* 100:523–531.
39. Fox, I.H., Palella, T.D., Thompson, D., and Herring, C. (1982). Adenosine metabolism: Modification of *S*-adenosylhomocysteine and 5′-methylthioadenosine. *Arch Biochem Biophys* 215:302–308.
40. Borchardt, R.T. (1980). *S*-Adenosyl-L-methionine-dependent macromolecule methyltransferases: Potential targets for the design of chemotherapeutic agents. *J Med Chem* 23:347–357.
41. Ueland, P.M. (1982). Pharmacological and biochemical aspects of *S*-adenosylhomocysteine and *S*-adenosylhomocysteine hydrolase. *Pharmacol Rev* 34:223–253.
42. Clarke, S., and Banfield, K. (2001). *S*-adenosylmethionine-dependent methyltransferases: Potential targets in homocysteine-linked pathology, in R. Carmel and D. Jacobsen (eds.), pp. 63–78, *Homocysteine in Health and Disease.*, Cambridge, England: Cambridge University Press.
43. Martens, J.H., Verlaan, M., Kalkhoven, E. and Zantema, A. (2003). Cascade of distinct histone modifications during collagenase gene activation. *Mol Cell Biol* 23:1808–1816.
44. Lee, S.H., and Cho, Y.D. (1998). Induction of apoptosis in leukemia U937 cells by 5′-deoxy-5′-methylthioadenosine, a potent inhibitor of protein carboxylmethyltransferase. *Exp Cell Res* 240:282–292.
45. Bottiglieri, T. (2002). *S*-Adenosyl-L-methionine (SAMe): From the bench to the bedside: Molecular basis of a pleiotropic molecule. *Am J Clin Nutr* 76:1151S–1157S.
46. Bressa, G.M. (1994). *S*-Adenosyl-L-methionine (SAMe) as antidepressant: meta-analysis of clinical studies. *Acta Neurol Scand* 154:7–14.
47. Mischoulon, D., and Fava, M. (2002). Role of *S*-adenosyl-L-methionine in the treatment of depression: A review of the evidence *Am J Clin Nutr* 76:1158S–1161S.

48. Papakostas, G.I., Alpert, J.E., and Fava, M. (2003). *S*-adenosyl-methionine in depression: A comprehensive review of the literature. *Curr Psychiatry Rep* 5:460–466.
49. Williams, A.L. Girard, C., Jui, D., Sabina, A., and Katz, D.L. (2005). *S*-adenosylmethionine (SAMe) as treatment for depression: A systematic review. *Clin Invest Med* 28:132–139.
50. di Padova, C. (1987). *S*-Adenosylmethionine in the treatment of osteoarthritis: Review of the clinical studies. *Am J Med* 83:60–65.
51. Lieber, C.S. (1999). Role of *S*-adenosyl-L-methionine in the treatment of liver diseases. *J Hepatol* 30:1155–1159.
52. Lieber, C.S. (2002). *S*-adenosyl-L-methionine: Its role in the treatment of liver disorders. *Amer J Clin Nutr* 76:1183S–1187S.
53. Chiang, P.K., Gordon, R.K., Tal, J., Zeng, G.C., Doctor, B.P., Pardhasaradhi, K., and McCann, P.P. (1996). *S*-Adenosylmethionine and methylation. *FASEB J* 10:471–480.
54. Stramentinoli, G. (1987). Pharmacologic aspects of *S*-adenosylmethionine. Pharmacokinetics and pharmacodynamics. *Am J Med* 83:35–42.
55. Goren, J.L., Stoll, A.L., Damico, K.E., Sarmiento, I.A., and Cohen, B.M. (2004). Bioavailability and lack of toxicity of *S*-adenosyl-L-methionine (SAMe) in humans. *Pharmacotherapy* 24:1501–1507.
56. Castagna, A., Le Grazie, C., Accordini, A., Giulidori, P., Cavalli, G., Bottiglieri, T., and Lazzarin, A. (1995). Cerebrospinal fluid *S*-adenosylmethionine (SAMe) and glutathione concentrations in HIV infection: Effect of parenteral treatment with SAMe. *Neurology* 45:1678–1683.
57. Baldessarini, R.J. (1987). Neuropharmacology of *S*-adenosyl-L-methionine. *Am J Med* 83:95–103.
58. Young, S.N., and Shalchi, M. (2005). The effect of methionine and *S*-adenosylmethionine on *S*-adenosylmethionine levels in the rat brain. *J Psychiatry Neurosci* 30:44–48.
59. Van Phi, L., and Soling. H.D. (1982). Methyl group transfer from exogenous *S*-adenosylmethionine to plasma-membrane phospholipids without cellular uptake in isolated hepatocytes. *Biochem J* 206:481–487.
60. Bontemps, F., and Van den Berghe, G. (1987). Metabolism of exogenous *S*-adenosylmethionine in isolated rat hepatocyte suspensions: Methylation of plasma-membrane phospholipids without intracellular uptake. *Biochem J* 327:383–389.
61. Stramentinoli, G., Pezzoli, C., Kienle, M.G. (1978). Uptake of *S*-adenosyl-L-methionine by rabbit erythrocytes. *Biochem Pharmacol* 27:1427–1430.
62. Pezzoli, C., Stramentinoli, G., Galli-Kienle, M., and Pfaff, E. (1978). Uptake and metabolism of *S*-adenosyl-L-methionine by isolated rat hepatocytes. *Biochem Biophys Res Commun* 85:1031–1038.
63. Zappia, V., Galletti, P., Porcelli, M., Ruggiero, G., and Andreana, A. (1978). Uptake of adenosylmethionine and related sulfur compounds by isolated rat liver. *FEBS Lett* 90:331–335.
64. Freitag, C., and Clarke, S. (1981). Reversible methylation of cytoskeletal and membrane proteins in intact human erythrocytes. *J Biol Chem* 256:6162–6108.
65. McMillan, J.M., Walle, U. K., and Walle, T. (2005). *S*-adenosyl-L-methionine: Transcellular transport and uptake by Caco-2 cells and hepatocytes. *J Pharma Pharmacol* 57:599–605.
66. Giulidori, P., Galli-Keinle, M., Catto, E. and Stramentinoli, G. (1984). Transmethylation, transsulfuration, and aminopropylation reactions of *S*-adenosyl-L-methionine *in vivo*. *J Biol Chem* 259:4205–4211.
67. Agrimi, G., Di Noia, M.A., Marobbio, C.M., Fiermonte, G., Lasorsa, F.M., and Palmieri, F. (2004). Identification of the human mitochondrial *S*-adenosylmethionine transporter: Bacterial expression, reconstitution, functional characterization and tissue distribution. *Biochem J* 379:183–190.
68. Horne, D.W., Holloway, R.S., and Wagner, C. (1997). Transport of *S*-adenosylmethionine in isolated rat liver mitochondria. *Arch Biochem Biophys* 343:201–206.
69. Marobbio, C.M., Agrimi, G., Lasorsa, F.M., and Palmieri, F. (2003). Identification and functional reconstitution of yeast mitochondrial carrier for *S*-adenosylmethionine. *EMBO J* 22:5975–5982.

70. Rouillon, A., Surdin-Kerjan, Y., and Thomas, D. (1999). Transport of sulfonium compounds. Characterization of the *S*-adenosylmethionine and *S*-methylmethionine permeases from the yeast *Saccharomyces cerevisiae*. *J Biol Chem* 274:28096–28105.
71. Caro, A.A., and Cederbaum, A.I. (2004). Antioxidant properties of *S*-adenosyl-ʟ-methionine in Fe(2+)-inititated oxidations. *Free Radic Biol Med* 36:1301–1316.
72. Katz, J.E., Dlakic, M., and Clarke, S. (2003). Automated identification of putative methyltransferases from genomic open reading frames. *Mol Cell Proteom* 2:525–540.
73. Pascale, R.M., Similie, M.M., Satta, G., Seddaiu, M.A., Daino, L., Pinna, G., Vinci, M.A., Gaspa, L., and Feo, F. (1991). Comparative effects of L-methionine, *S*-adenosyl-ʟ-methionine and 5′-methylthioadenosine on the growth of preneoplastic lesions and DNA methylation in rat liver during the early stages of hepatocarcinogenesis. *Anticancer Res* 11:1617–1624.
74. Pascale, R.M., Simile, M.M., De Miglio, M.R., and Feo, F. (2002). Chemoprevention of hepatocarcinogenesis: *S*-Adenosyl-ʟ-methionine. *Alcohol* 27:193–198.
75. Bedford, M.T., and Richard, S. (2005). Arginine methylation: An emerging regulator of protein function. *Mol Cell* 18:263–272.
76. Bedford, M.T. (2006). The family of protein arginine methyltransferases, in S.G. Clarke and F. Tamanoi (eds.), *The Enzymes* (3d Ed.), (in press).
77. McBride, A.E. (2006). Diverse roles of protein arginine methyltransferases, in S.G. Clarke and F. Tamanoi (eds.), *The Enzymes* (3d Ed.), (in press).
78. Zhang, X., and Cheng, X. (2006). Structure of protein arginine methyltransferases, in S.G. Clarke and F. Tamanoi (eds.), *The Enzymes* (3d Ed.), (in press).
79. Gualano, M., Berti, F., and Strammentinoli, G. (1985). Anti-inflammatory activity of *S*-adenosyl-ʟ-methionine in animal models: Possible interference with the eicosanoid system. *Int J Tissue React* 7:41–46.
80. Hevia, H., Varela-Rey, M., Corrales, F.J., Berasain, C., Martinez-Chantar, M.L., Latasa, M.U., Lu, S.C., Mato, J.M., Garcia-Trevijano, E.R., and Avila, M.A. (2004). 5′-Methylthioadenosine modulates the inflammatory response to endotoxin in mice and in rat hepatocytes. *Hepatology* 39:1088–1098.
81. Veal, N., Hsieh, C.-H., Xiong, S., Mato, J.M., Lu, S., and Tsukamotom H. (2004). Inhibition of lipopolysaccharide-stimulated TNA-alpha promoter activity by S-Adenosylmethionine and 5′-methylthioadenosine. *Am J Physiol Gastrointest Liver Physiol* 287:G353–G362.
82. Law, R.E., Stimmel, J.B., Damore, M.A., Carter, C., Clarke, S., and Wall, R. (1992). Lipopolysaccharide-induced NF-kappa B activation in mouse 70Z/3 pre-B lymphocytes is inhibited by mevinolin and 5′-methylthioadenosine: Roles of protein isoprenylation and carboxyl methylation reactions. *Mol Cell Biol* 12:103–111.
83. Cerri, M.A., Beltran-Nunez, A., Bernasconi, S., Dejana, E., Bassi, L., and Bazzoni, G. (1993). Inhibition of cytokine production and endothelial expression of adhesion antigens by 5′-methylthioadenosine. *Eur J Pharmacol* 232:291–294.
84. Song, Z., Uriarte, S., Sahoo, R., Chen, T., Barve, S., Hill, D., McClain, C. (2005). *S*-adenosylmethionine (SAMe) modulates interleukin-10 and interleukin-6, but not TNF, production via the adenosine (A2) receptor. *Biochem Biophys Acta* 1743:205–213.
85. Li, H., Park, S., Kilburn, B., Jelinek, M.A., Henschen-Edman, A., Aswad, D.W., Stallcup, M.R., and Laird-Offringa, I.A. (2005). Lipopolysaccharide-induced methylation of HuR, an mRNA-stabilizing protein, by CARM1. *J Biol Chem* 277:44623–44630.
86. Abramovich, C., Yakobson, B., Chebath, J., and Revel, M. (1997). A protein-arginine methyltransferase binds to the intracytoplasmic domain of the IFNAR1 chain in the type I interferon receptor. *EMBO J* 16:260–266.
87. Meissner, T., Krause, E., Lodige, I., and Vinkemeier, U. (2004). Arginine methylation of STAT1: A Reassessment. *Cell* 119:587–590.
88. Komyod, W., Bauer, U.-M., Heinrich, P.C., Haan, S., and Behrmann, I. (2005). Are STATS Arginine-methylated? *J Biol Chem* 280:21700–21705.

89. Richard, S., Morel, M., and Cleroux, P. (2005). Arginine methylation regulates IL-2 gene expression: A role for protein arginine methyltransferase 5 (PRMT5). *Biochem J* 388:379–386.

90. Maher, P.A. (1993). Inhibition of the tyrosine kinase activity of the fibroblast growth factor receptor by the methyltransferase inhibitor 5′-methylthioadenosine. *J Biol Chem* 268:4244–4249.

91. Miyaji, K., Tani, E., Nakano, A., Ikemoto, H., and Kaba, K. (1995). Inhibition by 5′-methylthioadenosine of cell growth and tyrosine kinase activity stimulated by fibroblast growth factor receptor in human gliomas. *J Neurosurg* 83:690–697.

92. Pintucci, G., Quarto, N., and Rifkin, D.B. (1996). Methylation of high molecular weight fibroblast growth factor-2 determines post-translation increases in molecular weight and affects its intracellular distribution. *Mol Biol Cell* 7:1249–1258.

93. Burgess, W.H., Bizik, J., Mehlman, T., Quarto, N., and Rifkin, D.B. (1991). Direct evidence for methylation of arginine residues in high molecular weight forms of basic fibroblast growth factor. *Cell Regulation* 2:87–93.

94. Klein, S., Carroll, J.A., Chen, Y., Henry, M.R., Henry, P.A., Ortonowski, E.E., Pintucci, G., Beavis, R.C., Burgess, W.H., and Rifkin, D.B. (2000). Biochemical analysis of the arginine methylation of high molecular weight fibroblast growth factor-2. *J Biol Chem* 275:3150–3157.

95. Reilly, J.F., Maher, P.A., and Kumari, V.G. (1998). Regulation of astrocyte GFAP expression by TGF-beta1 and FGF-2. *Glia* 22:202–120.

96. Seeley, P.J., Rukenstein, A., Connolly, J.L., and Greene, L.A. (1984). Differential inhibition of nerve growth factor and epidermal growth factor effects on the PC12 pheochromocytoma line. *J Cell Biol* 98:417–426.

97. Kujubu, D.A., Stimmel, J.B., Law, R.E., Herschman, H.R., and Clarke, S. (1993). Early responses of PC-12 cells to NGF and EFG: Effect of K252a and 5′-methylthioadenosine on gene expression and membrane protein methylation. *J Neurosci Res* 36:58–65.

98. Haklai, R., Lerner, S., and Kloog, Y. (1993). Nerve growth factor induces a succession of increases in isoprenylated methylated small GTP-binding proteins of PC-12 phyochromocytoma cells. *Neuropeptides* 24:11–25.

99. Smith, D.S., King, C.S., Pearson, E., Gittinger, C.K., Landreth, G.E. (1989). Selective inhibition of nerve growth factor-stimulated protein kinases by K-252a and 5′-S-methyladenosine in PC12 cells. *J Neurochem* 53:800–806.

100. Cimato, T.R., Ettinger, M.J., Zhou, X., and Aletta, J.M. (1997). Nerve growth factor-specific regulation of protein methylation during neuronal differentiation of PC12 cells. *J Cell Biol* 138:1089–1103.

101. Cimato, T.R., Tang, J., Xu, Y., Guarnaccia, C., Herschman, H.R., Pongor, S., and Aletta, J.M. (2002). Nerve growth factor-mediated increases in protein methylation occur predominantly at type I arginine methylation sites and involve protein arginine methyltransferase 1. *J Neurosci Res* 67:435–443.

102. Lee, J., Sayegh, J., Daniel, J., Clarke, S., and Bedford, M.T. (2005). PRMT8, a new membrane-bound tissue-specific member of the protein arginine methyltransferase family. *J Biol Chem* 280 (in press).

103. Acheson, A., Vogl, W., Huttner, W.B., and Thoenen, H. (1986). Methyltransferase inhibitors block NGF-regulated survival and protein phosphorylation in sympathetic neurons. *EMBO J* 5:2799–2803.

104. Vandenbark, A.A., Ferro, A.J., and Barney, C.L. (1980). Inhibition of lymphocyte transformation by a naturally occurring metabolite: 5′-Methylthioadenosine. *Cell Immunol* 49:26–33.

105. Pegg, A.E., Borchardt, R.T., and Coward, J.K. (1981). Effects of inhibitors of spermidine and spermine synthesis on polyamine concentrations and growth of transformed mouse fibroblasts. *Biochem J* 194:79–89.

106. Riscoe, M.K., Schwamborn, J., Ferro, A.J., Olson, K.D., and Fitchen, J.H. (1987). Inhibition of growth but not differentiation of normal and leukemic myeloid cells by methylthioadenosine. *Cancer Res* 47:3830–3834.

107. Riscoe, M.K., Tower, P.A., and Ferro, A.J. (1984). Mechanism of action of 5'-methylthioadenosine in S49 cells. *Biochem Pharmacol* 33:3639–3643.
108. Boisvert, F.-M., Cote, J., Boulanger, M.-C., Cleroux, P., Bachand, F., Autexier, C., and Richard, S. (2002). Symmetrical dimethylarginine methylation is required for the localization of SMN in Cajal bodies and pre-mRNA splicing. *J Cell Biol* 159:957–969.
109. Jenuwein, T., and Allis, C.D. (2001). Translating the histone code. *Science* 293:1074–1080.
110. Bernstein, B.E., Humphrey, E.L., Erlich, R.L., Schneider, R., Bouman, P., Liu, J.S., Kouzarides, T., and Schreiber, S.L. (2002). Methylation of histone H3 Lys 4 in coding regions of active genes. *Proc Natl Acad Sci USA* 99:8695–8700.
111. Young, S.G., Clarke, S., Bergo, M.O., Philips, M. and Fong, L.G. (2006). Genetic approaches for understanding the physiologic importance of the carboxyl methylation of isoprenylated proteins, in S.G. Clarke and F. Tamanoi (eds.). *The Enzymes* (3d ed.), (in press).
112. Ansorena, E., Garcia-Trevijano, E.R., Martinez-Chanar, M.L., Huang, Z.-Z., Chen, L., Mato, J.M., Iraburu, M., Lu, S.C., and Avila, M.A. (2002). *S*-Adenosylmethionine and methylthioadenosine are antiapoptotic in cultured rat hepatocytes but proapoptotic in human hepatoma cells. *Hepatology* 35:274–280.
113. Yang, H., Sadda, M.R., Li, M., Zeng, Y., Chen, L., Bae, W., Ou, X., Runnegar, M.T., Mato, J.M., and Lu, S.C. (2004). *S*-Adenosylmethionine and its metabolites induce apoptosis in HepG2 cells: Role of protein phosphatase 1 and Bcl-x$_s$. *Hepatology* 40:221–231.
114. O'Connor, C.M. (2006). Protein L-isoaspartyl, D-aspartyl *O*-methyltransferases: Catalysts for protein repair, in S.G. Clarke and F. Tamanoi (eds.). *The Enzymes* (3d ed.), (in press).
115. Galletti, P., Oliva, A., Manna, C., Della Ragione, F., and Carteni-Farina, M. (1981). Effect of 5'-methylthioadenosine on in vivo methyl esterification of human erythrocyte membrane proteins *FEBS Lett* 126:236–240.
116. Raina, A., Tuomi, K., and Pajula, R.L. (1982). Inhibition of the synthesis of polyamines and macromolecules by 5'-methylthioadenosine and 5'-alkylthiotubercidins in BHK21 cells. *Biochem J* 204:697–703.
117. Robinson-White, A., and Stratakis, C.A. (2002). Protein kinase A signaling: "Cross-talk" with other pathways in endocrine cells. *Ann NY Acad Sci* 968:256–270.
118. Munshi, R., Clanachan, A.S., and Baer, H.P. (1988). 5'-Deoxy-5'-methylthioadenosine: A nucleoside which differentiates between adenosine receptor subtypes. *Biochem Pharmacol* 37:2085–2090.
119. Lorenzen, A., Beukers, M.W., van der Graaf, P.H., Lang, H., van Muijlwijk-Koezen, J., de Groote, M., Menge, W., Schwabe, U., and Ijzerman, A.P. (2002). Modulation of agonist responses at the A$_1$ adenosine receptor by an irreversible antagonist, receptor-G protein uncoupling and by the G protein activation state. *Biochem Pharmacol* 64:1251–1265.
120. Nishida, Y., Suzuki, S., and Miyamoto, T. (1985). Pharmacological action of 5'-methyl-thioadenosine on isolated rabbit aorta strips. *Blood Vessels* 22:229–233.
121. El Yacoubi, M., Costentin, J., Vaugeois, J.M. (2003). Adenosine A2A receptors and depression. *Neurology* 61:S82–S87.
122. Kaster, M.P. Rosa, A.O., Rosso, M.M., Goulart, E.C., Santos, A.R.S., and Rodrigues, L.S. (2004). Adenosine administration produces an antidepressant-like effect in mice: Evidence for the involvement of A1 and A2A receptors. *Neurosci Lett* 355:21–24.
123. Tsvetnitsky, V., Campbell, I.C., Gibbons, W.A. (1995). *S*-adenosyl-L-homocysteine and 5'-methylthioadenosine inhibit binding of [^{3}H]flunitrazepam to rat brain membranes. *Eur J Pharmacol* 282:255–258.
124. Simile, M.M., Banni, S., Angioni, E., Carta, G., De Miglio, M.R., Muroni, M.R., Calvisi, D.F., Carru, A., Pascale, R.M., and Feo, F. (2001). 5'-methylthioadenosine administration prevents lipid peroxidation and fibrogenesis induced in rat liver by carbon-tetrachloride intoxication. *J Hepatol* 34:386–394.

125. Yamanaka, H., Kajander, E.O., Carson, D.A. (1986). Modulation of diphthamide synthesis by 5'-deoxy-5'-methylthioadenosine in murine lymphoma cells. *Biochem Biophys Acta* 888:157–162.
126. Riscoe, M.K., and Ferro, A.J. (1984). 5-Methylthioribose: Its effects and function in mammalian cells. *J Biol Chem* 259:5465–5471.
127. Kaneko, K., Fujimori, S., Kamatani, N., and Akaoka, I. (1984). 5'-Methylthioadenosine in urine from normal subjects and cancer patients. *Biochem Biophys Acta* 802:169–174.
128. Najm, W.I., Reinsch, S., Hoehler, F., Tobis, J.S., and Harvey, P.W. (2004). *S*-adenosylmethionine (SAMe) versus celecoxib for the treatment of osteoarthritis symptoms: a double-blind cross-over trial. *BMC Musculoskelet Disord* 5:6 (doi:10.1186/1471–2474–5–6).
129. Cheng, D., Yadav, N., King, R.W., Swanson, M.S., Weinstein, E.J., and Bedford, M.T. (2004). Small molecule regulators of protein arginine methyltransferases. *J Biol Chem* 279:23892–23899.

Author Index

Numbers in regular font are reference numbers and indicate that an author's work is referred to although the name is not cited in the text. Numbers in italics refer to the page numbers on which the complete reference appears.

W

Z

Index

A

a-factor, in carboxyl methylation, 274–276, 278
αβα domains, in PIMTs, 396, 398
Active sites
  in PrmC/glutamine residues, 443–446
  in PRMT structure, 114–116, 116f
Adenosine, MTA inhibition and, 485
aDMA. *See* Asymmetric dimethyl arginine
AdoHcy. *See* S-adenosyl-L-homocysteine
AdoMet. *See* S-adenosyl-L-methionine
AdOx, in PRMT roles, 58, 60, 67
Aer protein, in reversible methylation/
  glutamate residues, 329, 329n2, 330, 347
AFC. *See* N-acetyl-S-farnesyl-L-cysteine
AGGC. *See* N-acetyl-S-geranylgeranyl-
  L-cysteine
Aging, PIMTs and, 404–405, 408–411
Alzheimer's amyloid, PIMTs and, 410–411
Alzheimer's disease, PP2A reversible
  methylation and, 303, 315–316
AMI. *See* Arginine methyltransferase inhibitors
Amine oxidases, in demethylation pathways,
  230, 237
Antagonism, of arginine methylation, PRMT
  structure and, 116–117
Arginine methylation reactions, PRMT roles
  and, 52f
Arginine methyltransferase inhibitors (AMI),
  PRMTs and, 43–44
Asparagine (Asn) peptide degradation, in
  phycobiliprotein modification, 459, 460f
Asymmetric dimethyl arginine (aDMA). *See
  also* Symmetric dimethyl arginine
  in PRMT roles, 52, 67, 71, 82, 84–85, 88
  PRMT structure and, 114–116, 115f, 116f
  PRMTs and, 32–33, 39, 40

B

Benzodiazepine receptors, MTA inhibition
  and, 484
Bilins, donor/acceptor, in phycobiliprotein
  modification, 460–461
Biomineralization, non-histone PKMT in,
  188–189

C

CAAX motif
  in carboxyl methylation, 273–275
  in Icmts, 245, 246–248, 247f, 249, 261, 262
Caenorhabditis elegans, PIMTs and, 419–420
Calmodulin (CaM), PIMT and, 394, 396, 406
Calmodulin-lysine N-methyltransferase
  (CLNMT)
  calcium binding sites and, 198–199, 199f
  kinetic mechanism in, 197–198
  Lys-115 methylation in, 200–201
  in non-histone PKMT, 196–201
  substrate specificity in, 197–198, 199f
cAMP phosphodiesterases, MTA inhibition
  and, 483–485
Cancer
  carboxyl methylation and, 295, 296
  MTA inhibition and, 470, 471, 478
  PRMT roles and, 84
Carboxyl methylation, of isoprenylated proteins
  a-factor in, 274–276, 278
  AFC and, 277, 279, 279f, 283, 283f, 296
  AGGC and, 277, 279, 279f, 283, 283f, 296
  CAAX motif in, 273–275
  cancer and, 295, 296
  CXC motif in, 274, 275
  CXC-Rab binding in, 287, 288f

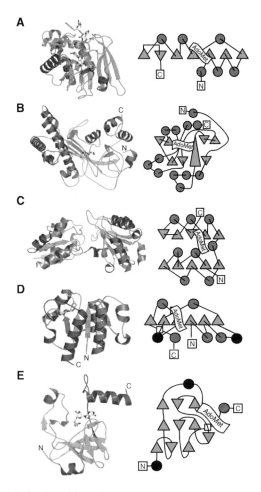

Chapter 1. Figure 8. (See legend in text.)

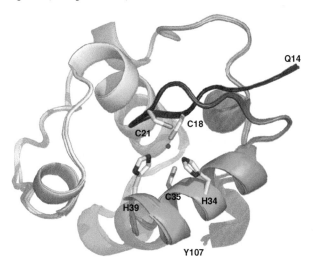

Chapter 4. Figure 4. (See legend in text.)

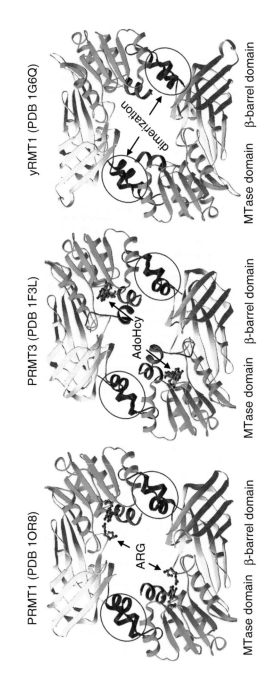

PRMT1 (PDB 1OR8)

MTase domain   β-barrel domain

PRMT3 (PDB 1F3L)

MTase domain   β-barrel domain

yRMT1 (PDB 1G6Q)

MTase domain   β-barrel domain

ARG

AdoHcy

dimerization

Chapter 4. Figure 5. (See legend in text.)

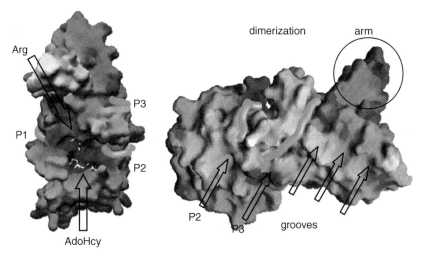

R3 peptide (19): GG<u>R</u>GGFGG<u>R</u>GGFGG<u>R</u>GGFG

Chapter 4. Figure 6. (See legend in text.)

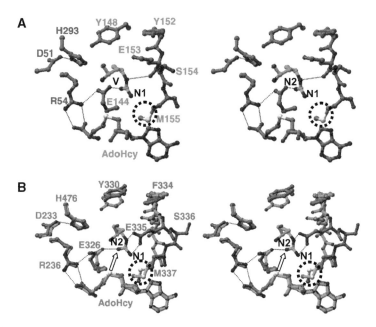

Chapter 4. Figure 7. (See legend in text.)

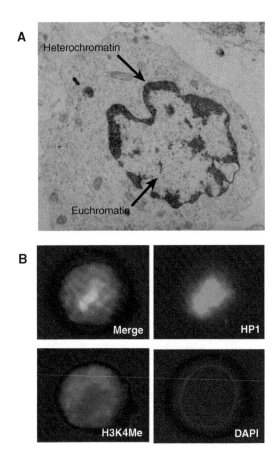

Chapter 5. Figure 1. (See legend in text.)

A HKMT and binding effectors in the same complex

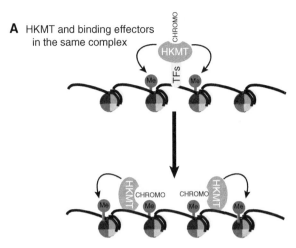

B HKMT and binding effectors in different complexes

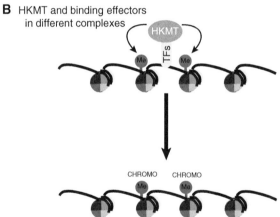

C Prevent binding of effectors proteins

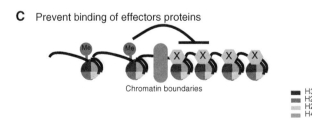

Chromatin boundaries

H3
H2B
H2A
H4

Chapter 5. Figure 3. (See legend in text.)

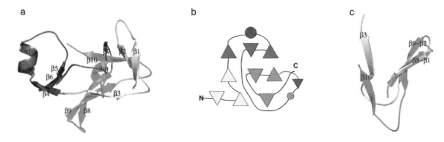

Chapter 6. Figure 3. (See legend in text.)

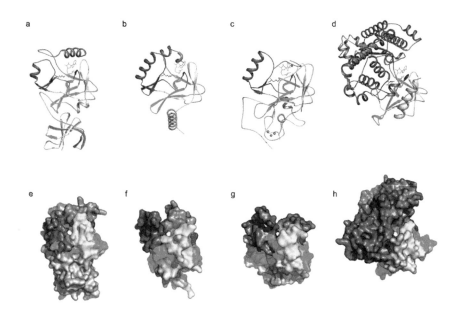

Chapter 6. Figure 4. (See legend in text.)

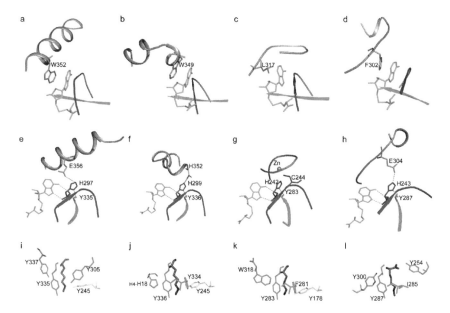

Chapter 6. Figure 5. (See legend in text.)

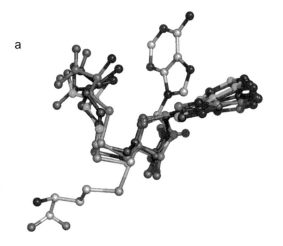

a

b. Set7/9

Asn-296 His-297

Glu-356

Ala-226

Lys-294 Glu-228

c. PR-Set7 (Set8)

Tyr-271 Asn-298 His-299

Lys-226

Arg-228

d. Dim-5

Tyr-204 Asn-241 His-242

NH₂

Lys-201 Leu-307

Arg-159

Trp-161

e. LSMT

Asn-242 His-243

Glu-80

Leu-82

Arg-222

Chapter 6. Figure 6. (See legend in text.)

a. Set7/9 H3 K4

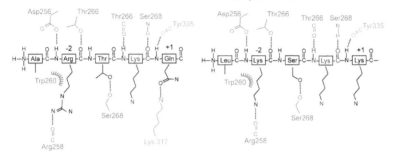

b. Set7/9 p53 K272

c. PR-Set7 (Set8) H4 K20

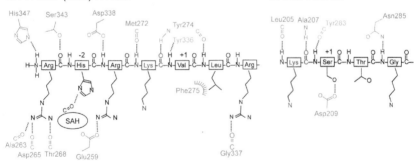

d. Dim-5 H3 K9

Chapter 6. Figure 7. (See legend in text.)

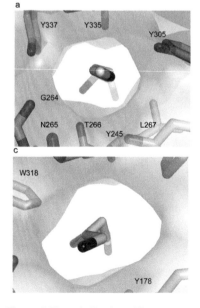

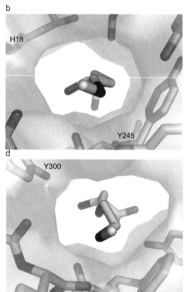

Tyr-305
(290)
Histone peptide
C=O
(295)
Tyr-245
177.2°
AdoHcy

a. Set7/9

Tyr-334
(297)
Histone peptide
C=O
(294)
Tyr-245
175.7°
AdoHcy

b. PR-Set7 (Set8)

Histone peptide
Tyr-178
AdoMet

c. Dim-5

Histone peptide
C=O
(222)
127°
AdoHcy

d. LSMT

Chapter 6. Figure 8. (See legend in text.)

a
Y337   Y335   Y305
G264
N265   T266   L267
Y245

b
H18
Y245

c
W318
Y178

d
Y300

Chapter 6. Figure 9. (See legend in text.)

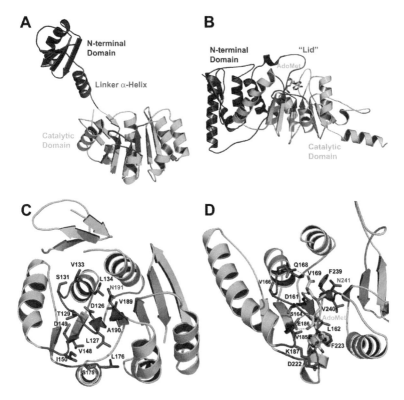

Chapter 7. Figure 2. (See legend in text.)

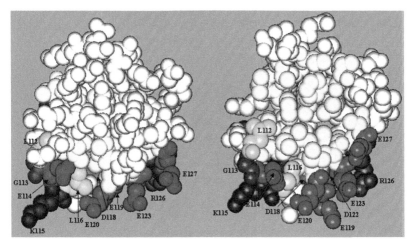

Chapter 7. Figure 3. (See legend in text.)

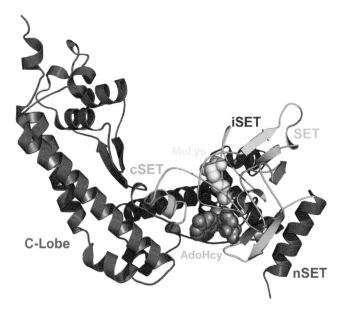

Chapter 7. Figure 5. (See legend in text.)

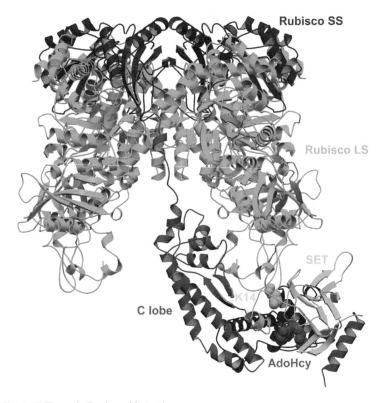

Chapter 7. Figure 6. (See legend in text.)

a — Rubisco Large Subunit N-Terminal Sequences

Arabidopsis thaliana: Ac-PQTETKASVGFK$_{14}$AGVKEYKLTYY
Pea (*Pisum sativum*):   Ac-PQTETKAKVGFK$_{14}$AGVKDYKLTYY
Tobacco (*Nicotiana tabacum*): Ac-PQTETKASVGFK$_{14}$AGVKEYKLTYY
Spinach (*Spinacia oleracea*): Ac-PQTETKASVGFK$_{14}$AGVKDYKLTYY
Consensus Sequence: VGFK$_{14}$AGV

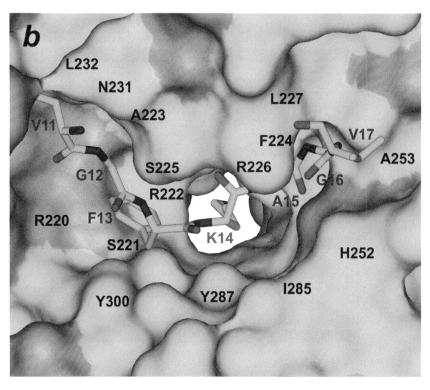

Chapter 7. Figure 7. (See legend in text.)

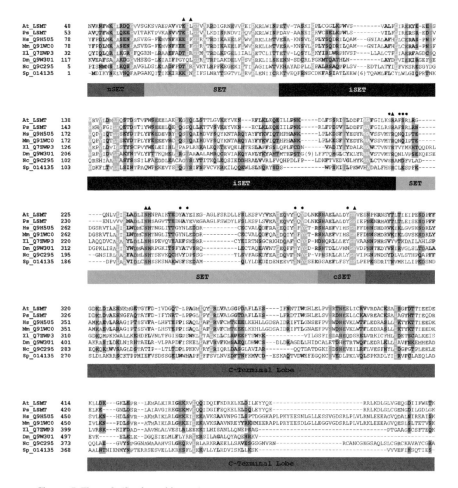

Chapter 7. Figure 8. (See legend in text.)

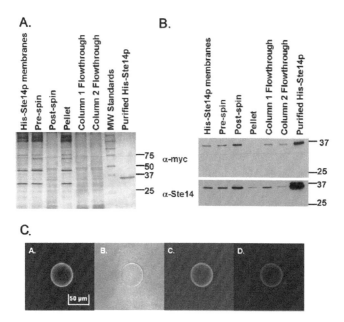

Chapter 9. Figure 4. (See legend in text.)

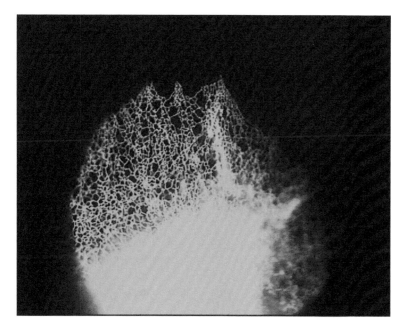

Chapter 10. Figure 1. (See legend in text.)

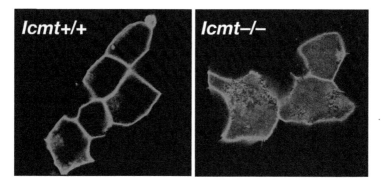

Chapter 10. Figure 5. (See legend in text.)

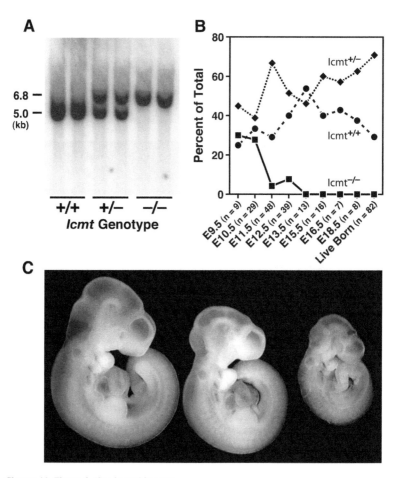

Chapter 10. Figure 6. (See legend in text.)

Printed and bound by CPI Group (UK) Ltd, Croydon, CR0 4YY

08/05/2025

01864951-0004